8th Scientific-Technical Conference on Material Problems in Civil Engineering (MATBUD'2018)

MATEC Web of Conferences Volume 163 (2018)

Cracow, Poland
25-27 June 2018

Editors:

I. Hager **T. Tracz**

ISBN: 978-1-5108-6800-7

Printed from e-media with permission by:

Curran Associates, Inc.
57 Morehouse Lane
Red Hook, NY 12571

Some format issues inherent in the e-media version may also appear in this print version.

Printed by Curran Associates, Inc. (2018)

For additional information, please contact EDP Sciences – Web of Conferences at the address below.

EDP Sciences – Web of Conferences
17, Avenue du Hoggar
Parc d'Activité de Courtabœuf
BP 112
F-91944 Les Ulis Cedex A
France

Phone: +33 (0) 1 69 18 75 75
Fax: +33 (0) 1 69 28 84 91

contact-edps@webofconferences.org

Additional copies of this publication are available from:

Curran Associates, Inc.
57 Morehouse Lane
Red Hook, NY 12571 USA
Phone: 845-758-0400
Fax: 845-758-2633
Email: curran@proceedings.com
Web: www.proceedings.com

TABLE OF CONTENTS

Preface

The Scientific-Technical Conference on Material Problems in Civil Engineering (MATBUD'2018) is organised for the 8th time by the Chair of Building Materials Engineering of the Faculty of Civil Engineering at Cracow University of Technology, under the auspices of the Building Materials Engineering Section, part of the Civil Engineering Committee of the Polish Academy of Sciences.

This year we have received about 75 paper submissions, which were reviewed by national and international experts. 55 of the received papers have been accepted by the Scientific Committee for oral presentation at the conference and publication in its proceedings. The papers have been grouped into 8 chapters: i) *new generation concretes*; ii) *modelling of concrete behaviour and test methods*; iii) *waste products as components of sustainable building materials*; iv) *cement pastes and mortars*; v) *durability and environmental impact*, *vi) geopolymers and soil-cement materials*; vii) *structural materials performance evaluation*, and viii) *energy-efficient building materials*, which correspond to the main conference areas and sessions.

The primary focus of this conference are several areas of building materials engineering: development, processing, evaluation of properties and applications. As usual, MATBUD'2018 was dominated by topics related to cement binders and aspects of durability, but we observe a definite shift towards new sustainable material solutions. Among them is the search for robust and sustainable mineral materials for the additive construction and alkali-activated binders that meet the demanding mechanical and durability performance requirements of civil engineering applications. One of the best examples of awareness and respect for principles of sustainable development is the widespread use of waste materials and by-products for the production of mineral binders presented in session iii) *waste products as components of sustainable building materials*.

For this MATBUD edition, we have invited Dr Philippe Poullain to be the Keynote Speaker presenting 3-D printing technology and the Yhnova^{TM} project. The additive manufacturing in construction is an emerging industry with numerous opportunities and challenges. Not only does this new paradigm for construction industry demand changes in construction processes, but it also requires new materials with specific rheological and setting properties. Watching with great interest the first successful applications, we cannot help asking: will the 3D printing technology be sufficiently efficient and affordable to be widely used in the construction industry? To what extent can 3D printing replace traditional building technologies?

Looking forward to seeing this technology on site, we would like to take the opportunity to express our gratitude to the originator and architect of these cyclic meetings – Professor Jacek Śliwiński, who has spared no time or effort in organizing MATBUD events with absolute commitment since its first edition in 1996, and also strongly supported the organization of this year's event. From the beginning of MATBUD, Prof. Lech Czarnecki has also played a key role in these events, devoting his time and sharing his knowledge with us as the Chairman of the Scientific Committee he successfully leads.

Warm thanks to all members of the Scientific Committee who have accepted papers for review on a voluntary basis as well as provided valuable suggestions during preparation of this conference. Thank you very much for your assistance!

We express our sincere gratitude to the chairperson of each conference session for smoothly conducting oral presentations of papers and coordinating discussions.

We are deeply indebted to EDP Sciences for excellent presentation of this conference proceedings through the MATEC Web of Conferences. The online publication of papers allows them to be distributed quickly and freely all over the world. We believe that this will significantly strengthen the impact of the MATBUD conference among the scientific community dealing with civil engineering materials.

On behalf of the organizing committee of MATBUD'2018, we would like to express our gratitude to all participants for their valuable contribution to the conference. We would also like to thank the authors for sending papers and for their patience with which they corrected them in compliance with the recommendations received from reviewers. Without your contribution, we would not have been able to appreciate so much this meeting and all the pleasant moments spent together in beautiful Cracow.

Izabela Hager, Tomasz Tracz

Chairpersons of MATBUD'2018

Statement of Peer review

In submitting conference proceedings to *Web of Conferences*, the editors of the proceedings certify to the Publisher that

1. They adhere to its **Policy on Publishing Integrity** in order to safeguard good scientific practice in publishing.
2. All articles have been subjected to peer review administered by the proceedings editors.
3. Reviews have been conducted by expert referees, who have been requested to provide unbiased and constructive comments aimed, whenever possible, at improving the work.
4. Proceedings editors have taken all reasonable steps to ensure the quality of the materials they publish and their decision to accept or reject a paper for publication has been based only on the merits of the work and the relevance to the journal.

Title, date and place of the conference

MATBUD' 2018 - 8th Scientific-Technical Conference on Material Problems in Civil Engineering
Cracow University of Technology
Cracow, Poland, 25-27 June 2018

Proceedings editor(s):

Izabela Hager and Tomasz Tracz

Date and editor's signature

03.04.2018

Izabela Hager

Tomasz Tracz

MATEC Web of Conferences **163**, 01001 (2018) https://doi.org/10.1051/matecconf/201816301001
MATBUD'2018

On site deployment of 3D printing for the building construction – The case of Yhnova^{TM}

Philippe Poullain[1,*], *Elodie* Paquet[2], *Sébastien* Garnier[2], and *Benoît* Furet[2]

[1]University of Nantes – GeM UMR CNRS 6183 – 58 rue Michel Ange – 44606 Saint Nazaire CEDEX – France
[2]University of Nantes – LS2N UMR CNRS 6004 – 2 Avenue du Professeur Jean Rouxel – 44475 Carquefou – France

Abstract. The University of Nantes has developed a 3D printing technique (BatiPrint3D^{TM}) dedicated to the construction of the walls of a house. This innovative on site construction technique is based on the deposition of two layers of expansive foam used as a formwork for a third concrete layer. It allows to build at the same time the structure and the insulation. This new construction technology has first been developed at the laboratory, but rapidly, we decided to deploy it on site, in order to demonstrate its technical viability. We present the technology Batiprint3D^{TM} and the demonstrator Yhnova^{TM}, a 95m² social dwelling built for the social landlord Nantes Métropole Habitat (NMH).

1 Introduction

According to Chaparro-Pelaez et al. [1], the crisis has caused a slowdown in the building industry and a way for small and medium enterprises to survive is to be flexible, specialized and innovative. This concept applies especially for services and relation to the client but could be extrapolated to specialization and innovation in the techniques of construction. Moreover, in a recent report published by the McKinsey Institute [2], it is shown that the building industry has known nearly zero gains of productivity during the last decades. The McKinsey report gives sevens areas to improve the construction productivity and boost it by 50-60%, such that [2]: rethink design of construction, improve onsite execution, infuse technology and innovation, reskill workers.

The burgeoning technology of large scale additive manufacturing could be a tool for a new impulse in the construction industry.

ASTM standards define additive manufacturing as a process of joining materials to make objects from 3D model data, usually layer upon layer [3]. It englobes various manufacturing techniques like binder jetting, direct energy deposition, material extrusion, powder bed fusion…

Large scale additive manufacturing presents potentially many advantages towards other construction technologies, as it is supposed to *reduce the environmental impact* of the

* Corresponding author: Philippe.Poullain@univ-nantes.fr

manufacturing industry as the amount of used material is adjusted to the need (no losses) or the shape of the object fits exactly the need (topologic optimization [4]), *produce complex parts or objects without increase in the cost of the process* [5–8], *increase building rates* [6], *reduce the arduousness* for workers on construction plants [7, 9], give the opportunity to architects to *release creativity* [9], *allow the functionalization of the printed parts* of the building (acoustic or thermal properties) [10].

During the last decade, many projects of additive manufacturing of buildings have arised particularly in the United States, China, Russia and Europe. In a recent comprehensive paper, Labonnote et al. [7] have carried out a review of the scientific works, publications, projects, web pages in the field of 3D printing, thanks to scientific search engines. The aim was to determine the evolution in the interest of the academia and industrials for this topic. The article shows that large scale additive manufacturing has known a growing interest since the development of the first 3D printing technique by Khoshnevis in the middle of the 1990's, the so-called Contour Crafting [11], and especially since five years.

Among these projects, we can cite for example Contour Crafting [11–13], WinSun [14], Loughborough University [15–16], D-Shape [17], Skanska [18], 3D-Construction [19], XTreeE [10, 20], Apis Core [21], Cazza Construction [22], BatiPrint3D™ [9, 23], Total Kustom [24], MIT [25], CyBe [26], BET Abram [27], 3D Printhuset [28], Ja-Wa [29]...

The technologies developed by these universities or enterprises split into three main techniques : *binder jetting* (a thin bed of sand is aspersed with a binder layer by layer [18]), *material extrusion* (a material – cementitious or clayey – is deposited and stiffens after deposition enough to support the upper layers [10–17, 19–22, 25, 27–29]), *filled formwork* (a formwork is 3D printed and a material is poured into it [9, 24, 26]).

Most of these techniques of 3D printing for building are based on cementitious material extrusion using cartesian printer, poly-articulated arms or cable robot [7]. The cementitious material has to exhibit particular properties, as well at early age (rheological or flow properties) as at late age (mechanical and durability properties) such that [17]:

• it is *pumpable* to the printer head in ducts

• it is *printable* : it can keep its shape once deposited and support the upper deposited layers

• its *open time* is long enough to fulfill the previous conditions

• it is *buildable* : its properties are adapted to the mechanical load of the wall and to the regulation requirements.

To match 3D printing requirements, the cementitious materials have to contain a large amount of cement and a low water content. Thus, the cementitious materials used have for now an unfavorable environmental impact as the production of cement involves the use of large amounts of energy. The expensiveness of these materials and of the technique is also still an obstacle for the development of building 3D printing. Moreover, the deposition of a cementitious material layer by layer produces a layered structure that is not taken into account by the Eurocode. As a consequence, the material extrusion techniques still need to be investigated to solve these issues.

To withdraw the technical and economical drawbacks of material extrusion applied to the construction of building walls, we developed a new technique through the project Batiprint3D™ funded by the Society for the Acceleration of Technology Transfer (SATT Ouest Valorisation). It is based on the deposition by a nozzle of poly-urethane foam that expanses and stiffens rapidly. The deposition of two parallel layers of polyurethane (PU) foam creates a FW in which the concrete is poured. This technique, first developed at laboratory, has been applied in september 2017 to build the walls of Yhnova™, a 95m² social dwelling for the social landlord Nantes Métropole Habitat.

This paper first presents the technology Batiprint3D™ by describing the principle of the three-layer wall and the way it is deployed on the construction site. Then, we present the demonstrator Yhnova™, the different stakeholders of the project, some pictures of the construction site and the main problems encountered during the construction.

2 BATIPRINT3D™

In this section, we describe the innovative technology of construction named Batiprint3D™. This technology was developed by two laboratories from the University of Nantes (France), the LS2N (Laboratory for the Sciences of Numerics) and the GeM (Research Institute in Civil Engineering and Mechanics). First, the research project involved only the laboratories but rapidly, other stakeholders joined the consortium such that institutionals, private firms, financing entities… The technology Batiprint3D™ has been developed to build the walls of building and has been applied for now for the construction of dwellings.

2.1 Composition of the wall

The wall created by the technique Batiprint3D™ is composed of a layer of self-compacting concrete between two layers of polyurethane (PU) foam (figure 1). The layers of polyurethane foam are first deposited to create a formwork (FW) in which the concrete is poured with a slight time delay to ensure that the foam has acquired sufficient stiffness to support the pressure exerted by the fresh concrete with deformations of the FW lower than the maximum value accepted by the standards of construction (5<mm).

2.2 Polyurethane foam

The polyurethane is a polymer bi component material obtained by mixing together isocyanate and polyol. Mixing is performed in a dynamic mixer and the compound is then deposited on the surface of the slab or on the sublayers (figure 1). The material then expanses (3s) and acquire stiffness in a few seconds (45s), depending on the temperature of the mix and on the reactivity of the compound.

The density of the foam is 35kg/m^3, its thermal conductivity is 0.027W/m.K and the Young modulus is 7MPa. Due to the low thermal conductivity of the PU foam, the thermal resistance of the wall (PU foam+concrete) is equal to approximately 6.5m^2.K/W.

The minimum width of one PU wall created by foam deposition is 80mm obtained by adapting the flow rates of isocyanate and polyol, the distance between the nozzle and the speed of displacement of the nozzle. Figure 2 shows the 3D scan of the surface acquired on a 300×300mm^2 sample extracted from a larger wall to ensure the representativity of the sample. The height of each layer is 35mm (10 layers for 350mm). The sample shows foam cushions distributed on the whole surface. The analysis of the surface shows that the mean height of the cushions is 15.7mm with a standard deviation of 3.7mm. The space periodic analysis of the cushions (obtained by picture autocorrelation) shows that the surface is periodic in the direction:

- *perpendicular to the displacement* of the nozzle corresponding to the height of a foam layer
- *parallel to the displacement* of the nozzle. This periodic scheme is due to instabilities in the deposition and in the expansion of the foam.

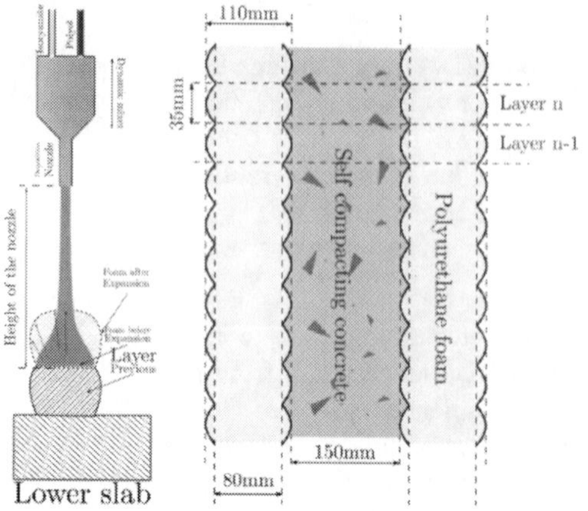

Fig. 1. Wall – vertical cut.

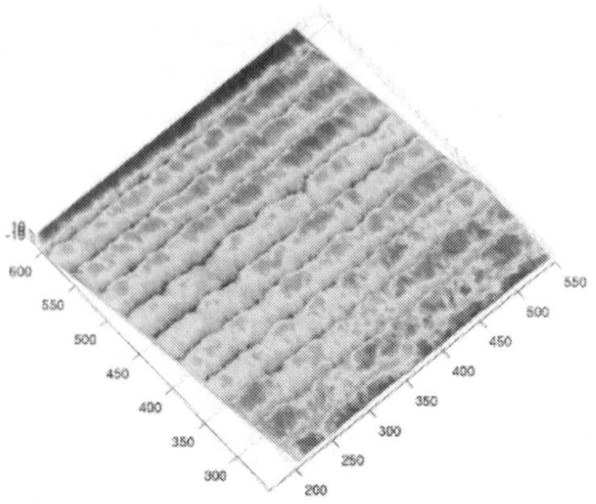

Fig. 2. 3D Scan of the polyurethane foam surface.

2.3 Concrete

The concrete is a self-compacting concrete (SCC) with a slump flow property SF=600mm. It is composed of cement CEM III 42.5, limestone filler, sand (0-4mm), gravel (4-10mm), water and set accelerator. The composition of the concrete allows it to be pumped and to flow in the FW in order to fill it with a plane horizontal free surface without the need of vibrations.

As the FW is made of an insulation material, it is important to design the concrete in such a way that the heat of hydration liberated yields a low temperature increase of the material. Indeed, if the cure temperature is too high, long term pathologies can arise and can lower the life time of the wall.

2.4 Robotics

The robotic part (see figure 4) is composed of a poly-articulated arm by Staübli (PAA) and an Automatic Guided Vehicle by BA Systèmes (AGV). The PAA is used for the deposition of the polyurethane foam and is equipped with a dynamic mixer linked to the isocyanate and polyol containers. It is carried by the AGV that can then move the PAA to different locations on the lower slab. The inaccurate positioning of the PAA induced by defaults in the flatness of the slab is corrected by first playing the deposition paths and measuring the true position of the nozzle. A correction matrix for each position is then calculated and stored.

A laser beam guidance system composed of a laser, a sensor and a set of 11 targets allows the AGV to track its location with an accuracy of 1mm on the construction site.

2.5 Process

The process Batiprint3D^{TM} is dedicated to the construction of the walls only and requires the casting of the lower concrete slab prior to the installation of the equipment necessary to the erection of the walls.

As mentioned in the ASTM standards [3], the 3D printing of the building requires the 3D model data relative to the project. They are obtained from the Building Information Model (BIM). These data are then converted into trajectories defining the displacements of the AGV and the deposition procedures of the PAA. This essential phase is performed at desk.

After casting of the lower slab, positioning targets are placed around the slab and the AGV is brought onto the slab. A calibration phase is necessary so that the AGV can track its location on the slab.

Then, the dynamic mixer and the nozzle mounted on the PAA are linked to the polyol and isocyanate containers. The AGV and the PAA deposit the polyurethane foam according to the deposition path. The number of layers N dropped down depends on different parameters : stiffness of the foam, set time of the SCC, density of the SCC, class of fluidity of the SCC, time necessary to drop down N layers of foam depending on the geometry of the building. Indeed, as the polyurethane foam exhibits weak mechanical properties, the height of concrete deposited must be determined in order to keep the deformation of the FW under a value depending on the local construction tolerance about the deformation of FWs. Once the N layers are deposited, the PAA is equipped with a duct linked to a concrete pump and the concrete is poured into the FW up to the predefined height. Then, the AGV together with the PAA, go on depositing the next N layers.

At the end of the whole casting, the polyurethane FW is kept in place as the external and internal insulation of the building.

The final rendering of the outer surface is obtained by applying a traditional coating like those used for external insulation techniques. For the inner surface, plaster boards are used.

Figure 3 presents the different phases necessary to build the wall.

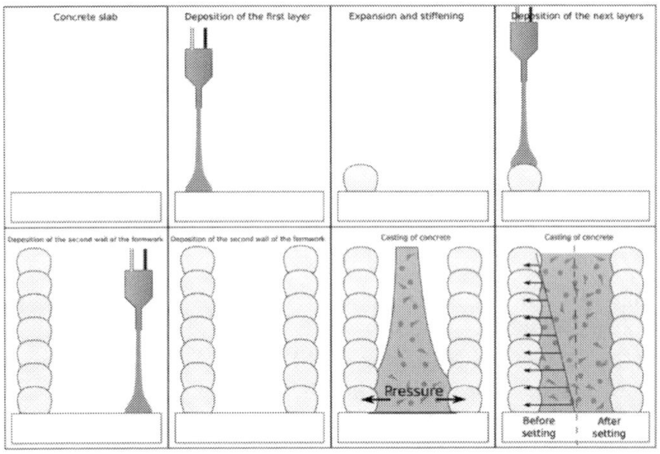

Fig. 3. – Different phases for the erection of the wall.

Fig. 4. – Different parts for the material deposition.

3 Yhnova^{TM}

3.1 A social house

The technique Batiprint3D^{TM} has been applied in september 2017 to the construction of a social house for NMH called Yhnova^{TM}. The name has been chosen as the house has the shape of a Y and because its construction is based on an in-NOVA-tive technique. It is a one level 95m^2 dwelling located in Nantes in a social housing neighbourhood. Figure 5 and 6 show different views of the house designed by TICA architecture. It is located on a parcel with protected trees (aerial part and root system). Consequently, TICA designed a house that does not invade the tree area. Therefore, Yhnova™ will respect not only the trees but also the environment of people who are used to benefit the proximity of the park. Tica tried to fulfil its credo of respect of both the environment and the social context of the building.

Yhnova™ is one of the three projects granted by Nantes City Lab, an action that aims to promote the innovation in the Metropole of Nantes and the collaboration between public and private partners. It encourages the partners (University, firms, SME, startups, associations, users and institutions) *to build together the city of tomorrow*.

Fig. 5. Views of Yhnova™.

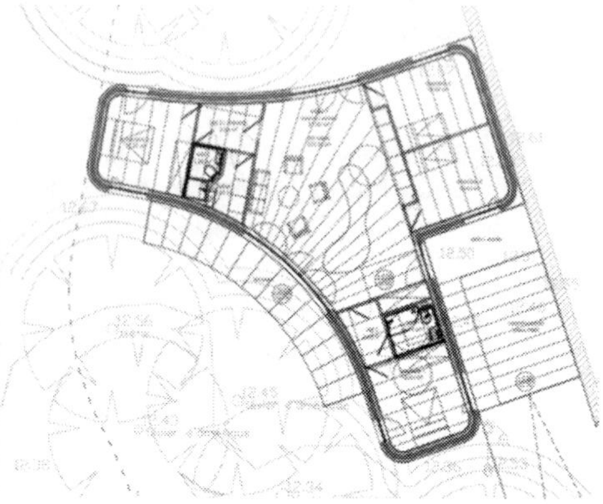

Fig. 6. Plan of Yhnova™.

3.2 Partners of the project

Yhnova™ involves many actors: academic partners, institutional partners, federations, private firms.

3.2.1 Academic partners

The technology Batiprint3D™ has been developed at the University of Nantes (France) by two research teams from:

• the GeM (Research Institute in Civil Engineering and Mechanics – UMR CNRS 6183 University of Nantes/Centrale Nantes/CNRS)

• the LS2N (Laboratory for the Sciences of Numerics – UMR CNRS 6004 University of Nantes/Centrale Nantes/Institut Mines Telecom Altlantique/INRIA/CNRS)

3.2.2 Institutional partners

CSTB – It is the French Building Research Institute. Its role is to guarantee the quality and the security of buildings. It is an expert of the techniques of construction and is in charge of evaluate the projects Batiprint3D™ and Yhnova™, and especially the impact of the technique on the other subtrades or its reliability regarding the seismic behaviour of Yhnova™.

Nantes Metropole (NM) – Nantes is the sixth metropole in France with 600.000 inhabitants. Its population is young as 36% of the population is less than 25 years old. The mayor of Nantes Johanna Rolland aims at having Nantes become a major actor of the evolution of the society by promoting social cohesion, sustainable development and innovation. In this latter aim, Nantes Metropole inaugurated in march 2017 the Nantes City Lab, a new manner to encourage innovation.

Nantes Metropole Habitat (NMH) – It is a social landlord located in Nantes. It owns around 25.000 dwellings (houses or apartments) in the Metropole of Nantes and dwells nearly 50.000 persons. NMH is always in search of innovation directly applicable to the dwellings and was interested in participating to the development of our technique of additive manufacturing for the construction. It will be the owner of Yhnova™.

SATT Ouest Valorisation – It is a public society dedicated to the transfer of technologies from the university laboratories to the industrial environment. It aims at helping the researchers to increase the TRL (Technology Readiness Level) of their developments and promote the intellectual protection of the discoveries.

3.2.3 Federations

French Building Federation (FFB) – It gathers 50.000 subscribers and 45.000 enterprises, acting in the sector of the construction of buildings.

Novabuild – It is a cluster in the Region Pays de la Loire (France), which promotes the eco-construction, the energy transition, the environmental transition, the numerical transition and the social transition in the context of the building industry

3.2.4 Private firms

BA Systèmes – It is specialized in the mobile robotics and in particular in the manufacturing and the development of AGV.

Bouygues Bâtiment – It is a global actor of the construction and is present in 80 countries. It claims that innovation is the first additional value for construction. It is part of several projects of additive manufacturing or robotization, applied as well to the construction of buildings as to the construction of infrastructures.

Carretero Meyer – It is a construction enterprise located in the region of Nantes. Even if it is a SME, it promotes innovation and has been part of the project since the beginning. The main interest of the enterprise for the development of additive manufacturing for the construction of dwellings is the fact that in the developed countries, it is more and more difficult to renew the retired employees, due to the arduousness of the jobs. Additive manufacturing could be a solution to this problem of unattractiveness.

Lafarge-Holcim – It is one of the worldwide leading provider of construction materials (cement, gravels, concrete). In the project, it supports the development of the concrete mix.

PRB – It is a manufacturer and supplier of building coatings.

TICA Architecture – It is the architect of the project Yhnova™. TICA is preoccupied by the social and the environmental aspects of a project of construction. TICA designed the house Yhnova™.

3.3 Deployment on the construction site

The figure 7 show different steps of the construction of Yhnova™. We can notice that the whole construction site is protected by a tent so that the robots can work in safe environmental conditions. In this section we point out the main problems that occurred during the construction of Yhnova™.

3.3.1 Construction site nuisance

During the construction of the walls, the robots worked a total of 54h. Nevertheless, the robot worked from 8AM to 7PM and did not work at the week end, as the owner did not want to disturb the neighbourhood with working noises. Especially, the need for an air compressor and an electrical power unit that are quite noisy, impeached the robot to work at night. Moreover, the concrete delivery could not occur at night because the concrete plant was closed and because this is also a noisy operation (engine noise, reverse beeper…). This shows that even if we have the means to make the robots work continuously, we cannot do so because of the neighbourhood and because of the opening time of the concrete plant. For now, the increase in the building rates allowed by 3D printing is thus still a mean term objective. But, even if the industrial context evolves to match the requirement of a continuous work of the robots, the concern in the quietness for the neighbourhood will be difficult to manage at night.

3.3.2 Interferences with the printhead

Prior to bring the robots on the construction plant, the lower slab have to be cast and set and the reinforcing bars must be placed, especially near the wood frames placed at the location of the openings.

When we brought the robots on the slab, we noticed that the slab had not the predefined shape. The tolerance on the staking out of the slab the workers are used to do not match the tolerance required by the robots. Indeed, as explained above, the robots can accurately position the print head with a tolerance of 1mm. This was not the case nor for the slab neither for the reinforcement bars. As a consequence, we had to make the masonry enterprise move the rebars prior to play the deposition paths, in order to avoid interference between the print head and the rebars.

Moreover, a ancient wall as close to the slab and the true position of the wall had not been determined accurately, in particular the vertical shape of this wall was irregular. There were thus differences between the true shape and the shape taken into account to generate the trajectories of the print head. This obliged us to redefine the trajectories on site.

The same problem occurred with the fluid networks (water and heat supply).

We can see that the tolerances in the staking out of the different elements of the building and the consideration of existing elements close to the robots operating zone is of major importance. This will oblige the construction sector to lower the tolerances to match those of an automated construction of the building.

4 Conclusions

In this paper, we presented briefly the projects Batiprint3D™ and Yhnova™. The first was dedicated to the development of a new technology of construction based on 3D printing. The second was dedicated to the use of this technology to build a social dwelling for a social landlord in Nantes.

In Batiprint3D™, to use a PU FW in which the SCC is poured allowed us to work with a concrete with simpler properties than those usually required by the extrusion of cementitious or clayey materials. Moreover, it allowed us to obtain more easily the authorizations to build our house as the justification of the structural safety of the walls relied on usual calculation procedures. Indeed, we had not to justify the structural safety of a layered concrete wall but of a wall very similar to those poured in classical formworks.

The fact to deploy our technology on a real construction site after development at the laboratory taught us a lot about the context of construction and how the 3D printing can match the standards of construction or how the standards of construction have to evolve to include this new construction technology.

Nevertheless, the development of 3D printing in order to increase the construction rates is for now still a potential objective as this requires to rethink the management of the construction phase, the delivery of concrete and the opening time of the concrete plant, but also to find a way to limit the noise nuisance for the neighbourhood.

Thus, even if we did the job and delivered the walls of Yhnova™ on time, a lot is still to do to make of 3D printing a mature construction technology.

Fig. 7. The construction site of Yhnova™ – a) Deposition of the 6 first layers on the slab showing the reinforcement bars – b) Deposition of PU showing the AGV, the PAA and the protective tent – c) Timber frame for the roof – d) Inner curved plaster boards – e) Outer view of the walls – f) View of the wall covered with external coating.

The authors would like to thank the project's research engineers who worked hard to make the project a success: Laurence Guiheneuf, Nicolas Houssais, Thomas Bressac, Romain Ponson.

References

1. J. Chaparro-Pelaez, A. Pereira-Rama, F.J. Pascual-Miguel, J. Bus. Res. **67**(5), 673–679 (2014)
2. *Reinventing Construction : a Route to Higher Productivity* (McKinsey&Company, 2017)
3. ASTM F2792-12a (2013)
4. B.H. Jared, M.A. Aguilo, L.L. Beghini, B.L. Boyce, B.W. Clark, A. Cook, B.J. Kaehr, J. Robbins, Scripta Materialia **135**, 141 – 147 (2017)
5. T. Friedman, *When complexity is free* (The New York Times, 2013)
6. I. Krimi, Z. Lafhaj, L. Ducoulombier, Additive Manufacturing, **16**, 107 – 114 (2017)
7. N. Labonnote, A. Rønnquist, B. Manum, P. Rüther, Automation in Construction **72**, 347– 366 (2016)
8. I. Hager, A. Golonka, R. Putanowicz, Procedia Eng. **151**, 292 – 299 (2016)
9. Ph. Poullain, B. Furet, E. Paquet, S. Garnier, *ACE Workshop - Lille* (2017)
10. C. Gosselin, R. Duballet, Ph. Roux, N.Gaudillière, J. Dirrenberger, Ph. Morel, Mat. and Designs **100**, 102–109 (2016)
11. B. Khoshnevis., Automation in Construction **13**(1), 5 – 19 (2004)
12. http://contourcrafting.com/ (January 2018)
13. J. Zhang, B. Khoshnevis, Automation in Construction **29**, 50 – 67 (2013)
14. http://www.yhbm.com/ (January 2018)
15. http://www.lboro.ac.uk/enterprise/case-studies/3d-concrete-printing/ (January, 2018)
16. S. Lim, R.A. Buswell, T.T. Le, S.A. Austin, A.G.F. Gibb, T. Thorpe, Automation in Construction **21**, 262 – 268 (2012)
17. https://d-shape.com/ (January, 2018)
18. http://www.skanska.co.uk/about-skanska/innovation-and-digital-engineering/innovation/3d-printing/ (January, 2018)
19. https://www.constructions-3d.com/ (January, 2018)
20. http://www.xtreee.eu/ (January, 2018)
21. http://apis-cor.com/en/ (January, 2018)
22. https://cazza.co/ (January, 2018)
23. http://www.batiprint3d.fr (January, 2018)
24. http://www.totalkustom.com/ (January, 2018)
25. http://news.mit.edu/2017/3-d-printing-buildings-0426 (January, 2018)
26. https://www.cybe.eu (January, 2018)
27. betabram.com (January, 2018)
28. https://3dprinthuset.dk (January, 2018)
29. A. Więckowski, Automation in Construction **83**, 19–28 (2017)

MATEC Web of Conferences **163**, 01002 (2018) https://doi.org/10.1051/matecconf/201816301002
MATBUD'2018

Evaluation of suitability for 3D printing of high performance concretes

Maria Kaszyńska[1], *Marcin* Hoffmann[1], *Szymon* Skibicki[1,*], *Adam* Zieliński[1], *Mateusz* Techman[1], *Norbert* Olczyk[1], and *Tomasz* Wróblewski[1]

[1]West Pomeranian University of Technology, al. Piastow 17, Szczecin, 70-310, Poland

Abstract. The article presents overview of additive manufacturing for concrete structures. Study focuses on specific tests used to determine suitability of high performance mixes for 3D printing. The tests include determination of compressive strength and evaluation of printing speed, extrudability and overall surface quality. Tests were performed on a High-Performance Concrete mix with fine natural aggregate up to 2 mm. Mineral additives such as silica fume and fly ash, and superplasticizer were added to obtain proper consistency. The tests were performed using specially designed site consisting of Cartesian robot and pumping module. The design of printed paths were tailored for specific tests. The evaluation of pump performance was made by measuring pumped mix volume in time. The determination of correct pumping speed was made based on the visual quality of the printed layers. Study determines also the ability to print multilayered structures and printability window of proposed mix.

1 Introduction

Additive manufacturing techniques for building construction have been intensively developed in the recent years. The beginning of this novel technology can be dated to mid '90s. The first technique known as Contour Crafting was invented at the University of Southern California [1–4]. The idea of additive manufacturing have also been studied at University of Loughborough [5–7]. The British research team has developed their own method of printing concrete, that does not utilize trowelling plates. Different approach called D-Shape, utilizes ideas from FDM (*Fused Deposition Modelling*) techniques to create structures using support materials [8]. Since mid-2012 many research teams and companies have taken interest in the additive manufacturing with concrete [3,9].

The major issue with the 3D concrete printing is designing a mix with proper rheological properties for a specific time window [3,4,6,10–12]. There are no standardized methods for determining required properties of the concrete mixes for 3D printing. Many research facilities have proposed their own frameworks for laboratory testing of printing concrete mixes [3–6].

* Corresponding author: szymon.skibicki@zut.edu.pl

The mixtures are evaluated in terms of extrudability, buildability, flowability, open time and, after curing, compressive strength. Those properties are necessary for proper extrusion and forming of 3D structures.

Extrudability (quality of being extrudable) is the mix ability to pass from the tank through the pump and hose to the nozzle. The concrete must be extrudable with structural integrity, without discontinuity and segregation, maintaining consistency throughout whole process. To determine the extrudability one can print several single layers with certain length and evaluate occuring blockage and tearing/fracturing [5,13]. Other method, called Print Quality, was proposed by the team from University of Southern California [10]. Printed layers are visually tested for surface defects, inadequate cohesion, rounded edges and changes in dimensions.

Workability/Flowability is a rheological parameter of mix consistency in time. There are several Standards dedicated to determination of this characteristic. Le et al. [6] used British test for soils BS1337-9 :1990, while Kazemian et al. [10] performed test according to American Standard ASTM C1437-15. Other methods include Hagermann flow table and mini cone test. Many research teams [4,12–16] correlate the flowability and extrudability with rheological properties. A team from Technische Universität Dresden modified the Sliding Pipe Rheometer to perform the tests according to their assumptions [12]. French research team [16] studied a time-dependant characteristics of the mix, proposing a new framework based on the plastic deformation of printed layer-by-layer structures.

Open time, also known as printability window, is a period in which the mix maintains proper pumpability. The mix must also maintaining proper quality and adhesion in a layer-by-layer structural build-up. Le et al. [5] and Secrieru et al. [12] defined the open time based on the shear stress test in accordance with British BS1377-9:1990 and German DIN 398. Ma et al. [13] proposed to test the open time with the V-Funnel method, Vicat apparatus, jump table, penetration tests and mini cone. Additionally, every couple minutes a layer is printed and measured to determine the dimensions and quality. A team from USA [10] prints 1.2 m long layers in a set period of time. Open time for concrete mixes used in 3D printing should not exceed 90 minutes from water-cement contact. The results of a study [6] showed that the time can be extended with additional use of superplasticizers. It needs to be remembered though that this decreases the quality of the concrete. A research team [13] has determined that the working time should be limited to 50-80 minutes.

Buildability is an important parameter of printed layers. It determines the feasibility of fresh mix for layer-by-layer printing and its resistance to deformation under the pressure of following layers. The parameter is strictly correlated with workability and extrudability of the mix. In order to test the buildability, research teams [5,7,10,13,14,17] have proposed printing several layers on top of each other and determine the shape stability and resistance to deformation. Other method, developed for sole purpose of evaluating shape stability, was proposed in [10]. The cylinder stability test proceeds by applying 5.5 kg load on a cylindrical Φ40x80 mm sample. Mix deformation determines the usability of the mix. Different rheological models are used to determine the buildability of fresh concrete mixes in time [4,15,16]. The buildability of the mix allows to determine the speed of layerwise construction.

There are several types of concrete mixes used in the additive manufacturing. First includes mixes with water-binder ratio of $w/b \approx 0.4\text{-}0.5$ and VMA and superplasticizers for regulating the consistency [10,18]. The compressive strength after 28 days reaches 50-60 MPa. Second type of mixes with $w/b \approx 0.23\text{-}0.3$ have addition of fly ashes and micro silica. The consistency is acquired by adding superplasticizer [6,14,17,19,20]. The compressive strength of those concretes at 28-day age reaches up to 125 MPa. Some studies [16,21], discusses use of mixes with 50-50 content of cement and microfillers like kaolin or limestone powder, with additional superplasticizer. Some mixes incorporate small addition (1.2 kg/m^3) of 12 mm long, 0.18 mm in diameter steel fibres [6,10,13,20].

The article presents printing process of concrete layers used for determination of geometrical characteristics, pump capacity, nozzle movement speed and multilayer buildability. This is a pilot research for testing technical capabilities of a 3D printer of own design.

2 Materials and experimental procedure

Based on the research conducted at Loughborough University [6] a new High-Performance Concrete was designed with w/b=0.23. The density of fresh mix was determined as 2168.25 kg/m^3 (CoV=0.24%, n=3). Preliminary tests determined the 28-day strength of 113.7 MPa (CoV=1.13%, n=4) on 100x100x100 mm samples. Preliminary tests were performed to qualify the mix for the 3D printing purposes. The mix composition is visible in Table 1.

Table 1. Concrete mix composition

Mix	[kg/m^3]
Cement CEM I 52.5R	579
Fly Ash	165
Silica Fume	83
Superplasticizer	8.33
Water	185
Aggregate 0-2 mm	1241

Mix preparation and printing were made in a laboratory at temperature of 20°C (± 2°) and relative humidity of RH=55% (± 5%). The printing was made using Cartesian robot of own design connected to a pumping module (Fig. 1a).

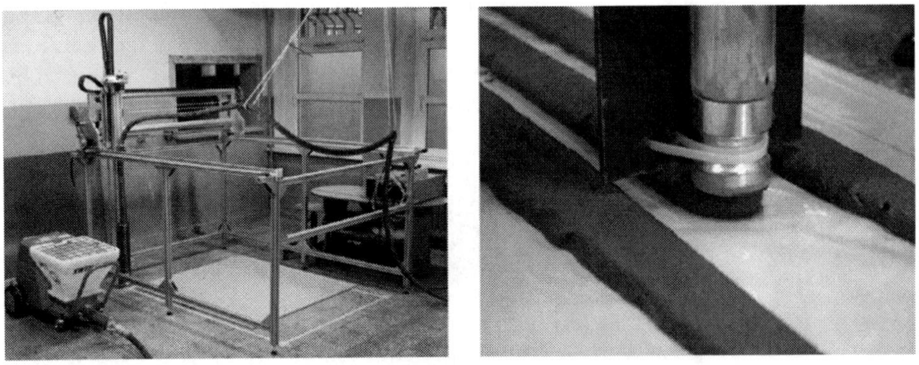

Fig. 1. Cartesian robot with pumping module during concrete printing.

The first test was conducted on printed layers with changing geometry. Test was performed to evaluate two things, the correlation between layer geometry and pump output and multilayer buildability (Fig. 2b). The nozzle with internal diameter of D=25 mm, moves 15 mm above from the printed surface. Test were performed for a constant movement speed of v=50 mm/s.

a)

b)

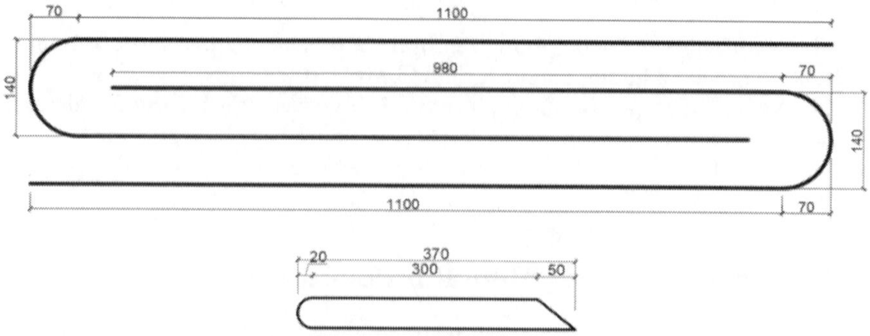

Fig. 2. Designed pathways for determination of layer properties: a) pump output; b) buildability – view on a single layer of printed wall.

3 Test results

3.1 Extrudability and quality issues of printed layers

During the first stage several layers were printed to evaluate the influence of pump output on geometrical characteristics of the layers and their visual quality. The quality of the layers was evaluated using classification presented in a study [10]: surface defects, inadequate cohesion, edge shape and printed layer dimensions. Oval shape of the layer (with increased edge radius) affects negatively the adhesion zone between two consecutive layers, thus decreasing bearing capacity of printed structure. Proper cross section should be symmetrical without rounded edges. The top surface should be flat with width similar to nozzle diameter. Figure 3 presents different cross-sections of layers printed with different pump output. At 2.88 l/min output (Fig. 3a) the layer is 61 mm wide and its edges are rounded significantly. Increased amount of air bubbles is visible in the cross section. With a lower output of 2.12 l/min (Fig. 3b) the layer is 45 mm wide, while at the output of 1.33 l/min the width met assumed requirements. The latter layer had also the least rounded upper edges. Increase in the layer width was noticed on the curves. This was caused by the decrease of the nozzle linear movement speed on curves, without simultaneous reduction of pump output.

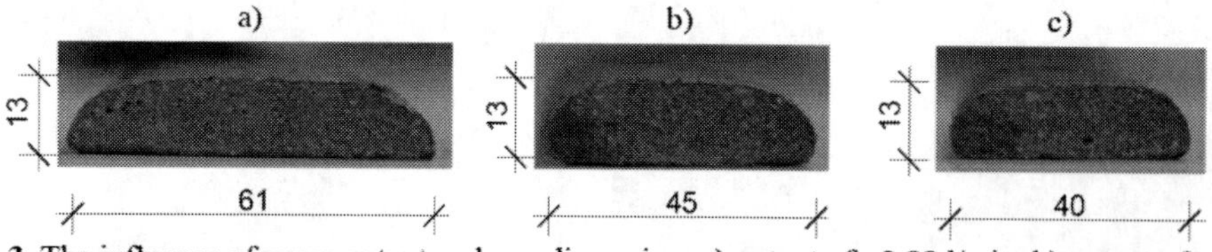

Fig. 3. The influence of pump output on layer dimension: a) output of =2.88 l/min; b) output of = 2.12 l/min; c) output of =1.33 l/min .

The extrudability of concrete is directly correlated with rheological characteristics of the mix and has to be controlled throughout the whole process of additive manufacturing. Based on the results [6] the open window was assumed between 15-60 minutes for designed mix. Exceeding the printability window causes the layer to lose continuity and consistency, exhibiting tearing (Fig. 4a). The surface of printed layer exhibit multiple defects and does not meet the requirements of the extrudability evaluation from [5,10]. The tears decrease significantly the mechanical properties of printed structure. Printing during the open time resulted in a continuous mix with stable geometry and without air bubbles (Fig. 4b). The mix presented in this study met the requirements of the extrudability evaluation.

a) b)

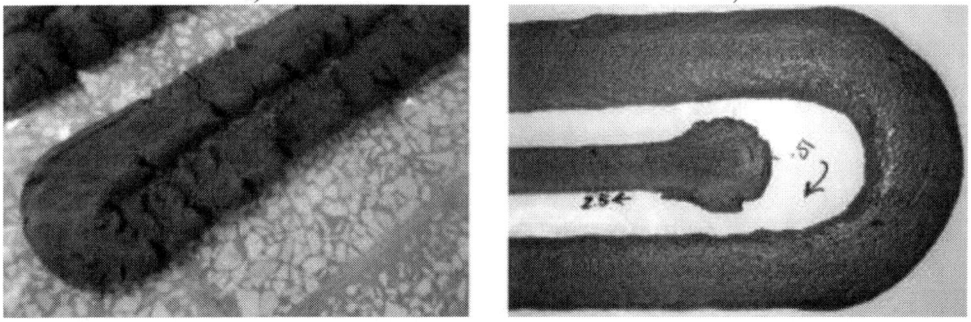

Fig. 4. The influence of open time on the print quality a) open time over 60 min, b) open time between 15 – 60 min.

3.2 Test buildability

The property of the concrete mix to retain its shape under own load and following layers is tested by printing multilayer structure. In this study the test was performed by printing 13 layer on top of each other as a continuous path, with constant 1.33 l/min pump output (Fig. 5a). After visual evaluation it was noticed that consecutive layers were not deformed, and the surface did not exhibit any defects or air bubbles. There was a visible decrease in a linear movement speed of the printing nozzle at the turns of designed path. As the pump output remains the same, the amount of material at the turns has increased causing thickening of the layer, thus additional load. In result there was a visible outflow of laitance (Fig. 5b).

a) b)

Fig. 5. Buildability test: a) printed multilayer structure, b) laitance at the turn of printed structure.

4 Preparation of control code for 3D printer

Previous tests allowed to determine various characteristics necessary for designing of 3D structures with use of additive manufacturing. To prove the findings additional 3D structure, in which the nozzle moved along conical spiral was designed. Designed cone was 345 mm high, with base diameter of 200 mm. Assumed layers were 15 mm high and 40 mm wide (Fig. 6a). Based on proposed geometry a 3D structural model was prepared (Fig. 6b). The visualization of G-Code is visible in Figure 6c [22]. Printing speed was set to v=39 mm/s and pump output to Q=1.33 l/min. Figure 6d shows the printed cone structure. As seen in Figure 6d the cone did not collapse, proving proper characteristics of designed mix and assumed printing setup.

 https://doi.org/10.1051/matecconf/201816301002

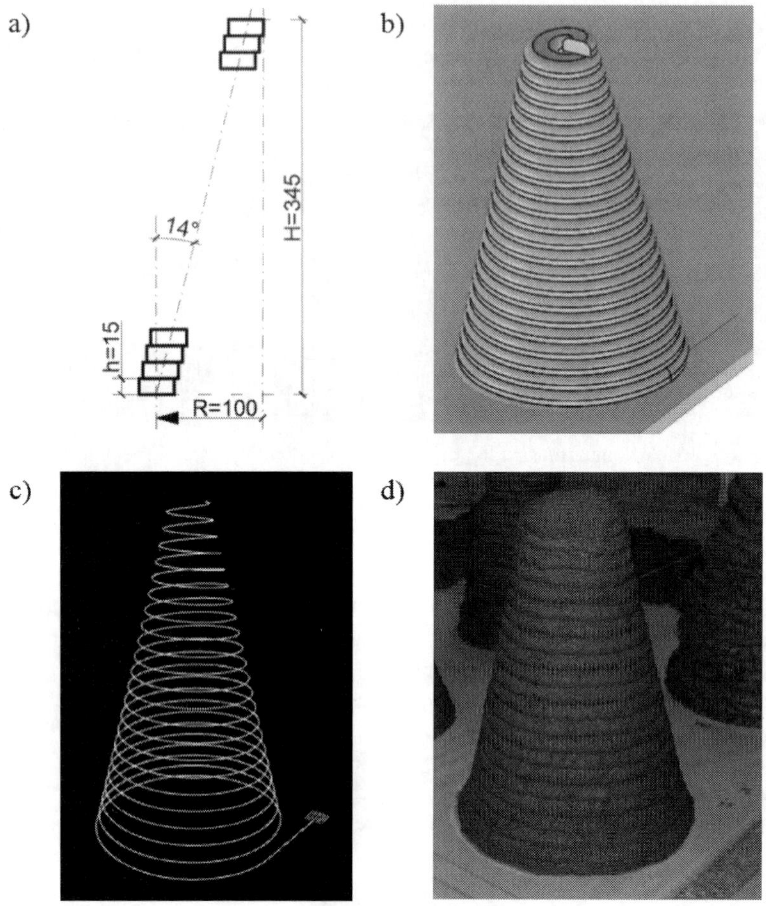

Fig. 6. (a) conical spiral geometry, (b) 3D model, (c) G-code simulation [22], (d) printed structure.

5 Conclusion

Performed tests allowed to correlate the nozzle movement speed and pump output in a manner that produced homogenous layers with constant geometry. Produced layers were evaluated in accordance to different classifications found in the literature. Increasing pump output affected the shape and geometry of the layers. Even though the cross-sections were symmetrical, produced layers had higher width with rounded upper edges.

Study have showed that the open time factor is one of the most significant in 3D printing of concrete. Exceeding the assumed time caused the drop in consistency of the mix, limiting its extrudability and lowering its surface quality.

Proposed design methods and tests allowed to produce structures that met all of the assumed criteria.

References

1. B. Khoshnevis, D. Hwang, K.-T. Yao, Y. Zhenghao, Int. J. Industrial and Systems Engineering **1**, 3 (2006)

2. B. Khoshnevis, Automation in Construction **13**, 1 (2004)
3. F. Bos, R. Wolfs, Z. Ahmed, T. Salet, Virtual and Physical Prototyping **11**, 3 (2016)
4. T. Wangler, E. Lloret, L. Reiter, N. Hack, F. Gramazio, M. Kohler, M. Bernhard, B. Dillenburger, J. Buchli, N. Roussel, et al., Digital Concrete: Opportunities and Challenges. RILEM Letters **1** (2016)
5. T. Le, S. Austin, S. Lim, R. Buswell, A. Gibb, T. Thorpe, Mater. Struct. **45** (2012)
6. T. Le, S. Austin, S. Lim, R. Buswell, R. Law, A.G.F. Gibb, T. Thorpe, Cem. Concr. Res. **42**, 3 (2012)
7. S. Lim, R. Buswell, T. Le, S. Austin, A. Gibb, T. Thorpe, Automation in Construction **21** (2012)
8. G. Cesaretti, E. Dini, X. Kestelier, V. Colla, L. Pambaguian, Acta Astronautica **93** (2014)
9. http://www.3dprinting architecture.net/?p=601 (January 2018)
10. A. Kazemian, X. Yuan, E. Cochran, B. Khoshnevis, Constr. Build. Mater. **145** (2017)
11. N. Labonnote, A. Rønnquist, B. Manum, P. Rüther, Automation in Constr. **72** (2016)
12. E. Secrieru, S. Fataei, C. Schröfl, V. Mechtcherine, Constr. Build. Mater. **144** (2017)
13. G. Ma, Z. Li, L. Wang, Constr. Build. Mater. **162** (2018)
14. V. Nerella, M. Krause, M. Näther, V. Mechtcherine, *25th Conference on Rheology of Building Materials* (2016)
15. N. Roussel, F. Cussigh, Cem. Concr. Res. **38**, 5 (2008)
16. A. Perrot, D. Rangeard, A. Pierre, Mater. Struct. **49**, 4 (2016)
17. Z. Malaeb, H. Hachem, A. Tourbah, T. Maalouf, N. El Zarwi, F. Hamzeh, Int. J. Civ. Eng. Tech. **6**, 6 (2015)
18. B. Khoshnevis, S. Bukkapatnam, H. Kwon, J. Saito, Rapid Prototyping Journal **7**, 1 (2001)
19. L. Anell, *Concrete 3d printer* (Master Thesis, Lund University, 2015)
20. J. Nylund, A. Järf, K. Kekäle, J. Rönnskog, F. Al-Neshawy, P. Kiviluoma, P. Kuosmanen, *10th International DAAAM Baltic Conference "INDUSTRIAL ENGINEERING"* (May 2015)
21. C. Gosselin, R. Duballet, P. Roux, N. Gaudillière, J. Dirrenberger, P. Morel, Mater. Des. **100** (2016)
22. https://nraynaud.github.io/webgcode/ (January, 2018)

Influence of the w/c ratio and aggregate composition on the autogenous shrinkage in self-consolidating concrete

Adam Zielinski[1,*], and *Maria* Kaszynska[1]

[1] Faculty of Civil Engineering and Architecture, West Pomeranian University of Technology, Piastów 50, Szczecin, Poland

Abstract. Paper presents the analysis of shrinkage development in self-consolidating concretes (SCC). Concretes were cured for 28 day in an insulated environment. The shrinkage was tested on 35x150x1150 mm samples. The specimen had the same volume as ring specimen proposed in the ASTM method of testing the susceptibility of concretes to cracking caused by the restricted radial shrinkage. Linear deformation of concrete samples was measured in constant periods of 500 s using dial gauges with digital data loggers. Conducted tests showed the influence of w/c ratio of 0.28, 0.34, 0.42 and aggregate composition on the development of the autogenous shrinkage in self-consolidating concretes. Additionally, rheological properties of the concrete mixes were tested and compressive and splitting strength was determined. Conducted research allowed to evaluate the influence of w/c ratio and natural aggregate composition on the development of autogenous shrinkage in different stages of curing of SCC.

1 Introduction

The early age cracking of High-Performance Concretes (HPC) is a result of increasing internal stresses caused by the restraining free deformation of the element. Deformation can be limited by various internal and external factors. External factors include: surrounding formwork, casting on existing, rigid structure or improper expansion joints. Internal stresses are mostly caused by the improper rebar deployment or concrete mix design, including the water-cement ratio, non-deformable natural aggregate and high cement amount [1].

Autogenous shrinkage of concrete is a dangerous phenomenon, especially for flat structural elements such as bridge slabs, floors and walls. Use of the High Performance Self-Consolidating Concretes (HPSCC) with w/c<0.4 speeds up the construction process, simultaneously improving the strength and durability [2,3]. However, such elements exhibit unintentional cracking. The cause of the cracking of concrete slabs lies not in the hydration heat or expansion factors, but in the simultaneous action of two shrinkages – drying and autogenous. In case of the flat concrete elements with a high surface area, the drying shrinkage will also cause higher deformation. Drying shrinkage is reversible and its effect

* Corresponding author: adam.zielinski@zut.edu.pl

can be minimized with a proper external curing. In case of the autogenous shrinkage, which is irreversible, the external curing does not significantly improve the condition of the element. There are several methods of the internal curing that can limit the autogenous shrinkage [4,5]. One of the most basic methods is to increase the w/c ratio.

The article presents the analysis of the influence of the water-cement and water-aggregate ratios on the autogenous shrinkage of cement composites with constant binder composition. The analysis of the shrinkage development was performed after 1 and 28 days. Mixes used in the study exhibited similar rheological properties and were analyzed based on their material ratios. Rheological properties were acquired with use of a superplasticizer.

2 Mixture proportions and specimen

The study was performed on four Self-Consolidating mixes. Half of the mixes were made using only fine natural aggregate, the other using fine and coarse natural aggregates. Three mixes had water-cement ratio of w/c<0.35 that is usually found in HPC. The fourth concrete had w/c=0.42, typical for ordinary concretes. The latter was used as a reference for evaluation of the autogenous shrinkage in HPC.

Authors used Portland Cement CEM I 42.5 R, mineral additives of fly ash and silica ash and superplasticizer. The amount of mineral additives in the mixes is constant in relation to cement mass. The superplasticizer was added in amount between 0.5% and 2.5% to maintain the assumed consistency. Mix designs are shown in Table 1.

Table 1. The mix design.

Mix	Cement [kg/m^3]	SCM [kg/m^3]		Water [kg/m^3]	SP [kg/m^3]	Natural aggregate [kg/m^3]	
		Fly Ash	Silica Fume			0-2 mm	2-8 mm
M-1	725	116	61	305	3.4	1004	-
M-2	794	127	67	222	18	1100	-
M-3	450	72	38	155	8	624	1072
M-4	458	73	39	128	11.5	636	1092

Table 2 presents the material ratios in prepared mixes: water-cement, water-binder, cement-aggregate and binder/aggregate.

The autogenous shrinkage of prepared concretes was measured in a test bench of own design. The test bench was situated in a climatic room with constant environmental conditions of T=20 °C ± 2 °C and RH = 50% ± 5%. Three slab samples 35x150x1150 mm were made from each mix. The volume of the samples was equal to the volume of the ring samples proposed by the ASTM C 1581-16. The displacement was measured with digital gauges on both ends of the insulated samples. The deformation was registered every 500 s and recorded in a database.

Table 2. Material ratios of studied mixes.

Mix	w/c [-]	w/b [-]	c/a [-]	b/a [-]
M-1	0.42	0.338	0.722	0.898
M-2	0.28	0.225	0.722	0.898
M-3	0.34	0.28	0.265	0.330
M-4	0.28	0.225	0.265	0.330

Four times longer samples as in the ASTM C 1698-09, high precision of the gauges and short measuring intervals allowed to determine the development of the autogenous shrinkage. The test allowed to precisely determine the inflection points on the shrinkage development curve in the first 24 hours. Stable surrounding conditions and automatic measurement constituted to the low standard deviation of the values of autogenous shrinkage in concrete samples. Figure 1 presents the test bench during the shrinkage test of M-1 mix.

a) b)

Fig. 1 Test bench for registration of linear shrinkage: a) dial gauges, b) general view

3 Test results and discussion

The first step was to obtain the assumed consistency in rheological tests. In case of insufficient flowability of the mix, additional amounts of the superplasticizer were added. Moulds 10x10x10 cm were casted to acquire samples for strength tests. Prepared mix was then taken to the climatic chamber and casted into the test site.

3.1 Rheological and strength properties

The evaluation of self-consolidating properties was made based on the Slum Flow test in accordance with EN 12350-8 – for mixes with fine and coarse aggregate; in accordance with [6] for mixes with fine aggregate only. The study assumed the slump flow between 600-750 mm for mixtures with fine and coarse aggregate and 240-260 mm for mix with fine aggregate only. Additionally, in accordance with ACI 237R-07 authors evaluated the resistance to segregation using Visual Stability Index (VSI). Results of the rheological and strength tests are visible in Table 3.

Table 3. Rheological and strength properties

Mix / Composite	M-1	M-2	M-3	M-4
Slump-flow, D_{max}[mm]	250	260	720	610
Visual Stability Index	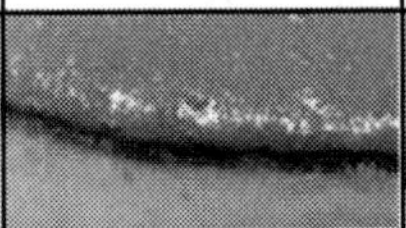 VSI0	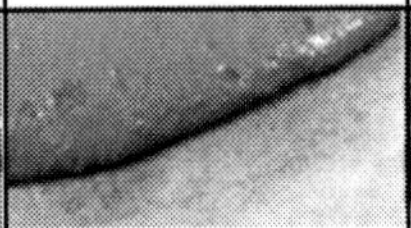 VSI0	 VSI0	 VSI0
Compressive strength (28 d.) $f_{c,10x10}$ [MPa]	65.2	102.9	99.6	100.5
Splitting tensile strength (28 d.) $f_{ct,sp}$ [MPa]	4.47	4.73	4.51	5.21

3.2 Free linear autogenous shrinkage test

3.2.1 Water/cement ratio

The influence of the w/c ratio on the development and values of the autogenous deformation was studied for all mixes.

Figure 2 presents the development of autogenous deformation in M-1 and M-2 mixes during the first 24 hours. Figure 3 presents the shrinkage development in the same mixes in a longer, 28-day period. Figure 4 shows the daily development of autogenous shrinkage as percentage of the overall shrinkage after 28 days.

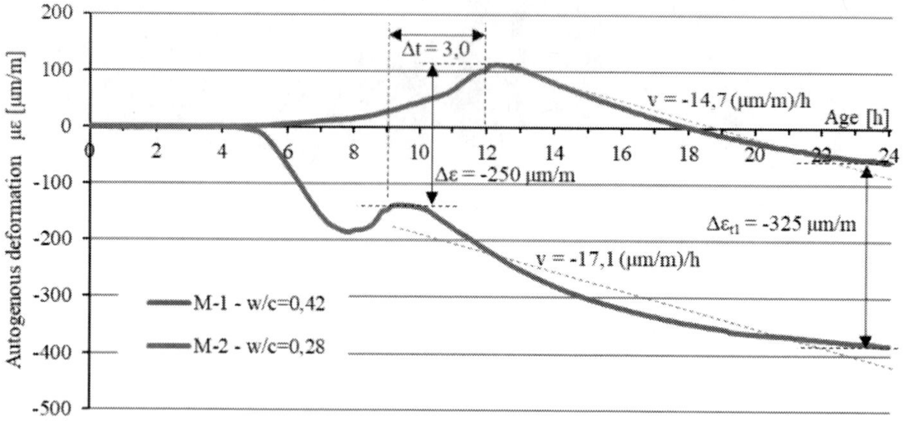

Fig. 2 The influence of the w/c ratio on the value of autogenous deformation of M-1 and M-2 composites in the first 24 h

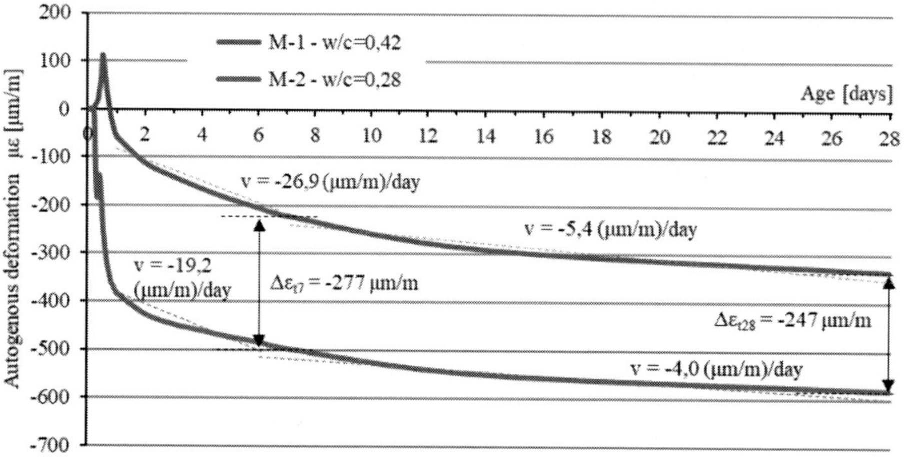

Fig. 3 The influence of the w/c ratio on the value of autogenous deformation of M-1 and M-2 composites during 28 days

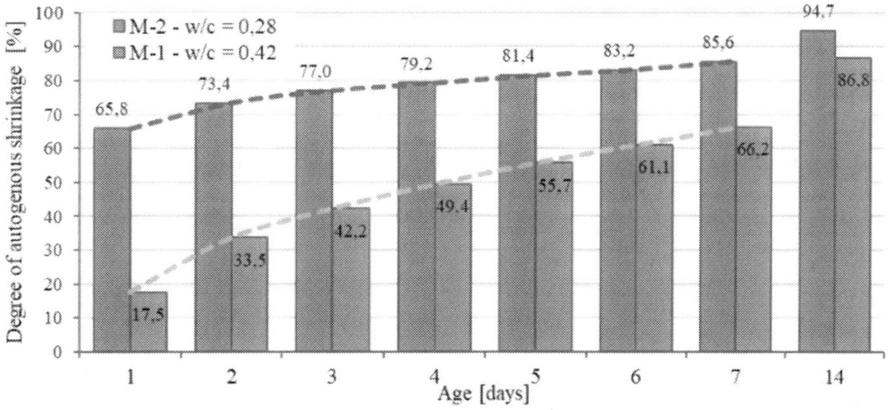

Fig. 4 Daily development of autogenous shrinkage as percentage of the overall shrinkage after 28 days

Following conclusions can be drawn from the aforementioned Figures:

- The reduction of the w/c ratio from 0.42 to 0.28 changed the mechanism behind the development of autogenous deformation. The M-1 composite exhibited swelling, while M-2 exhibit autogenous shrinkage. Reduction of the w/c ratio to 0.28 causes a shortage in mixing water;
- Increase of the w/c ratio causes increment of the autogenous shrinkage by about 15% after a day, 55% after 7 days and 40% after 28 days comparing to composite with the lower w/c ratio;
- Reduction of the w/c ratio from 0.42 to 0.28 affects the development speed of the autogenous shrinkage. Composite with the lower w/c has higher rate by about 16% in the first 24 hours, but exhibit lower rate during the next 6 days by about 40%. Between 7th and 28th day shrinkage development rate for the composite with higher w/c ratio is higher by about 30%.

Figure 5 presents the development of the autogenous deformation in M-3 and M-4 mixes during the first 24 hours. Figure 6 presents the shrinkage development in the same mixes in a longer, 28-day period. Figure 7 shows the daily development of the autogenous shrinkage as percentage of the overall shrinkage after 28 days.

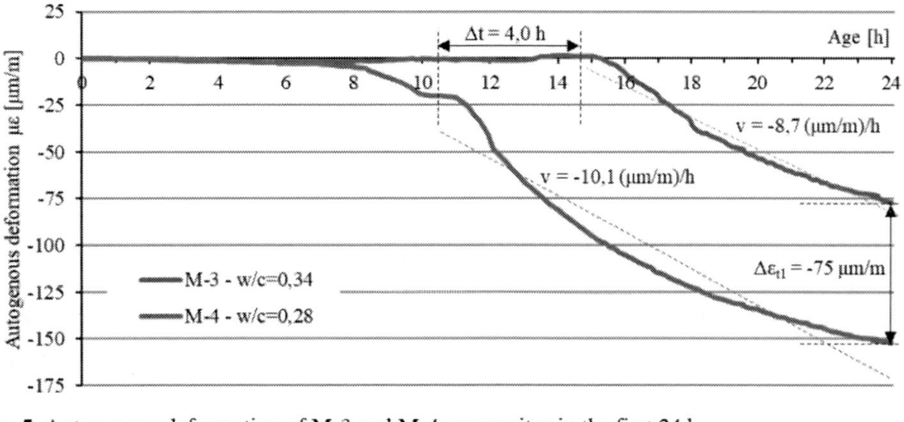

Fig. 5 Autogenous deformation of M-3 and M-4 composites in the first 24 h

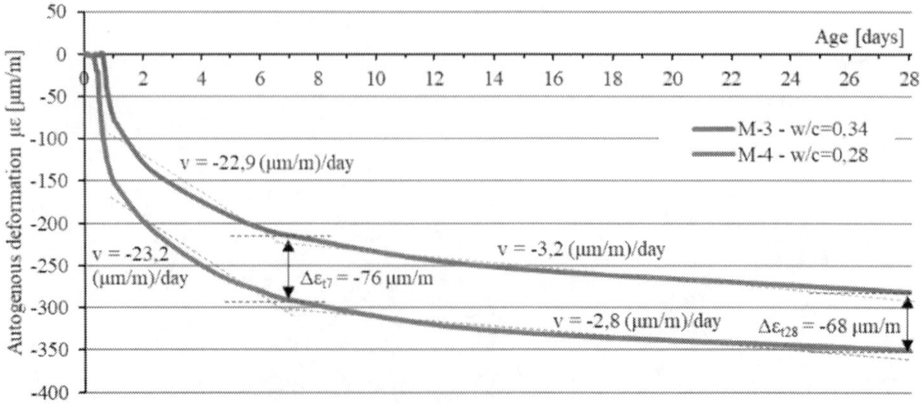

Fig. 6 Autogenous deformation of M-3 and M-4 composites during 28 days

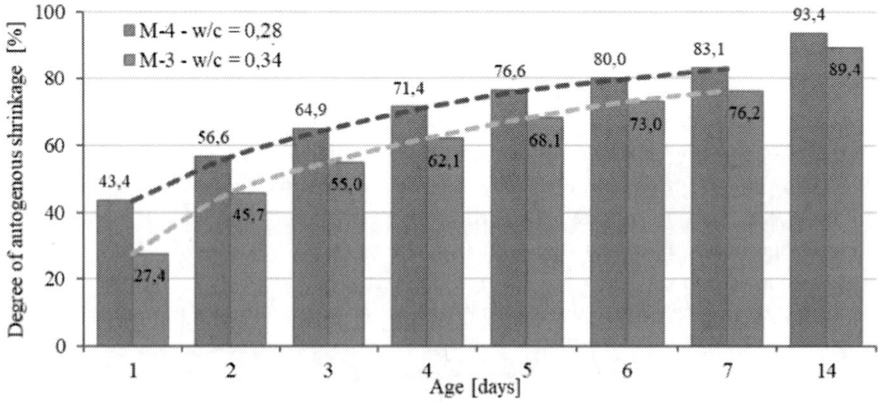

Fig. 7 Daily development of autogenous shrinkage as percentage of the overall shrinkage after 28 days

Following conclusions can be drawn from the aforementioned Figures:

- Reducing the w/c ratio from 0.34 to 0.28 caused the autogenous shrinkage value to rise by about 50% after a day, 25% after 7 days and 20% after 28 days;
- Reducing the w/c ratio from 0.34 to 0.28 increased the development rate of the autogenous shrinkage by 14% in the first 24 hours. The difference in the development rate however decreases to about 1% during following 6 days. After 7 days higher shrinkage development rate (12%) was visible in a concrete with higher w/c;
- Autogenous shrinkage of M-4 composite begins to develop just after water-cement contact, while in M-3 only after hardening.

Presented analysis shows that the composites with lower w/c ratio have higher autogenous shrinkage which develops at higher rate in the first 24 hours. However, during the hardening phase the development slows down and begins to stabilize. In case of concretes with higher water content, in the first 24 hours the value of autogenous shrinkage is lower, but its development rate during hardening phase is higher.

The development of the autogenous shrinkage in composites with the higher w/c ratio is slower but progresses for a longer time up till the later stages of hardening. The effect is caused by the increased amount and size of the capillary pores. Composites are thus more susceptible to volume changes during the hardening.

3.2.2 Aggregate composition

The analysis of the influence of the aggregate composition on the development and value of autogenous deformation was conducted on M-2 and M-4 composites with w/c ratio of 0.28. The composites had respectively cement-aggregate ratios of 0.722 and 0.265.

Figure 8 presents the development of the autogenous deformation in M-3 and M-4 mixes during the first 24 hours. Figure 9 presents the shrinkage development in the same mixes during a longer, 28-day period. Figure 10 shows the daily development of autogenous shrinkage as percentage of the overall shrinkage after 28 days.

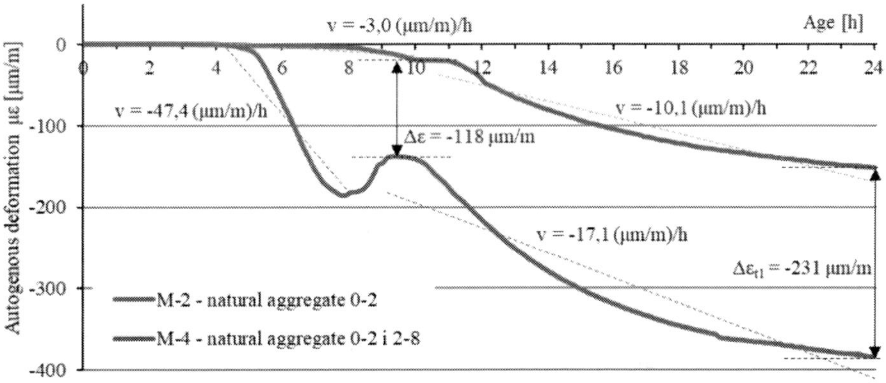

Fig. 8 Autogenous deformation of M-2 and M-4 composites in the first 24 h

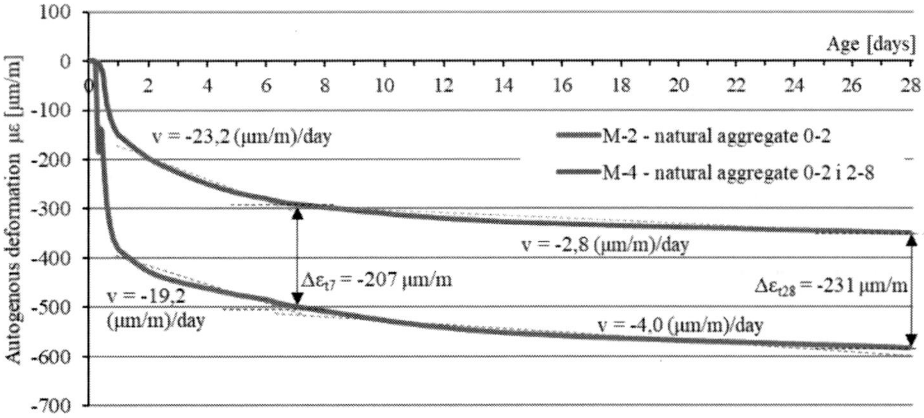

Fig. 9 Autogenous deformation of M-2 and M-4 composites in the first 28 h

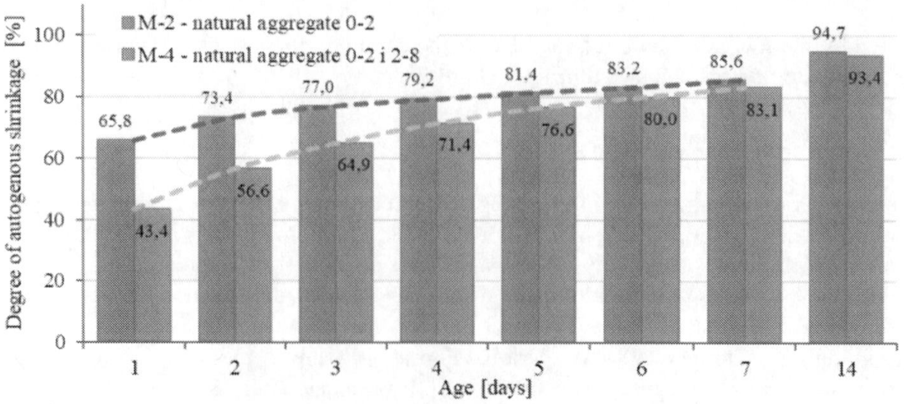

Fig. 10 Daily values of autogenous shrinkage as percentage of the total shrinkage

Following conclusions can be drawn from the aforementioned Figures:

- Increasing the amount of aggregate by replacing the cement in M-4 aggregate decreased the autogenous shrinkage by 60% after a day, 40% after 7 days and 40% after 28 days compared to M-2 composite;
- Increasing the amount of non-deformable aggregate reduces the shrinkage during the first stage of hydration process. It also decreases the development rate of shrinkage by 40% right after the end of setting time;
- Decreasing the volume of aggregate influenced the development rate of shrinkage. The M-2 concrete exhibited 50% of the overall shrinkage value after first 24 hours, while M-4 composite reached about 30%. The development of the early age shrinkage is correlated with the chemical reactions during setting. The higher the amount of the cement, the higher the autogenous shrinkage;
- Development rate of the early age shrinkage (from day 1 to 7) is higher by about 20% in composite with higher amount of aggregate. The increase is caused by a higher capillary pressure in the cement matrix. The pressure is increased due to water deficit, higher in M-4 composite. Higher aggregate amount requires higher water demand, which, if not included, might limit the bonding water;
- Similar rates of the shrinkage development are visible in M-2 and M-4 concretes after 7 days of curing. It can be noticed that hardened cement matrix is more resistant to autogenous deformation.

4 Conclusions

Performed tests and analysis allowed to draw following conclusions:

- Reduction of the w/c ratio increases the value of autogenous shrinkage and causes it to stabilize faster as the result of less developed network of capillary pores in hardening concrete;
- Composites with lower w/c ratio exhibit intensive and high autogenous shrinkage during setting time, which decreases and stabilizes in the hardening phase;
- Composites with higher w/c ratio exhibit lower shrinkage or swelling in the early stages of setting. During the hardening phase the shrinkage develops at a faster rate than in composites with lower w/c ratio;
- Natural aggregate reduces the value and development rate of autogenous shrinkage;

- Development of autogenous deformation is more intensive in the first 24 hours of concrete curing.

References

1. P.C. Aïtcin, *High Performance Concrete* (1998)
2. M. Kaszynska, A. Zielinski, Procedia Engineering **108**, 608-615 (2015)
3. M. Kaszynska, A. Zielinski, Brittle Matrix Composites **10**, 265-274 (2012)
4. D. Bentz, J. Weiss, Technical Report (2011)
5. K. Kovler, O.M. Jensen, RILEM Report **41** (2007)
6. G. de Schutter, P. Bartos, P. Domone, J. Gibbs, *Self-Compacting Concrete* (2008)

MATEC Web of Conferences **163**, 01004 (2018) https://doi.org/10.1051/matecconf/201816301004
MATBUD'2018

Low cement content SCC (Eco-SCC) – the alternative for ready-mix traditional concrete

Maciej Urban[1,*]

[1]Cracow University of Technology, 24 Warszawska St. PL-31-155 Cracow, Poland

Abstract. Self-Compacting Concrete (SCC) is often regarded for in-situ applications as a more expensive and less environmentally friendly version of traditional concrete for use in places where the latter is highly inconvenient for various reasons. To counteract this belief, a new type of SCC has been proposed, called Eco-SCC or green-SCC. In comparison to typical SCC this building material has lower powder content resulting in low cement content and a slightly lower paste volume. The disadvantage of this approach is high water content and low viscosity resulting in high vulnerability of fresh concrete to sedimentation. Additionally, comparing to EN-206 limits, mixes with too low cement content are often obtained. This is why a research program has been undertaken to check the possibility of obtaining fresh Eco-SCCs of higher viscosity, thereby fulfilling all EN-206 requirements. It was possible to obtain concretes of C25/30-C35/45, SF2, VF2, PL2, containing 265-300 kg/m^3 of cement and 165-190 kg/m^3 of water owing to the use of limestone powder and typical rounded aggregate of local origin with very low sand to a total aggregate ratio of s/a = 0.4. Based on Global Warming Potential (GWP) analysis, a modification of Eco-SCC definition is proposed: Eco-SCC is an SCC having cementitious materials volumetric content no greater than 100 dm^3/m^3 (or mass content no greater than 315 kg/m^3).

1 Introduction

The building industry often treats Self-Compacting Concrete (SCC) as a more expensive and less environmentally friendly alternative to traditional concrete for places where the use of traditional concrete is, for a variety reasons, highly questionable (e.g. complicated formwork, very dense reinforcement, limited access etc.) [1]. To overcome this belief, a new type of SCC has been proposed, called Eco-SCC or green-SCC [2, 3]. According to the definition proposed by Wallevik [1, 4], this material should contain a maximum 315 kg/m^2 of binder. Binder here is defined as the sum of particles lower than 1/8 mm, excluding aggregate particles. In many respects this is a doubtful assumption. Firstly, the paste volume needed to obtain the self-compactibility limit is correlated to the loose voids volume in the aggregate skeleton [5]. This means that such a kind of concrete has to contain high water doses resulting in low viscosity and low segregation-resistance. An example of this problem is shown in [2]. It was possible to obtain only SF1, VF1 consistence classes and C16/20-C25/30 strength

* Corresponding author: maurban@pk.edu.pl

classes (acc. to EN-206) using c.a. 200 kg/m^3 of water, resulting in $w/c > 0.75$ so were too high for use in practice (nominally with exception for X0 exposure class). Then, balancing 'on the edge' with durability requirements is the second important drawback of this approach. Thirdly, judging from the point of view of water and superplasticizer demand, the author in [5] has shown that aggregate fumes act in the same way as other additions. Thus, there is no reason to distinguish between these materials and treat them differently, especially as even inert additions, due to their densifying effect, play a significant role in enhancing both mechanical and durability properties of SCC [6-8]. Fourthly, according to common practice (see e.g. [9,10]), SF1, VS1 SCCs have a very limited usability. Lastly, according to ERMCO contemporary data [11], more than 50% of ready mix concrete production is within the range C25/30-C30/37 with a tendency to rise. The market share of lower strength classes (and so X0 exposure class) is lower than 20% with a tendency to regress. Additionally, the mean value of cementitious materials' equivalent content is within the range 290-320 kg/m^3. For comparison, typical exposure classes in the C25/30-C30/37 strength range require at least 280 to 300 kg/m^3 of cement equivalent, maintaining a maximum w/c limit in the range 0.55 - 0.65. This means that these ranges have to be used as the target for Eco-SCC design, assuming that this kind of concrete should be used more often than accidentally (as, according to [11], SCC shares c.a. 3-4% of the ready-mix concrete market). This is why the Eco-SCC limit should be altered so as not to exceed 315 kg/m^3 of cement equivalent materials, treating all inert additions (as defined in EN 430-1) as a part of aggregate. Minimizing the content of inert additions is also important, but only from the point of view of lowering the overall cost of concrete, as its environmental cost (see Table 1) is low.

This is why in the following part of the article this target range is used to check the possibility of Eco-SCC design in practice, additionally assuming that at least a VS2 bottom consistence limit and VF2 viscosity are to be obtained. These two last requirements were chosen to widen the resulting Eco-SCC applicability.

2 Materials

Considering the aforementioned target values, a binary binder made of cement and an inert addition was chosen. This is why a commercial limestone powder (lp) dedicated for the use in SCC was selected. According to [1] the lowest values for carbon footprint are typically reached for cements containing ground granulated blast-furnace slag, therefore commercial cement of CEM II/B-S 32.5R type was selected. Its mean strength value measured according to EN 196-1 was 47.5 MPa. To fluidize the mixes, commercial superplasticizer (sp) of polycarboxylate type was used (density: 1.07 kg/dm^3).

Aggregate was composed of river material obtained from the same local quarry. It consisted of mid-sand and sorted rounded coarse aggregate (Carpathian flysch). All granular compositions used are shown in fig. 1.

3 Design method

SCC compositions were designed according to the original method shown in [5], and then modified for the purpose of this research. Aggregate composition was designed following the original procedure (to obtain minimum voids in a loose state). Additionally, PL2 passing ability class (i.e. L-box, 3 bars) was assumed. To assess this feature, Van and Montgomery criterion [12] was used. The target value was $V_a = 660 - 670$ dm^3/m^3. The final composition is shown in Fig. 1. The sand to total aggregate ratio (s/a) of this composition is only 40% (compare to e.g. [2,13], where much higher values were used).

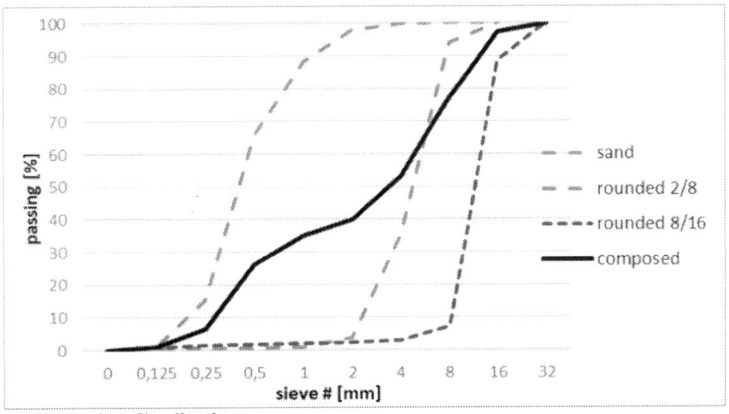

Fig. 1. Aggregate sieve size distribution.

Paste compositions were prepared assuming a *w/c* ratio on 0.5, 0.55, 0.6, 0.65 and 0.70 levels. To obtain higher viscosity (preferably VF2) water-to-binder ratio (*w/b*) was set in the range 0.33-0.41. This range was chosen to secure water content not higher than 185 kg/m^3, following typical Japanese recommendations [7]. Next, superplasticizer dose was optimized using the following procedure. Firstly, plain cement paste containing a maximum allowable dosage of superplasticizer (2.0% of binder mass *b*) was prepared and tested using standard Haegermann cone with rising *w/c* up to sedimentation limit D_{sed} (following the original procedure). Fluidity of this plain paste was treated afterwards as the target for testing paste with addition. Next, paste compositions with binder proportions fulfilling the aforementioned criteria were calculated. Then, following the original procedure, *w/b* ratio correction due to aggregate fumes (a_f= 2.5% of aggregate mass, measured by washing acc. to EN 933-1) was calculated assuming paste content (i.e. the sum of $V_w + V_c + V_{lp}$) equal to 330 dm^3 per 1m^3 of concrete. Final paste composition for testing was calculated by the equation (1)

$$\frac{w}{w+lp+a_f} = \frac{w_{corr}}{c+lp} \tag{1}$$

The paste (containing w_{corr}, *c and lp*) was tested with steeply rising *sp* dose to reach D_{sed}. Established this way *sp* content was treated as the final one for fresh concrete composition design, treating aggregate water demand as anti-sedimentation reserve for the paste.

Additionally, based on the same aggregate composition, five mixes treated as the reference ones were designed. Two of them were standard concretes (S2 consistence) with $w/c = 0.5$, $c = 345$ kg/m^3 and $w/c = 0.7$, $c = 268$ kg/m^3 (denoted in Tables 1-3 as 50S and 70S, accordingly). They were prepared for strength comparison with compliant SCC using the same cement content. Another three mixes were chosen to represent typical binder-rich SCC based on paste of the highest obtainable viscosity. The design procedure strictly followed the one shown in [5]. Maximum viscosity was obtained using $lp/c = 0.33$ and $sp = 1.8\%b$. This composition was used as the bottom extreme, denoted 33R. Next, following the same procedure and *lp/c* ratio, *w/b* ratios for $sp = 1.2$ and $0.6\%b$ were found. These mixes were denoted 40R and 55R, accordingly.

4 Testing

Dry aggregate mixture was prepared in a counter-current mixer. Ready-mixed paste was prepared in a bucket using a mechanical hand mixer. Superplasticizer was added premixed with water without delay. The mixing time was c.a. 75 s. Concrete mixes were obtained by adding the paste (immediately after preparation) to dry aggregate mix in an amount enabling the assumed consistence level to be obtained. The mixing time was c.a. 2 min. Then, fresh concrete was tested using slump flow, V-funnel and L-box PL2 tests. If the consistence was too viscous, the batch was remixed with additional paste portion, and the tests were repeated. Both compositions are shown in Table 1. In two cases, for *w/c*=0.65 and 0.70 (denoted 65-2, 70-3 and 70-4, accordingly), the mixes were redesigned: the first one because the passing ability was too low for the given cement content, and the second one because the paste and *sp* contents were too high. Lastly, 15 cubes of 100 mm rib were formed for compressive strength testing. Strength was evaluated after 1 day (3 samples), 28 days (6 samples) and 90 days (6 samples).

Table 1. Mix compositions.

Mix no.	c *	w *	lp *	sp *	a *	V_b **	V_{paste} **	V_c **	w/c	w/b
33R	503	165	165	12	1,542	223	388	162	0.33	0.25
40R	454	175	149	7.3	1,581	201.5	376.5	146.5	0.385	0.29
50S	345	173	0	1.75	1,835	113	286	113	0.5	0.5
50-1	336	167	168	6.5	1,689	170.5	337.5	108.5	0.5	0.33
50-2	345	172	171	8	1,664	174.5	346.5	111.5	0.5	0.33
55R	369	200	122	3	1,628	164	364	119	0.55	0.41
55-1	314	174	190	5	1,671	171.5	345.5	101.5	0.55	0.34
55-2	314	174	190	5.8	1,671	171.5	345.5	101.5	0.55	0.34
60-1	290	174	173	3.5	1,710	157.5	331.5	93.5	0.6	0.375
60-2	302	182	182	3.6	1,674	164.5	346.5	97. 5	0.6	0.375
65-1	278	182	165	2.6	1,710	150.5	332.5	89.5	0.65	0.41
65-2	278	181	176	3.4	1,703	154.5	335.5	89.5	0.65	0.37
70-1	256	177	209	4.2	1,698	159.5	336.5	82.5	0.69	0.38
70-2	268	187	220	4.4	1,651	167.5	354.5	86.5	0.69	0.38
70-3	253	176	187	3.3	1,724	150.5	326.5	81.5	0.7	0.40
70-4	268	187	198	3.5	1,674	159.5	346.5	86.5	0.7	0.4
70S	268	188	0	0.7	1,866	87	276	87	0.7	0.7
GWP ***	0.55	0.00034	0.032	0.72	0.00345	-	-	-	-	-

* Units: [kg/m^3]
** Units: [dm^3/m^3]
*** Units: [kg CO_2/kg]

5 Analysis

All results are shown in Tables 2-3. The strength results obtained on the same *w/c* ratio for SCCs and traditional concretes were between ± 4% range for all testing times, so that the difference was negligible. This means that in this particular case limestone powder was inert, without the dilution effect reported e.g. in [8,14], and might be treated as a part of aggregate.

PL2 passing ability class was obtained in the range of 335-350 dm^3/m^3 of paste, so that Van and Montgomery formula (see [12]) predicted this feature with reasonable accuracy.

GWP (Global Warming Potential, in kg CO_2 per $1m^3$ of concrete) values in Table 2 were calculated using data shown in Table 1 (taken from [1]). Additionally, a hypothetical substitution of used cement by CEMI or CEMIII/B by simple 1:1 method was assumed. The adequate GWP values are shown in the last two columns of Table 2. Next, correlations between this parameter and other material characteristics were checked. Only two significant relationships of GWP were found: with *w/c* ratio, see Fig. 2, and with cement content, see Fig. 3. For comparison, see also Fig. 4. This means that cement content is an absolutely dominating factor in terms of GWP assessment for SCC containing rounded aggregate and inert addition(s).

The Eco-concrete limit proposed in [1, 4] is 125 $kgCO_2/m^3$. The results shown in Table 2 and fig. 3 reveal that cement type (strictly: the level of Portland clinker replacement by low-GWP active addition) dominates all other aforementioned factors: only concrete compositions recalculated assuming CEM III/B usage fulfilled this condition (up to over 350 kg/m^3 content, thereby far exceeding the top limit of eco-concrete requirement given in the Introduction). The results for CEM II/B-S are c.a. 75% higher. In the case of CEM I the difference is c.a. 260%. For comparison: rising cement content from 250 to 450 kg/m^3 (mixes 70-3 and 40R, accordingly) resulted in c.a. 70% GWP increase (Fig. 3); rising of *w/c* ratio by c.a. 0.2 resulted in 26% GWP rise for Eco-SCC and 38% for binder-rich SCC (Fig. 2).

Table 2. Test results – Global Warming Potential (GWP) [kg CO_2/m^3] and consistence.

Mix no.	GWP	D_0 [cm]	t_{500} [s]	t_v [s]	L-box PL2	GWP** (CEMIII/B)	GWP** (CEMI)
33R	296	82	6.5	20.8	100%	171	433
40R	265	76	7	15.4	100%	152	388
50S	200	6*	-	-	-	113	295
50-1	201	70	7.5	34	60%	116	291
50-2	207	74	4.5	13	84%	120	299
55R	215	71	3	6	91%	122	314
55-1	188	70	7	25	78%	110	275
55-2	189	70	8	18	85%	111	275
60-1	174	68	5.5	16.6	70%	101	253
60-2	180	72	2.8	11	89%	105	264
65-1	166	66	3.5	15.1	52%	97	242
65-2	167	67	5	14.8	89%	99	244
70-1	156	65	7.5	33	80%	93	226
70-2	163	74	2.6	12	87%	97	237
70-3	154	63	8	46	44%	91	223
70-4	162	68	3.3	12	87%	96	236
70S	155	5*	-	-	-	88	228

* slump
** calculated assuming hypothetic substitution of used cement by CEMI (GWP=0.82) and CEMIII/B (GWP=0.3)

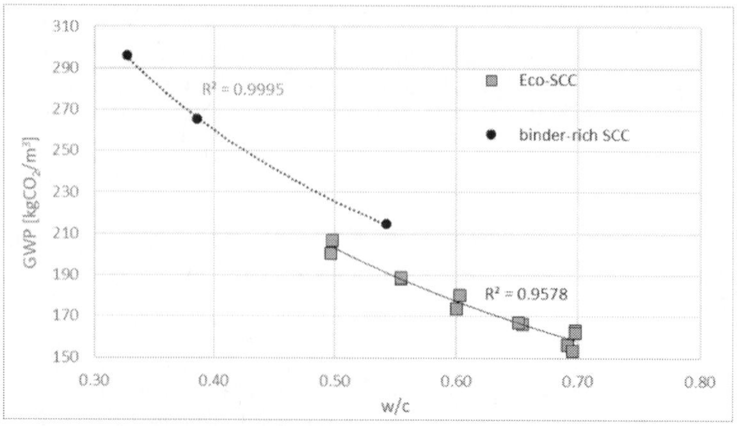

Fig. 2. Global Warming Potential (GWP) as a function of *w/c* for binder-rich and Eco-SCC.

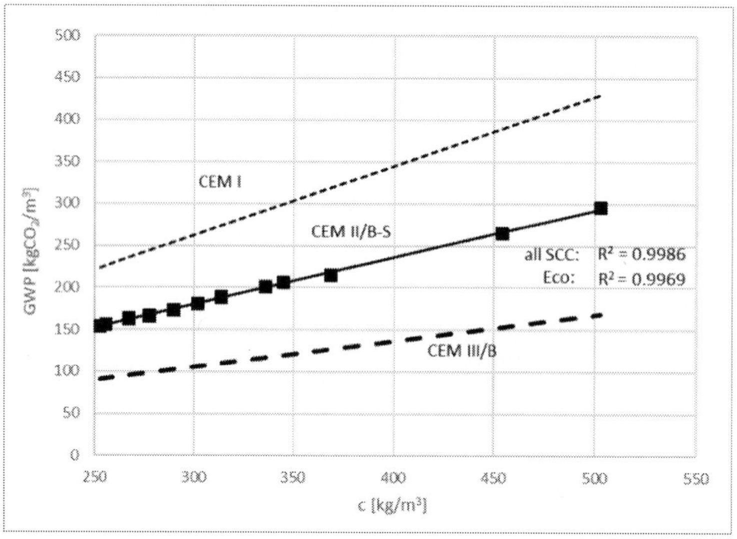

Fig. 3. Global Warming Potential (GWP) as a function of cement content. Solid line with data points – test results. Dashed lines –calculations for hypothetical cement 1:1 replacement.

Consequently, taking into consideration SCC stability requirements, it is clear that the limiting criterion for Eco-SCC should be cement (or cementitious materials) volume (or mass), not binder content, as proposed in [1, 4] (for comparison see Fig. 4). The presented data shows that the bottom limit of Eco-SCC is c.a. 80 dm^3/m^3 and the top limit should be established at c.a. 100 dm^3/m^3 (depending on exposure class requirements). Taking all of the above into consideration, the altered definition of Eco-SCC given in the Introduction (maximum 315 kg/m^3 of cementitious materials) is positively verified.

CEMII/B-S produced SCC with GWP of c.a. 165 $kgCO_2/m^3$ minimum, only 6% higher than comparable traditional concrete. For $w/c = 0.5$ this difference is almost negligible (3%). The difference between binder–rich SCC and Eco-SCC in terms of GWP is c.a. 15%, which is also acceptable. To sum up, the design proposed in [5] (with the alterations presented above) is well-suited for Eco-SCC design.

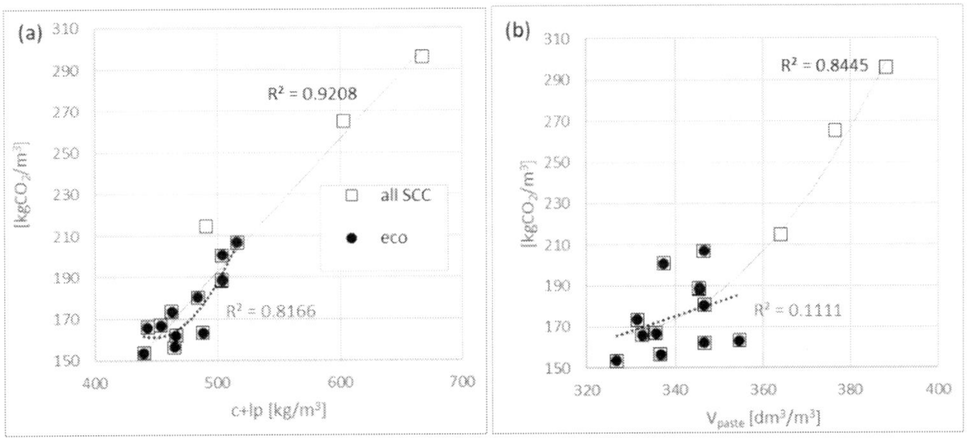

Fig. 4. Global Warming Potential (GWP) as a function of binder (a) and paste content (b).

Table 3 summarises compressive strength results, and strength and exposure classes assessment according to EN-206. The comparison of this data with the matching one concerning mix compositions (Table 1) demonstrates that the given exposure class was obtained using cement content close to the minimum allowed. In contrast, the strength class is typically higher, most probably due to the high strength reserve of the cement used (c.a. 15 MPa measured on standard mortar beams).

Table 3. Test results – Strength and exposure classes.

Mix no.	f_{cm} 1day [MPa]	f_{cm} 28 day [MPa]	f_{cm} 90 day [MPa]	$f_{ck,cube}$ [MPa]	Strength class	Exposure class
33R	15.1	85.4	93.5	77.1	C60/75	XC4,XS3,XD3,XF1,XA3
40R	12.8	66.2	74.6	58.9	C45/55	XC4,XS3,XD3,XF1,XA3
50S	11.4	51.8	58	45.2	C35/45	XC4,XS3,XD3,XF1,XA3
50-1	-	-	-	-	-	XC4,XS1,XD2,XF1,XA2
50-2	11.8	53.2	57.1	46.5	C35/45	XC4,XS1,XD2,XF1,XA2
55R	9.5	52.7	64.7	46.1	C35/45	XC3,XS1,XD2,XF1,XA1
55-1	-	-	-	-	-	XC3,XD1,XF1,XA1
55-2	12.1	52.5	60.5	45.9	C35/45	XC3,XD1,XF1,XA1
60-1	-	-	-	-	-	XC2
60-2	10.5	47.2	55.2	40.8	C30/37	XC2
65-1	8.5	44.5	52	38.3	C30/37	XC1
65-2	9.4	46.1	55	39.8	C30/37	XC1
70-1	-	-	-	-	-	X0
70-2	7.6	39.3	48	33.3	C25/30	X0
70-3	-	-	-	-	-	X0
70-4	7	39.5	49.3	33.5	C25/30	X0
70S	7.2	38	50.2	32.1	C25/30	X0

6 Conclusions

1) There is a possibility to obtain Eco-SCC of excellent fluid properties with only slightly higher GWP (Global Warming Potential) than comparable traditional concretes.
2) It was possible to reach the target specification using the minimum cement content and *w/c* ratio allowed by EN 206. The only exception is for the concrete strength class, where sometimes results at least one class higher than the minimum were obtained, this is most probably due to the fact that the cement had a very significant strength reserve (15 MPa) in comparison to the minimum standard requirement (acc. to EN 197-1).
3) Aggregate composed using minimum voids in a loose state criterion allowed to obtain a low sand content SCC fulfilling the PL2 requirement. The PL2 passing ability class was predicted owing to Van and Montgomery formula [12] with reasonable accuracy.
4) The presented data show that GWP value in SCC is governed by cement type. Other factors are of lower importance. Two such factors have been recognized. The first is *w/c* ratio and the second is cement content. This means that cement type and content is an absolutely dominating factor in terms of GWP assessment for SCC containing rounded aggregate and inert additions.
5) Properly chosen inert addition (here: limestone powder) may, in fact, be treated as a powder aggregate (of high GWP in comparison with a natural aggregate). This means that powder SCC always has a higher GWP than traditional concrete of comparable composition, but the difference could be small (here: 3 – 6%).
6) The obtained results lead to the conclusion that the definition of Eco-SCC proposed in literature should be altered, as following: Eco-SCC is an SCC having a cementitious materials content not greater than 315 kg/m^3 or 100 dm^3/m^3, depending on exposure class.

References

1. S. Grunewald, G. De Shutter, Concr. Plant. Intern. **10**(5), 34-38 (2016)
2. B. Esmaeilkhanian, K.H. Khayat, O.H. Wallevik, Mater, Struct, **50**, 124 (2017)
3. M. Urban, *Energy Efficient Sustainable Building Materials and Products,* 119-136, (*ed. I. Hager,* Cracow University of Technology Publishing, Kraków, 2017)
4. O.H. Wallevik, F.V. Mueller, B. Hjartarson, S. Kubens, *XVII IBAUSIL 1*, 1105–1116 (2009)
5. M. Urban, Mater. Struct. **48,** 1321-1341(2015)
6. Y. Xie, Y.Liu, G. Long, RILEM PRO054, 189-198 (2007)
7. G. De Shutter, P.J.M. Bartos, P. Domone, J. Gibbs, *Self-Compacting Concrete.* (Whittless Publ., 2008)
8. M. Urban, Czas. Techn. Politechniki Krakowskiej, seria 'Budownictwo' **14-B/2004**, 189-202 (2004) *(in Polish)*
9. J. Walraven, RILEM PRO033, 15-22 (2003)
10. *EFNARC Guidelines for SCC.* (EFNARC 2005)
11. *ERMCO ready-mix concrete industry statistics, year 2015.* (Ed. ERMCO, 2016)
12. B.K. Van, D. Montgomery RILEM PRO007, 373-384(1999)
13. F.V. Mueller, O.H. Wallevik, K.H. Khayat, RILEM PRO100, 25-34 (2016)
14. M. Tokyay, *Cement and concrete mineral admixtures* (CRC Press, 2016)

Influence of tensile cracking and of aggregate size on concrete permeability

Marta Choinska [1,*], *Hayder* Al-Khazraji [1,2], *Nathan* Benkemoun [1], and *Abdelhafid* Khelidj[1]

[1] Institut de Recherche en Génie Civil et Mécanique GeM - UMR CNRS 6183, Université Bretagne Loire, Université de Nantes, IUT Saint Nazaire, 58 rue Michel Ange, 44600 Saint-Nazaire, France
[2] Missan University, Engineering College, Iraq

Abstract. The aim of this study is to investigate the interaction between crack opening (COD), aggregate size and gas transfer in concrete submitted to mechanical loading in the Brazilian splitting tensile test. The lab-made devices have been developed to investigate physical phenomena during loading and to provide data to validate a developed mesoscale hydromechanical model by the same authors, based upon a 3D lattice approach to represent the heterogeneity of the material [1]. Experimental studies has been carried out on five materials with different aggregate sizes. The results emphasize that permeability of mortar increases with cracking following a sigmoid law, with the most important kinetics due to passing-through connected crack growth, after nonsymmetric one-face crack initiation. Furthermore, the obtained results highlight that permeability increase, due to aggregate size, may be separated from permeability increase due to tensile cracking: for all the five materials tested results fall on the same master sigmoid curve. This behaviour law represents a strong advantage for concrete modelling.

1 Introduction

Permeability is a parameter that may indirectly influence the durability of concrete structures by governing the penetration rate of aggressive agents responsible for degradation under a pressure gradient, or directly in the case of structures with tightness role. Hence, there is a clear motivation to investigate the variation of the permeability with cracking of concrete-like materials during loading.

On the one hand, existing experimental data on tensile behavior provide some correlations between material damage and permeability, but they are in general limited by the residual permeability measurements performed after complete unloading [2-5]. Some studies [6-7] have been performed during loading but concern only water permeability, which remains delicate for interpretation as far as autogenous healing phenomena and water flow reduction are met [8-9]. Therefore, there is a clear need of data on intrinsic permeability evolution during tensile loading.

On the other hand, aggregate size should influence mechanical behaviour and permeability. Bisschop and van Mier [10] observed drying shrinkage microcracking increase

* Corresponding author: Marta.Choinska@univ-nantes.fr

in cement-based composites with spherical glass aggregates size and with volume percentage increases. Grassl et al. [11] pointed out that the concrete permeability increases with the increasing of crack width and with the aggregate diameter. Grabiec et al. [12] highlighted the influence of the type and maximum aggregate size on some properties of high strength concrete. Fabien et al. [13] emphasized permeability increase with limestone aggregate size and compressive damage: higher aggregate specific surface of smaller aggregates impacts on higher cohesion and results in lower damage and lower permeability, independently of cement nature (CEM I or CEM III).

This work has therefore two main objectives: the first one concerns the evolution of permeability under tensile loading and the second one is focused on the effects of aggregate size on mechanical behavior and permeability. The authors use experimental methods to investigate these issues. The experimental results will be discussed in the light of comparison with numerical simulations based on a mesoscale model, developed by the same authors [1].

2 Experimental program

For the sake of simplicity, experimental study has been restricted to dry materials with mix proportions given in Tables 1 and 2.

Table 1. Mortar constituents and mix proportions.

Ingredient	Quantity (kg/m^3)
Cement (CEM I 52.5R PM-ES)	451
Sand (alluvial rolled Quartz, Loire) 0-4mm	1517
Water	248

Table 2. Microconcrete, concretes and macroconcrete constituents and mix proportions.

Ingredient	Quantity (kg/m^3)		
	Microconcrete	**Concrete**	**Macroconcrete**
Cement (CEM I 52.5R PM-ES or CEM III/A 52.5 PM ES)	350	350 or 350	350
Sand (Limestone) 0-4mm	868		
Medium aggregate (Limestone) 4-12mm	1009	432	-
Coarse aggregate (Limestone) 12-20mm	-	561	981
Superplasticizer (solid content 20%)	1.71		
Water	168		

Material presented in Table 1 corresponds to a mortar, with water-to-cement ratio equal to 0.55 and aggregate volume fraction equal to 60%.

All the materials presented in Table 2 have the same water-to-cement ratio, equal to 0.48, as well as the same aggregate volume fraction, equal to 66% of the total volume. All the aggregates are crushed limestone. In order to investigate aggregate size effect on mechanical and transport properties, three types of formulations have been proposed: microconcrete, concrete and macroconrete. Microconcrete contains sand and medium size aggregates, two concretes contain sand, medium and coarse size aggregates and macroconcrete contains sand and coarse size aggregates only [13].

Cylindrical specimens (diameter of 110 mm, length of 220 mm) of mortar, microconcrete, two concretes and macroconcrete have been prepared from a single batch and cast in steel or PVC rigid moulds. 24 h after casting, they have been stored in a room maintained at 20°C with a 95% relative humidity, and then cured in water at 20°C for 4 weeks. After the storage in water, the cylinders have been cut using a diamond blade saw in order to obtain three "disc" specimens (diameter of 110 mm, length of 47 mm) within each cylinder. Mortar "disc" specimens have then been dried at 80°C down to constant mass, and then placed in a desiccator to keep their dry state. Microconcrete, concretes and macroconcrete "disc" specimens have been dried at 80°C over 2 weeks, then at 105°C down to constant mass.

2.1 Tensile loading - gas permeability set-up

For many years, our lab (GeM) has been working and developing coupled mechanical and permeability tests. After a compressive loading - gas permeability set-up [14], we propose a tensile loading - gas permeability set-up [15], based on indirect tension in the Brazilian splitting test (compression in the diameter plane). The principle of this test is to perform gas permeability measurements during loading, for different levels of controlled cracking in the pre-peak and post-peak phases, using a "disc" specimen loaded in the Brazilian splitting test. This test presents the advantage of forming a single crack, parallel to the loading direction [2-5, 7, 16], where openings may be measured at one or two sides of a specimen, using f.ex. a COD sensor (crack-mouth opening displacement) as a feedback signal during the test [16]. Two devices, called a "simple" and a "double" one, have been developed in our lab and are presented in the next section.

2.1.1 Developed experimental devices

A simplified diagram of the "simple" loading-permeability device is shown in Fig. 1.

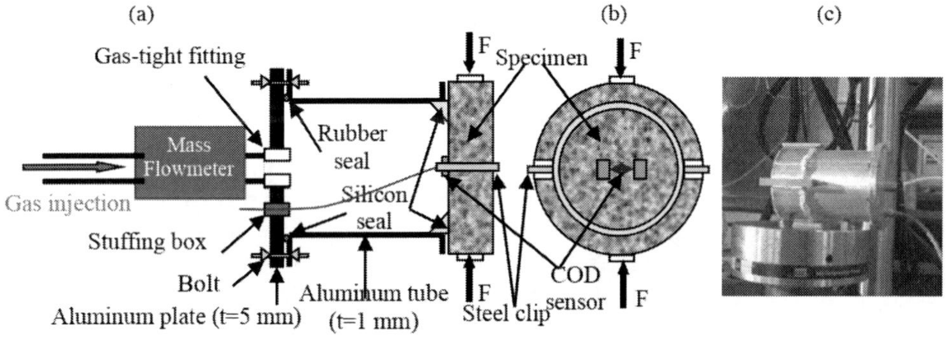

Fig. 1. Diagram of the "simple" experimental device to measure crack opening displacement (COD) and gas permeability during loading in the Brazilian splitting test: (a) side section view (b) front section view (c) overall view (GeM, Ecole Centrale de Nantes).

An inert gas (dry nitrogen) is injected into the cylindrical lightweight aluminium compartment, sealed to the specimen and maintained by two steel clips (see Fig. 1). The gas flow occurs in the direction perpendicular to the disc and thus parallel to the crack plane. In order to ensure a uniaxial gas flow and to prevent any parasitic gas leakage, the lateral face and the outer ring part of the up-stream side of a cylindrical specimen are protected by an adhesive aluminium film. The gas is injected at the up-stream side of a specimen at a pressure Pi, equal here to 0.2 MPa (absolute), the down-stream side being at atmospheric pressure (P_{atm}). Injection pressure is applied and maintained until gas flow stabilization. This pressure is recorded during the test, as well as the flow rate, parallel to the recording of mechanical data (COD, axial force F, see Fig. 1). The relative pressure and mass flow rate are measured up-stream of the specimen by a thermal mass flowmeter which converts a mass flow rate to an equivalent volumetric gas flow rate at normalized conditions (P_0, T_0) [14]. Using Darcy's law, mass balance equation and the thermal mass flowmeters principle, an apparent permeability k (m²) may be calculated as follows (1):

$$k(T_a) = \frac{2Q_i(T_0, P_0)P_0 \frac{T_a}{T_0} \mu(T_a)L}{S(P_i^2 - P_{atm}^2)} \quad (1)$$

where P_i : injection pressure (Pa), T_a : room temperature (K), $P_0 = 1.013 \cdot 10^5$ Pa, $T_0 = 273$ K, Q_i (T_0, P_0) : equivalent volumetric gas flow rate at normalized conditions (m³/s), μ : nitrogen dynamic viscosity (Pa·s), L : specimen length (m), S : specimen surface (m²).

In this study, a COD sensor, used as a feedback signal of the tensile test, is placed in the aluminium compartment, between two points, determined by the position of the steel plates stuck on one flat face of a specimen, at mid-height of the latter and across the potential failure plane (see Fig. 1(b)). The two points are distanced by 10 mm, sufficiently close to integrate only the area around the crack and sufficiently elongated to let the crack, not exactly rectilinear, pass through.

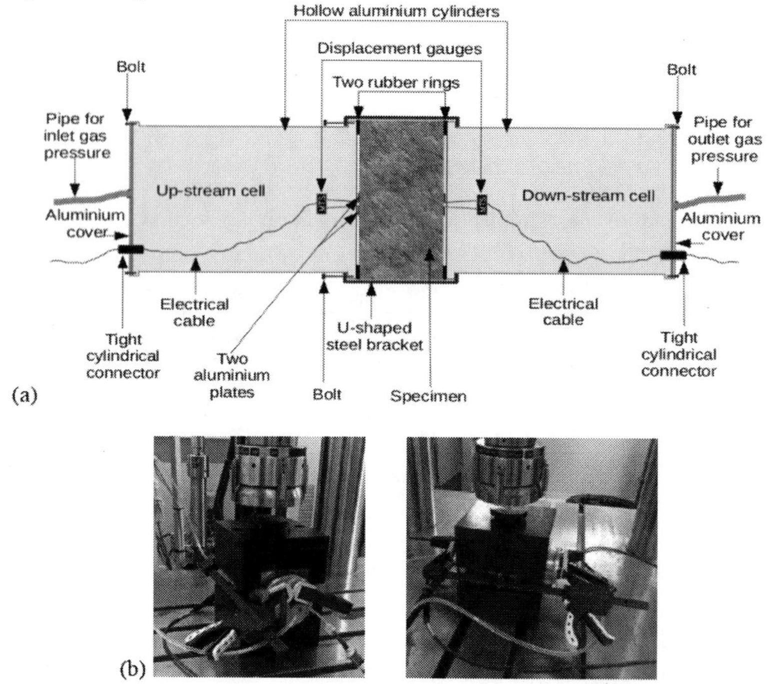

Fig. 2. (a) Side section view of the "double" loading-permeability device, (b) overall views of the "double" experimental device (GeM, Nantes University, IUT Saint-Nazaire).

The tests within the "simple" device, COD-controlled with a constant rate for loading and unloading, are performed within a hydraulic loading frame MTS 500 kN. A specimen, supplied with an aluminium compartment, is placed between two rigid metal platens of the loading frame equipped with plywood strips, in order to redistribute the load and prevent crushing at the loading points [2]. The second flat face of a specimen is left voluntarily free to allow the measurements of the displacement field by a Digital Image Correlation method, but the results won't be presented in this paper.

The second "double" device (see Fig. 2 (a)) includes two parts: one placed on the up-stream side and the second one on the down-stream side, each one fitted with a COD sensor:

COD 1 and COD 2. The principle is common to the first device, except the use of the incurved shape loading platens (see Fig. 2 (b)), instead of flat ones with plywood strips. Tightness and security elements are also modified: up-stream and down-stream cells are clamped in the perpendicular direction to the loading (see Fig. 2 (b)) and a 250 kN MTS loading cell is employed within a hydraulic loading frame CFM / MTS 1000 kN. Gas pressure and flow rates are measured here at the up-stream and at the down-stream sides and the feedback signal of the tensile test corresponds to the mean COD value between COD 1 and COD 2.

2.1.2 Validation and Testing procedure

The external diameter of the aluminium compartments is 10 mm lower than the diameter of a "disc" specimen in order to not interfere with the boundary conditions of the mechanical test. Once attached to a specimen using two steel clips, it doesn't impact the mechanical behaviour: global rigidity of a specimen is not impacted by the permeability device, even under gas pressure.

As a Brazilian splitting test is unstable in the post-peak phase [16], use of a COD sensor on the only one flat face specimen is highly risky. For this reason, before a test within the "simple" device, each specimen needs to be geometrically investigated and the larger diameter face needs to be chosen as the COD-driven to avoid loss of control. In fact, a slight difference of about 0.05-0.1mm in diameter between two faces, due to mould imperfections, is sufficient to perform tests successfully, despite nonsymmetric crack initiation. On the other hand, as the injected gas pressure occurs some flexural effects and therefore may promote crack propagation on the non-driven face within the "simple" device, some tests may not succeed by loss of control. For this reason, results of four mortar specimens will be presented, instead of six initially desired.

As the "double" device permits to control the tests by a mean COD value, there is no risk of collapse: all the tests, realised on micoconcrete, two concretes and macroconcrete succeded.

Testing procedure: Before a test, a specimen, supplied with loading-permeability device(s), is positioned on the lower platen of the loading frame and alignment control is performed (force driven test) when the upper platen approaches a specimen to contact it. After this stage, the controlled variable is switched and mechanical loading becomes COD-driven ("simple" device) or mean COD-driven ("double" device) with a constant rate of 0,02μm/s. In order to avoid any specimen failure within the "simple" device, due to coupled mechanical and gas pressure effects, after each loading step, a specimen is slightly unloaded, and then gas pressure imposed and permeability test performed. This procedure is maintained until cracking propagates to the non-driven face. The latter occurs well after the peak load and demonstrates like a step in F-COD response (see Fig. 3 (a)). Once the step is exceeded, permeability tests may be performed under full loading, nevertheless accompanied by stress relaxation effects, shown in Fig. 3(a).

3 Results and discussion

The initial permeability of the mortar specimens varies between $9.8 \cdot 10^{-17}$ m^2 and $12.3 \cdot 10^{-17}$ m^2. This variation is relatively small and illustrates the good reproducibility of the material in each specimen.

The variation of gas permeability with the COD during loading within the "simple" device is plotted in Fig. 3(b). One may say that, for all the four mortar specimens tested, results fall on the same master sigmoid curve. The latter presents three characteristic phases, delayed to the mechanical ones: the first (1) one (up to COD of 30/40 microns) where the permeability increases relatively slightly, the second (2) one (COD of 30/40-80 microns) where the permeability increases rapidly and the third (3) one (COD beyond 80 microns) which is characterized by a slower rate than the latter (see Fig. 3(b)). The first phase corresponds to the gas transfer through both the mortar pores and the incompletely developed crack on the stronger diameter face, which propagates to the second face. The second phase corresponds probably to the gas flow by the crack "passing-through" with increasing opening and length, while the third phase (3) is probably related to the localized gas flow through the crack for which only the opening increases. These observations on sigmoid-like tendencies are in qualitative agreement with the numerical purposes of Pijaudier-Cabot et al. [17] as well as with the experimental results obtained by other researchers performing indirect tensile-permeability tests [3, 6-7] as well as compressive tests [13]. For the latter, permeability increase is more important around the peak-load in comparison to the Brazilian tensile test. This effect is probably due to numerous micro-cracks and their connectivity development in comparison to only one main crack formation and propagation in the Brazilian test. Indeed, recent modelling, carried out by the authors of this study [1], based upon a 3D lattice approach to represent the heterogeneity of the material and the failure mechanism, has highlighted good quantitative agreement between the numerical simulations and the experimental results for the range [25 mm ; 125 mm]. This comparison has to be seen as a starting point for a future coupled approach between numerical and experimental investigations and not as a complete validation considering the fact that only one realization has been regarded for the numerical simulations.

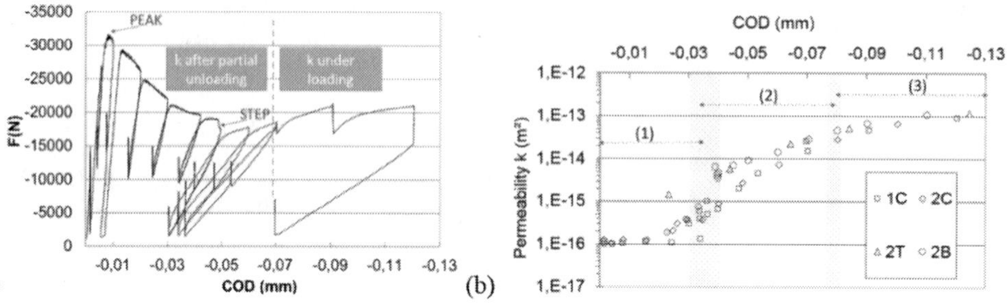

Fig. 3. (a) F-COD typical relation for a mortar specimen with two phases corresponding to different permeability test procedures, (b) variation of gas permeability with COD for four mortar specimens during tensile loading in the Brazilian splitting test.

Fig. 4(a) plots the experimental results obtained on microconrete, concrete and macroconcrete, within the "double" device. The tendencies concerning the sigmoidal evolutions, observed for mortar, are still confirmed. Here, permeabilities are reported to the permeability of the concrete (equal to $3.6\cdot10^{-17}m^2$). Moreover, one may observe that permeability increases not only with cracking, but also with aggregate size. These observations are in qualitative agreement with recent numerical simulations based on the already mentioned 3D lattice approach (see Fig. 4(b)). Besides, these results confirm other observations [13] concerning permeability increase with limestone aggregate size. By the way, permeability increase with aggregate size has also been recently confirmed for other materials, containing limestone sand and glass spheres as a model aggregate [18].

Fig. 5 plots permeability evolution for all the tested specimens and materials (with 2 different cements) within the "simple" and the "double" devices. One may observe that all the results fall on the same master curve, function of the permeability reported to the initial one of each material, and to the COD reported to the peak COD at the failure side. These results assume that a relative permeability evolution follows a single sigmoid law, independently to other parameters. Therefore, one may say that the overall evolution of the permeability may follow a multiplicative format: k=f(tensile cracking)*g(aggregate size). The increase of permeability is the product of two functions (f and g) which reflect each contribution separately. In other words, permeability increase due to tensile cracking seems to be decoupled from the aggregate size effect.

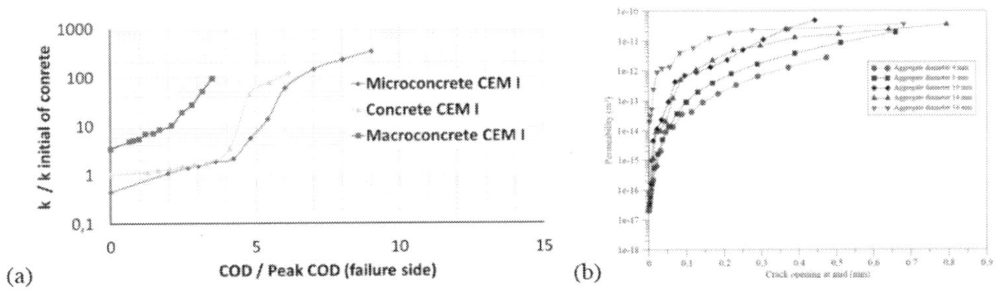

Fig. 4. (a) Relative permeability evolution with crack opening and aggregate size, (b) numerical permeability evolution with cracking and aggregate size (mono-size inclusions: 4,8,10,14,16mm)[18].

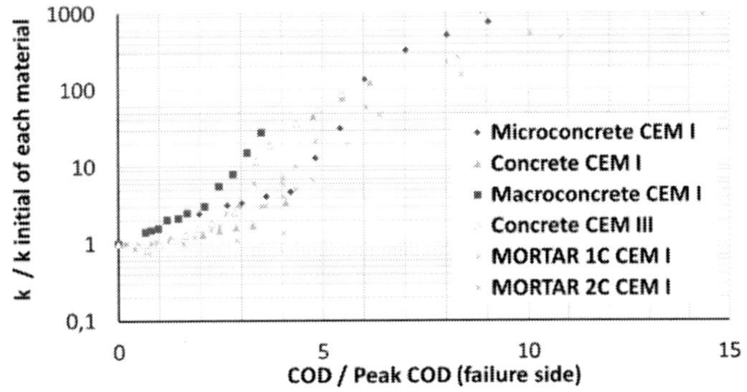

Fig. 5. Relative permeability evolution for all the tested materials and specimens within the "simple" and the "double" devices.

4 Concluding Remarks

The results reported in this study can be summarised as follows:

- The developed testing method has several advantages: cracking is produced by tensile stress and crack opening displacement is measured during loading simultaneously to gas permeability.
- Permeability increase depending on the crack opening (COD) follows a sigmoid curve, already observed for the compressive tests [14]. The experimental results are in quantitative agreement for the range [25 μm ; 125 μm] with numerical simulations, based upon a 3D lattice approach to represent the heterogeneity, developed by the authors of this study [1].
- Three regimes with different rates may be distinguished for the sigmoidal evolution, depending on crack connectivity and geometry. The first one corresponds to a slight increase of permeability which starts beyond the peak load due to nonsymmetric crack initiation. Then, an important increase of permeability is observed, corresponding to the gas flow by the passing-through connected crack with increasing opening and length. The third phase with moderate rate is probably related to the localized gas flow through the crack for which only the opening increases.

- Permeability increases not only with the COD, but also with the aggregate size, independently of their nature and shape (limestone crushed aggregates, glass spheres). The experimental results are in qualitative agreement with numerical simulation results based on the mesoscopic modelling [1, 18].
- Experimental results highlight that the overall evolution of a relative permeability depending on the COD follows a single sigmoid law for all the materials tested. The effects of cracking and of aggregate size may therefore be separated by two distinct functions. Further studies with different concretes and a wider range of aggregate size, nature and volume fractions are needed to expand this relationship.

References

1. N. Benkemoun, H. Al-Khazraji, P. Poullain, M. Choinska, A. Khelidj, Int. J. Numer. Anal. Methods Geomech. 1–20 (2017)
2. K. Wang, J. Daniel, S. Surendra, K. Alan, Cem Concr Res. **27**(3), 381–93 (1997)
3. C.M. Aldea, S.P. Shah, A. Karr, Mater. Struct. **32**(219), 370–76 (1999)
4. A. Djerbi, S. Bonnet, A. Khelidj, V. Baroghel-Bouny, Cem Concr Res. **38(6)**, 877-883 (2008)
5. V. Picandet, A. Khelidj, H. Bellegou, Cem. Concr. Res. **39**(6), 537–547 (2009)
6. B. Gerard, *Contribution des couplages mécanique-chimie-transfert dans la tenue à long terme des ouvrages de stockage de déchets radioactifs* (PhD Thesis, Ecole Normale Supérieure de Cachan, 1996)
7. G. Rastiello, C. Boulay, S. Dal Pont, J.L. Tailhan, P. Rossi, Cem. Concr. Res. **56**, 20-28, (2014)
8. S. Granger, A. Loukili, G. Pijaudier-Cabot, G. Chanvillard, Cem. Concr. Res. **37**(4), 519–527 (2007)
9. Z. A. Kameche, F. Ghomari, M. Choinska, A. Khelidj, Constr. Build. Mater. **65**, 551-565 (2014)
10. J. Bisschop, J. Van Mier, J. Mater. Struct. **35**(8), 453-461 (2002)
11. P. Grassl, H.S. Wong, N.R. Buenfeld, Cem. Concr. Res. **40**(1), 85-93 (2010)
12. A.M. Grabiec, D. Zawal, J. Szulc, Constr. Build. Mater. **98**, 17-24 (2015)
13. A. Fabien, M. Choinska, S. Bonnet, A. Khelidj, Key Eng. Mater. **729**, 115-121 (2017)
14. M. Choinska, A. Khelidj, G. Chatzigeorgiou, G. Pijaudier-Cabot, Cem. Concr. Res. **37**, 79-88 (2007)
15. M. Choinska, F. Dufour, G. Pijaudier-Cabot, A. Huerta, A. Khelidj, *How to extract a crack opening from a continuous damage finite element computation? Application for the estimation of permeability* (ISTE - Wiley. GeoProc 2008, Lille, France, J.F. Shao & N. Burlion Eds., 2008)
16. R. Gettu, B. Mobasher, S. Carmona, D. Jansen, Cem. Based Mater. **3**(2), 54-71 (1996)
17. G. Pijaudier-Cabot, F. Dufour, M. Choinska, J. Eng. Mech. **135**, 1022-1028 (2009)
18. H. Al-Khazraji, *Influence of crack opening, aggregates size and volume fraction on hydro-mechanical properties of concrete in a Brazilian splitting test: 3D meso-macro scale modeling and experimental work* (PhD Thesis, Université de Nantes, 2017)

MATEC Web of Conferences **163**, 02002 (2018) https://doi.org/10.1051/matecconf/201816302002
MATBUD'2018

Application of the nanoindentation method in assessing of properties of cement composites modified with silica-magnetite nanostructures

Elżbieta Horszczaruk[1,*], *Roman* Jedrzejewski[2], *Jolanta* Baranowska[2], and *Ewa* Mijowska[3]

[1]Faculty of Civil Engineering and Architecture, West Pomeranian University of Technology, Szczecin, Al. Piastow 50, 70-311 Szczecin, Poland
[2]Faculty of Mechanical Engineering and Mechatronics, West Pomeranian University of Technology, Szczecin, Al. Piastow 19, 70-310 Szczecin, Poland
[3]Faculty of Chemical Technology and Engineering, West Pomeranian University of Technology, Szczecin, Al. Piastow 45, 70-311 Szczecin, Poland

Abstract. The results of investigation of the cement composites modified with 5% of silica-magnetite nanostructures of the core-shell type are presented in the paper. The nanoindentation method employing three-sided pyramidal Berkovich indenter was used in the research. The mechanical properties and microstructure of the modified cement composites were evaluated on the basis of the values of hardness and indentation modulus measured inside the cement matrix and in the aggregate-paste interfacial zone. The results were compared with those obtained for the reference composites without nanostructures. The positive influence of the presence of silica-magnetite nanoparticles on the tested properties was found out.

1 Introduction

Recently, new testing techniques have been efficiently used for the control of properties and durability of the cement-based materials, particularly the cement composites modified with addition of various types of nanoparticles. In the cement composites, like concrete or mortar, the weakest place is still the interfacial transition zone between the cement matrix and the aggregate or other filler, as well as between the components of the matrix itself. The transition zone affects significantly the macro-properties of the cement concretes, like strength, modulus of elasticity or crack propagation. One of the reason for use of nanomaterials in the cement composites is strengthening of the transition zone. The nanoindentation method is increasingly used in testing of nanoproperties of composites modified in this way [1-6].

The basic difficulty in verification of the values of elasticity modulus and hardness of the cement composites, obtained using the nanoindentation method, is the proper selection of the scale of lattice, pressing force and type of indenter. The authors of [7] have observed the high variability of the obtained results of measurements of the elasticity modulus for the phases of the cement paste. This variability is a result of use of the different loadings and indenters

[*] Corresponding author: elzbieta.horszczaruk@zut.edu.pl

as well as the variable assumptions regarding to the scale of lattice, often maladjusted to the structure of the tested composite.

The results of investigation of the mechanical properties of cement composites modified with admixture of silica-magnetite nanostructures are presented in the paper. The basic goal in the use of such nanostructures was, apart from improvement of shielding properties of the composites on gamma-rays, achieving of better resistance of the composite on high temperature [8]. Therefore, the subject of the detailed investigation was the characteristic of the aggregate-paste contact zone. For this aim, the nanoindentation method was used.

2 Materials and methods

2.1 Materials and specimens preparation

The cement composites were made using portland cement CEM I 42,5R; the composition of the cement is presented in the Table 1. Two types of the cement pastes were prepared with water to cement ratio w/c=0.5. The first paste (R) was the reference one and contained no nanoparticles, while the second paste (N5) contained 5% (refering to the cement mass) of the silica-magnetite nanostructures of the core-shell type (Table 2).

The Fe_3O_4 particles used as a core for the nanostructures were purchased from Sigma Aldrich (637106). The nanomagnetite structures were in the size of 50–100 nm. In order to create a silica shell on the surface of the nanomagnetite ($Fe_3O_4/nSiO_2$), 2 g of Fe_3O_4 were dispersed in 200.0 mL of ethanol (EtOH), mixed together with 2.5 mL of ammonia solution ($NH_3 \cdot H_2O$). Next, 1.5 mL of tetraethyl orthosilicate (TEOS) was added. The resultant mixture was stirred for 18 h at room temperature. The particles were separated with a magnet, washed with distilled water and air-dried. Detailed description of synthesis process is reported in [9]. Transmission electron micrograph (TEM) of the pristine (Fe_3O_4) and silica coated ($Fe_3O_4/nSiO_2$) magnetite nanoparticles are shown in the Figure 1.

Table 1. Chemical composition of the portland cement (wt.%).

Material	SiO_2	Al_2O_3	Fe_2O_3	CaO	MgO	SO_3	Na_2O	K_2O	Cl^-
CEM I 42.5R	20.2	5.0	2.6	64.1	1.58	2.5	0.2	0.7	0.06

Table 2. Proportions of the cement pastes.

Sample designation	Cement	Nano-particles	Water
R	1.0	0	0.5
N5	1.0	0.05	0.5

The specimens for investigation of the aggregate-paste contact zone were prepared in the steel cubic moulds with side length of 20 mm. For this aim, the grain of the river gravel with dimensions 10-12 mm was placed on the bottom of the mould and flooded with the cement paste. The specimens were demoulded after 24 hours and then stored for 28 days in the container filled with water at the temperature 20±2°C. After removing from water, the specimens were cut in half, and the surfaces with the cut aggregate were additionally treated for nanoindentation testing of the interfacial zone. The method of surface treatment is described in 2.2.

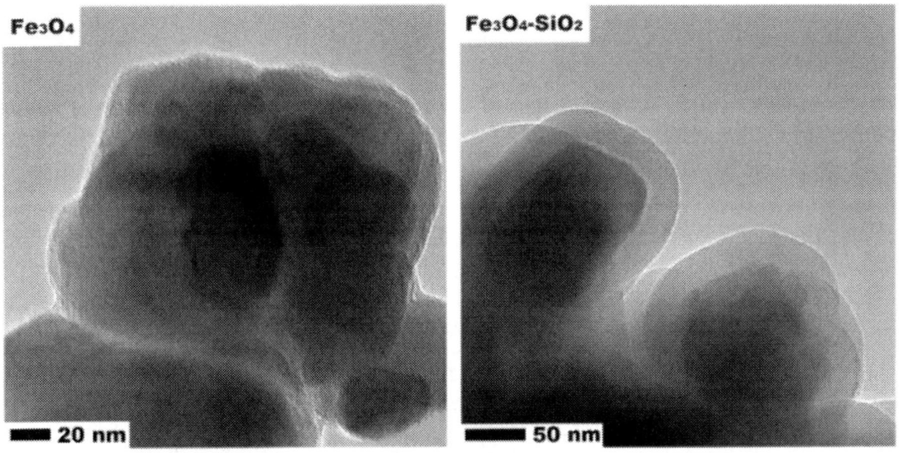

Fig. 1. TEM images of Fe_3O_4 and $Fe_3O_4/nSiO_2$ nanoparticles.

2.2 Nanoindentation test

The surface of the concrete specimens needs special preparation for nanoindentation testing. For ensuring of the flat-parallelism of the surfaces, the specimens were fixed to the handle using thermal glue of the crystalbond type. Every specimen was polished using the sandpapers (the grain size: 200, 400, 800, 1000, 1200, 1500, 2000, in sequence) in water for max. 5 min. Then the specimens were polished using the cloths with diamond suspensions with the grain size 9, 6 and 3 μm, in sequence. Finally, the specimens were polished on the diamond films with the grain size 1 and 0.5 μm (in ethanol, for 10-15 min.). After polishing the specimens were washed in the ultrasonic washer with ethanol for 5 min.

The hardness and indentation modulus were measured by nanoindentation method, using Nanoindenter XP produced by Agilent Place and three-sided pyramidal Berkovich indenter. At least 15 series, each with 8 imprints in the row, was performed for every specimen. The first imprint was placed about 10–20 μm from the edge of the aggregate (Fig. 2). The distance between the imprints was about 20 μm.

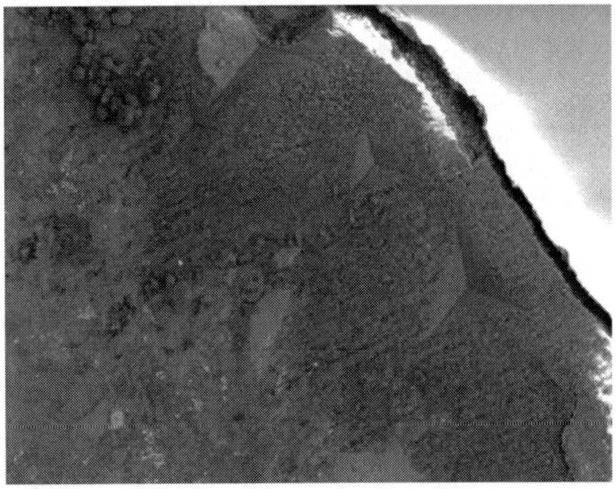

Fig. 2. SEM image of the imprints during the nanoindentation test.

3 Results and discussion

The obtained results of nanoindentation tests were selected according to the measurement correctness. The values of hardness, H, and indentation modulus, M, were determined and statistically analyzed. The final results of hardness and modulus measurements have been averaged within the range of the values determined at the same distance from the aggregate. The mean values for the tested specimens are presented in the Table 3.

Table 3. The mean values of hardness and indentation modulus determined using nanoindentation method.

Area		**R**	**N5**
Interfacial zone	H [GPa]	0.52	0.59
	M [GPa]	14.21	17.72
Cement matrix	H [GPa]	0.57	0.74
	M [GPa]	17.53	24.11

In the composite with nanoparticles admixture (N5), the hardness in the aggregate-paste contact zone was by 13% higher and inside the cement matrix by 29% higher than in the reference material (R). Similarly, the indentation modulus in the N5 composite was by 22% higher in the transition zone and by 37% higher inside the cement matrix, as compared to the reference sample (Fig. 3).

It has been found on the basis of the test results analysis that the width of the transition zone in the composite with nanoparticles (N5) is, on average, half as much as in the reference composite (R). This has been confirmed by measurements of the transition zone width performed using scanning microscope (SEM analysis). This width in the sample N5 was 25-40 μm, while in the case of the reference sample it was 40-70 μm.

Analysis of the results of nanoindentation testing shows that the admixture of silica-magnetite nanoparticles improves the properties of the cement composites inside the cement matrix and causes a decrease of the width of the transition zone, improving its mechanical nanoproperties.

The nanoindentation method makes possible to carry out testing on the relatively small specimens, what is a big advantage of this method, taking into consideration the high cost of nanomodifiers as well as long and often complicated production process. However, in the case of cement composites, the method requires very careful preparation of the samples and large number of the measurements as well as application of statistical analysis for the interpretaion of the test results

The improvement of the mechanical properties in the transition zone of the composite N5 can be explained by the presence of a shell of nano-SiO_2. As it was confirmed in a number of researches [10-16], application of nanosilica in the cement composites accelerates the hydration process, improving the strength and microstructure of the composite. Investigation by Y. Qing et al. [17] showed that the use of nano-SiO_2 leads to improvement of the bond in the aggregate-paste contact zone.

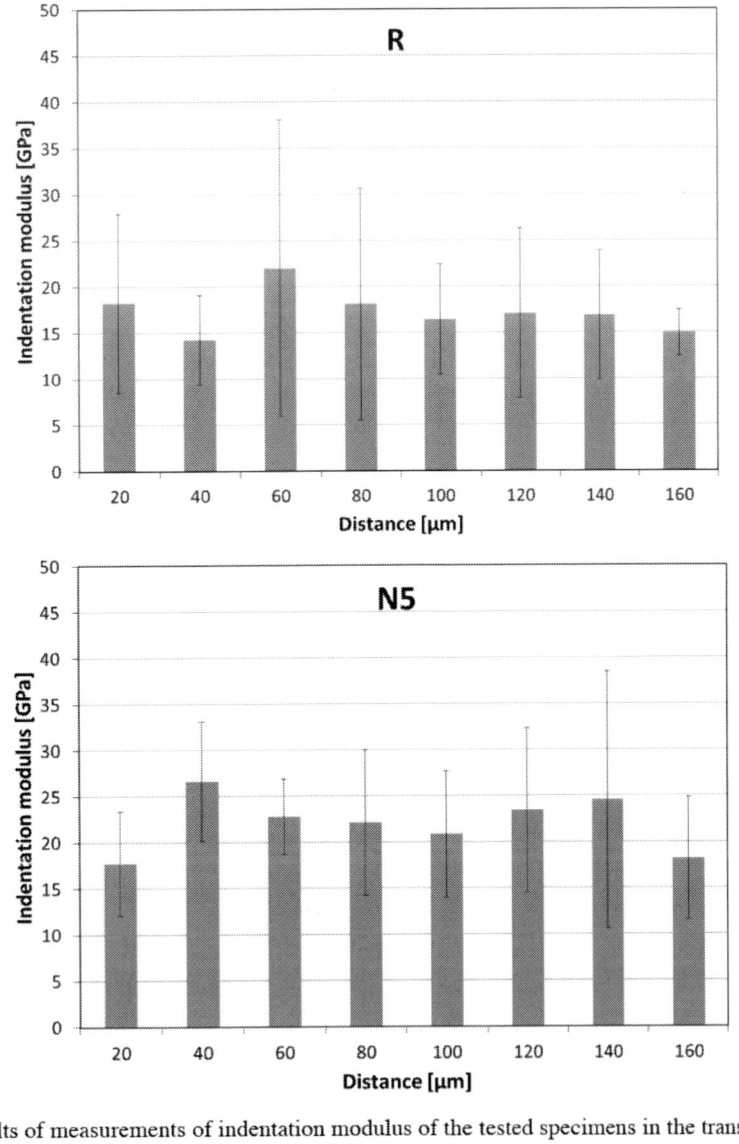

Fig. 3. Results of measurements of indentation modulus of the tested specimens in the transition zone.

4 Summary

Application of the nanoindentation method and SEM analysis made possible to describe the mechanical properties and structure of the aggregate-paste contact zone in the cement composites modified with silica-magnetite nanoparticles. The improvement of the mechanical properties of modified composite was observed both in the transition zone and inside the cement matrix itself, as compared to the reference composite. The reason for the decrease of the width of interfacial transition zone is application of the silica shell in the used nanostructures.

The nanoindentation method, when applied for the testing of cement composites, requires very careful preparation of the specimens. The roughness of the surfaces of the tested specimens should be as low as possible.

This research was funded by the National Science Centre within the project No. 2014/13/B/ST8/03875 (OPUS 7).

References

1. G. Constantinides, F.-J. Ulm, K. Van Vliet, Mater. Struc. **36**, 7 (2003)
2. M. Sebastiani, R. Moscatelli F. Ridi P. Baglioni, F. Carassiti, Mater. Des. **97**, 9 (2016)
3. W.R.L. da Silva, J. Nemecek , P. Štemberk, Cem. Concr. Compos. **45**, 12 (2014)
4. P. Trtik, B. Munch, P. Lura, Cem. Concr. Compos. **31**, 10 (2009)
5. V. Zadeh, Ch. P. Bobko, Cem. Concr. Compos. **52**, 9 (2014)
6. I. Zyganitidis,M. Stefanidou, N. Kalfagiannis, S. Logothetidis, Mater. Sci. Eng. B. **176**, 5 (2011)
7. G. Constantinides G, F-J Ulm, J. Mech. Phys. Sol. **55**, 26 (2007)
8. E. Horszczaruk, P. Sikora, K. Cendrowski, E. Mijowska, Constr. Build. Mater. **137**, 12 (2017)
9. K. Cendrowski, P. Sikora, B. Zielinska, E. Horszczaruk, E. Mijowska, Appl. Surf. Sci. **407**, 7 (2017)
10. B.W. Jo, C. H. Kim, Tae. G-h, J.-B. Park, Constr. Build. Mater. **21**, 6 (2007)
11. E. Horszczaruk, E, Mijowska, K. Cendrowski, S. Mijowska, P. Sikora, Cement Lime Concrete **1**, 9 (2013)
12. E. Horszczaruk, E, Mijowska, K. Cendrowski, S. Mijowska, P. Sikora, Constr. Build. Mater. **66** (2014)
13. K.L. Lin, W.C. Chang, D.F. Lin, H.L. Luo, M.C. Tsai , J. Environ. Manage. **88**, 7 (2008)
14. M. Nili, A. Ehsani, Mater. Des.**75**, 10 (2015)
15. J. Schoepfer, A. Maji, ACI Spec. Publ. **267**, 18 (2009)
16. P. Sikora, P. Lukowski, K. Cendrowski, E. Horszczaruk, E. Mijowska, Proced. Eng. **108**, 7 (2015)
17. Y. Qing, Z. Zenan, K. Deyu, C. Rongshen, Constr. Build. Mater. **21**, 7 (2007)

Failure mode in shear of steel fiber reinforced concrete beams

Julita Krassowska[1, *], *Marta* Kosior-Kazberuk[1]

[1] Faculty of Civil and Environmental Engineering, Bialystok University of Technology, Bialystok, Poland

Abstract. Experimental tests were carried out to assess the failure model of steel fiber reinforced concrete beams. Experimental research was focused on observing changes in the behavior of the tested elements depending on the amount of shear reinforcement and the fiber. Model two-span beams with a cross-section of 80x180 mm and a length of 2000 mm were tested. The beams had varied stirrup spacing. The following amounts of steel fibres in concrete were used: 78.5 kg/m^3 (1.0%) i 118 kg/m^3 (1.5%). Concrete beams without fibres were examined at the same time. The beams were loaded in a five-point bending test until they were destroyed. Shear or bending capacity of the element was observed. Fibre reinforced concrete beams were not destroyed rapidly, but they kept their shape consistent under load. Larger number of diagonal cracks with a smaller width were observed in fibre reinforced concrete beams. Failure of concrete beams without fibres was rapid, with a characteristic brittle cracking. Steel fibres revealed the ability to transfer significant shear stress after cracking in comparison to plain concrete.

1 Introduction

In recent years, the use of fibers as dispersed reinforcement in concrete is constantly growing. The increased interest the use of fibers exhibited by both concrete technologists and structure designers is related to the development of experimental studies confirming the beneficial effect that different types of fibers have on the properties of cement concrete [1, 2, 3] Fibers used as dispersed reinforcement can be made of steel, glass, plastic, basalt, etc. The introduction of this type of reinforcement changes the properties of both fresh concrete mix and hardened concrete. A wide description of the research on the characteristics of fiber reinforced concrete can be found, among others, in the works of Lee and Jacobsena [4], Topçu and Canbaza [5], Fischera and Li [6].

In the case of structural concrete, steel fibers are more and more popular addition nowadays, because their presence changes the behavior of the concrete element under the load - from brittle to quasi-plastic [7, 8]. In comparison to ordinary concrete, concrete with steel fibers is characterized by higher bending tensile strength, and often also shear strength; paired with increased fatigue and impact resistance.

[*] Corresponding author: j.krassowska@pb.edu.pl

Research on the application of concrete with steel fibers in structural elements have begun in the 1960s [9], and resulted in proposals for procedures and standards in testing methods and calculations of structures made of fiber reinforced concrete were created, including RILEM TC162-TDF [10] and *fib* Model Code 2010 [11]. In these standard recommendations the contribution of concrete in the shear stress transfer is included as a component of the load-bearing capacity of reinforced element, making the issue of estimating the load-bearing capacity of concrete fiber elements subjected to bending more important.-The design of fiber concrete structure is based mainly on the use of residual strength after cracking, provided by dispersed reinforcement [12]. The fibers can be used to improve the behavior of structure at the ultimate limit state, or to improve service conditions at the serviceability limit state. In the ultimate limit state, the addition of steel fibers can partially or completely replace the traditional reinforcement for tensile or shear [13, 14]. However, material engineering issues related to the use of fiber reinforced concrete in structural elements are still insufficiently recognized, which makes it difficult to disseminate design methods that take the presence of dispersed reinforcement in concrete into account.

Shear failure can occur in reinforced concrete elements subjected to bending, mainly in support zones. This type of failure may cause a reduction of the load bearing capacity below the value of full bending capacity resulting from the longitudinal reinforcement used, and is particularly dangerous in elements without shear reinforcement [15]. The shear capacity is a complex issue that depends on many parameters. The main factors affecting shear capacity are: shear span-depth ratio, concrete strength characteristics, longitudinal reinforcement ratio, element dimensions and method of load [12, 14, 15]. The parameters recently listed in the literature include: type and grain diameters of aggregate, diameter of longitudinal reinforcement bars, spacing and inclination of cracks, conditions of the element support.

The aim of the work was to assess the impact of steel fibers on the shear resistance of a reinforced concrete beam. Two-span reinforced concrete beams with various shear reinforcement and different content of steel fibers in concrete were tested.

2 Experimental programme

2.1 Materials

Portland cement CEM I 42.5R was used to make concrete for structural elements. The cement content was 320 kg/m^3. A mixture of sand and gravel with grain size up to 4 mm was used as aggregate. The maximum size of aggregate was limited to reduce its influence on toughening mechanism and to provide the homogenous fiber distribution in concrete. Table 1 gives the mix proportions for reference concrete. The modified polycarboxylate and phosphonate based super-plasticizer (1% related to cement mass) was used to minimize fiber clumping and enhance fiber dispersion in concrete.

Table 1. Mix proportions of the reference concrete.

w/c	Cement [kg/m^3]	Water [kg/m^3]	Aggregate 0.125-2 mm [kg/m^3]	Aggregate 2-4 mm [kg/m^3]	Superplasticizer [kg/m^3]
0.5	320	160	1329	626	3.0

The main reinforcement bars were made of RB500 steel. As a dispersed reinforcement, round hooked-end steel fibers with a length of 50 mm and a diameter of 1 mm were used. The tensile strength of steel of fibers was equal to 800 MPa.

2.2 Methods

The strength properties of concretes with fibers and reference concretes were determined. The test of compressive strength was carried out in accordance with PN-EN 12390-3 [16] using cubic samples with 100 mm side size. The flexural strength was determined on specimens of size 100×100×400 mm according to PN-EN 12390-5 [17]. The elastic modulus was determined in accordance with PN-EN 12390-13 [18]. using cylindrical samples with a diameter of 150 mm and a depth of 300 mm. Each series consisted of 3 samples.

In order to evaluate the failure mode of concrete beam with steel fibers, experimental studies were carried out, in which the emphasis was placed on analysing the behavior of tested elements depending on the amount of shear reinforcement and the amount of fibers. Fig. 1 shows the reinforcement schema for tested beams.

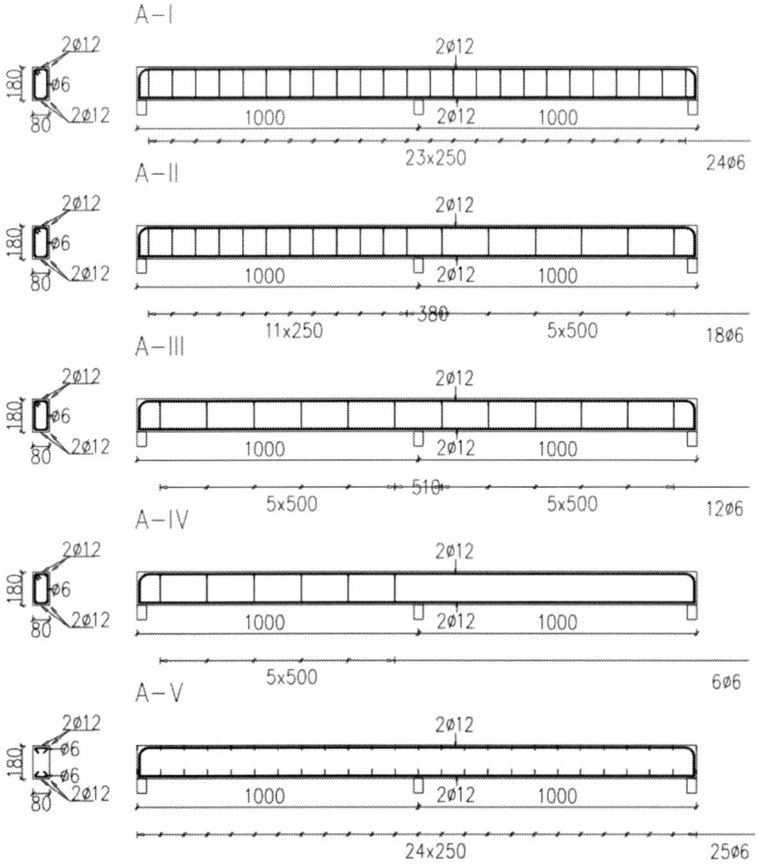

Fig. 1. Schema of reinforcement in beams tested.

Five series of two-span model beams with cross-sectional dimensions of 80×180 mm and length of 2000 mm were made, each containing 5 elements with various dosage of steel fiber additions and with different stirrup spacing. Steel fibers were introduced into concrete in an amount of 78.5 kg/m^3, i.e. 1.0% by volume (WS1.0) and 118 kg/m^3, i.e. 1.5% by volume (WS1.5). The amounts of fibers were selected based on test results of post-cracking behavior of fine grained concrete, described in [20]. Comparably, reinforced concrete beams made of concrete without fiber (W0) were also tested. The amount of main longitudinal reinforcement and shear reinforcement were calculated in accordance with PN-EN 1992-1-1 [19], assuming a load in the form of concentrated force in the middle of each span. In each test series, the top and bottom reinforcements were identical and consisted of two ϕ12 mm bars. The stirrups with diameter of ϕ6 mm were used. The A-I series was made with stirrups spacing designed in accordance with the standard [19], then for each series the stirrups spacing was reduced. No stirrups were used in the A-V series. The shear span-depth ratio a_v/d was equal to 2.7.

The beams were loaded in a five-point system. During the tests, the shear capacity and/or bending capacity of the beam were determined. The peak load and support reactions were also measured.

3 Analysis of the test results of concrete properties

Table 2 presents the results of strength properties tests: compressive strength f_{cm}, flexural strength f_{ctm} and modulus of elasticity E_s in dependence on the fiber content in concrete (V_f) and changes in parameter mean values in relation to the reference concrete.

Table 2. Strength properties of concretes (average values and range of accuracy).

Series	V_f [%]	f_{cm} [MPa]	Δf_{cm} [%]	f_{ctm} [MPa]	Δf_{ctm} [%]	E_s [GPa]	ΔE_s [%]
W0	-	38.95 (±1.64)	-	3.56 (±0.33)	-	33.10 (±2.25)	-
WS1.0	1.0	41.23 (±1.64)	5.86	4.45 (±0.12)	25.05	33.40 (±3.72)	0.91
WS1.5	1.5	39.87 (±1.21)	2.37	4.92 (±0.3)	38.37	34.50 (±3.64)	4.23

The addition of steel fibers did not significantly affect the compressive strength f_{cm} and the modulus of elasticity E_s. The increase in flexural strength Δf_{ctm} for concretes with fibers in the amount of 1.0% and 1.5% was respectively 25% and 38%.

4 Analysis of two-span beams test results

4.1 Failure mode and cracking

Table 3 shows the values of destructive load P_{cr} and their increase in comparison to the values of destructive loads for the reference beams (without fibers). In the case of concrete beams with steel fibers, the increase in peak load ranged from 36.5% to 100%.

The maximum load increase was observed for A-IV-WS1.5 series beams. Comparing the value of the destructive load in the A-I-W0 series with a typical longitudinal and shear reinforcement in accordance with the guidelines [19] with the force value in series A-III-WS1.5, it can be concluded that reducing the number of stirrups by half while using concrete with the addition of steel fibers in 1.5% is enough to achieve required value of destructive force P_{cr}.

Table 3. Destructive load P_{cr} and their increase ΔP_{cr} in particular series with steel fibers.

Series	P_{cr} [kN]		P_{cr} [kN]		ΔP_{cr} [%]	P_{cr} [kN]		ΔP_{cr} [%]
A-I	A-I-W0	80.6	A-I-WS1.0	110.0	36	A-I-WS1.5	114.71	42
A-II	A-II-W0	57.8	A-II- WS1.0	93.4	62	A-II-WS1.5	109.7	90
A-III	A-III-W0	60.7	A-III- WS1.0	64.1	6	A-III-WS1.5	102.7	69
A-IV	A-IV-W0	29.9	A-IV- WS1.0	47.6	59	A-IV-WS1.5	60.0	101
A-V	A-V-W0	33.1	A-V- WS1.0	47.0	42	A-V-WS1.5	52.8	60

The A-I series beams, in which the shear reinforcement provided the transfer of transverse forces, were destroyed in the compression zone by concrete crushing with wide cracks perpendicular to the element axis, caused by bending. The first crack appeared in cross-section with maximum value of tensile stress due to the bending moment. The structural element carried the loads as long as the adhesion of concrete and steel was not exhausted - the elements reached the phase of successive cracks formation, in which the number of cracks increased. Cracks developed up to the neutral axis of beam and the longitudinal steel bars were responsible for stress transfer. At the limit state, the reinforcement in the cracked sections reached the yield point and the element was destroyed. The stirrups in the support zones ensured the transfer of shear stresses, and prevented the beams from destruction due to shear.

In the other series (A-II - A-V), in the majority of cases the failure due to shear occurred in the span, where the spacing of stirrups was reduced. Figs 2-4 present the model of failure of selected beams. There were diagonal cracks due to shear in the support zone. The morphology of these cracks formation is varied. At first, the perpendicular cracks appeared as a result of exceeding the tensile strength of concrete. Perpendicular cracks appeared much later in the case of elements made of concrete with steel fibers, due to their greater tensile strength. Rapid loss of the bearing capacity occurred when the shear cracks were accompanied by cracks associated with the loss of bond between steel and concrete. In some of the tested elements, a phenomenon of reinforcement slip occurred, as evidenced by splitting cracks (series A-II-WS1.0; A-IV-W0; A-IV-WS1.0; A-IV-WS1.5; A-V-WS1.0; A-V-WS1.5). The phenomenon was caused by the lack or reduced number of stirrups in a given zone. In most cases, the destruction of the elements is a typical shear compression failure or shear tension failure. The cause of failure is the destruction of concrete in the compressive zone at the tip of the diagonal crack, where kind of pin joint appeared, around which two parts of the beams, separated by a crack, rotated with a simultaneous tangential movement. In the case of a beam series in a slip of reinforcement occurred, the opening of the first inclined crack significantly increased, and a series of small diagonal cracks appeared at the level of the main reinforcement, which indicated the bond failure between the main reinforcement and concrete. What is typical for this kind of failure, is that when the stresses in the main reinforcement are lower than the yield stress of steel, the complete loss of load-bearing capacity of reinforced concrete element can occur without warning.

The beams of concrete with steel fibers showed quasi-plastic features. The cross-section still carried full load after the crack appeared. In the reinforced beams of concrete with steel fiber, after exceeding the limit stresses there was no case of element failure due to concrete crushing. The presence of fibers allowed the development of multiple diagonal cracks and widening of at least one of them prior to shear failure, which provided some warning about the imminence of failure. The elements were destroyed by the extreme crack opening caused by pulling out steel fibers from cement matrix. The rate of element destruction depended on the content and shape of the fibers in the concrete.

Fig. 2. Failure mode of A-I-W0 beam.

Fig. 3. Failure mode of A-IV-WS1.5 beam.

Fig. 4. Failure mode of A-V-WS1.0 beam.

4.2 Shear stress analysis

The dependencies presented in Figs 5 and 6 make the influence of steel fiber content on the shear stress $v_u = V_{ult}/bd$ clearly visible. The values of stress were determined at three supports. The graphs show the dependence of shear stress on the content of steel fibers in concrete. The shear stress values were calculated on the basis of the shear force at failure V_{ult}.

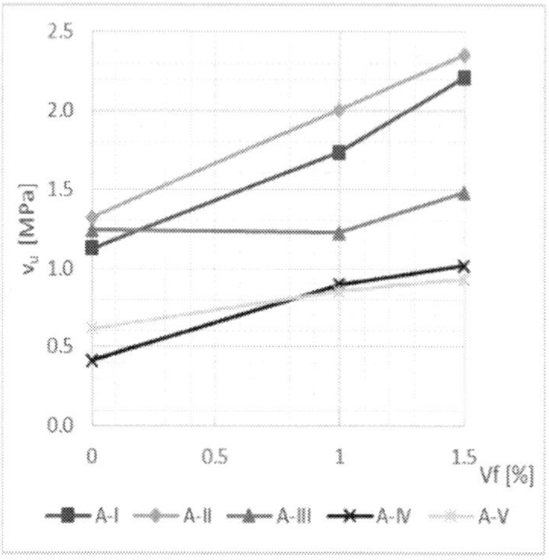

Fig. 5. Shear stress v_u versus fibre content V_f at the central support.

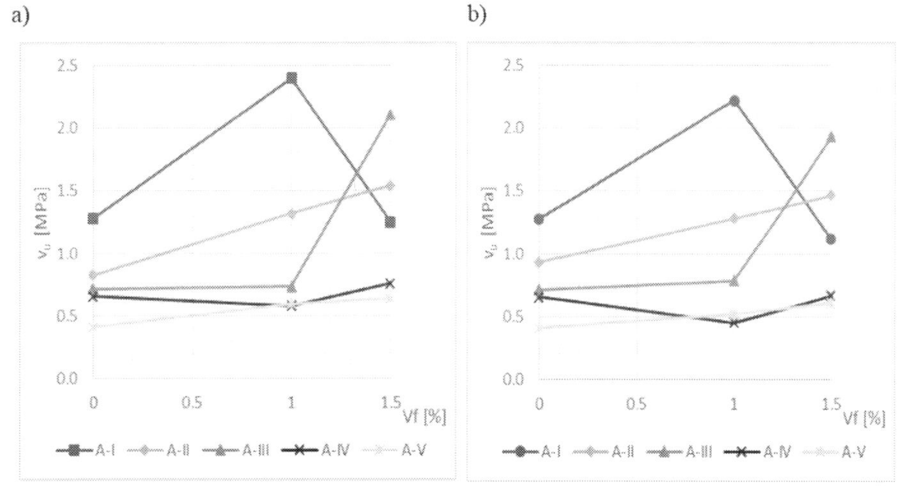

Fig. 6. Shear stress v_u versus fiber content V_f at the left support (a) and right support (b).

Analyzing the shear stresses at the central support, it can be concluded that in the case of a series of concrete beams with steel fiber, the higher the fiber content in the concrete, the higher the stresses at failure. The highest shear stress was recorded in the case of the A-II series, where in the right span the stirrup spacing was reduced by half in relation to that required by the standard [19]. The largest increase in shear stress due to the effect of dispersed steel reinforcement was obtained for the A-IV series.

The effect of dispersed steel reinforcement on both edge supports was similar. The highest stresses were obtained in beam series where stirrups and dispersed reinforcement were used. Compared to beams without dispersed reinforcement, the increase in shear stresses at failure was significant. In series A-III-WS1.5, the reactions during tests reached similar values on all supports due to the quasi-plastic behavior of fiber reinforced concrete. The failure process was running smoothly as long as the bond between steel fibers and concrete was transferring shear stress. The fiber bridging effect contributed to uniform redistribution of stresses in the cross-section weakened by the crack. In the series A-I-WS1.5, the stresses on the edge supports decreased significantly, which resulted from the failure mode due to bending. In this series, the reaction at the central support achieved the greatest value.

5 Conclusions

The following conclusions can be drawn from the results of the experimental investigation.

- The stiff, hooked-end steel fibres (incorporated in the volume fraction of 1.0 and 1.5%) influenced concrete performance effectively. The flexural tensile strength and the ductility increased in line with increasing fibre volume as expected.
- Steel fibers directly influenced the shear capacity of concrete beams. Inclusion of fibers can transform brittle shear mechanism into ductile flexural mechanism. The obtained results confirm the potential possibility of using fibers to partially replace the stirrups. The combination of stirrups and steel fibers can improve both the ultimate and shear cracking strength. Analyzing the shear stresses on the central support of two-span beams, it can be concluded that in the case of steel fiber series, the higher the fiber content in the concrete, the higher the stresses at failure. The largest increase in shear stress at failure due to the effect of dispersed steel reinforcement was obtained in the A-IV series with a highly reduced number of stirrups.

- Awareness of the dispersed reinforcement influence on the shear capacity of element subjected to bending can also be helpful in determining the minimum shear reinforcement needed to protect the element against a sudden brittle failure.

This research work was financially supported by National Centre for Research and Development, Poland; project number PBS3/A2/20/2015 (ID 245084).

References

1. T. Domański, A. Czkwianianc, Przegląd Budowlany **6**, 32-36 (2006) (in Polish)
2. W. Głodkowska, J. Kobaka, Constr. Build. Mater. **44**, 645-653 (2013)
3. A. Khaloo, E.M. Raisi, P. Hosseini, H. Tahsiri, Constr. Build. Mater. **51**, 179-186 (2014)
4. S.F. Lee, S. Jacobsen, Mater. Struct. **44**, 1451-1465 (2011)
5. I.B. Topçu, M. Canbaza, Constr. Build. Mater. **44**, 1486-1491 (2007)
6. G. Fischer, V.C. Li, Eng. Fracture Mechanics **74**, 258-272 (2007)
7. J. Michels, R. Christen, D. Waldmann, Eng. Fracture Mechanics **98,** 326-349 (2013)
8. F.A. Farhat, D. Nicolaides, A. Kanellopoulos, B.L. Karihaloo, Eng. Fracture Mechanics **74**, 151-167 (2007)
9. J. Romualdi, J. Mandel, ACI J. Proc. **61**, 657-672 (1964)
10. RILEM TC 162 – TDF **36**, 560 – 567 (2003)
11. A. Ajdukiewicz, J. Walraven, *fib Model Code 2010. Pre – norma Konstrukcji Betonowych. Tom 1* (SPC, Kraków 2014)
12. M. Kamiński, C. Bywalski, M. Kaźmierowski, Materiały Budowlane **6**, 76-77 (2014)
13. Q. Chuxiang, I. Patnaikuni, Cem. Concr. Comp. **21**, 73-81 (1999)
14. H.H. Dinh, G.J. Parra-Montesinos, J.K. Wight, ACI Structural J. **107**, 597-606 (2010)
15. M. Słowik, *Nośność na ścinanie zginanych elementów żelbetowych bez zbrojenia poprzecznego*, (Wyd. Politechnika Lubelska, Lublin, 2016).
16. PN-EN 12390-3 (2011)
17. PN-EN 12390-5 (2011)
18. PN-EN 12390-13 (2014)
19. PN-EN 1992-1-1 (2008)
20. M. Kosior-Kazberuk, J. Krassowska, Budownictwo i Inżynieria Środowiska (Civil and Environmental Engineering) **6**(2), 73-80 (2015)

Concrete propensity to fire spalling: testing and observations

Izabela Hager[1], *Katarzyna* Mróz[1,*], and *Tomasz* Tracz[1]

[1]Cracow University of Technology, Warszawska St. 24, Cracow 31-155, Poland

Abstract. This article presents results of fire spalling tests on small concrete slabs and studies of material parameters that may increase its occurrence. Experimental techniques enabling to study and determination of material features are presented and discussed. Experimental studies on spalling behaviour of elements were carried out on seven different concrete mixes with constant content of cement paste and mortar. Research aimed at determining influence of the following parameters: w/c ratio (0.30; 0.45; 0.60), cement type (CEM I, CEM III) and type of aggregates (riverbed gravel, granite, basalt) on fire concrete spalling. Paper discusses also the influence of cold rim that forms while testing slab-like element is subjected to one-side heating.

1 Introduction

Recent achievements in concrete mix design have led to new types of concretes which, besides an increased mechanical performance, have also shown enhanced behaviour when exposed to harsh environment conditions. The concrete properties improvement is obtained by reducing the water cement ratio and application of pozzolanic materials. However, when exposed to high temperature those dense materials may be prone to explosive spalling when exposed to high temperature. The sensitivity towards spalling of concrete is investigated by numerous research teams, nevertheless it is until now not fully understood phenomenon and more research is needed to predict risk of spalling. One of the issues that seems to be of particular importance is the question if spalling is a material property or spalling propensity is governed by structural effects?

There is a wide range of factors that may induce spalling among which we can distinguish both mechanical and physical ones. Since concrete is a composite material, in which dispersed phase is represented by aggregate and continuous one by cement paste, both aggregate and cement paste may have an influence on propensity of spalling. It is already known that aggregates and moisture play an important role in the behaviour of concrete during fire [1-7]. However, the concrete structures, ex. tunnels, off-shore platforms, nuclear power plant concrete structures, work in different conditions (curing, load, size, environment, etc.), and each of these conditions may also have a great impact on the thermal stability of concrete.

* Corresponding author: kmroz@pk.edu.pl

Numerous experimental tests results can be found in the literature [8-16]. All attempting to investigate the fire spalling phenomenon by identifying different parameters that may enhance spalling risk, such as the mix composition, the heating scenario, the initial water content and the geometry of specimen or mechanical boundary conditions. Unfortunately, the influence of specific spalling parameters is difficult to assess and to compare with others due to differences in the testing procedures.

While testing unloaded slab-like elements subjected to one-side heating, tested slab specimen is generally placed horizontally on the top of a furnace. Concrete element lies on the furnace edges, which provide an insulation to fire load. As a result, the unheated part of the tested material does not experience thermal expansion and hence limits the thermal expansion of the heated central part of specimen. Therefore, the cold rim of specimen creates a kind of restraint to thermal expansion of the inner part of the concrete slab and contribute to development of compressive stresses in heated part.

High compressive stresses, caused by restraint to thermal expansion develop when the rate of heating is such high that the stresses cannot be relieved by creep quickly enough [17]. The importance of cold rim presence was discussed in [18]. While testing a slab subjected to central heating, the spalling events stopped as soon as the internal cracks of unheated part were developed [18].

In order to facilitate the comparison of results obtained within fire concrete spalling tests a multiparameter assessment has been carried out with the use of unified procedures for seven different concrete mixes. The objective of presented research was to identify most significant parameters that are believed to induce spalling in concrete and investigate them within unified research procedures.

2 Tested materials

2.1 Mix composition

Experimental tests were performed with specimens manufactured with high strength concrete (HSC) of various mix compositions, as indicated in Table 1. The individual compositions are designated by three indicators: type of aggregate / w/c - ratio / cement type.

Within 7 mix compositions, three types of coarse aggregates, namely: basalt (B) crushed aggregate, granite (G) crushed aggregate and riverbed gravel (O) were used in order to study the influence of aggregate nature on fire spalling propensity.

Also the effect of presence of Portland cement clinker was observed by fabricating concrete by use of two types of cement: CEM I 42.5 and CEM III/A 42,5 N for basalt concrete with w/c ratio of 0.30.

Finally, the influence of water to cement ratio is investigated by considering three levels of w/c = 0.30, 0.45 and 0.60 for riverbed gravel concrete being manufactured with Portland cement CEM I 42.5 R.

It shall be emphasized that all tested compositions have the same content of cement paste and mortar, which facilitates the comparison of particular results.

2.2 Production and curing of test specimen

The slabs (1.2 x 1.0 x 0.3 m) and laboratory testing specimen of standard size were produced with the aforementioned concrete mixtures without reinforcement.

Directly after casting, all the test specimens were stored in moulds for 24 hours. Subsequently, concrete is protected from water evaporation by covering it with plastic lids for next 7 days. After demoulding, all the specimens were stored in air-drying conditions:

temperature of 20 ± 5 °C and relative humidity of 50 ± 5 %. The reference test and test concerning fire properties were carried out after 90 days.

Table 1. Mix proportions of the tested concretes

Component \ Concrete	Unit	B/0.30/ CEM I	G/0.30/ CEM I	O/0.30/ CEM I	B/0.30/ CEM III	O/0.30/ CEM III	O/0.45/ CEM I	O/0.60/ CEM I
Cement CEM I 42,5 R Małogoszcz	kg / m^3	482			-		388	325
Cement CEM III/A 42,5 N Małogoszcz	kg / m^3	-			482		-	
Water	dm^3 / m^3	145					175	195
w/c ratio	-	0.30					0.45	0.60
Aggregates								
Dwudniaki river sand 0 - 2 mm	kg / m^3	662	663	663	662	663	663	663
Dwudniaki gravel 2 - 8 mm	kg / m^3	-	-	610	-	610	610	610
Dwudniaki gravel 8 - 16 mm	kg / m^3	-	-	558	-	558	558	558
Gracze basalt 2 - 8 mm	kg / m^3	709	-	-	709	-	-	-
Gracze basalt 8 - 16 mm	kg / m^3	648	-	-	648	-	-	-
Strzegom granite 2 - 8 mm	kg / m^3	-	635	-	-	-	-	-
Strzegom granite 8 - 16 mm	kg / m^3	-	580	-	-	-	-	-
Admixtures:								
plastyfier BASF BV 18	% mc	0.90	0.90	0.90	0.90	0.90	0.80	0.30
superplastyfier BASF SKY 591	% mc	2.20	2.20	2.10	2.35	2.35	-	-
Content:								
Cement paste	dm^3 / m^3	300						
Mortar	dm^3 / m^3	550						

2.3 Material properties

The reference properties of manufactured concretes were determined 90 days after casting and curing in air-drying conditions. Physical (density, water content, gas permeability (Cembureau method)) and mechanical properties (compressive strength, splitting tensile strength, modulus of elasticity) are tested using standard concrete specimens. The obtained values for reference properties are presented in Table 2.

Thanks to the diversity of mix compositions, an evident spread of the obtained characteristic values of physical and mechanical properties can be noticed within tested concretes. Therefore, it is expected to observe different spalling propensity for the individual tests.

Table 2. Mean values of reference properties of tested concretes

Property	Unit	B/0,30/ CEM I	G/0,30/ CEM I	O/0,30/ CEM I	O/0,45/ CEM I	O/0,60/ CEM I	B/0,30/ CEM III	O/0,30/ CEM III
Density	kg/m^3	2558.8	2376.7	2300.7	2268.6	2177.4	2533.2	2315.6
Water content	%	2.6	2.4	2.3	2.2	2.3	2.7	2.8
Permeability	m^2	7.00 E-18	2.58 E-17	1.20 E-17	3.39 E-17	2.22 E-16	5.23 E-18	9.96 E-18
Compressive strength	MPa	84.9	73.3	77.0	64.4	51.5	96.2	87.4

Spliting tensile strength	MPa	6.2	4.9	6.0	4.7	4.0	6.9	5.6
Elastic modulus	GPa	44.4	30.6	29.7	27.3	24.7	48.9	30.9

3 Heating and testing procedure

The slab is placed horizontally on chamber of the DRAGON laboratory furnace. During the fire test, the slab is subjected to one side heating for 120 minutes by ISO 834 fire scenario from the bottom with the use of a 140 kW gas burner.

The specimen is freely supported by thermally insulated furnace external walls of 0.125 m in thickness. As a result, the directly exposed surface has an area of 0.95 x 0.75 m and the unheated support may work as cold rim and limit the thermal dilation of heated part. Slab is tested under unloaded conditions.

During the test, the temperature *(T)* is measured near to the heated surface with a plate thermocouple and in the concrete at different depths (1 cm to 22 cm with spacing of 3 cm) by aid of of thermocouples type K placed in the specimen before casting. Additionally, the time of spalling initiation (t_0) and its end (t_1) are reported.

After heating, the specimens could freely cooled down. Before taking measurements, the fire damaged surface has not been brushed or cleaned. The spalling propensity is assessed by investigation of fire-exposed area. Measurements of spalling depth are taken manually with spacing of 25 mm. Having the data gathered, the maximum spalling depth (d_{max}), volume of spalled material (V_s) and volume V_s related to the initial volume of slab (V_{rel}) have been derived.

4 Test results

4.1 Observations during heating

Fire spalling events: if a fire spalling event occurs, the time of its initiation and its duration is listed. In six of seven slabs, fire spalling was observed. The concrete that did not experience an explosive behaviour was the one being manufactured with riverbed gravel aggregate, w/c ratio equal 0.6 and cement type CEM I 42.5 R, designated as O/0.6/CEMI. In the remaining six concretes, the popcorn-like spalling started between 1st and 4th minute of fire exposure, when temperature of surface was between 350 °C and 550 °C. Afterwards, between 5th and 7th minute of testing, when temperature of surface was ca. 500 °C - 600 °C, the explosive behaviour occurs with large pieces spalled off the surface, Figure 1. Spalling continued for 30 - 40 minutes and its duration did not change significantly for concretes of different compositions.

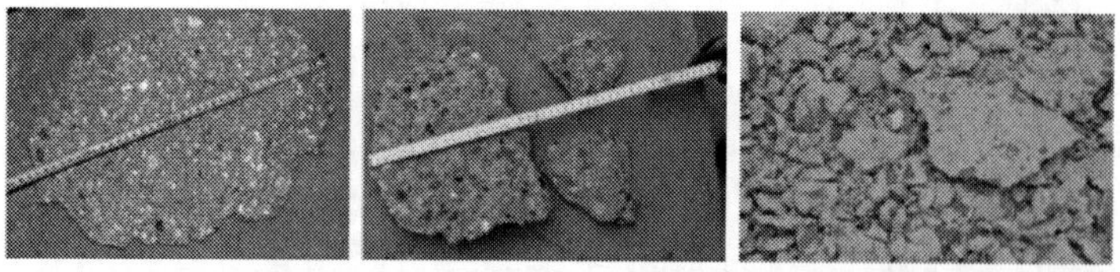

Fig. 1. Spalled pieces of concrete, from left: G/0.3/CEM I, O/0.45/CEM I, B/0.3/CEM I.

Lateral cracks: During fire tests the propagation of lateral cracks in unheated rim took place. It was observed that the lateral cracks formation (Figure 2) effectively allowed to release a large amount of water vapor during heating, thus reducing gas pore pressure and limiting the spalling. In the tested slabs, two to three lateral cracks appeared between 5th and

15th minute after start of fire exposure. In most cases, the spalling stopped when two or three lateral cracks became visible at each side of the tested element. It has been noticed that greater spalling intensity corresponds to more lateral cracking, irrespective of initial moisture content of the concrete. In case of specimen O/0.6/CEMI that did not spall, one lateral crack of significant width (0.5 mm to 2.0 mm) has been formed after ca. 6 minutes of fire exposure and intense release of water vapor has been observed for ca. 30 minutes.

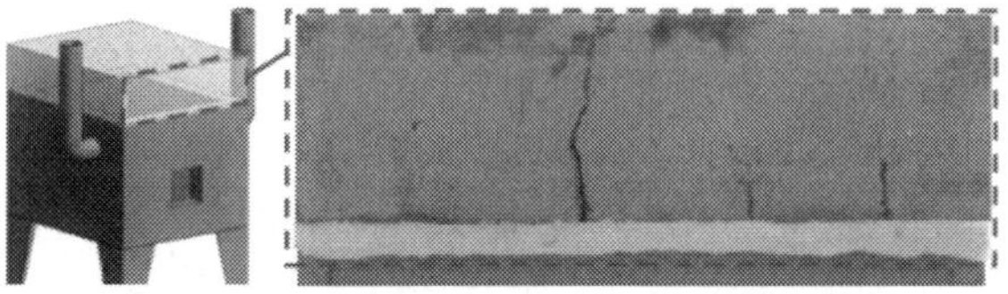

Fig. 2. External side of specimen of O/0.45/CEMI during heating: lateral cracking and release of water vapor.

4.2 Post-fire investigations

After cooling the specimens down, the slab were removed from the furnace and the post-fire investigations took place.

No post-cooling sloughing off was observed, probably due to lack of aggregate of calcareous nature in the compared mixes. The post-fire observations consist of the following determination:

- Spalling topography: post-cooling investigations of tested slabs consisted in measurements of spalling depth at grid with mesh of 25 mm and visual inspections of cracking of the unheated part of specimen (cold rim). Spalling topography was prepared in the same scale for all tested concretes. Blue color correspond to 0 mm of spalling depth while red one to 65 mm, which was the maximum depth measured within all the specimens. The maps of spalling topography along with the corresponding measurements of maximum depth, volume and relative volume of spalling are presented in Figure 3 with the increasing order. It has been noticed that concretes made with riverbed aggregate are generally more susceptible to spall, and the measured maximum depth and volume of spalled material are greater in comparison to concretes made with basalt and granite aggregates.
- Cement type effect: The effect of cement type on spalling has been analyzed. Investigated concretes were manufactured with cement types CEM I and CEM III. It was observed that concretes made with CEM I had greater permeability and lower compressive strength in comparison to those made with CEM III. However, more permeable concretes of lower strength made with CEM I indicated greater volume and depth of spalling, Figure 4a.
- Water / cement ratio effect: the effect of w/c ratio has been analysed by comparison of measured maximum spalling depth for concretes made with riverbed aggregate and cement type CEM I. For manufacturing concretes with w/c ratio of 0.3, 0.45 and 0.6, different amount of water has been used: 145 dm^3 / m^3, 175 dm^3 / m^3 and 195 dm^3 / m^3, respectively. As a result, concrete with less amount of water, presents higher density and compressive strength and lower permeability. The experimental results show that for concrete with w/c ratio of 0.6 spalling does not occur. However, for concretes of w/c = 0.45 and 0.3 spalling takes place and its depth and volume is greater for more dense and less permeable material, Figure 4b.

Mix	d_{max} (mm)	V_s (cm^3)	V_{rel} (%)
O/0.60/CEM I	0	0	0
B/0.30/CEM III	40	6043	1.7
G/0.3/CEM I	35	7260	2.0
B/0.3/CEM I	45	12378	3.4
O/0.30/CEM III	50	12903	3.6
O/0.45/CEM I	55	14594	4.0
O/0.3/CEM I	65	14859	4.1

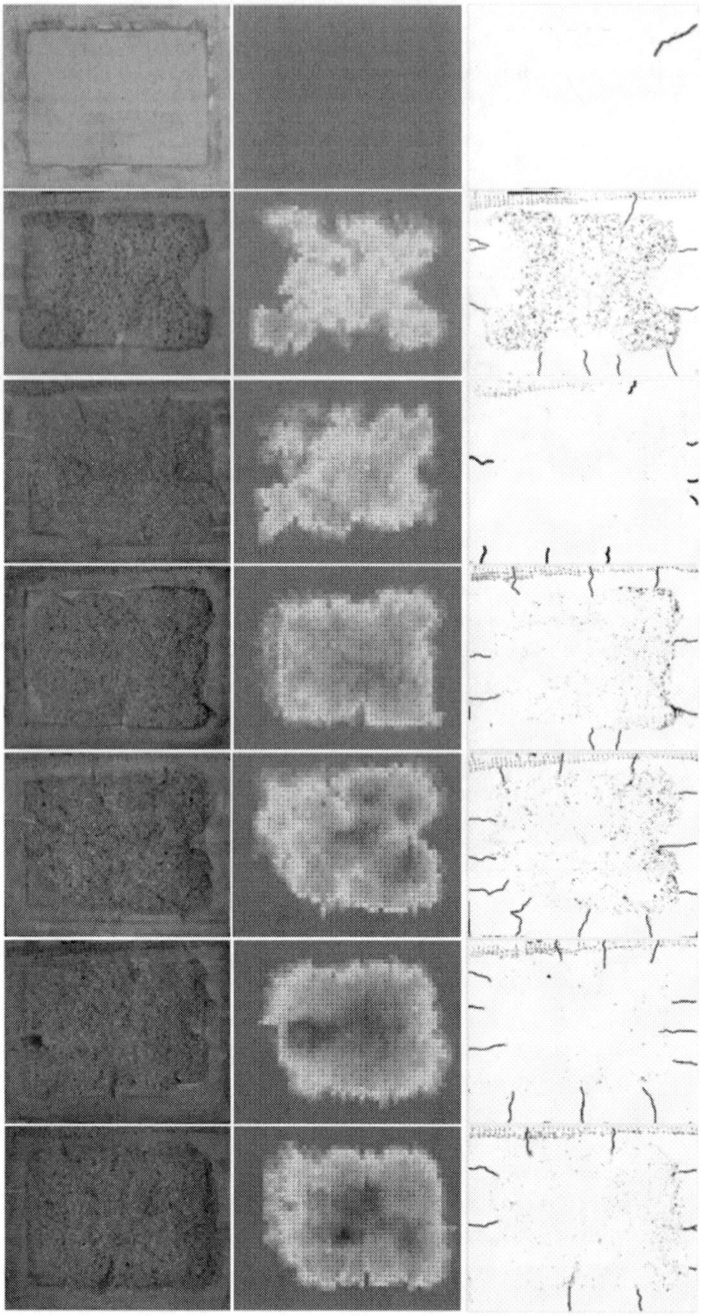

Fig. 3. Spalling observations, from left: measured data, picture of slab after test, spalling topography (blue = 0 mm, red = 65 mm), cracking of cold rim.

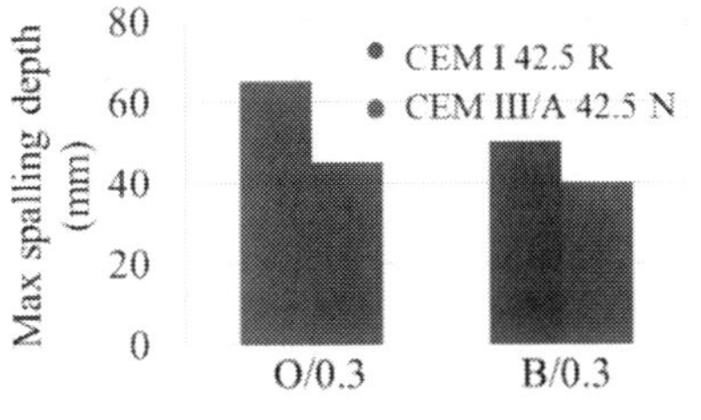

a)

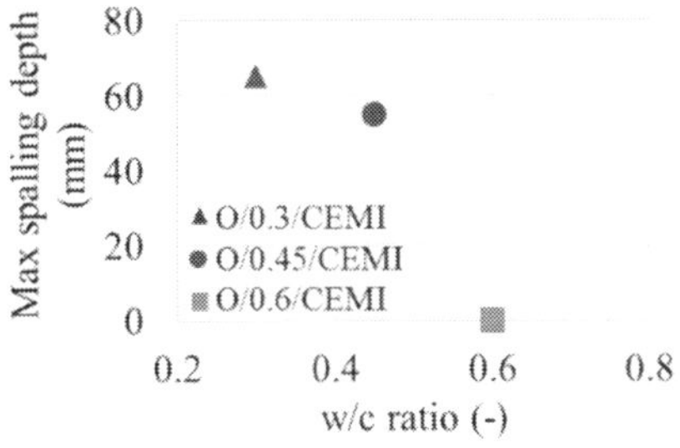

b)

Fig. 4. Effect of a) cement type, b) water / cement ratio on maximum spalling depth.

5 Conclusions

The presented results have been gathered with concrete specimens exposed to one side heating in the fire testing DRAGON setup. Research program includes test of seven concretes of different mix proportions. Within all designed concrete mixture the same content of cement paste and mortar has been assumed. Therefore, the comparison of the influence of individual factors on concrete susceptibility to spalling is more evident.

On the basis of gathered results, following conclusions and observations can be stated:

- One of seven concretes, designated as O/0.6/CEMIII did not indicate spalling. Due to significantly greater amount of water used for production (195 dm^3 / m^3) in comparison to the other six concretes (145 - 175 dm^3 / m^3), its initial properties (density, permeability, strength) greatly stand apart the results obtained for other compositions.
- For remaining six concretes, spalling occurred when the temperature of surface was between 350 °C and 550 °C. When temperature increased to ca. 500 °C - 600 °C, the explosive behaviour occurred with large pieces spalled off the surface.
- Within spalled specimens, the greatest value of spalling depth was measured as 65 mm (for concrete O/0.3/CEM I). It corresponded to 22 % of the entire thickness of tested specimen.
- Significantly more intense spalling was observed for concretes made with riverbed gravel aggregate. Basalt and granite concretes spalled less of ca. 50 % in volume in average.
- More dense and less permeable concretes are more prone to spall. However, the results show that such conclusion holds only for concretes made with aggregate of the same nature and with the same type of cement. While comparing the results for all the tested concretes, such conclusion seems not to be such evident.
- It has been shown that spalling is more likely to occur for concrete made with cement type CEM I 42.5 R than for CEM III 42.5 N. It is despite greater permeability of concrete made with CEM I than CEM III. It indicates that the initial permeability of concrete may not be a decisive factor inducing spalling.
- During heating, two to three lateral cracks occurred in unheated part of concrete specimens, irrespective of spalling intensity. However, it seems that greater spalling volume corresponds to lateral cracks of greater width. It is probably due to more severe spalling and therefore higher release of water vapor is needed.
- Observations of development of lateral cracks during heating confirm that the unheated part of specimen work as a cold rim and limits thermal dilation of concrete.
- In case of specimen O/0.6/CEMI that did not spall, one lateral crack of significant (0.5 mm to 2.0 mm) width has created and intense release of water vapor has been

observed. It confirms that unheated part of concrete retraints the thermal dilation of heated one and experience tension in a transverse plane.

Lack of significant differences in spalling mechanisms within the tested concretes provides the presumptions that material properties may not be a decisive factors influencing the behavior of concrete element under fire attack. In other words, it seems that most of High Performance Concretes is prone to spall, regardless of its properties. Therefore, further comprehensive experimental studies on the effect of cold rim and other structural aspects are highly needed in order to clarify the effect of boundary conditions on spalling severity.

This work was supported by the Polish National Science Centre [grant number N N506 045040].

References

1. D.R. Lankard, D.L. Birkimer, F.F. Fondriest, M.J. Snyder, *Temperature and Concrete*, 59–102, (ACI, Detroit, 1971)
2. R. Jansson, L. Boström, *1st International Workshop on Concrete Spalling due to Fire Exposure*, 120-129 (2009)
3. I. Netinger, I. Kesegic, I. Guljas, Fire Saf. J. **46**, 425–430 (2011)
4. J. Mindeguia, P. Pimienta, C. La Borderie, Eur. J. Environ. Civ. Eng. **16**, 236–253 (2012)
5. R. Jansson, L. Boström, *MATEC Web Conf.* **6**, 3003 (2013)
6. K. Krzemień, P. Pimienta, N. Pinoteau, I. Hager, *4th International Workshop on Concrete Spalling due to Fire Exposure*, 165–176 (2015)
7. I. Hager, T. Tracz, J. Śliwiński, K. Krzemień, Fire Mater. **40**, 668–682 (2016)
8. R. Jansson, *Fire spalling of concrete: theoretical and experimental studies*, 154 (KTH, Stockholm, 2013)
9. K. Krzemień, I. Hager, Procedia Eng. **108**, 285–292 (2015)
10. H. Carré, P. Pimienta, C. La Borderie, F. Pereira, J.C. Mindeguia, MATEC Web Conf. **6**, 1007 (2013)
11. I. Hager, P. Pimienta, *Workshop Fire Design of Concretes: What now? What next?*, (2004)
12. J.C. Mindeguia, *Contribution Expérimentale A La Compréhension Des Risques D'instabilité Thermique Des Bétons*, (Universitè de Pau et des Pays de l'adour, 2009)
13. L. Boström, R. Jansson, *Self-compacting concrete exposed to fire*, (SP, Borås, 2008)
14. R. Felicetti, F. Lo Monte, P. Pimienta, Cem. Concr. Res. **94**, 13–23 (2017)
15. S. Huismann, F. Weise, B. Meng, U. Schneider, Mater. Struct. Constr. **45**, 793–801 (2012)
16. W. Lindlbauer, M. Zeiml, A. Heel, W. Kusterle, *Fire resistance of fiber-reinforced, reinforced, and prestressed concrete* (in German) (Straßenforschung 3.269, Vienna, 2004)
17. G.A. Khoury, *Spalling Review* (Fire Safety Design, 2005)
18. K.D. Hertz, Fire Saf. J. **38**, 103–116 (2003)
19. EN 1992-1-2 (2004)

High frequency impulse ground penetrating radar application in assessment of interlayer connections

Jacek Sudyka[1,*], *Lech* Krysiński[1], *Adam* Zofka[1], *Marek* Pszczoła[2], and *Piotr* Jaskuła[2]

[1]Road and Bridge Research Institute, Warsaw 03-302, Poland
[2]Gdansk University of Technology, Gdansk 80-233, Poland

Abstract. Ground Penetrating Radar (GPR) technique is commonly used in the nondestructive evaluation of pavement structures. In particular, this method is used to estimate thicknesses of pavement layers as well as it can be utilized in advanced studies of pavement structures. The device presented in this paper comprise the high frequency impulse antennas that allow for investigating the interlayer zones in terms of their electromagnetic properties (e.g. dielectric constant). In some cases these electromagnetic responses can be suitable in the assessment of layer bonding in the pavement structure. This paper discusses the assessment of the quality of asphalt pavement interlayer bonding with the use of high frequency GPR techniques. The preliminary laboratory measurements were performed using an impulse antenna in the zero-offset configuration combined with the large-scale models simulating an idealized horizontal delamination. These measurements allowed to estimate the antenna sensitivity to detect interlayer connection under dry and wet conditions. Analysis of collected results led to formulating practical conclusions regarding critical limitations of the measuring system and adequate methods of signal processing and interpretation.
The field investigations consisted of the GPR measurements along selected road sections and collection of the core samples at the locations associated with the specific reflexes. Inspection of the cores provided some real insights into the structure of different delaminations associated with characteristic reflexes. Analysis showed the reflection properties are able to expresses some important features of the interlayer zone, such as delaminations, presence of alien material at the interface, insufficient compaction occurring at the base of layer, and water penetration.

1 Introduction

The bonding of asphalt layers is a significant factor that directly influences the strength and durability of pavement structures. Due to insufficient bonding between asphalt layers, individual layers can move horizontally independently from each other that weakens the entire structure and leads to premature surface and structural failures [1]. In particular,

* Corresponding author: jsudyka@ibdim.edu.pl

debonded asphalt layers can lead to corrugation, slippage and transverse cracking of pavement surfaces that significantly affect traffic safety [2]. In the case of weak bonding predicted pavement life may drop to as low as 40% [3]. According to Judycki [4] fatigue life decreases rapidly when the interlayer connection is weakened. When there is a total lack of bonding the fatigue life of pavement decreases several times. While such an extreme condition is unlikely to occur, even with the partial bonding of 50% the decrease in fatigue life is clearly evident. In another study, Raab and Partl [2] estimated that a defective connection between asphalt layers may, in extreme cases, reduce the expected fatigue life of pavement by 30%. According to Raposerias and Castro-Fernas [5] the combined effect of high temperature and traffic can improve the interlayer connection provided however that the bond has not been totally disconnected. Vaitkus [6] has demonstrated that insufficient bonding between asphalt layers can cause corrugation and rutting of pavement. He states that cracks in asphalt pavement occur because of insufficient bonding of asphalt layers and reports also that an optimal bonding is necessary in order to obtain a sufficient asphalt pavement strength to prevent rutting. Raposeiras [7] have determined that the deformation of asphalt pavement layers increases significantly if the layer bonding decreases from 100% to 90%, which results in a premature deterioration of asphalt pavement.

There are three types of asphalt layer bonding [8]:

- sufficiently bonded – asphalt layers act as a monolithic structure. A large shear stress is created and no deformations (displacements) are developed,
- partially bonded – depends on the interlocking strength. The shear stress and deformations (displacements) of various sizes occur between layers. In case of strong interlocking, the large shear stress and small deformations are registered. Conversely, if the interlocking is weak, the shear stress is small and the deformation is substantial,
- insufficiently bonded – the friction and bonding occur only as a result of the load and self-weight of layers.

As can be seen an appropriate bonding plays significant role in the functioning of the entire pavement structure and contributes to its quality and durability. Sufficiently bonded asphalt layers give improvement of the pavement structure strength and prevent surface deformation and cracks, as well as water penetration between asphalt layers [9].

There are a number of laboratory methods for determining the strength of bond between layers, but this type of research is limited in terms of the number of sampling points and the nature of information obtained. Among non-destructive tests, one could mention Impulse Hammer Test [10], the test using an FWD (Falling Weight Deflectometer) [11], or examinations with the use of acoustic methods [12].

Other very common non-destructive technique is the Ground Penetrating Radar method (GPR) [13-16]. This method is used routinely to assess the continuity of pavement layers, or for measuring the thickness of the asphalt structure. The GPR technique also offers a great potential in examining the detailed architecture of the asphalt packet and qualitative evaluation of the road's condition, including diagnoses of internal defects such as interlayer delaminations [17] or cracking [18]. These non-standard tasks are far more challenging as a study task, hidden with problems inherent in the nature and limitations of the method.

GPR measuring systems emit electromagnetic pulses towards the test object and receive the reflected signals. Emitted electromagnetic pulses passing through successive layers of materials with different dielectric properties are partly reflected on their boundaries. The primary recorded data is the return time and the amplitude of the signal reflected from the boundary.

The interpretation of echograms recorded is essential in the GPR methodology. One of the ways to identify the faulty interlayer connection was tested earlier using laboratory model and calculative examples [19]. This study presented methodology to identify the connections between layers of asphalt where delamination has occurred. This methodology is based on identifying the type of reflection, i.e. a single reflection $w(t)$ or double reflection $dw(t)/dt$, with one leading extreme or two adjacent leading extremes of inverse signs, respectively (Fig. 1). The simple single reflection is generated by the commonly known type of the planar boundary between a two homogeneous layers having mutual contrast of impedance. Depending on the sign of contrast the single reflection can be positive, as it is in case of boundary between successive mediums of low and high permittivity, or negative, as between mediums of high and low permittivity. When a crevice is very wide (in comparison with wavelength) the resulting signals are single type with positive or negative sign. When crevice becomes thinner and thinner the signal becomes similar to the derivative of the incident signal, its amplitude becomes proportional to the decreasing crevice thickness and then it is called the double reflection. The presence of strong positive double reflection marks the existence of serious damage at the joints of asphalt layers, e.g. delamination fissures filled with water. Strong negative double reflections may be the sign of sever dry delaminations, connections strongly sprayed with binder or cracks filled with foreign material of low permittivity.

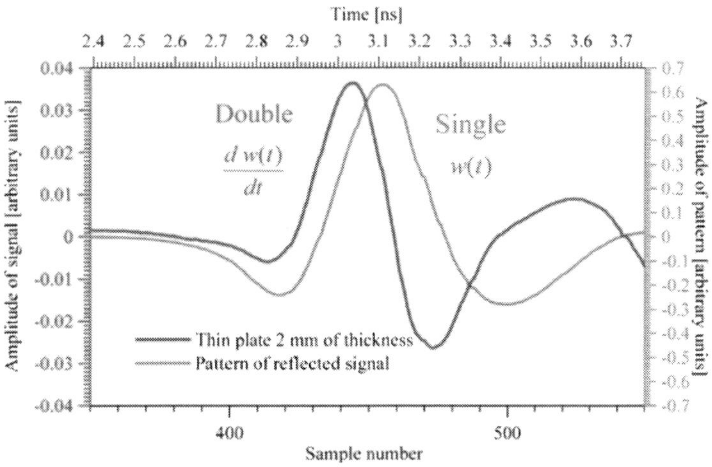

Fig. 1. Comparison of the head package shapes of the single w(t) and the double reflection dw(t)/dt measured on laboratory model of the crevice simulated by thin glass plate [9]

The double reflections have large diagnostic importance showing delaminations developed usually on the interlayer connections [20] and its amplitude can be a base of the effective thickness estimation, for example by use of direct waveform modeling. But in the field practice, a variety of complications should be expected when using this terminology like occurrence at mixed types of reflections and further complication of their shape in zones of substantial thickness (comparable to the local wavelength in the medium) and a gradient structure. Complications of the shape of reflections also occur as a result of diffraction where the pavement structure is characterized by the major local deviations from the plane-parallel structure. In the case of reflections of a small amplitude, serious difficulties of interpretation are also present in the identification of the type due to their observed shape being disturbed by masking background which is difficult to remove.

This paper presents the results of GPR surveys carried out on a road section where a problem with improper interlayer connection (bonding) between the layers of asphalt was found. The measurement was done using horn-type pulse antennae with centre frequencies of 2 GHz. Results showed a lack of layer connection in more than ten sections and sometimes over the entire width of the pavement. The article presents selected cases, confirmed by cores extracted at the specified locations. These cases are therefore an excellent illustration of the issue of recognizing the condition of interlayer connections in field practice. The term "interlayer connection" is used in this article instead of bonding just to underline the difference between laboratory and on test site evaluation of asphalt layer bonding.

According to the authors' opinion the following causes and phenomena, simultaneously present in the studied section of pavement are responsible for the lack of connection between the layers:

- insufficient compaction of upper layers; adverse weather conditions (wind and low temperature) during the laying of asphalt concrete in the base layers (dual layer) and the bonding layer, causing premature increase in viscosity of asphalt in the mix and hence hindrance to compaction; increased number of voids at the bottom of the layer (strong porosity of the floor of the top layer near the contact)
- no effect of interlock of contacting layers due to inefficient compaction of upper layer,
- the segregation of the mixture in the upper layer; reduction of the contact area of layers due to the presence of clusters of coarse aggregate particles at the bottom of the top layer,
- lack of or inadequate spraying of the interlayer surface, the lack of bonding between layers,
- increasing insufficiency of bond or an increase of the area of lack of bond between the layers resulting from the presence of water penetrating through micro fissures, insufficient compaction of the bottom layers, leaky longitudinal and transverse joints; intensified penetration of water due to interruption of the watercourse in the surrounding area,
- dirt in the interlayer space, failure to clean layers during the incorporation of successive layers.

The knowledge about these phenomena is very useful in explaining presence of strong, specific reflections on the recorded echograms. Opened delaminations, possibility of water penetration, presence of foreign material at the interlayer connection, porous base of the layer and change of particle size at the boundary have crucial importance in interpretation of the type, sign and amplitude of the resulting reflections.

2 Field studies

The results presented here are the effect of field studies carried out on one of the new sections of road where it passes through a deep trench with deficient hydrological conditions. GPR scans were taken along four parallel sections at four wheel tracks (two on each lane). The primary objective of this study was to perform a quantitative evaluation of the total thickness of layers built in, but the results also showed areas where the lack of a proper connection between layers of asphalt was found. The bad condition of the interlayer connection was confirmed by extracting cores at the specified locations of pavement.

Recorded echograms (Fig. 2) clearly show the layered structure of a packet of asphalt (wearing course, binding and base course asphalt layer). Even the very clear visibility of boundaries between the layers is intriguing, because the contrasts between the electrical properties of the lower layers of the packet (binding layer and base course) are usually minor and reflections corresponding to the boundaries should be hardly distinguishable from the remaining undeleted masking background. Significant contrast is present here only on the border between the wearing course and the binding layer. However, the phenomena occurring at interlayer interfaces in the course of applying layers lead to considerable material differences in the medium of layer near the interface (insufficient compaction, segregation or introduction of a foreign material). As a result, boundaries are often very well visible on echograms, but the corresponding reflections are rarely simple single reflections and their shape and amplitude reflect the complex structure of the contact. Thus, in interpretation practice, it is necessary to semi quantitatively evaluate the amplitude of reflections and to comment the amplitude using assumed properties of the medium and contact.

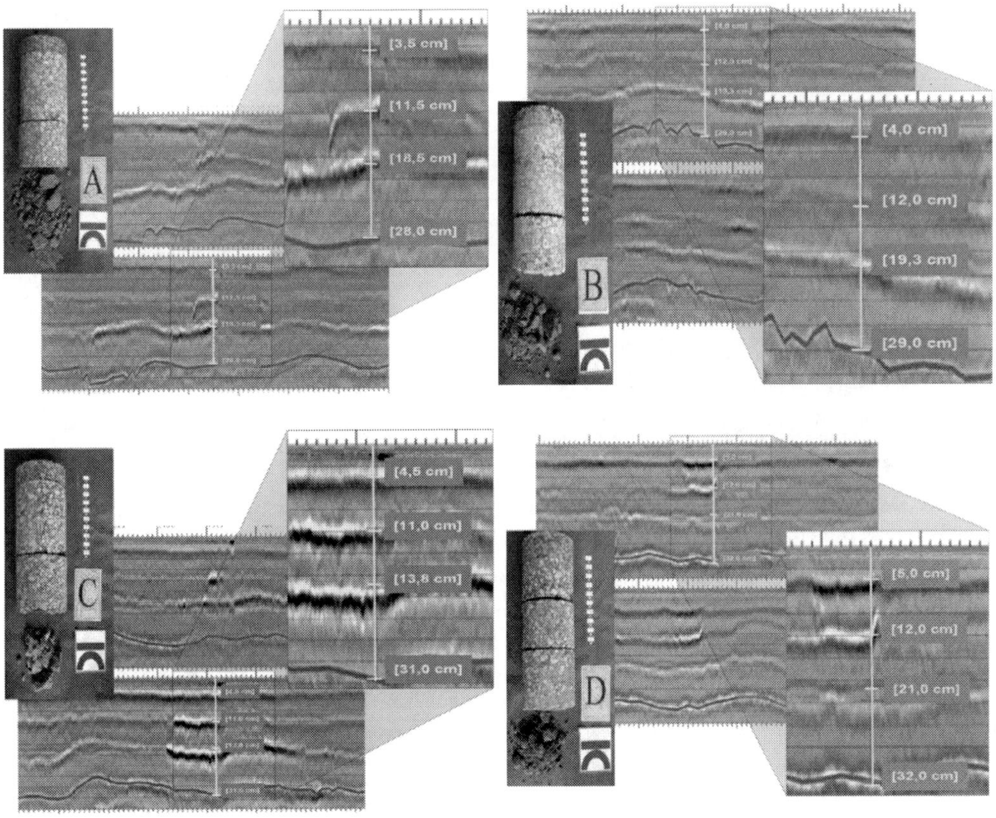

Fig. 2. Four sections (cases A, B, C and D) where the lack of a proper connection between asphalt layers was found. The cores ware taken at the specified locations of pavement on the basis of GPR echograms analyse. The yellow markings indicate the thicknesses of asphalt layers.

On the background of radar image of asphalt layers numerous, extremely strong positive double reflections are highlighted. At least four sites of particular accumulation of these strong reflections can be distinguished along the test sections. In addition to numerous local reflections, widespread reflections are also visible, occurring at lengths of several meters or even several tens of meters, often in all the profiles synchronously. Their depth position here always corresponds to interlayer surfaces, and their very large amplitude makes them stand out spectacularly from their environment in the adjacent areas of the profile. The positive sign of the double reflections and their high amplitude strongly suggest that they correspond to delamination fissures filled with water. Assuming a simple idealized model of such a fissure [9], and assuming its thickness d, it is possible to estimate very approximately the amplitude $g_R^{Double}|_{\max}$ of the corresponding double reflection:

$$g_R^{Double}\,|_{\max} \approx \gamma \cdot 4\pi \frac{d}{\lambda'_c} \frac{r}{1-r^2} \cdot w_I\,|_{\max} \tag{1}$$

Here $w_I|_{max}$ is a characteristic amplitude of the incident signal, r is the water-asphalt reflection coefficient, λ'_c is the central wavelength of the signal in water, γ is an amplitude correcting factor corresponding to weakening of the signal as a result of the double transition through the road surface and geometrical divergence of the wave front. These estimates show that the strongest reflections in the test case correspond to the fissures, whose equivalent thickness d reaches up to 0.3 mm.

The strongest extensive reflections are present in Case C, in which hydrological problems are dramatically demonstrated at the roadside (Fig. 3a), and water infiltrated into the borehole from below (Fig. 3b).

Fig. 3. a) section of road on which the drainage system is not functional; b) core extraction spot cleaned of dirt and water, spontaneously filling with water coming up from below.

Similarly, in case B there is a serious hydrological problem near the road. These observations are important because they confirm that water is the factor causing delamination cracks to show up in the echograms. Water will probably also intensify the development of defects, but as examination of cores demonstrated, the problematic interlayer connection has probably been present on the road ever since it was built. Strong double reflections probably correspond here to those places where the water entered (often extensively) into the faulty interfaces. The phenomenon of faulty interfaces (not necessarily wet), however, appears to be much more common. In echogram analysis and interpretation practice, it relates to clear visibility of interlayer horizons (large amplitude) and the type of the corresponding reflections, often more double (positive or negative) than single.

Careful quantitative analysis of layers is needed, especially in such cases described above. Observed changes in reflections, sometimes drastic (e.g. case D) do not necessarily indicate the same change in layer thickness. It should always be remembered that the radar echogram is a graphical presentation of the electromagnetic wave transition, the speed of which depends on the dielectric properties of the materials forming tested object.

3 Summary

The example of using GPR in pavement diagnosis, presented in this article, shows that this technique has great potential and described cases provide a very good illustration of diagnostic method in field conditions in terms of identifying pavement defects. Identification of interlayer connections at such early stage of road service often allows for more accurate planning of repairs, as well as seeking compensation payable under the guarantee by the contractor building the road.

A particular advantage of this method is the ability to identify and observe the subsurface damage. Sophisticated GPR observations of interlayer connection allows to evaluate bonding in more detailed way. Krysinski and Sudyka [9] describe response signals considering that boundaries of the interlayer connection are frequently not thin zones associated with a single or double reflection, but are usually thick, transitional zones manifested by the elongated response signals. Such blurred zones would correspond to:

- undulated boundary surface,
- the belt where interlocking of the aggregate grains of the two contacting layers occurs,
- zone of changing porosity (e.g. insufficient compaction),
- zone of intense material degradation e.g. aggregate crumbling.

The most serious limitation of the method is caused by the difficulty in removing the background that masks the structural signals. Insufficiently effective removal of background, leads, among others, to significant problems in distinguishing a single reflection from a double one.

In the case of the investigated road the key phenomenon that cause a faulty connection, show up on an echogram (strong double positive reflection), is an intense infiltration of water into an empty or porous zone of interlayer connection. The additional phenomena at the interlayer connections investigated on echograms having evidence in the drilling cores are:
- full delaminations between upper and lower base layer (sometimes between binding and base layer also),
- insufficient compaction at the base of binding layer and upper base layer (porous zones at the foot of these layers),
- presence of an foreign material at the connection.

References

1. P. Jaskula, *9th International Conference Environmental Engineering*, 7 (2014)
2. C. Raab, M.N. Partl, Constr. Build. Mater. **23**(8), 2926–2931 (2009)
3. S. Romanoschi, J. Metcalf, Transportation Res. Record **1778**, 132–139 (2001)
4. J. Judycki, Drogownictwo, 275–279 (2003)
5. A.C. Raposeiras, D. Castro-Fresno, A. Vega-Zamanillo, J. Rodriguez-Hernandez, Constr. Build. Mater. **43**, 372–381 (2013)
6. A. Vaitkus, D. Žilionienė, S. Paulauskaitė, F. Tuminienė, L. Žiliūtė, The Baltic J. of Road and Bridge Eng. **6**, 210–218 (2011)
7. A.C. Raposeiras, A. Vega-Zamanillo, M.A. Calzada-Pérez, D. Castro-Fresno, Constr. Build. Mater. **44**, 228–235 (2013)
8. A. Vaitkus, D. Cygas, A. Laurinavicius, V. Vorobjovas, R. Kleiziene, Građevinar **64**, 915–920 (2012)
9. J. Sudyka, L. Krysiński, Int. J. Pavement Res. Technol. **4**(3), 176–184 (2011)
10. C. Sangiorgi, A.C. Collop, N.H. Thom, *International Symposium: Non-Destructive Testing in Civil Engineering* (2008)
11. B. Al Hakim, L.W. Cheung, R.J. Armitage, Int. J. Pavement Eng. **1**, 49–59 (1999)
12. A. Saeed, J.W. Hall, Pavement evaluation **601** (2002)
13. L. Laguerre, L. Laguerre, F. Liu, F. Liu, O. Durand, O. Durand, X. Dérobert, X. Dérobert, Symp. A Q. J. Mod. Foreign Lit., 26–28 (200)
14. J.J. Daniels, *Ground Penetrating Radar Fundamentals* (USEPA Publ., 2000)
15. S. Lahouar, I.L. Al-Qadi, NDT and E International **41**, 69–81 (2008)
16. X. Dérobert X, *Ground Penetrating Radar 2nd Edition* (J. Daniels, Ed. 2009)
17. L.F. Walubita, T. Scullion, J. Leidy, W. Liu, Road Mater. Pav. Des. **10**, 259–286 (2009)
18. L. Krysiński, J. Sudyka, J. Appl. Geophys. **97**, 27–36 (2013)
19. L. Krysiński, J. Sudyka, Nondestruct. Test. Eval. **27**(May), 219–227 (2012)

20. A. Van Der Wielen, L. Courard, F. Nguyen, *Proceedings of the 13th International Conference on Ground Penetrating Radar* (GPR 2010, 2010)

MATEC Web of Conferences **163**, 02006 (2018) https://doi.org/10.1051/matecconf/201816302006
MATBUD'2018

Assessment of the compressive strength of lime mortar in the joints of brick walls - case study

Dawid Łątka[1,*], and *Piotr* Matysek[1]

[1]Faculty of Civil Engineering, Cracow University of Technology, 24 Warszawska Str., 31-155, Cracow, Poland

Abstract. The paper presents the test results of lime mortar compressive strength in the joints of brick walls. The tests were carried out with the Double Punch Test (DPT) method and with the use of an impact penetrometer (penetrometric test PT) on original samples taken from the structure of a building erected in the 1880s. The obtained results have shown that the predictions of the mortar compressive strength using both methods were very consistent (the difference 6%). The penetrometric method also made it possible to assess the homogeneity of the mortar in the direction of the wall thickness.

1 Introduction

The most reliable methods of assessing the masonry compressive strength are destructive tests carried out on masonry samples cut out from the existing structure. For the purpose of the study, extracted samples should have dimensions appropriate for the massif of the wall, also sample size should allow for statistical evaluation. However, it is a costly method and cause a significant damage to the structure and therefore can rarely be used. An additional problem is extracting intact samples from the walls. It is especially difficult when samples are taken from the walls on weak mortars. Masonry samples are often stratified and moistened during cutting by the use of water as cooling agent. The transport of larger masonry samples to the laboratory is also problematic. For these reasons, other methods of assessing the masonry strength in existing structure are developed. For example, the masonry compression strength is calculated from empirical formulas on the basis of mortar and brick strength determined in the tests. Taking bricks for destructive testing is usually not a problem. On the other hand, reliable determination of mortar strength in masonry joints is more difficult. First of all, due to the thickness of joints it is not possible taking mortar samples $40x40x80mm^3$ compliant with the requirements of EN 1015-11[1]. Therefore, a number of minor destructive methods (MDT) have been developed to assess the mortar strength in masonry joints [2-7]. The penetrometric method presented in this article is used for mortars of low strength, whereas DPT method is used for mortars of cohesion, which allows samples to be cut out of the masonry joint. Other common

* Corresponding author: dawid.latka@pk.edu.pl

diagnostic methods worth mentioning are the Windsor Pin Test and the Helix pull-out test [3],[4]. However, these methods are developed for medium and high strength mortars. The advantage of minor destructive methods is the ability to perform a large number of in-situ tests, which is crucial in statistical evaluation. So far, verification of minor destructive methods of testing mortar strength in the joints of masonry walls has been carried out mainly on samples made in the laboratory using modern materials. On the other hand, there are very few test results conducted on real objects. This fact strongly limits the practical application of MDT methods. In order to increase the reliability of the test results, the authors of this article suggested to carry out research using 2 minor destructive methods.

2 Description of test procedures

2.1 Penetrometric test (PT)

The penetrometric test consists of measuring the immersion of a steel needle in the mortar due to the repeated loading of the needle with a constant force. The view of the penetrometer used in the studies is presented in the figure 1. The penetrometric test, carried out according to the full incremental procedure, allows assessment of the mortar compressive strength as well as its homogeneity in the masonry joints. The sequence of activities in the case of testing mortar in joints on the external side of the wall (face side) starts with the designation of test points. Then, after cleaning the external surface of the tested joint and possibly removing the repointing layer, the reference needle position is measured. Next, 5-10-15-20 consecutive needle strokes are made (4 series of 5 strokes each) measuring immersion of the needle. On the basis of these actions, the mortar strength can be determined in the joint zone located closest to the external surface of the masonry wall. The testing of mortar in the deeper layers of the masonry starts with the drilling at the tested point a 12mm diameter and 100mm depth to remove the previously damaged joint material. Then, the reference measurement and steps described above are realized.
The compressive strength of f_{PT} is determined on the basis of a needle penetration expressed in mm, after the tenth impact. The correlation curve given by the penetrometer manufacturer [8] is used.

The homogeneity of the mortar in the direction of the wall thickness is determined using the graphical method, comparing the penetration of the penetrometer needle among the four measuring series. Similar immersion of the needles for series 2 to 4 means satisfactory uniformity of mortar.

Fig. 1. Penetrometer of RSM-15 version 1.0 of DRC Srl used for testing of mortar in joints of the masonry walls.

2.2 Double punch test (DPT)

The DPT method was first described by Henzel & Karl in [9]. This method carried out in accordance with DIN 18555-9 [10] and UIC 778-3R [11] has many advantages that have made it very popular. The tested material is taken directly from the structure. The mortar joints are separated from the bricks. Then, irregular mortar samples are cut to regular square form of a side shape of at least 40mm (or a circle of identical diameter). Thickness of samples is equal to the thickness of the bed joints. The compressive force is applied by means of steel punches, usually with a diameter of 20mm. The surface of the sample is levelled with a layer of material (gypsum, mortar) with the strength approximately similar to the strength of tested mortar. The compression load should be applied in a static manner. The test result is the maximum load transmitted to the sample before its destruction. The mortar strength in the DPT test is denoted as f_{DPT} and is determined from the relation (1):

$$f_{DPT} = F_{max} / A \tag{1}$$

where, F_{max} is the max. load in kN, whereas A is the area of the steel punch.

Due to the fact that the load area is smaller than the surface area of the specimen and friction forces are present at the contact of steel punches with the specimen, the mortar tested by DPT is in a complex stress state. For this reason, the strength of the mortar f_{DPT} calculated from formula (1) may be significantly higher than that determined according to EN 1015-11 [1].

3 Experimental and results

3.1 Samples for tests

The tests were carried out on 250x260x330mm^3 brick pillar (PT tests) and 26 mortar samples (DPT tests) taken from external walls of a masonry building erected in the 1880s. The building structure was made of bricks 280x130x68mm^3 bonded with lime mortar. The thickness of joints in the masonry was characterized by a great variety. The bed joints were between 12mm and 25mm thick, while the thickness of the vertical joints was between 7mm and 30mm. The quality of bed joints was assessed as satisfactory. Vertical joints showed numerous voids and local lack of adhesion between mortar and bricks. The brick pillar cutting stage, preparation for testing and heterogeneity of the material in the area of vertical joints are presented in figure 2.

On the two sides of the brick pillar (the ones that were the external sides of the wall) all joints contain repointing with cement mortar to the depth 25-30mm. The extraction of the brick pillar made it possible to conduct penetrometric tests of the bed joints located inside the masonry wall and comparison with the results for the measurements of the joints in the facing layer.

Samples for DPT tests were taken from different parts of the structure. The specimens with a height equal to the thickness of the bed joints had a square size of about 50mm. Prior to the test, gypsum layers with a diameter of approx. 30mm were made on all mortar samples.

Fig. 2. Tested brick pillar: (a) cutting the pillar from the structure; (b) load induction - 20kN; (c) losses of mortar in vertical joints, local lack of adhesion between mortar and bricks.

3.2 Penetrometric test on masonry brick pillar

After seasoning in air-dry conditions, the brick pillar with a 20kN compression force (corresponding to a compressive stress of approximately 0.3MPa) was loaded. The EDU 400 testing machine was used to test. The penetrometric tests were carried out on the bed joints in places far from voids, cracks or intersections of vertical and bed joints. The tests were carried out using RSM-15 penetrometer, according to the methodology described in point 2.1. Tests on external surfaces (2 measuring points marked A-B) and on internal surfaces (8 measuring points marked C-J) of the brick pillar were performed. The method of testing was shown in figure 3.

In figure 4a the penetration depth increment functions for all measurement series are presented, the results of which did not raise any objections. Two measurement series (marked A and B) conducted on the external surface of the pillar were omitted. Tests have shown that under the repointing cement mortar layer, a degraded original mortar with a loosened structure is located, which resulted in significant penetrometer needle immersions. The remaining 8 series of measurements were carried out on the internal (cut) sides of the brick pillar and thus on the mortar, which was located inside the wall and was not degraded. The mean global value of the needle penetration depth after one impact for all measurements was 1,5mm/impact. The greater uniformity of test results for surface measurements (marked on the graph with a continuous line) was observed (see figure 4a). This is due to the fact that surface points were selected in such a way that the results were not affected by any defects as cracks or voids in mortar. Such conditions for downhole measurements could not be observed, thus the results of these measurements have larger scattering (dashed line in figure 4a vs table 1).

Fig. 3. Testing methodology: (a) penetrometric measurements; (b) location of the needle of the penetrometer after surface examination; (c) dredging of the hole in the test point; (d) penetrometric deep measurements; (e) location of the needle of the penetrometer after the depth test.

Analyzing the penetration distribution (see figure 4b), it should be recognized that the mortar in the bed joints is characterized by high homogeneity, except for zones under repointing mortar. The average compressive strength estimated on the basis of penetrometric tests (table 1) was 2.41MPa at a standard deviation of 0.48MPa (coefficient of variation equal 20%).

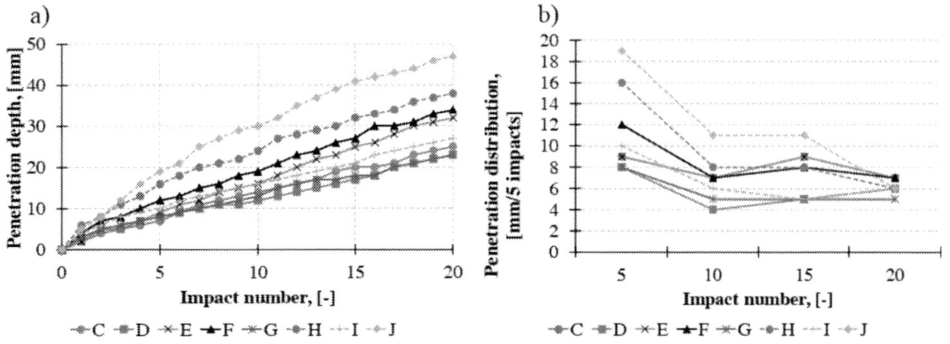

Fig. 4. Results of penetrometric tests (surface measurements - continuous line; downhole measurements - dashed line): a) increase in penetration depth for 20 consecutive impacts; b) penetration depth distribution (4 groups of 5 impacts each).

Table 1. PT results (10 series).

Measurements	Strength f_{PT} min. - max. [MPa]	Average [MPa]	Std. deviation [MPa]	CoV [%]
Surface	2.32 – 3.26	2.26	0.39	15
Downhole	1.55 – 2.58	2.07	0.42	20
Total	1.55 – 3.26	2.41	0.48	20

3.3 Double punch tests on mortar samples cut out from masonry

The tests were carried out on the Zwick Roell Z100 in 0.5kN/min force control mode. 20mm diameter steel punches were used for the test. The tested sample in figure 5a is presented. The damage of the mortar sample was caused typically by the appearance of radial cracks outside the load area, as shown in figure 5b.

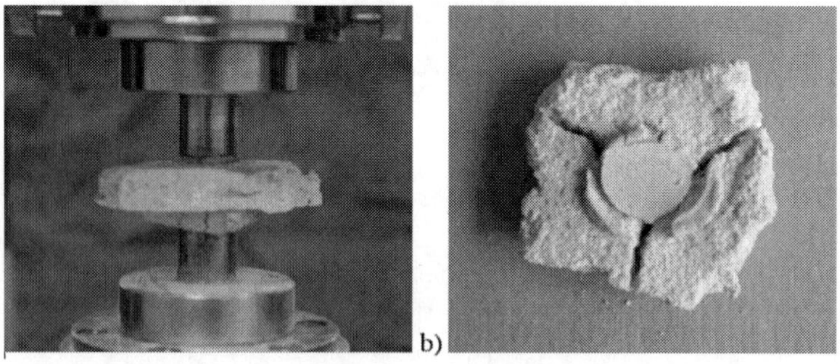

a) b)

Fig. 5. Testing methodology: (a) DPT test; (b) typical failure mode of mortar sample.

The average mortar strength (f_{DPT}) obtained according to the formula (1) was 2.28MPa, while the standard deviation – 0.62MPa (table 2). The considerable heterogeneity of mortar in joints is a typical feature of historical structures. The difference in thicknesses between the samples also affected the results. These issues were already the subject of our own research [12] as well as other authors' studies [13],[14].

Table 2. DPT results (26 samples).

Feature	Strength f_{DPT} min. - max.	Average	Standard deviation	CoV [%]
Compressive strength, f_{DPT}	1.34MPa – 3.92MPa	2.28MPa	0.62MPa	27%
Thickness of samples	13.4mm – 25.6mm	20.3mm	2.6mm	13%

4 Comparison of DPT and PT results

Comparison of the results of the DPT and PT tests for lime mortar strength are presented in table 3. It is necessary to notice the similar values of all strength parameters. The difference in compressive strength of mortar determined by DPT and PT methods was 0.13MPa, which is a small value from an engineering point of view. A greater difference was made in the standard deviation of mortar strength. This is due to the fact that samples for the DPT test were taken from different points of construction and therefore there is a greater dispersion of the test results. Significant dispersions of research results are also caused by the specificity of historical masonry materials. The internal structure of mortar and bricks of the 19th century is definitely different from those used today. The characteristic feature of historical bricks and mortars is the lower homogeneity expressed by the occurrence of numerous structural cracks, as well as inclusions of various materials and aggregates of considerable diameter (figure 6).

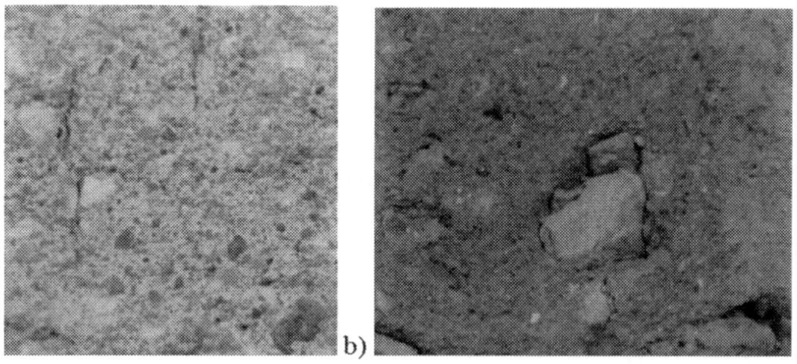

a) b)

Fig. 6. The internal structure of the tested historical materials: a) typical mortar structure in a bed joint - a stack of aggregates with multiple fractions, including thick fractions; c) typical brick structure - numerous interjections.

In paper [2] it was shown that the compressive strength determined in the DPT test on samples taken from joints with a thickness of 20mm is approximately equal to the compressive strength of mortar determined on 40x40x80mm3 prism specimens. However, this matter requires further research and analysis.

Table 3. Comparing PT and DPT results.

Test	Strenght		Standard deviation		CoV	
	Value [MPa]	Difference [%]	Value [MPa]	Difference [%]	Value [%]	Difference [%]
DPT	2.28	-6	0.62	23	27	26
PT	2.41		0.48		20	

5 Conclusions

The paper presents the test results of lime mortar strength in joints of brick walls. The tests were carried out on an original brick masonry erected in the 19th century. This is particularly important because the previous verification tests in the field of minor destructive methods were carried out mainly on masonry made of modern materials and therefore they omitted important features distinguishing historical structures.

The compression strength of the lime mortar determined in the penetrometric test (PT) was similar to that of the Double Punch Test (DPT). Differences in mortar strength according to DPT and PT did not exceed 6%.

Both test methods, the DPT method and the PT method, seem to be useful in estimating the compressive strength and homogeneity of mortars in the joints of masonry. Good results, as shown in the article, can be achieved mainly by compilation of research methods.

References

1. EN 1015-11 (2007)
2. E. Sassoni, E. Franzoni, C. Mazzotti, Key Eng. Mat., **624** (2015)
3. L. Pela, P. Roca, A. Aprile, *Structural Analysis of Historical Constructions: Anamnesis, Diagnosis, Therapy, Controls*.10 (CRC Press, 2016)
4. RILEM Recommendation MS-D.9, Mater. Struct. **30** (1997)
5. RILEM Recommendation MS-D.7, Mater. Struct. **30** (1997)
6. N. Gucci, R. Barsotti, Mater. Struct. **28,** 276–283 (1995)
7. D. Liberatore, N. Masini, L. Sorrentino, V. Racina, M. Sileo, O. Alshawa, L. Frezza, Constr. Build. Mater. **122**, 810-822 (2016)
8. *Penetrometro malta SRM 1.0 user manual - 2015*. www.drcitalia.it/en, (February 5, 2017)
9. J. Henzel, S. Karl, Darcon., **2** (1987)
10. DIN 18555-9 (1999)
11. UIC code 778-3 (2011)
12. D. Łątka, S. Seręga, P. Matysek, *Structural Analysis of Historical Constructions: An Interdisciplinary Approach* 11 (2018) (in revision)
13. M. Drdácký, D. Makšin, M.D. Mekonone, Z. Slížková, *Historical Mortars Conference: Characterization, Diagnosis, Conservation, Repair and Compatibility*, (2008)
14. M. Drdácký, Int. J. Archit. Herit. **5**, 383-394 (2011)

Towards the utilization of waste glass powder in sustainable cement based overlays

Łukasz Sadowski[1,*]

[1]Wroclaw University of Science and Technology, Wybrzeże Wyspiańskiego 27, 50-370 Wroclaw, Poland

Abstract. The article focuses on the possibility of utilizing waste glass powder in cement based overlays for sustainable construction. In the work, a study was carried out on the evaluation of the influence of waste glass powder, which modifies the overlay, on its adhesion to a concrete substrate. The waste glass powder was added in various amounts in relation to the binder mass. The adhesion tests were conducted using the commonly known pull-off method. The preliminary test results are presented and discussed in the article.

1 Introduction

Innovations in civil engineering and materials science have recently not been possible without considering sustainable development [1]. Some parts of these innovations may be successfully applied to the overlay industry. More attempts have recently been focused on the modification of the material of an overlay with different additives. These attempts can be made during the construction or repair of a concrete structure [2-3]. As estimated in [4], around 50% of Europe's annual construction budget is spent on refurbishment and repair. To achieve sustainable repair, waste material should be utilized.

Waste mineral powders, such as fly ash, blast-furnace slag or waste perlite powder are successfully used for the production of cement and concrete [5]. As pointed out in [6], waste glass powder can also further contribute to sustainability in construction. According to [7], waste glass can mainly be sourced from the production of container glass (bottles, jars), flat glass (windows, windscreens), bulb glass (light globes) or cathode ray tube glass (TV screens, monitors, etc). Waste glass powder has recently become popular as a supplementary cementitious material in glass-based concrete [8], architectural cement mortars [9], cement paste [10] or geopolymer paste [11]. To the best of my knowledge, there is no application of waste glass powder in the overlay industry. Thus, there is a need to perform research towards the utilization of waste glass powder in sustainable cement based overlays.

The durability of multi-layered concrete elements depends on the proper adhesion between an overlay and a concrete substrate [12]. According to literature reports, some researchers try to modify normal-strength concrete with various additives [13-17]. These tests can be aimed at achieving the required adhesion value between an overlay and substrate without the need for special treatments to prepare the surface of the concrete substrate. For

* Corresponding author: lukasz.sadowski@pwr.edu.pl

example, *Moghtadaei, Mohammadi et al.* [18, 19] conducted research on the adhesion between an overlay with the addition of metakaolin and a concrete substrate, and showed that this additive increases this adhesion. In turn, *Nagaonkar et al.* [20] showed an increase in the adhesion of an overlay made with the addition of silica fume to a concrete substrate. On the other hand, *Li* [21], in order to increase the adhesion between concrete layers, modified them with the addition of fly ash derived from the combustion of hard coal. These additives have recently been more frequently sourced from industrial wastes.

There are no reports in literature of whether the modification of an overlay with additives in the form of glass powder, which is a waste material in construction, favourably affects the adhesion of this overlay to a concrete substrate. This work is an initial contribution to filling this knowledge gap. The preliminary attempts of this utilization will be presented.

2 Materials and methods

The research involved two model two-layer concrete test elements with dimensions of 600 x 1200 mm and a total thickness of 150 mm (Fig. 1). The substrate was 125 mm thick and the overlay was 25 mm thick.

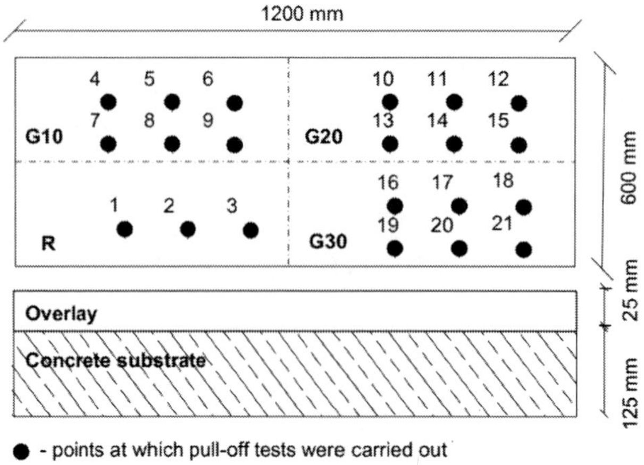

Fig. 1. Scheme of the model research element with the location of the points where the pull-off tests were carried out.

Table 1 shows the weight composition of the concrete mix used to make the substrate. Concrete basalt aggregate with a maximum grain size of up to 8 mm and a volume density of 2.60 g/cm^3 was used to make the concrete substrate.

Table 1. Weight composition of the concrete mix used to make the substrate.

Portland cement CEM II A-LL 42,5 R	Fly ash (FA)	Sand	Coarse aggregate (up to 8 mm)	Water	Plasticizer Visco Flow 6920
[kg/m^3]					
352.0	40.0	724.4	1086.6	165.0	2.0

In turn, Table 2 presents the weight compositions of the concrete mixes used to make the overlay. In the case of the overlay concrete, it was made with natural quartz aggregate with a maximum grain size of up to 8 mm and a bulk density of 2.64 g/cm^3. To make the concrete of both layers, quartz sand with a maximum grain size of up to 2 mm and a volume density of 2.62 g/cm^3 was used. It should be clarified that the surface of both elements was divided into four parts, each of which had an overlay with a different composition, which is given in Table 2. In the case of the first element, as presented in Figure 1, on the part marked with the letter R (reference) the top layer was made of the mixture marked R. In turn, on the part marked G, the top layer was made of waste glass powder G10, G20 and G30 containing 10, 20 and 30% of glass powder, respectively in relation to the binder mass. According to [22], Portland cement CEM 42.5 R was used.

Table 2. Weight composition of the concrete mix used to make the overlay.

Designation of concrete mix	Component of concrete mix [kg/m^3]					
	Portland cement CEM I 42.5 R	**Waste glass powder**	**Sand**	**Coarse aggregate (up to 8 mm)**	**Water**	**Plasticizer Master Pozzolith 18BV**
R	352.0	0	676.0	1205.0	141.0	2.1
G10	316.8	35.2	676.0	1205.0	141.0	2.1
G20	281.6	70.4	676.0	1205.0	141.0	2.1
G30	246.4	105.6	676.0	1205.0	141.0	2.1

The concrete of the surface and overlay matured naturally at an air temperature of + 20°C (± 3°C) and relative air humidity of 60% (± 5%). Table 3 shows the particle size distribution of waste glass powder and Table 4 shows the chemical composition of waste glass powder. For comparison, the chemical composition of the Portland cement used in the testing is also presented.

Table 3. Grain size distribution of waste glass powder.

Test sieve (μm)	Fraction passing (%)
1400	0.0
1000	0.2
430	2.0
150	8.0
100	15.0
63	40.0
< 63	90.0

It can be seen from Table 4 that the waste glass powder and cement CEM I 42.5R differ significantly in all the main components. The waste glass powder is mainly composed of SiO_2 and Na_2O+K_2O, while the dominant component of the cement was CaO and SiO_2.

Table 4. Chemical composition of waste glass powder and cement CEM I 42.5R used in the testing (data from producers)

Component	Content in waste glass powder, (mass %)	Content in cement, (mass %)
SiO_2	71.50	22.18
Al_2O_3	1.25	5.98
CaO	9.00	61.13
MgO	4.00	1.12
Na_2O+K_2O	14.00	0.65
TiO_2	max. 0.1	max. 0.1
Other	0.15	8.84

28 days after concreting the substrate, the overlay was made on it. Then, after 7 and 28 days, pull-off adhesion tests were carried out using the pull-off method. The location of points where the pull-off tests were performed was previously presented in Figure 1.

According to [23], the pull-off tests involved the drilling of holes in the overlay with a diameter of 50 mm and the gluing of steel discs, followed by detaching them from the surface of the substrate with a special actuator with the registration of the pull-off force F_b. Knowing the average dimension of the diameter of the steel disc D_f, the value of pull-off adhesion f_b is obtained from the following dependence (1):

$$f_b = \frac{4 \cdot F_b}{\pi \cdot D_f^2} \quad (1)$$

3 Preliminary test results

Figure 2 shows the influence of various amounts of waste glass powder, which modify the concrete of the overlay, on the pull-off adhesion f_b to the concrete substrate. The analysis of Figure 2 shows that a 10 and 20% addition of waste glass powder in relation to binder mass does not cause the increase of the value of the pull-off adhesion f_b to the concrete substrate after 7 and 28 days. However, a 30% addition of this waste reduces the pull-off adhesion by about 5% after 7 days, and by about 20% after 28 days.

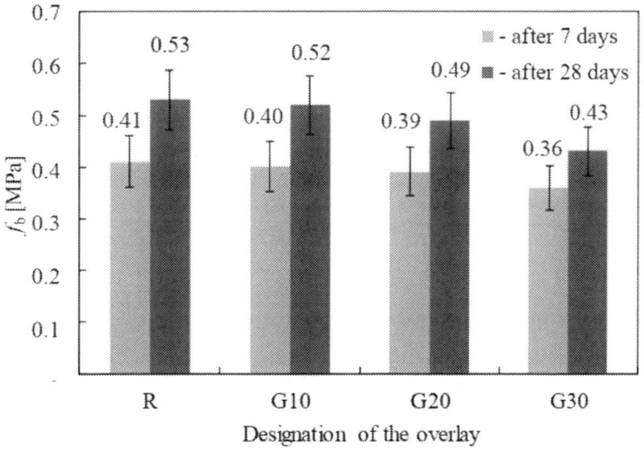

Fig. 2. Influence of different amounts of waste glass powder, which modify the concrete of the overlay, on the pull-off adhesion f_b to the concrete substrate.

4 Conclusions

The paper evaluates the influence of selected waste additives, which modify the concrete of the overlay, on its bond to the concrete substrate. This additive was glass powder.

The preliminary results show that the modification of the concrete of the overlay with the addition of waste glass powder in amounts of 10 and 20% of the binder mass results in the pull-off adhesion f_b value of the overlay to the substrate remaining unchanged. Increasing the amount of this powder in the concrete composition by up to 20% of the binder mass reduces the pull-off adhesion f_b value by about 5% after 7 days, and by about 8% after 28 days.

The presented preliminary results are promising when taking into account the route towards the utilization of waste glass powder in sustainable cement based overlays. Future studies will be needed and will be related mainly to the testing of the mechanical properties of modified overlays, as well as some microscopic observations.

References

1. L. Czarnecki, D. Van Gemert, Bull. Pol. Acad. Scie. Techn. Scie. **65**, 6 (2017)
2. F. Leonelli, P. Di Mascio, A. Germinario, F. Picarella, L. Moretti, M. Cassata, A. De Rubeis, Appl. Scie. **7**, 11 (2017)
3. A. I. Ortega, T. M. Pellicer, J. M. Adam, P. A. Calderón, Const. Build. Mater. **161** (2018)
4. G. P. Tilly, J. Jacobs, *Concrete repairs: Observations on performance in service and current practice* (CONREPNET Project Report, IHS BRE Press, Watford, UK, 2007)
5. P. Łukowski, Mater., **9**, 10 (2016)
6. A.M. Matos, J. Sousa-Coutinho, Const. Build. Mater. **36** (2012)
7. A. Shayan, A. Xu, Cem. Concr. Res. **34**, 1, (2004)
8. H. Lee, A. Hanif, M. Usman, J. Sim, H. Oh, J. Cle. Pro. **170** (2018)
9. J.X. Lu, Z.H. Duan, C.S. Poon, Const. Build. Mater. **131** (2017)
10. M. Kamali, A. Ghahremaninezhad, Const. Build. Mater. **149** (2017)
11. T. Tho-In, V. Sata, K. Boonserm, P. Chindaprasirt, J. Cle. Pro. **172** (2018)

12. L. Courard, B. Bissonnette, A. Garbacz, *Concrete Surface Engineering* (CRC Press, Boca Raton, 2015)
13. Y. Shi, I. Matsui, J. Guo, Cem. Concr. Res. **34**, 8 (2004)
14. A. Galińska, S. Czarnecki, IOP Conf. Ser.: Mater. Sci. Eng. **245**, 3 (2017)
15. M. Nepomuceno, L. Oliveira, S. Lopes, Const. Build. Mater. **26**, 1 (2012)
16. J. Tikkanen, V. Penttala, A. Cwirzen, Mag. Concr. Res. **63**, 12 (2011)
17. J. Tikkanen, Const. Build. Mater. **72** (2014)
18. R. Moghtadaei, M. Mohammadi, N. Samani, S. Mousavi, Const. Build. Mater. **80** (2015)
19. M. Mohammadi, R. Moghtadaei, N. Samani, Const. Build. Mater. **51** (2014)
20. D. Nagaonkar, J. Bhusari, Int. J. Sci. Eng. Res. **5**, 5 (2014)
21. G. Li, Cem. Concr. Res. **33**, 6 (2003)
22. EN 197-1 (2011)
23. ASTM D7234-05 (2005)

Assessing the Performance of Concrete Containing Recycled Glass on Tensile Splitting Strength and Resistance to Alkali Silica Reaction

Chun-Hsing Ho[1,*], *Anas Almonnieay*[1], *Khin Kyaw*[1], and *Phoo Myat Sandy Maung*[1]

[1]Department of Civil Engineering, Construction Management, and Environmental Engineering Northern Arizona University, PO Box 15600, Flagstaff, Arizona, 86011, USA

Abstract. Using recycled glass in concrete applications decreases the amount of glass in landfills and substitutes for expensive aggregates in the concrete mix. However, there has been a concern on recycled glass with smooth surfaces that would result in a drop in strength and in particular a reduction of an already low ductility. Thus, in many design aspects, the use of recycled glass in concrete is limited up to 30% by weight due to concern on concrete strength reduction. The current manufacturing technology in the recycling glass has been grown and evolved through which recycled glass has been processed to exhibit the following features: basically zero water absorption, excellent hardness (great abrasion resistance), high durability to resist extreme weather conditions, etc. The paper challenges the currently used recycled glass mixtures and presents new mix design principles for concrete mixed with 10%, 20%, 30%, 50%, and 100% recycled glass as replacements of nature sand and Portland cement to assess (1) strength changes and (2) resistance to alkali silica reaction (ASR). Aggregate, water reducer, hydration stabilizer, mid-range water reducer, fiber, and viscosity modifier were prepared with varying dosages of recycled glass. A series of scanning electron microscope (SEM) imaging were performed to evaluate the resistance of recycled glass specimens to ASR. The paper concludes that the use of recycled glass as an alternative aggregate and cement binder in the concrete mixtures show promising performance in both tensile splitting strength and ASR.

1 Introduction

Conventional or typical concrete is designed to have a high compression strength that can be applied to a diverse spectrum of real world projects. Approximately 7 million cubic yards of concrete are produced per year, making it the second most consumed substance on earth (next to the water) [1]. Cement production is growing by 2.5% annually, and is expected to rise from 2.55 billion tons in 2006 to 3.7-4.4 billion tons by 2050 [2]. However, convectional cement production has created substantial volumes of greenhouse gas (GHG) that have influenced the environment. Based on a report done by Portland Cement Association [3], cement plants in the U.S. accounted for about 1.1 percent of U.S. GHG emissions, 82 million

* Corresponding author: chun-hsing.ho@nau.edu

tons out of a total of 7.28 billion tons. Recently, the increasing interest in strategies to reduce (GHG) emissions has led to the examination options for reducing emissions from the production of Portland cement.

In addition to cement in the concrete mix, aggregates known as granular materials collected from natural resources, can also be used in an environmentally-friendly way for concrete construction. Aggregates typically account for 70-85 percent, by weight, of the materials used in concrete. Using recycled materials for replacement of nature aggregates would substantially lessen the disruption of the environment. One of methods is to use alternative materials to replace manmade Portland cement such as recycled glass. Pulverized post-consumer glass is eco-friendly alternative to be considered in concrete, specifically to replace the Portland cement, natural sand and gravel aggregate. However, there has been a concern on recycled glass with smooth surfaces that would result in a drop in strength and in particular a reduction of an already low ductility. Thus, in many design aspects, the use of recycled glass in concrete is limited up to 30% by weight due to concern on concrete strength reduction [4]. However, the current manufacturing technology in the recycling glass has been grown and evolved through which recycled glass has been processed to exhibit the following features: basically zero water absorption, excellent hardness (great abrasion resistance), high durability to resist extreme weather conditions, etc. In addition, many researchers have investigated the use of recycled glass as fine and coarse aggregate of the concrete in the past years [5]-[7]. More recently, research works have been started using recycled materials in concrete mixtures in 2012 [8]-[9]. If recycled glass can be used as a replacement of aggregate and cement, the effort to reduce the carbon dioxide due to production of virgin aggregate and cement will be significant. Furthermore, recycled glass (RG) is a product that acts as a pozzolan material. The presence of pozzolan in a concrete mix can help consume the alkali that are produced from the cement and the aggregate in the mix which in overall helps to reduce the amount of ASR within the concrete structure. Thus, the purposes of this paper is to present a list of concrete mixtures using recycled glass as an alternative aggregate and cement binder and evaluate the effectiveness of recycled concrete mixtures in the performance of strength and resistance to alkali-silica reaction.

2 Material preparation

Two types of recycled glass were used in experiments including coarse recycled glass (RG sand) and fine recycled glass (RG powder as a replacement of cement binder). Based on literature findings, we would like to challenge the currently used recycled glass mixtures and present new mix design principles for concrete mixed with 10%, 20%, 30%, 50%, and 100% recycled glass as replacements of nature sand and Portland cement. Based on the goal of achieving desired high strength of recycled glass concrete, we used a number of mix formulas to produce a volume of 600 cubic inch/9,832 cubic centimeter concrete mixed with and without fiber. The mix design formula contained recycled glass, coarse aggregate (1/2", 3/8" and No.4), admixture (Water Reducer, VMA900, Air Entrainment and polymer), water, cement, and sand. Recycled glass powder was used to replace cement while coarse recycled glass was used to replace nature sand. The coarse aggregate (1/2", 3/8" and No.4) remain unchanged in the mixing process. Table 1 shows the design mix matrix containing a variety of ranges using recycled glass. It should be noted that Control 1 mix design in Table 1 containing 100% cement and 100% natural aggregate (sand) was used as a control group in order to compare strength difference with the rest of mix designs containing different percentages of recycled glass.

Table 1. Mix design matrix containing recycled glass

Mix Designs	Cement (kg. /m^3)	RG Powder	Sand (kg. /m^3)	RG Sand
Control 1	470.27	0%	654.61	0%
MD# 1.1	470.27	0%	458.23	30%
MD# 1.2	329.19	30%	654.61	0%
MD# 1.3	470.27	0%	523.69	20%
MD# 1.4	376.21	20%	654.61	0%
MD# 1.5	470.27	0%	589.15	10%
MD# 1.6	423.24	10%	654.61	0%
MD# 1.7	470.27	0%	327.30	50%
MD# 1.8	235.13	50%	654.61	0%
MD# 1.9	470.27	0%	0.00	100%

Note: RG represents recycled glass

In addition to aggregate/cement or aggregate/cement replacements, admixtures were also an important resource in the concrete mixture used to improve strength and to provide the design mix with resistance against thermal and fatigue cracks. The admixtures used in the all mix designs are mid-range water reducer, air entrainment (Micro Air), viscosity modifier (VMA), and polymer. Water reducer is a chemical that reduces the amount of water needed for the mix to reach its high strength. This chemical improves the strength of the concrete mix (compression and tensile), as well as the workability of the concrete. Air entrainment creates air bubbles in the concrete structure to facilitate resistance to thermal expansion and contraction of the concrete. Viscosity modifier also helps to increase the strength and the workability of the concrete mix. All specimens were produced using the mix design matrix depicted in Table 1 and placed in a 4 x 8 inches (15 x 30 cm) cylinder. All demolded specimens were subsequently submerged in a water tank for curing until the 7th days and the 28th days for testing.

3 Material testing and result discussions

3.1 Tensile splitting test

A tensile strength test was used to evaluate the behavior of the concrete under applied load. The testing procedure follows the American Society for Testing and Materials ASTM C496: Standard Test Method for Splitting Tensile Strength of Cylindrical Concrete Specimens [10]. The testing device, a specimen set up, and a close view of a tested specimen are shown in Figure 1. The primary reason for testing splitting strength instead of compression was to review the usefulness of recycled glass in the concrete pavement structure (both airfield and highway systems). So, the interest of research was focused on tensile splitting performance than compressive strength of recycled specimens.

Fig. 1. Splitting test device and specimen set up

The test results were categorized in to two groups for comparison. Mixtures containing less than 50% RG replacement are shown in Table 2 while mixtures containing greater than 50% RG replacement are shown in Table 3. In comparison with the specimens containing less than 50% RG replacement, the specimens containing more than 50% RG replacement matrix (MD# 1.7-MD# 1.9) have better splitting strength. When the amount of RG increases and it is more consistent in distribution throughout the matrix in the concrete, the strength of concrete specimens increased in the twenty-eight (28) days with the average strength of 4,71 MPa to 5.47 MPa. In the design mixes that contain less than 50% RG replacement, MD# 1.4 with a 20% RG as a cement binder replacement has the least tension strength tested at 28 days. On the other hand, for the design mix with RG replacement higher than 50%, the MD# 1.7 specimen (50% RG sand) has the highest strength (28 days) among all specimens. From Table 2 and Table 3, it seemed that the role of recycled glass with a replacement of coarse aggregate (sand) could exhibit better strength improvement than the fine RG power (as an alternative cement binder). Regardless of designs with RG higher and lower than 50%, both tensile strength values did not surpass the control group's strength (28 days). However, the overall strength of recycled glass specimens still meet the desired strength requirement of 2.76 MPa, a desired level of splitting strength for most concrete pavement structure.

Table 2. Tensile Tension Strength for mixes less than 50% RG replacement

Experimental Number	Experimental Detail	7 days average (MPa)	28 days average (MPa)
Control	0% RG	4.38	6.14
MD# 1.1	30% RG Sand	5.31	5.58
MD# 1.2	30% RG Cement	4.69	4.86
MD# 1.3	20% RG Sand	3.88	4.72
MD# 1.4	20% RG Cement	5.48	3.90
MD# 1.5	10% RG Sand	4.03	4.83
MD# 1.6	10% RG Cement	4.41	4.38
Average strength (#1.1 ~1.6)		4.63	4.71

Table 3. Tensile Tension Strength for mixes greater than 50% RG replacement

Experimental Number	Experimental Detail	7 days average (MPa)	28 days average (MPa)
MD# 1.7	50% RG Sand	4.79	5.72
MD# 1.8	50% RG Cement	4.31	5.03
MD# 1.9	100% RG Sand	4.48	5.65
Average strength (#1.7 ~1.9)		4.53	5.47

3.2 Evaluating resistance to alkali-silica reaction

Alkali-silica reaction (ASR) is a reaction that caused by the silica within the aggregates and cement that over time turn into a gel. As time passes, more gel is formed causing it to take more space, causing micro cracks and reducing its tensile strength [11]. Specifically, the ASR mechanism can be divided into two parts: (1) critical components for creating ASR, and (2) the actual chemical reaction and ASR distress expansion.

(1) Three critical components are essentially to produce ASR:

i. Aggregates that contain reactive silica or silicate chemicals.
ii. Sufficient alkali ions such as sodium (Na) and/or potassium (K).
iii. Sufficient water/moisture inside the Portland Cement Concrete pavement.

(2) The actual chemical reaction and distress mechanism of ASR can be formed by the following steps:

i. The actual chemical reaction:
Siliceous (aggregates) + alkali (mortar/cement paste) + water (intrusion/moisture environment) = ASR gel, a viscous material.
ii. ASR gel product + moisture (absorption of water) → expansion surrounding cement paste.
iii. Stresses inside pavements created by expansive forces eventually result in cracking and spalling distress.

It is recognized that the product of the chemical reaction is the ASR gel which subsequently absorbs moisture, expands, and eventually causes concrete to crack and weakens the integrity of a concrete structure.

3.3 Scanning electron microscope imaging

Scanning electron microscope (SEM) imaging was performed using specimens that were prepared for alkali silica reaction to observe the silica gel resistance. The SEM imaging is implemented to evaluate the resistance of recycled glass to silica gel. The SEM was done in the Microanalysis Core Facility located at Northern Arizona University. Before SEM imagining, all specimens were coated with plutonium (Pu) and gold (Au) to enhance the resolution of the imaging on the specimens. The mix designs that the test was performed on were selected from Control 1, MD#1.1, MD#1.2, MD#1.3, MD#1.4, and MD#1.5 in which % of RG was less than 30%. The specimens were added silica to enable ASR affection in individual concrete specimen.

In order to observe how ASR gels are produced within specimens, all ASR specimens were undergone a series of SEM screening to allow us observing the appearance of ASR gels on the specimens. The electron microscope device is capable of filtering out the amount of individual chemical element and showing in different colors on the screen (Figure 2). This helps us better understand the effectiveness of recycled glass in mitigating ASR gels among all specimens.

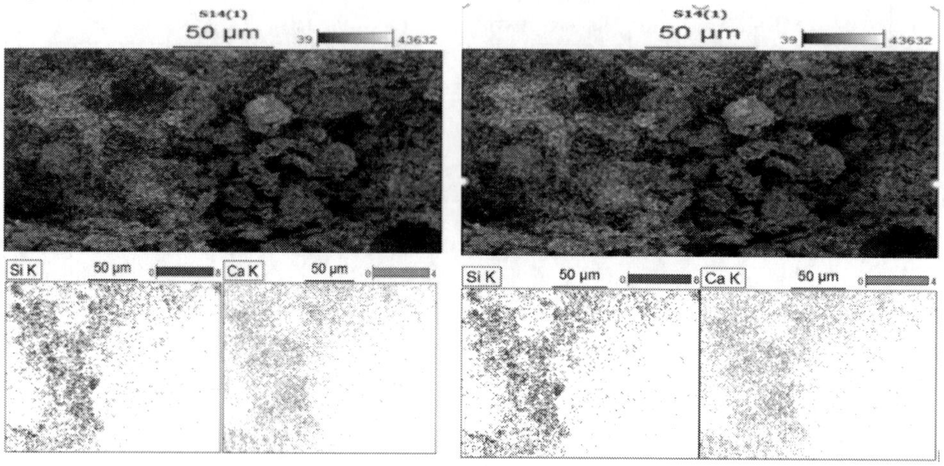

Fig. 2. Silica gel represented in MD 1.3 (L) and MD 1.4 (R).

After ASR occurred on all specimens, SEMs were able to measure a percentage of individual chemical contribution to the total weight in each specimen. The measure data allowed us to review the level of ASR affection on specimens based on the percentage of recycled glass used in the concrete mix. The results are presented in Table 4. Since ASR is primarily triggered by alkali, silica, and calcium elements (highlighted in yellow in Table 4), the main focus was to pay attention on the % weight that was generated by ASR on each specimen so enabling us to evaluate the effectiveness of recycled glass in mitigating ASR in the concrete mix. As shown in Table 4, it is clear that control group (without the use of recycled glass) generated the highest ASR gels among other specimens mixed with recycled glass. The MD-S 1.3 specimen (20% RG sand) has the least ASR gels remained on the specimen. In comparison with the use of coarse recycled glass in all specimens, the addition of fine recycled glass power in the concrete mix as an alternative cement binder seemed not to have better resistance to ASR affection. This finding suggests that the addition of coarse recycled glass in the conventional concrete mix could increase the ability to resist ASR occurrence in the concrete structure. However, more research works should be done to further verify the results done in the research.

Table 4. Measure results of individual chemical element within specimens.

Specimen #	Individual element, weight %							
	O-K	Mg-K	Al-K	Si-K	Ca-K	Pd	Au	Total
Control	0	0	0	48.45	31.55	0	0	100
MD-S 1.1	35.57	0.38	3.03	8.77	30.78	10.27	11.19	100
MD-S 1.2	0.64	0	0	25.00	2.16	51.67	20.53	100
MD-S 1.3	3.80	0	1.80	0.79	8.0	31.51	54.91	100
MD-S 1.4	21.21	1.01	1.30	9.32	31.65	13.51	22.01	100
MD-S 1.5	1.66	0	0	4.48	18.15	25.20	50.51	100

Note:

- O-K= oxidized ion; Mg-K: magnesium ion; Al-K: alkali ion; Si-K: silica ion; Ca-K: calcium ion; Pd: palladium and Au: gold (coating material)
- MD-S where S stands for adding silica in the mixtures

4 Conclusions

The paper presents the use of recycled glass in the concrete mix to promote the eco-environmental application of alternative materials in concrete production. The paper has the following conclusions:

1) Even though all specimens mixed with varying amounts of recycled glass did not exceed the control group (without recycled glass), however, the tensile splitting strength of all recycled glass specimens still show promising results ranging from 3.90 MPa to 5.72 MPa, higher than the desired level of 2.76 MPa (400 psi) splitting strength for most concrete pavement structure.
2) Based on SEM analysis, the resistance of recycled concrete mixtures to ASR is considered satisfactory as compared with the control group.
3) The use of coarse recycled glass as a sand replacement shows better performance than the fine recycled glass (as a cement binder replacement) in strength and ASR resistance.
4) The replacement of recycled glass with #4 or higher coarse aggregate sizes is worth testing for future research work.

References

1. R. Copp, *Proceeding of Automated pavement distress data collection Seminar*, 12–15 (1990)
2. http://blogs.ei.columbia.edu/2012/05/09/emissions-from-the-cement-industry/ (October 2016)
3. PCA. Portland Cement Association Summary of Portland Cement Plant Carbon Dioxide Emissions and U.S. Department of Energy, Emissions of Greenhouse Gases report: http://www.eia.doe.gov/oiaf/1605/ggrpt/ (2008)
4. M. Adaway, Y. Wang, Electronic J. Struct. Eng. **14**(1), 116-122 (2015)
5. W. Jin, C. Meyer, S. Baxter, ACI Mater. J. **97,** 208-213 (2000)
6. M. Arabani. Constr. Build. Mater. **25,** 1181-1185 (2011)
7. E.J.P. De Miranda Júnior. Mater. Res. UFSCar - Dep. De Engenharia De Materiais **17**(May) (2014)
8. M. Berry, J. Stephens, D. Cross, ACI Mater. J. **108**(4), 378-384 (2011)
9. C. Roskos, M. Berry, J. Stephens. *Proceedings of 2012 Annual Meeting of Transportation Research Board* (2012)
10. ASTM C496 (2017)
11. American Concrete Pavement Association (ACPA). R&T Update Concrete Pavement Research & Technology, September (2005).

Chemical testing of post-process ashes from municipal waste combustion

Teresa Stryszewska[1,*], and *Marta* Dudek[2]

[1]Cracow University of Technology, Faculty of Civil Engineering, ul. Warszawska 24, 31-155 Kraków, Poland
[2]Student of Cracow University of Technology, Faculty of Civil Engineering, ul. Warszawska 24, 31-155 Kraków, Poland

Abstract. The results of the study presented in the paper are a part of the research project about possibility of utilization of fly ashes obtained from municipal waste combustion in production of cement binders. The main goal was to determine possibility reduction of Cl^- and SO_4^{2-} contents in ashes by leaching. The scope of the examination included SEM observations along with an EDS analysis of chemical composition. Preliminary results show that the major problem is the presence of heavy metals such as Cr, Ni, Cu etc.

1 Introduction

The 'Waste Management Plan of the Małopolskie Province for 2016-2022' places special emphasis on reducing and preventing generation of waste and also on its recycling. In the aspect of methods of recycling and neutralising waste, one of the courses of action mentioned is thermal processing of waste [1]. In view of the introduction in Poland, as of the beginning of 2016, of a regulation that prohibits storing high-calorific waste, as well as the EU policy aimed at reducing storage of all sorts of waste, waste incineration for energy recovery makes something of an alternative. Incineration plants produce electricity and heat delivered to municipal heat distribution networks. In the case of the Waste Thermal Treatment Plant in Krakow, the power produced is capable of illuminating the city and the heat generated provides for approx. 10% of the demand [2]. This incineration plant has a maximum capacity of 220,000 t/year, which involves continuous incineration of 14.1 tonnes per hour in two process lines. As a result of this process, the waste generated makes approx. 25% of the input products and these are bottom ash and slag, which are not hazardous waste and can be reused as building material following the process of slag treatment, as well as boiler dust, fly ash and flue gas cleaning residues. These three other types of waste are treated in stabilisation and solidification processes in order to be transformed into inert or non-hazardous waste [3]. Fly ash that is generated from combustion of coal dust is utilised in the construction industry as a concrete additive while fly ash resulting from combustion of municipal waste makes post-processing waste. In order to determine the possibility of utilisation in the construction industry it is necessary to identify its physical and chemical properties. Pursuant to PN-EN

* Corresponding author: tstryszewska@pk.edu.pl

450-1 [4] fly ash intended for a concrete additive must comply with specific chemical requirements (Table 1) or have a technical approval certificate.

Table 1. Chemical requirements for fly ash as a concrete additive.

Component	Requirements
Loss of ignition	Category A: ≤ 5% Category B: ≤ 7% Category C: ≤ 9%
Chlorides	≤ 0.1%
Sulphates (VI)	≤ 3%
Free calcium oxide	≤ 1.5%*
Reactive calcium oxide	≤ 10%
Reactive silicon dioxide	Fly ash from co-combustion: ≥ 25%
	Fly ash from carbon combustion: met
Silicon dioxide, Alumina, Iron oxide (III)	Fly ash from co-combustion: ≥ 70%
	Fly ash from carbon combustion: met
Total alkali content	Fly ash from co-combustion: ≤ 5%
	Fly ash from carbon combustion: met
Magnesium oxide	Fly ash from co-combustion: ≤ 4%
	Fly ash from carbon combustion: met
Phosphates (V)	Fly ash from co-combustion: ≤ 100 mg/kg
	Fly ash from carbon combustion: met

One of the fundamental problems hindering utilisation of the said ash is the presence of heavy metals as well as other constituents, such as chlorides and sulphates [5], which are dangerous with regard to the use and durability of concrete. This necessitate the application of various technological treatments enabling their neutralisation already at the stage of manufacturing or at the stage of application. The most popular methods include the immobilisation process or, in other words, stabilisation and solidification, which consists in transforming hazardous waste into non-hazardous to the environment by means of elution of metals and transformation of soluble sulphide and chloride compound into insoluble ones. In addition to the hydraulic binders that are characteristic for the traditional technologies, the new modern technologies also use a specially compiled set of chemicals that act in a permanent manner on the ash. An example of such a technology is the Geodur, used in Poland by the company CDF - Technologie dla środowiska s.c. [6]. Authors of the study [7] describe options of immobilising heavy metals (Cr, Zn, Cd, Mn, Pb and Cu) by means of 4 types of binders with Portland cement as well as binders containing additives of silica fly ash, ash from a fluidal boiler and ground granulated blast-furnace slag. Heavy metals were introduced in the binder as constituents of fly ash from incineration of medical waste and dust from a dedusting system of an arc furnace. It has been shown that immobilisation of these materials has been achieved at a very high level (in the range of 99.82%-99.99%); only Cr has been neutralised with a slightly lower effect (85.97%-93.33%).

In the literature there are some examples of the use of fly ashes from municipal waste combustion and researches done on them. In the article [8], the author suggests that ashes can be used in production of powder cement, ceramic tiles or concrete as a partial replacement for cement or aggregates. They can also be used in building embankments or in road building where they could replace the sand or cement in pavements. Author confirms that by cited many publications about fly ashes from municipal waste combustion. Similar proposals for the use of ashes can be found in the article entitled "Use of Incineration MSW Ash" [9]. Also the authors of the article [10] described the possibility of using fly ashes in concrete. They focused on researches concerned information such as physical properties, durability (compressive strength) and ability to leaching. Based on the results they stated that fly ashes can be used in concrete. In turn Z. Pavlík and others in the article [11] presented the results of researches carried out on three ashes used as a filler in cement mortar, and in the conclusion they stated that, it is possible to use fly ashes in concrete, but these ashes have deteriorated the properties of concrete and first and foremost, it is necessary to focus on getting rid of the harmful elements from the ashes.

1.1 The scope and objectives of the study

The study used three types of ash from the Waste Thermal Treatment Plant in Krakow. The first type is fly ash from the flue gas cleaning system (FA 1), the second one is fly ash containing hazardous substances (FA 2), and the third one is a mixture of the other two (FA 3). All types contain heavy metals, such as mercury, lead, copper, nickel and others (Table 2). Another inconvenience is the very high content of chlorides, up to approx. 25% and of sulphates, up to several % of the mass. For this reason, an attempt was made to remove them by washing the ashes in water.

Table 2. Characteristics of the ash investigated from the incineration of municipal waste in Krakow.

Tested parameter [mg/kg]	FA 1	FA 2
Barium	4	0,34
Cadmium	<0.005	<0.005
Chromium	0.05	0.42
Cupper, mercury, molibdenium, nickel	<0.04	<0.04
Lead	13	<0.10
Antymony	0.029	<0.01
Selenium	0.038	<0.01
Zinc	4.2	2.7
Chloride	201382	18081
Fluorides	21	72
Sulphate (VI)	121732	30730
Arsenic [mg/kg]	<0.01	<0.01
Dissolved organic carbon	606	162
Soluble substances	498839	62920
pH [-]	12.1	12.6

The objective of the investigation presented in the paper was to determine the possibilities of removing soluble compounds (chlorides and sulphates) by washing ashes in water. For this purpose, an analysis was performed with regard to the contents of sulphates and chlorides in the water extracts, as well as the pH value. And the identification of the types of metal cations in chlorides and sulphates was performed based on SEM observations of the ashes concerned together with an EDS analysis.

When considering the applicability of the ashes being tested as an alternative additive to cement binders one should bear in mind the variability of composition over time. A further problem is the presence of heavy metals for which an effective method of immobilisation must be developed, as well as a very high content of chlorides and sulphates, which are a hazard to the durability of cement-based materials.

2 The investigation and methods

2.1 Chemical analysis of chlorides and sulphates by spectrophotometry

Leaching is one of the methods of treating the ash that is generated at municipal waste incineration plants. In order to determine the possibility of removing soluble compounds (especially chlorides and sulphates) by washing, 2 water extracts were prepared for each type of ash. The first was with 1:10 ash-to-distilled water ratio (by mass); the second one was with 1:100 ratio by mass. Both were left for 24 hours. After this time, chemical analyses were performed for the first set to determine Cl^- and SO_4^{2-} contents and the pH value. The second set was filtered and the remaining residues were mixed again with distilled water at 1:10 ash-to-water ratio by mass. After 24 hours another measurement of Cl^- and SO_4^{2-} contents and the pH value was made. The chemical analyses were performed by spectrophotometry using an AQUALYTICA AL800 spectrophotometer and a CP-411 Elmetron pH-meter.

2.2 SEM observations of the ash microstructure

SEM examinations were conducted in variable vacuum on non-sputtered samples using a VPSE detector. Additionally, a chemical EDS analysis of the elemental composition was performed in selected areas. The observations along with the analysis were performed using a Zeiss EVO-MA 10 Scanning Electron Microscope equipped with a Bruker EDS XFLASH 6/30 detector. The examinations were performed for three types of ashes in the dry and free flowing condition.

3 Results and discussion

3.1 Results of determination of the content of water-soluble chlorides and sulphates and the pH value

Results of the chemical tests concerning determination of the content of water-soluble chlorides and sulphates in the ashes before and after the process of washing have been presented in Table 3.

Table 3. Results of spectrophotometric tests.

Fly ash	Content before leaching [% mass]		Content after leaching [% mass]	
	Cl^-	SO_4^{2-}	Cl^-	SO_4^{2-}
FA 1	27	0.66	0.2	0.12
FA 2	21	0.71	0.26	0.12
FA 3	6	1.16	0.05	0.35

The results that were obtained indicate a possibility of removing considerable amounts of Cl^- and SO_4^{2-} from the test ashes by washing in water. Washing enables removing over 90% of chlorides present in the fly ashes examined. Their pH value both before and after washing is strongly alkaline and is in the ranges of 11.8÷12.3 and 11.1÷11.6 respectively.

3.2 Results of SEM examinations with an EDS analysis

Results of SEM observations along with mapping images of selected elements (O, Ca, Cl, Na, K, S, Si, Al and Ba) have been shown in Figures 1÷3.

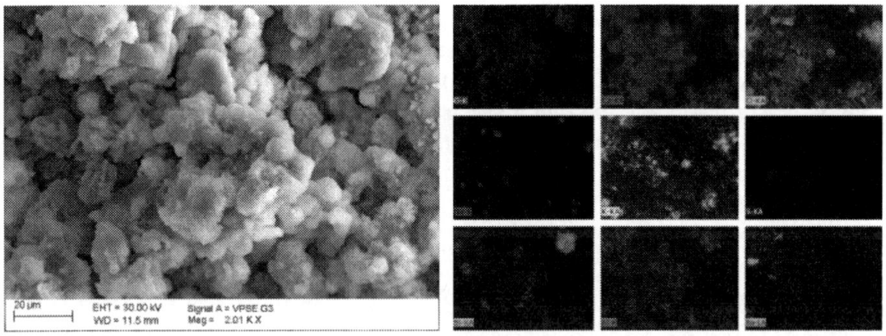

Fig. 1. SEM image of fly ash 1 with mapping of chosen elements.

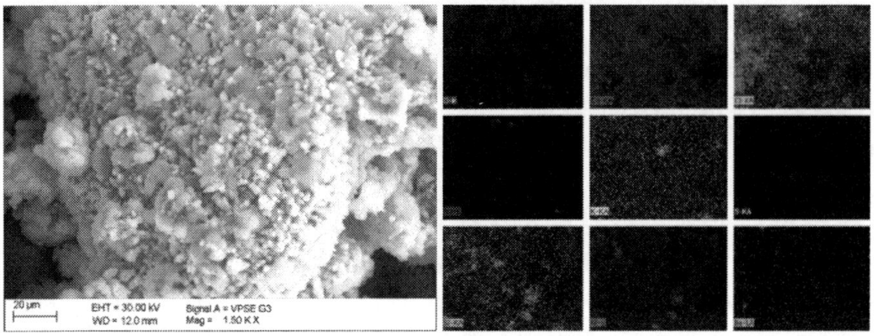

Fig. 2. SEM image of fly ash 2 with mapping of chosen elements.

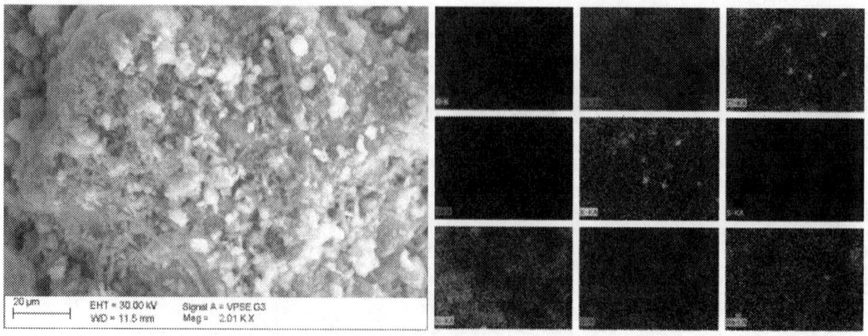

Fig. 3. SEM image of fly ash 3 with mapping of chosen elements.

In the microstructural examinations special focus was placed on the crystalline form of chlorides and sulphates. These are predominantly KCl crystals, more seldom NaCl, which take the form of fine and thick crystals. Sulphates occur most probably in the form of gypsum but, due to the high content of calcium compounds, this is difficult to determine based on EDS analysis alone.

4 Conclusion

The results presented in the paper form a part of a preliminary investigation the purpose of which was to determine the possibilities of utilising the ash in question in the construction industry. Due to a very high content of chlorides and sulphates, the examination was commenced with an attempt at removing these compounds using the method of washing in water. The results obtained give reasons to think that this is an effective method that enables reduction of Cl^- and SO_4^{2-} contents below a few % of the mass. However, it is necessary to conduct supplementary investigation concerning all the substances that are simultaneously washed out from the ash.

References

1. Waste Management Plan for the Małopolska Region for 2016-2022 (in Polish) https://www.malopolska.pl/_userfiles/uploads/PROJEKT%20PGOWM_2016-2022_30_08_2016.pdf
2. T. Pająk, Inżynier budownictwa **11**, 64-69 (2017)
3. Ekospalarnia Kraków (Eco-incinerator Krakow) https://khk.krakow.pl/pl/ekospalarnia/
4. PN-EN 450-1 (2012
5. B. Kabat, (Master thesis, Cracow University of Technology, Krakow, 2017) (in Polish)
6. M. Fengler, Piece Przemysłowe & Kotły **11-12**, 38-44 (2012)
7. Z. Giergiczny, A. Król, J. Hazardous Mater. **160**, 247-255 (2008)
8. R. Siddique, Resources, Conservation and Recycling **55**, 853-91 (2010)
9. Ch.H.K. Lam, W.M. Alvin, J.P. Barford, G. McKay, Sustainability **2**, 1943-1968 (2010)
10. J.E. Aubert, B. Husson, A. Vaquier, Cem. Concr. Res. **34**, 957–963 (2004)
11. Z. Pavlík, M. Keppert, M. Pavlíková, P. Volfová, R. Černý, WIT Transactions on Ecology and The Environment **148**, 325-334 (2011)

Physical properties of post-process fly ash from TWTP Krakow

Aneta Nowak-Michta[1,*], and *Brygida* Kabat[1]

[1]Chair of Building Materials Engineering, Cracow University of Technology, Warszawska 24, Krakow 31-155, Poland

Abstract. Fly ashes from eco-incinerator according to the Art.7 of the Act of December 14, 2012 on waste are a hazardous waste. The growing amount of this type of waste generates a problem in terms of their management. They are a post-process waste, so their chemical composition is not constant and their quality is not very high. In addition, a high content of heavy metals with different concentrations may adversely affect the properties of these ashes. The physical properties of fly ashes originating from the TWTP in Krakow in the light of the requirements of PN-EN 450-1 are analysed in the article. The aim of the paper is a preliminary assessment, based on the tests carried out on three random ashes, to check if there's a possibility of utilisation of this type of waste in concretes. The results of physical properties examination showed that they are coarse ashes with a several percent moisture content. They meet the requirements of PN-EN 450-1 in terms of the impact on the initial setting time and soundness. They cause an increase in the mortar's water demand. Their compression activity index after 28 days is 63-70%, and after 90 days 56-67% and do not meet the standard requirements (75% after 28 days and 85% after 90 days).

1 Introduction

An increasing amount of municipal solid waste generates a problem in terms of their management [1]. In order to reduce the amount of waste deposited, the process of thermal treatment of municipal waste is being disseminated. The third incinerator - The Thermal Waste Treatment Plant (TWTP) in Krakow - in the country was launched in December 2015. There are already 8 eco-incineration plants in Poland. Further investments are planned and by 2020 10 such establishments will be operating in Poland. It is estimated that this will allow for a thermal transformation of about 20% of waste per year [2-4].

As a result of the thermal transformation of municipal waste, in the waste-incineration plant in Krakow, post-process waste is produced in the form of:

- slag and bottom ash,
- boiler dust and fly ashes as well as solid residues from aftertreatment.

* Corresponding author: a_nowak@pk.edu.pl

All the listed residues represent 25% of the input stream. Slag and bottom ashes, which do not belong to hazardous waste, are used as a building material [4].

Boiler dusts and fly ashes as well as solid residues from the exhaust gas treatment are specified in the group 19 01 and in accordance with the Act [5] on waste are a hazardous waste. Subjecting them to the stabilization and solidification process is aimed at converting them into inert other non-hazardous waste. This effect is achieved by mixing them with appropriate additives and hydraulic binders. Stabilization and solidification consists of two stages. The first stage (stabilization) is used to obtain an internal barrier that prevents the solubilization of soluble heavy metal compounds. At the second stage (solidification) an external barrier is created using various types of hydraulic binders (eg cement). Waste that is a product of this process meets the conditions that allow its storage [4]. The Waste Management Plan for the Małopolska Voivodship for the years 2016-2022 recommends reduction in the amount of landfilled waste which is to be managed. One of the ways of its management is its use in construction, e.g. as a substructure for roads, filling of mining remains or by adding fly ash to concretes [1].

The aim of the paper is to analyze physical properties of fly ashes originating from the Department of Thermal Waste Transformation in Krakow based on the requirements of PN-EN 450-1 [6]. This work aims to determine, based on the tests carried out, whether it is possible to utilise of this type of waste in concretes.

2 Experimental part

2.1 The purpose and scope of research

The purpose of the research was to determine the impact of randomly selected fly ashes obtained from the thermal treatment of municipal waste in TWTP in Krakow on the physical properties of pastes and mortars made with their addition. The scope of research included physical properties required in accordance with PN-EN 450-1: 2012 [6]:

- fineness,
- initial and final setting time,
- soundness,
- water demand,
- activity index,

and also:

- sieve analysis,
- humidity.

2.2 Testing of fly ashes

Three different fly ashes from the thermal treatment of municipal waste in the TWTP in Krakow were obtained from a company "EKO SMOK" dealing with the collection of waste from public institutions. Ashes differ in the content of heavy metals in the chemical composition. The first of them (FA1) is a waste with a code 19 01 07 (fly ash from a flue gas cleaning system), the second one (FA2) has a code 19 01 13 (fly ashes containing dangerous substances), while the last one (FA3) is a mixture of the two previous ashes. All of these ashes are a hazardous waste due to the content of heavy metals: barium, cadmium, chromium, copper, nickel, mercury, lead and others contained in the Annex 4 of the Act of 14 December 2012 on waste [5].
The contents of heavy metals in the ashes and a detailed description of the test procedures are given in [7].

2.2.1 Fineness

Fineness of ashes was determined in the process of wet sieving according to PN-EN 451-2:2017 [8].

2.2.2 Sieve analysis

Sieve analysis of fly ashes was carried out according to PN-EN 933-1:2012 [9].

2.2.3 Humidity

The humidity of the ashes was determined by drying to a constant mass at a temperature of 105°C.

2.3 Testing of pastes

Pastes with a standard consistency were prepared in accordance with PN-EN 196-3: 2006 [10]. Four pastes were made, one using only cement and three being a mixture of 25% of ash and 75% of cement. Dry ashes, sieved through a 2mm sieve, were used for the pastes. The amounts of water needed to obtain the pastes with a standard consistency are given in Table 1.

Table 1. The amounts of water needed to obtain the pastes with a standard consistency.

	Cement	Fly ash	Water
	(g)		
C	500	-	147
FA1	375	125	173
FA2	375	125	180
FA3	375	125	140

2.3.1 Initial and final setting time

Initial and final setting time was marked on pastes with a standard consistency for cement and for three tested ashes FA1, FA2 and FA3 according to PN-EN 196-3:2006 [10].

2.3.2 Soundness

The soundness was determined on pastes with a standard consistency for cement and for three ashes FA1, FA2 and FA3 according to PN-EN 196-3: 2006 [10].

2.4 Testing of mortars

Four standard mortars were prepared according to PN-EN 196-1: 2006 [11]. One of the comparative cement itself and three, in which 30% of cement was successively replaced with FA1, FA2 and FA3. Dry ashes, sieved through a 2mm sieve, were used for the mortars.

2.4.1 Water demand

Water demand of fly ashes FA1, FA2 i FA3 were marked in accordance with Annex B of the standard PN-EN 450-1:2012 [6]. The test consisted of determining the amount of water with which the spreading of the mortar with a 30% ash content in the binder is equal to ± 10mm spreading of the control (cement) mortar. The water demand of ash is calculated by comparing the amount of water in the mortar with 30% ash content in the binder with the amount of water in the control mortar.

2.4.2 Activity index

In order to determine the activity indexes of ashes FA1, FA2 and FA3 four sets of bars with dimensions of 40x40x160mm were prepared for each of the ashes and one for the control mortar. Activity indexes were determined during flexing and compressing after 7, 14, 28 and 90 days according to PN-EN 450-1: 2012 [6]. In order to determine the indexes of flexural and compressive activity, the flexural strength and compressive strength of the mortar with the addition of ashes should be compared with the strength of the control mortar.

3 Research results and their analysis

3.1 Humidity of fly ashes

Humidity contents in FA1 and FA2 were at a similar level of around 18%. Fly ash FA3 has the least humidity (4.3%). The ashes for further tests were dried to a constant weight at 105°C.

3.2 Fly Ashes sieving analysis

Fig. 1 contains the curves of ash graining. Ashes FA1 and FA2 are characterized by a similar grain size, whereas FA3 is finer. For further tests, ashes with fractions up to 2mm were used.

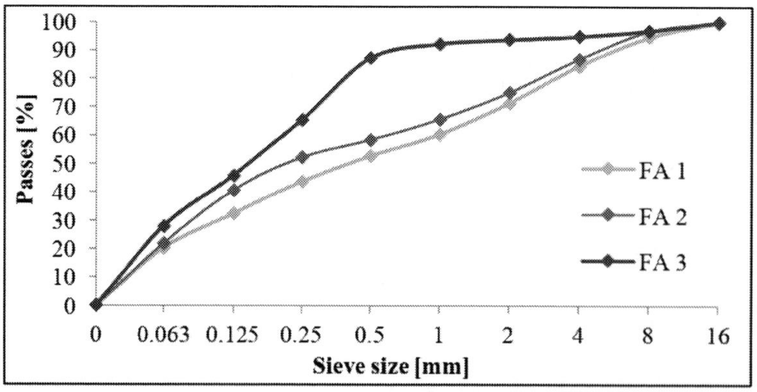

Fig. 1. Fly ashes grain curves.

3.3 Flexural strength of mortars

Fig. 2 presents graphs of flexural strength of mortars: the cement one and three with the addition of all ashes. The flexural strength of cement mortar was increasing in a standard way over a period of time. The mortars with the addition of all three ashes after 7 days gained 67-80% of the 90 day strength, which was about 8MPa. There was no increase in strength between 7 and 28 days. The activities of all three ashes during flexing are similar. After 7 days the activity was the highest and fluctuated for individual ashes from 69-84%, while at later dates it fluctuated around 60%.

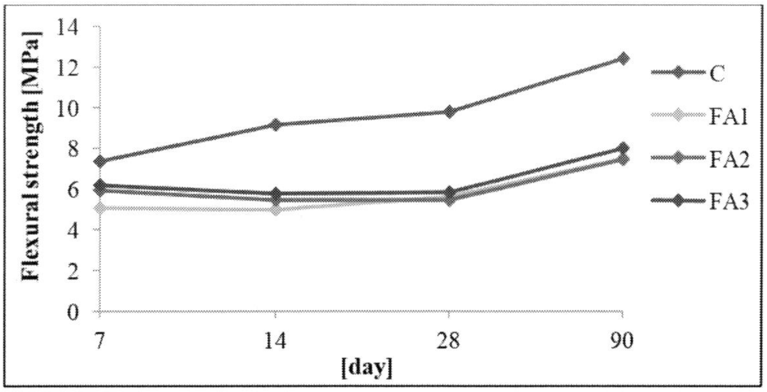

Fig. 2. Flexural strength of mortars.

3.4 Compressive strength of mortars

Fig. 3 presents the graphs of compressive strength of mortars: the cement one and three with the addition of all ashes. The kinetics of the compressive strength of cement mortar was a standard one. Mortars with the addition of all three ashes had similar compressive strengths and kinetics of their growth. After 7 days, the compressive strength of mortars with the addition of ashes was about 22MPa, after 14 days 28-30MPa, after 28 days 34- 38MPa, and after 90 days 40-46MPa. Thus, the activity of ashes after 7 days was 45%, and at later stages it fluctuated around 60%.

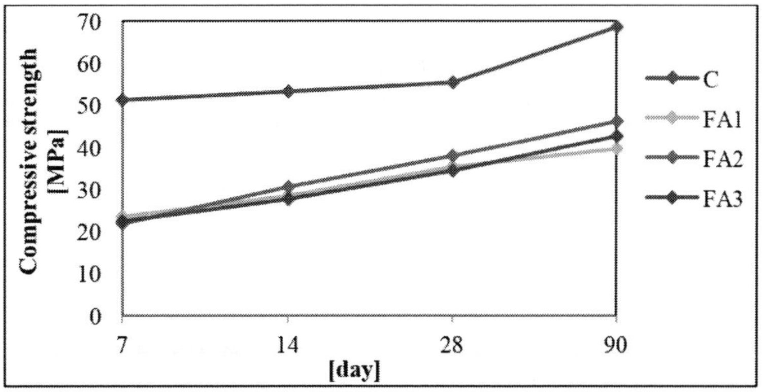

Fig. 3. Compressive strength of mortars.

3.5 Physical properties of fly ashes consistent with PN-EN 450-1:2012

Table 2 contains the requirements of PN-EN 450-1 [6] regarding physical properties for fly ashes, which can be used as an addition to concretes as well as the obtained results for ashes FA1, FA2 and FA3 from the incineration of municipal waste in TWTP in Krakow and their assessment in the light of the standard requirements. Fineness of the ash FA2 fulfilled the requirement [6] for the N category. All ashes met the requirements regarding the effect on the initial setting time as well as on the soundness. Ashes FA1 and FA2 had significantly shorter initial setting times than cement. Moreover, FA2 ash did not meet the requirements of PN-EN 197 [12] in terms of the minimum time for the cement class 42.5 which is 60 minutes. The requirements regarding the influence on the water demand and the compression activity index were not met by any of the tested ashes. The increase in water demand for FA1 and FA3 ashes was 18%, while for FA3 ash - 36%. Activity indexes neither after 28 nor 90 days reached the required levels. After 28 days, they ranged from 62.9-69.5% (required at least 75%), and after 90 days from 56.4 to 66.7% (required at least 85%).

Table 2. Physical properties of ashes and their assessment according to PN-EN 450-1 [6].

Feature	Requirements of PN-EN 450-1:2012	Tests results		Assesment
Fineness [%]	Cathegory S: ≤ 12 Cathegory N: ≤ 40	FA1: 42.2		does not meet
		FA2: 29.6		Cathegory N
		FA : 67.2		does not meet
Initial setting time [min]	The initial setting time not more than twice as long as the start of the leaven paste with cement alone	C: 158		-
		FA1: 70		meet
		FA2: 28		does not meet [13]
		FA3: 198		meet
Final setting time [min]	no requirements	C: 361		-
		FA 1: 233		-
		FA 2: 193		-
		FA 3: 325		-
Soundness [mm]	≤ 10mm	C: 0.5		meet
		FA 1: 2.4		meet
		FA 2: 0.2		meet
		FA 3: 0.8		meet
Water demand [%]	≤ 95% water demand for mortar from cement alone CEM I	FA 1: 118		does not meet
		FA 2: 136		does not meet
		FA 3: 118		does not meet
Flexural activity index [%]	no requirements	FA1	28: 57.6	-
			90: 60.4	
		FA 2	28: 56.0	-
			90: 60.0	
		FA 3	28: 60.0	-
			90: 64.3	

Compressive activity index [%]	After 28 days: ≥ 75% After 90 days: ≥ 85%	FA 1	28: 64.6	does not meet
			90: 56.4	
		FA 2	28: 69.5	does not meet
			90: 66.7	
		FA 3	28: 62.9	does not meet
			90: 61.9	

4 Conclusions

Fly ashes from the eco-incineration plant are a post-process waste and because of that their chemical composition is not constant and in terms of properties they do not meet all the standard requirements for ashes used as an addition to concretes [13]. Moreover, a high content of heavy metals with different concentrations may adversely affect the properties of these ashes. However, these ashes also show positive features, i.e.: such as soundness, setting time and some degree of pozzolanic activity, which open up the chances of utilizing this type of waste in concretes. In order to effectively assess the possibility of using these ashes in concretes, their chemical analysis is required, especially regarding the content of heavy metals and unwanted soluble compounds as well as the possibility of their immobilization [14].

References

1. Waste Management Plan for the Małopolska Region for 2016-2022 (Kraków, 2016) in Polish
2. G. Wielgosiński, O. Namiecińska, Nowa Energia **2** (2016)
3. R. Ziaja, *Ogólnopolska Konferencja Młodych Energetyków*, 5-16 (2016) in Polish
4. https://khk.krakow.pl/en/eco-incinerator/project/ (June 2017)
5. The Act of 14 December 2012 on waste (Journal of Laws, 2013 item 21)
6. PN-EN 450-1 (2012)
7. B. Kabat, *Preliminary assessment of municipal waste incineration ash disposal possibilities in concrete production*, (Cracow University of Technology, 2017) in Polish
8. PN-EN 451-2 (2017)
9. PN-EN 933-1 (2012)
10. PN-EN 196-3 (2016)
11. PN-EN 196-1 (2006)
12. PN-EN 197-1 (2012)
13. PN-EN 206+A1 (2016)
14. T. Stryszewska, M. Dudek, *8th Scientific-Technical Conference on Material Problems in Civil Engineering MATBUD'2018 Cracow* (2018)

Influence of cement substitution by calcareous fly ash on the mechanical properties of polymer-cement composites

Beata Jaworska[1,*], *Paweł* Łukowski[1], and *Jerzy* Jaworski[1]

[1]Warsaw University of Technology, Faculty of Civil Engineering, Al. Armii Ludowej 16, 00-637 Warsaw, Poland

Abstract. The aim of the presented research was to determine the influence of cement substitution with calcareous fly ash on the mechanical properties of polymer-cement composites. Coal combustion products such as calcareous fly ash have been already used in traditional cement composites as a part of cement and considered potential additions for concrete but its introduction into polymer-cement composites is still under preliminary investigation. The morphology of fly ash causes problems with its compatibility with polymer-cement binders but its insertion into those building materials is another way to utilize mineral combustion products that are cumbersome in storage and recycling. The influence of the mineral addition on polymer-cement composites containing 20% of polymer was especially taken into consideration. Mechanical properties of polymer-cement mortars modified with calcareous fly ash were tested after 28 and 90 days of curing. As a part of preliminary study, activity index of mineral addition was determined. Polymer-cement composites containing calcareous fly ash were characterized by higher flexural and tensile strength comparing to standardized mortar, even for the mortars containing 40% of mineral addition. The negative effect of the polymer-cement composites modification with calcareous fly ash was especially observed on the compressive strength of this composites.

1 Introduction

Polymer-cement concretes and mortars are part of the concrete-like composites group in which a polymer modifier is present in addition to the cement binder. The polymer-cement composites production technology is similar to the technology of obtaining conventional cement concrete. The content of the polymer modifier is usually in the range of 10 ÷ 20% by cement mass, but only a small number of polymers are suitable for the modification of cement binder. The presence of a polymer modifier causes the increase of flexural and tensile strength of the composite and the reduction of its permeability [1]. Chemical resistance and freezing/thawing resistance, and – therefore - durability of the material are improved.

* Corresponding author: b.jaworska@il.pw.edu.pl

Polymer-cement composites are used is such areas as repairs and protection of concrete structure [2], production of industrial floors and pavements [3] as well as production of pre-cast elements [4].

The main obstacle to using polymer-cement composites is their high material cost, although lower than the cement less resin composites, due to the smaller amount of the polymer needed to produce them. The need to reduce the material cost while maintaining the material properties leads to a search for new material solutions [1]. One of the methods to reduce the material cost is the replacement of some components with cheaper substitutes, both natural resources and waste materials. Examples of such materials are fly ashes – coal combustion products (CCP).

Whereas siliceous fly ash is commonly used as addition to concretes according to PN-EN 206 [5] and common cement according to PN-EN 197-1 [6], the calcareous fly ash is not such an obvious choice. This lignite combustion product differs from siliceous ashes both in chemical composition and in properties [7, 8]. Most of all, its irregular grains with a porous surface [9] are the reason for the increased water demand of composites containing this ash, thus negatively affecting the workability of the concrete mix, and consequently the properties of the hardened material.

The utilization of mineral wastes, also including coal combustion products, is a part of a sustainable development strategy, according to which, inter alia, it is necessary to minimize the consumption of matter, re-use elements and materials and effectively manage waste.

The purpose of this work was to solve the scientific problem regarding the influence of the presence of coal combustion product – calcareous fly ash – on the mechanical properties of polymer-cement composites.

2 Experimental

2.1 Methodology of the research program

The aim of the presented research was to determine the influence of cement substitution by calcareous fly ash on the mechanical properties of polymer-cement composites. Mixtures were designed with the following assumptions:

- dosage levels of calcareous fly ash: 0÷40% of cement mass,
- constant polymer content: 20% of cement mass,
- constant level of water-cement ratio equal to 0.5.

Polymer-cement mortars were prepared using Portland cement CEM I 42.5R, standard sand fulfilling requirements of PN-EN 196-1 [10], a water dispersion of styrene-acrylic copolymer and calcareous fly ash from one of the Polish power plants. The reference material used in this research was standardized cement mortar according to PN-EN 196-1.

2.2 Testing procedure

Mechanical properties of polymer-cement composites were tested after 28 and 90 days of curing in mixed wet conditions. For the first day of curing specimens were covered with polyethylene sheets and, after demoulding, immersed for five days in water at 20-22°C. These are favourable conditions for the hydration of cement binder. After removing specimens from water, until testing, they were held in air-dry conditions at temperature of 20-22°C and relative humidity of 60±5°C. The air-dry condition is adequate for the formation of a continuous polymer film. Standardized mortars were stored in wet conditions as recommended in PN-EN 196-1. As a part of preliminary study, the activity index of calcareous fly ash was tested.

2.3 Activity index of mineral addition

The activity index of calcareous fly ash which was used in the research as a substitution of cement in polymer-cement composites was tested according to PN-EN 450-1 [11] and is presented in Table 1, Figure 1.

Table 1. Compressive strength (f_c) and activity index (AI) of specimens after 28 and 90 days of curing (CFA - calcareous fly ash, SM – standardized mortar – reference level).

No.	Specimen	f_{c28}, MPa	AI_{28}, %	f_{c90}, MPa	AI_{90}, %
1	CFA	41.4	**83.0**	49.5	**85.9**
2	SM	49.9	-	57.6	-

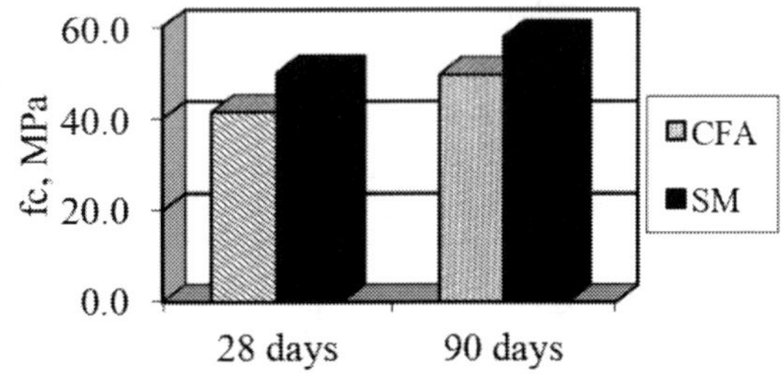

Fig. 1. Compressive strength of specimens after 28 and 90 days of curing (CFA - calcareous fly ash, SM – standardized mortar – reference level).

Calcareous fly ash was characterized by high activity index, reaching a value above 75% after 28 days and above 85% after 90 days of curing the specimens.

2.4 Mechanical properties of polymer-cement composites containing calcareous fly ash

Mechanical properties of polymer-cement composites containing calcareous fly ash were tested after 28 and 90 days of curing. The investigation was carried out for composites modified with 20% content of polymer and up to 40% content of mineral addition (by cement mass). Mechanical properties of these composites were tested in accordance with PN-85/B-04500 [12].

2.4.1 Flexural strength

Cement mortars containing 40.0% of mineral addition were characterized by the smallest flexural strength - 7.5 MPa after 28 days of curing and 8.3 MPa after 90 days of curing (Table 2). The cement substitution by calcareous fly ash in polymer-cement composites did not result in larger differences in the obtained flexural strength results. Polymer-cement composites containing calcareous fly ash were characterized by flexural strength of up to 12.0 MPa. Polymer-cement composites containing 40% of mineral addition were characterized by the smallest flexural strength (9.8 and 10.5 MPa after 28 and 90 days of curing respectively). Cement mortars containing 40% of calcareous fly ash were characterized by 1.3% and 4.6% decrease of flexural strength and polymer-cement composites with 40% of calcareous fly ash were characterized by 17.6% and 14.6% decrease of flexural strength, after 28 and 90 days of curing respectively, comparing to the mortars without fly ash (Table 2, Figure 2).

Table 2. Flexural strength of PCC mortars with different content of calcareous fly ash – after different time of curing (p/c – polymer content by cement mass, m/c – mineral addition by cement mass, SD – standard deviation, CV – variability index)

No.	p/c, %	m/c, %	Flexural strength, MPa					
			28 days			90 days		
			Average value	SD	CV, %	Average value	SD	CV, %
1	0.0	0.0	7.6	0.3	3.8	8.7	0.4	4.6
2	0.0	5.0	8.6	0.8	8.7	9.2	0.2	2.7
3	0.0	10.0	8.1	0.8	9.9	8.6	0.2	2.9
4	0.0	15.0	7.7	06	8.2	8.4	0.2	2.9
5	0.0	20.0	8.0	0.1	0.8	8.8	0.3	3.7
6	0.0	25.0	7.6	0.4	5.6	8.5	0.2	2.8
7	0.0	30.0	7.5	0.4	5.6	8.5	0.0	0.0
8	0.0	40.0	7.5	0.5	6.2	8.3	0.4	4.2
9	20.0	0.0	11.9	0.1	0.7	12.3	0.9	7.4
10	20.0	5.0	10.6	0.1	1.1	12.1	0.1	0.9
11	20.0	10.0	10.4	0.3	3.1	12.2	0.0	0.0
12	20.0	15.0	9.6	0.1	1.3	11.9	0.3	2.5
13	20.0	20.0	9.8	0.3	3.2	11.4	0.3	2.9
14	20.0	25.0	9.8	0.1	0.9	11.4	0.4	3.2
15	20.0	30.0	9.9	0.2	1.6	10.9	0.1	1.1
16	20.0	40.0	9.8	0.0	0.0	10.5	0.1	0.6

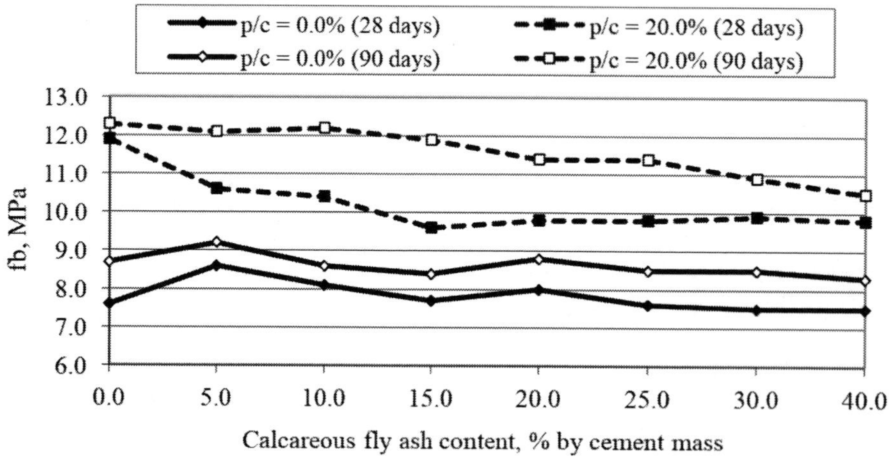

Fig. 2. Flexural strength of polymer-cement composites containing different content of calcareous fly ash after different time of curing.

2.4.2 Tensile strength

Cement mortars containing 40.0% of mineral addition were characterized by the smallest tensile strength (3.1 MPa). The cement substitution by calcareous fly ash in polymer-cement composites did not result in larger differences in the obtained tensile strength results. Polymer-cement composites containing calcareous fly ash were characterized by tensile strength of up to 6.0 MPa. Polymer-cement composites modified with 40% of calcareous fly ash were characterized by 11.1% and 15.8% decrease of tensile strength, after 28 and 90 days of curing respectively (Table 3, Figure 3).

Table 3. Tensile strength of PCC mortars with different content of calcareous fly ash – after different time of curing (p/c – polymer content by cement mass, m/c – mineral addition by cement mass, SD – standard deviation, CV – variability index).

No.	p/c, %	m/c, %	Tensile strength, MPa					
			28 days			90 days		
			Average value	SD	CV, %	Average value	SD	CV, %
1	0.0	0.0	3.1	0.3	8.0	3.2	0.2	6.3
2	0.0	5.0	4.2	0.2	4.8	4.6	0.3	6.5
3	0.0	10.0	4.1	0.1	2.4	4.9	0.2	3.1
4	0.0	15.0	3.9	0.3	8.1	4.6	0.1	1.5
5	0.0	20.0	3.6	0.0	0.0	4.4	0.1	1.3
6	0.0	25.0	3.5	0.1	2.4	4.3	0.0	0.0
7	0.0	30.0	3.5	0.3	8.6	4.0	0.1	2.5
8	0.0	40.0	3.1	0.1	1.9	4.1	0.1	2.4
9	20.0	0.0	5.4	0.3	4.6	5.7	0.2	3.5
10	20.0	5.0	5.6	0.4	7.1	6.1	0.0	0.0
11	20.0	10.0	5.1	0.1	1.1	5.9	0.1	1.7
12	20.0	15.0	5.1	0.0	0.0	5.6	0.1	1.1
13	20.0	20.0	5.2	0.2	3.8	5.4	0.1	1.9
14	20.0	25.0	5.2	0.1	3.3	5.2	0.1	1.0
15	20.0	30.0	5.4	0.2	3.7	5.1	0.1	2.0
16	20.0	40.0	4.8	0.1	1.2	4.8	0.1	1.2

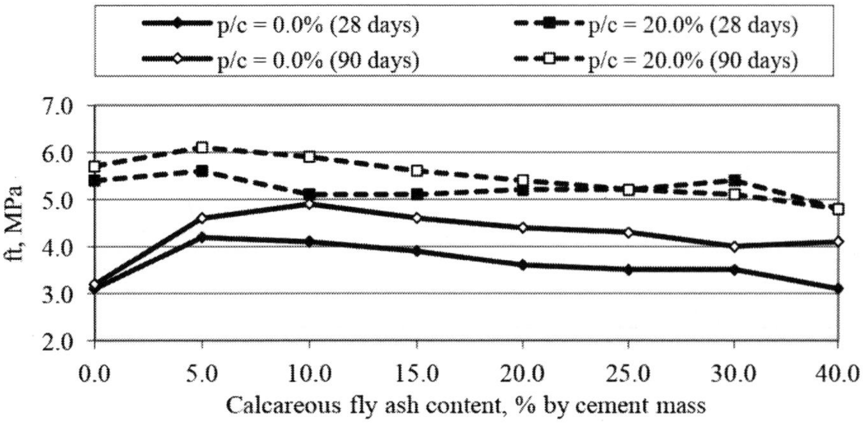

Fig. 3. Tensile strength of polymer-cement composites containing different content of calcareous fly ash after different time of curing.

2.4.3 Compressive strength

Polymer-cement composites containing calcareous fly ash were characterized by compressive strength of up to 46.0 MPa. Polymer-cement composites containing 40.0% of mineral addition were characterized by the smallest compressive strength (30.4 MPa) even comparing to cement mortar modified with the same amount of fly ash. Polymer-cement composites modified with 40% of calcareous fly ash were characterized by 29.0% and 25.7% decrease of compressive strength, after 28 and 90 days of curing respectively (Table 4, Figure 4).

Table 4. Compressive strength of PCC mortars with different content of calcareous fly ash – after different time of curing (p/c – polymer content by cement mass, m/c – mineral addition by cement mass, SD – standard deviation, CV – variability index).

No.	p/c, %	m/c, %	Compressive strength, MPa					
			28 days			90 days		
			Average value	SD	CV, %	Average value	SD	CV, %
1	0.0	0.0	48.3	1.4	2.9	59.6	1.4	2.4
2	0.0	5.0	50.6	0.7	1.4	59.2	1.8	3.1
3	0.0	10.0	46.7	0.3	0.7	55.1	1.1	2.0
4	0.0	15.0	42.3	1.0	2.4	53.6	1.3	2.3
5	0.0	20.0	41.7	1.3	3.1	52.2	0.7	1.4
6	0.0	25.0	38.7	1.6	4.0	49.7	1.8	3.5
7	0.0	30.0	35.2	0.6	1.7	48.1	2.4	5.0
8	0.0	40.0	31.7	1.4	4.3	40.2	0.7	1.6
9	20.0	0.0	42.8	0.4	1.0	47.1	0.7	1.4
10	20.0	5.0	39.5	1.4	3.6	46.2	0.8	1.7
11	20.0	10.0	39.4	0.9	2.2	44.8	1.3	2.9
12	20.0	15.0	37.6	1.4	3.8	43.4	0.8	1.8
13	20.0	20.0	37.0	0.7	1.9	43.4	0.2	0.5
14	20.0	25.0	35.1	0.6	1.6	40.2	0.9	2.3
15	20.0	30.0	31.6	0.6	1.8	37.2	0.3	0.8
16	20.0	40.0	30.4	0.4	1.2	35.0	0.5	1.5

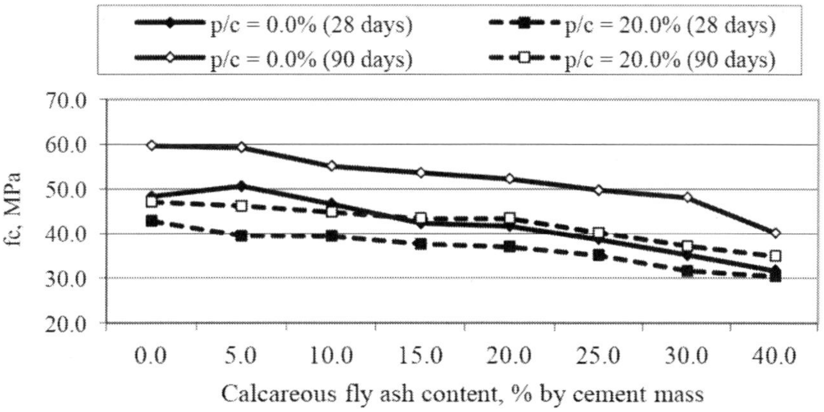

Fig. 4. Compressive strength of polymer-cement composites containing different content of calcareous fly ash after different time of curing.

3 Conclusions

Adding calcareous fly ash to the polymer-cement composites had more influence on the flexural strength of this composites than on the flexural strength of cement mortars. Polymer-cement composites containing 40% of mineral addition were characterized by almost 18% and 15% (after 28 and 90 days of curing respectively) lower flexural strength comparing to polymer-cement composites without ash. Despite this, their flexural strength was higher by about 20% than the flexural strength of standardized cement mortar.

Also, when taking into consideration the influence of cement substitution with calcareous fly ash on the tensile strength of these composites, it can be seen that the addition up to 40% of coal combustion product caused its reduction, but still all the polymer-cement composites were characterized by higher tensile strength comparing to cement mortars.

The biggest influence of calcareous fly ash content was observed for compressive strength. Polymer-cement composites modified with 40% of calcareous fly ash were characterized by 29.0% and 25.7% decrease of compressive strength. Despite this, the decrease was smaller than in the case of cement mortars, where substitution of cement with 40% of calcareous fly ash resulted in 34.4% and 32.6% decrease in compressive strength, after 28 and 90 days of curing respectively.

The studies presented in this paper showed that it is possible to use calcareous fly ash (even in large quantities, up to 40% of cement mass) in polymer-cement composites. The obtained composites were characterized by similar or slightly worse (decrease in strength ≤ 30%) mechanical properties comparing to polymer-cement composites without fly ash.

References

1. D.W. Fowler, Cem.Con.Com. **21**, 449-452 (1999)
2. F. Giustozzi, Con. Buil. Mat. **111**, 502-512 (2016)
3. D. Jeongyun, S. Yangseob, Cem. Con. Res. **33**, 1497-1505 (2003)
4. L. Czarnecki, P. Łukowski, Cem. Wap. Bet. **5**, 243-258 (2010)
5. PN-EN 206+A1 (2016-12)

6. PN-EN 197-1 (2012)
7. J. Małolepszy, *MATBUD'2011*, 17- 48 (2011)
8. B. Jaworska, J.J. Sokołowska, P. Łukowski, J. Jaworski, ACE, **LXI, 4** (2015)
9. Z. Giergiczny, *Role of high calcium and low calcium fly ashes in shaping properties of construction binders and cement materials* (Kraków, 2006)
10. PN-EN 196-1 (2006)
11. PN-EN 450-1 (2012)
12. PN-85/B-04500 (1985)

Lightweight concrete with copper slag waste as sand substitution

Roman Jaskulski[1,*], and *Wojciech* Kubissa[1]

[1]Faculty of Civil Engineering, Mechanics and Petrochemistry, Warsaw University of Technology, ul. Łukasiewicza 17, 09-400 Płock, Poland

Abstract. The paper presents results of tests of lightweight concrete made with copper slag waste material as fine aggregate (fraction 0 – 2 mm) and two types of light aggregate (fractions > 2 mm). Six series of concrete with three w/c ratios (0.50, 0.55 and 0.60) and two variants of cement content (200 kg/m^3 and 300 kg/m^3) were manufactured. Reference concretes were also prepared in which copper slag waste was replaced by the same mass of river sand. To maintain similar densities of mixtures with copper slag in comparison to those with sand mutual proportions of both lightweight aggregates were modified. On concrete specimens compressive strength was tested as well as the open porosity was determined by drying to stable mass at 65°C. The concrete bulk density in saturated and dried state was also determined using hydrostatic weighing. The obtained results were compared with the results of the reference concrete tests. Results have shown that replacing sand with copper slag waste in lightweight concrete does not negatively affect their strength but may slightly increase their porosity.

1 Introduction

Recycling of waste materials became necessary. In recent years much attention is paid, beside reduction of CO_2 emission, to rational management of waste. This is caused by necessity of the transition to an energy-efficient economy with frugal natural resources management policy, which became one of the main priorities of the EU 2020 strategy [1,2]. Cement and concrete industry try to fit in with the strategy and manufacturing of concrete with limited amount of cement clinker substituted by other binding materials and with waste components becomes one of the most important directions of development of concrete technology [3,4,5]. It should be emphasized, however, that simultaneously with these priorities, the demands on durability of concrete structures during its whole life cycle are increasing and impact of construction on the environment is becoming increasingly important [6].

The use of copper slag waste in the production of concrete can be a part of the sustainable development strategy. Copper slag is a by-product, produced besides mill tailing, during the process of copper extraction by smelting. Its physical properties are similar to natural sand [7] and some potential applications which have been described already [8,9]. The source of the waste is the surface blast-cleaning process in which copper slag is used as an abradant.

* Corresponding author: Roman.Jaskulski@pw.edu.pl

After the cleaning process the material is considered as waste because its gradation becomes too fine to use it again for the same purpose. The content of 0 – 0.125 mm and 0.125 – 0.25 mm fractions is significantly increased compared to river quartz sand. The waste also contains a small amount of corrosion products and corrosion protection coatings 10. Huge amount of its annual production is still disposed in landfills or stockpiles.

One of the most promising potential application for using original copper slag waste is concrete production [11,12,13]. It is possible to manufacture high-quality concrete, improving its properties compared to concrete with sand [14,15]. It has been found that copper slag used as sand substitution significantly improves consistency and compressive strength of concrete without changing the amount of mixing water [12,16]. It is also possible to retain the same consistency of the concrete mixture and reduce the amount of water by 22% with the significant compressive strength increase. It is very likely that copper slag waste, despite its higher fineness and some content of impurities, will also work well in the same applications. However it can affect workability of the concrete mixtures in a different way than original copper slag.

One of the fastest growing areas of concrete technology, which also allows to achieve a positive impact on the environment, are lightweight concretes, especially those made with the use of lightweight aggregates produced from waste materials (granules of recycled glass, fly ash and adjustment components, waste from sewage treatment plants and other).

The use of lightweight concrete can be traced to as early as 3000 B.C. when the towns of Mohenjo - Daro and Harappa in Pakistan were built with porous clay bricks. Later the Greeks and Romans used pumice when building their constructions, which some of them still exist like St. Sofia Cathedral (Hagia Sofia) in Istanbul and the Roman temple, the Pantheon and the Colosseum in Rome [17]. LWA (ie. expanded clay aggregate) was first manufactured from clay and shale sintered in a rotary kiln in 1917. Lightweight aggregate concrete (LWC) has been increasingly applied in high buildings (Shard London Bridge, Commerzbank in Frankfurt) and long-span bridges (Great Belt Bridge, Øresund Bridge, Wabash River Bridge, repairs of Tower Bridge), Concrete Offshore Platforms (Heidrun Floating Concrete Offshore Platform, CIDS Island Drilling System) and other applications [18].

LWC has many virtues. Mainly due to its 20-30% lower self-weight in comparison to normal weight concrete. It usually considered as an average- and low-strength material used mainly for subsidiary or non-bearing construction elements, but it is possible to manufacture lightweight concrete of high compressive strength [19]. The impact of high temperature on the strength of the lightweight concrete with aggregates fully saturated with water showed strength losses that were comparable to those observed in the concrete with standard aggregate 20. Another advantage of the material is wide use of waste materials in the production of lightweight aggregates [21,22,23] including hazardous waste which is difficult to dispose of in any other way, such as petroleum soaked absorbents [24] or industrial waste containing heavy metals [25]. Therefore, there is every reason to believe that LWC has strong market competitiveness and it will be widely used for the civil engineering structure in the near future [26].

The aim of this work is to compare selected mechanical and structural properties of LWC manufactured with two kinds of different fractions of lightweight aggregate (Certyd and expanded clay aggregate) and copper slag waste as sand substitution. Received results of compressive strength, bulk density and open porosity were compared with values obtained for LWC manufactured with using similar recipes and lightweight aggregate but with sand replacing copper slag waste. Due to the difference in density of sand and copper slag, in order to obtain concretes of similar density, the same mass share of both fine aggregates and different mutual proportions of light aggregates in mixtures were assumed. also the amount of additional water due to the difference in absorbability of both light aggregates has been corrected.

2 Materials and methods

2.1 Materials and mix compositions

Two lightweight aggregates as coarse aggregate were used to make lightweight concrete. The first, with a grain size of 4 - 8 mm, is an aggregate under the trade name Certyd produced in the process of high-temperature sintering of ash from the combustion of coal in fine coal industrial boilers [27]. The other, with grain size 8 - 16 mm, was expanded clay aggregate - keramzite. Both of the lightweight aggregates used are presented in the illustrative form in Fig. 1.

Fig. 1. Lightweight aggregate utilized in the tested concrete mixtures: a) Certyd, b) keramzite [28].

As fine aggregate, alternatively river quartz sand and copper slag waste were used. The second is shown in Fig. 2.

Fig. 2. Copper slag waste [28].

Portland-composite cement CEM II/B-M (V-LL) 32.5 R was used as a binder and the ordinary tap water as mix water. Total of twelve concrete mixture compositions were designed. Six of them were manufactured with quartz river sand as a fine aggregate and six more with copper slag waste as the sand substitution. The concrete mixtures with the same kind of fine aggregate were additionally varied by assuming different values of w/c ratio (0.50, 0.55, 0.60) as well as different cement content: 200 kg/m^3 and 300 kg/m^3.

In order to take into account the lightweight aggregate absorbability on the effective value of the w/c ratio of the concrete mixtures [29], the aggregate water absorption capacity was determined. The amount of water absorbed by the aggregate during 24 hours was respectively: 16.2% by mass in the case of Certyd aggregate and 21.6% in the case of keramzite. In the mix design, additional water was included in the amount of 80% of the water absorption capacity of both lightweight aggregates.

Two assumptions were made in concrete mixtures design. First, the mass fraction of fine aggregate in the corresponding mixtures with sand and copper slag waste was to be approximately the same. Second, bulk density of the hardened concrete was to be similar. Due to higher bulk density of copper slag waste in comparison to sand (approx. 2.99 g/cm^3 according to own tests) it was necessary to correct the mutual proportions of coarse lightweight aggregates, which also differed in bulk density (according to own tests: Certyd – 1.75 g/cm^3 and keramzite – 1.27 g/cm^3). As a result in the mixtures with sand the mutual proportions of lightweight aggregates were 40% of Certyd and 60% of keramzite, and in case of the mixtures with copper slag waste there were 30% of Certyd and 70% of keramzite. Concrete mix proportion based on the above described assumptions are presented in Table 1.

Table 1. Concrete mix proportions.

Mix designation	Component [kg/m^3]						w/c ratio
	Cement	Water	River sand	Copper slag waste	Certyd	Keramzite	
LCs-2/50	200	230	664	---	409	446	0.50
LCc-2/50	200	236	---	662	322	546	
LCs-3/50	300	267	599	---	369	402	
LCc-3/50	300	272	---	597	291	492	
LCs-2/55	200	238	656	---	404	440	0.55
LCc-2/55	200	244	---	654	319	539	
LCs-3/55	300	280	587	---	362	394	
LCc-3/55	300	285	---	585	285	482	
LCs-2/60	200	247	648	---	400	435	0.60
LCc-2/60	200	252	---	646	315	533	
LCs-3/60	300	293	575	---	354	386	
LCc-3/60	300	297	---	573	279	473	

2.2 Specimens preparation

The concrete mixtures were prepared in a forced action laboratory mixer. First the dry components were mixed and then water was added in two portions. Concrete specimens were formed in plastic moulds in two layers. Each of the layers was compacted on vibration table. The specimens were demoulded after two days and were put into water containers until the testing but not shorter than the concrete reached the age of 28 days.

Two kinds of specimens were prepared. For compressive strength test 100 mm cube specimens, while to measure the bulk density and open porosity 150 x 300 mm cylinders were casted, which were subsequently cut into approximately 25 mm thick discs on which the measurements were performed.

2.3 Performed tests

2.3.1 Compressive strength test

Compressive strength was tested on 100 mm cubic specimens 28 days after casting. Until the test the specimens were stored in a container with water in laboratory. To perform the test compression machine Toni Technik ToniPACT II with 3000 kN capacity was used.

The test procedure was performed in accordance with PN-EN 12390-3. The rate of loading was set at 0.5 MPa/s. For each concrete series eight specimens were tested. The results were statistically analysed in order to find possible outliers and then averaged. Besides the mean value the measurement uncertainty was also calculated with use of t Student distribution.

2.3.2 Bulk density and open porosity determination

Bulk density was determined on specimens in two conditions: fully saturated with water and oven dried in 65°C to stable mass. The procedure consisted of three weighing. Two of them were performed on saturated specimens and the third on dried ones. First the specimens saturated mass m_s was determined. Then the hydrostatic weighing was performed and the mass of the specimen submerged in water m_w was determined. The last, but not least, mass m_d was determined after specimens drying.

Bulk density ρ_s of saturated hardened concrete mixtures was calculated using Eq (1).

$$\rho_s = m_s / (m_s - m_w) \tag{1}$$

For calculations of density ρ_d of dried hardened concrete mixtures Eq. (2) was used

$$\rho_d = m_d / (m_s - m_w) \tag{2}$$

and the open porosity p was calculated using Eq. (3)

$$p = (m_s - m_d) / (m_s - m_w) \tag{3}$$

For each concrete series nine specimens were tested. The results were, as in case of compressive strength, statistically analysed in order to find possible outliers and then averaged. Besides the mean value the measurement uncertainty was also calculated with use of t Student distribution.

3 Results and conclusions

3.1 Compressive strength

Results of compressive strength test are presented on Fig. 3. For better illustration of the results of similar concrete series differing in fine aggregate type (sand or copper slag waste) they were arranged next to each other and described by one common mixture designation.

The highest compressive strength was obtained for LC-3/50 with sand and the lowest for LC-2/50 with sand. Analysing the results some more observations can be easily made. First of all it should be mentioned, that only in case of one concrete mixture (LC-3/50) the results differ more from each other than the measurement uncertainty. For this mixture the higher compressive strength is observed for series with sand as fine aggregate. If the mean values are analysed, without taking into account the measurement uncertainty, than we can see some another interesting trend. Increasing w/c ratio changes the differences in compressive strength between similar series in favour of those with copper slag waste. However for mixtures with 300 kg/m^3 the compressive strength is higher for series with sand, in the contrary to mixtures with 200 kg/m^3 of cement.

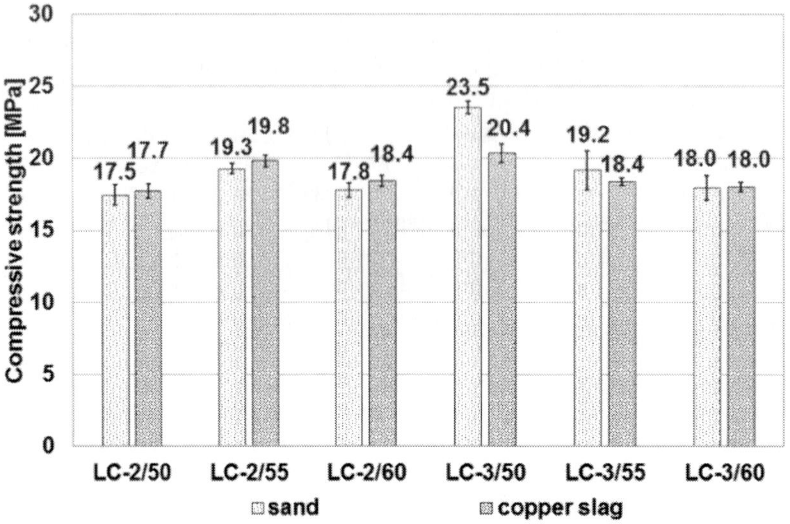

Fig. 3. Compressive strength test results.

3.2 Bulk density and open porosity

Results of bulk density determination are shown in Tab. 2. Obtained values of bulk density vary from 1.76 g/cm^3 to 1.88 g/cm^3 for saturated specimens and from 1.53 g/cm^3 to 1.63 g/cm^3 for dried specimens. In most cases bulk density of concrete series with copper slag waste are higher than of the series with sand, but the differences between determined values are not higher than 3%. This means that the initial assumption for achieving a similar density of the concrete mixes has been met.

Table 2. Values of bulk density.

Mixture	LC-2/50	LC-2/55	LC-2/60	LC-3/50	LC-3/55	LC-3/60
Bulk density determined on saturated specimens [g/cm^3]						
with sand	1.76±0.01	1.80±0.02	1.82±0.01	1.84±0.01	1.84±0.01	1.83±0.05
with copper slag	1.77±0.04	1.83±0.02	1.88±0.01	1.83±0.04	1.88±0.02	1.87±0.03
Bulk density determined on dried specimens [g/cm^3]						
with sand	1.55±0.01	1.58±0.02	1.59±0.01	1.59±0.02	1.58±0.01	1.54±0.06
with copper slag	1.53±0.05	1.60±0.02	1.63±0.01	1.59±0.04	1.60±0.02	1.58±0.03

Open porosity determination results are presented on Fig. 4. They are arranged in the same way as the compressive strength results. As it can be seen the porosity increases with increasing value of w/c ratio. What is more, the porosity is higher for mixtures with higher cement content. It strongly suggests that the porosity of cement matrix is higher than the porosity of lightweight aggregate. In addition, in all cases the porosity of the copper slag waste concrete series is higher than the porosity of the series with sand. This is probably the effect of higher content of lightweight aggregate and its different proportions (higher content of more porous aggregate). In most cases, however, the differences are not statistically significant.

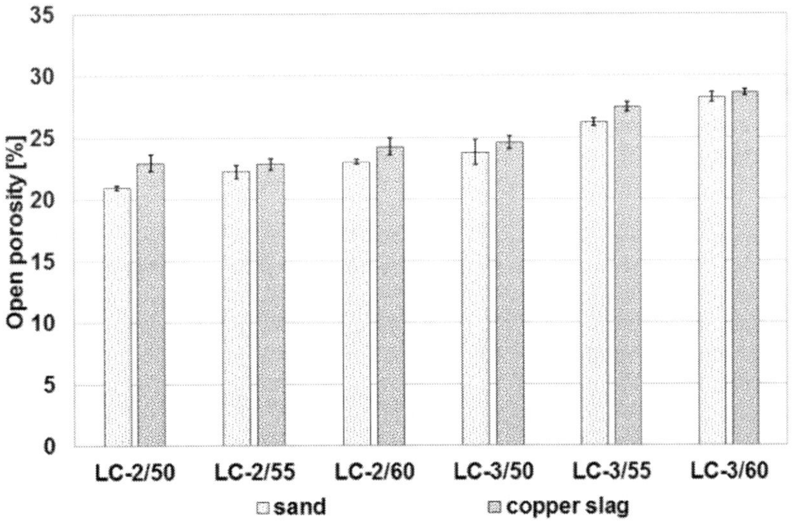

Fig. 4. Open porosity determination results.

3.3 Summary and conclusions

The obtained results indicate that it is possible to produce lightweight concrete with aggregate heavier than sand (copper slag waste) with similar density and without significant deterioration of mechanical properties. The resultant compressive strength of concrete with copper slag waste in range from 17.7 MPa to 20.4 MPa is sufficient in most applications of lightweight concrete. And in most cases of tested concrete mixtures the results of strength test are not significantly different from strength of concrete mixtures with sand.

The resultant higher porosity of the concrete produced with copper slag waste can be an advantage in many applications, because higher porosity is likely to result in a lower thermal conductivity. It may be important if such a material is used in external wall of buildings. The differences in porosity varies from 1.3% in case of LC-3/60 concrete mixture to 9.7% in case of LC-2/50 concrete mixture.

In the face of depletion of stocks of natural resources, the more often use of waste materials such as of copper slag waste is necessary and it would be a good news if it was used more often in manufacturing of lightweight concretes.

It seems advisable to continue research on lightweight concrete with copper slag waste to determine the thermal properties of the material, as well as other mechanical properties (eg. tensile strength, modulus of elasticity) and also some durability properties (eg. sorptivity, frost resistance, etc.).

Authors wish to thank Artur Sieciński for his help in preparing specimens and carrying out tests.

References

1. E. Holden, K. Linnerud, and D. Banister (2016)
2. I. Broman, K. Rob, J. Clean. Prod. J. **140**, 17–31 (2015)
3. L. Czarnecki, W. Kurdowski, Technol. Archit. **37**, 50–55 (2007)

4. M.A. Glinicki, Przegląd Bud. **12**, 24–30 (2007)
5. S. Bentur, A. Katz, S. Mindess, Cem. Wapno Beton **11**, 73, 102–121 (2006)
6. A. Ajdukiewicz, Przegląd Bud. **2**, 20–29 (2011)
7. S. Meenakshi Sudarvizhi, R. Ilangovan, Int. J. Civ. Struct. Eng. **1**(4), 918–927 (2011)
8. B. Gorai, R.K. Jana, Premchand, Resour. Conserv. Recycl. **39**(4), 299–313 (2003)
9. R.K. Dhir, J. de Brito, R. Mangabhai, C.Q. Lye, *Sustainable Construction Materials-Copper Slag* (Woodhead Publishing, Cambridge, 2017)
10. J. Rzechuła, *Fizykochem. Probl. Miner. Zesz. 28*207–218, A. Łuszczkiewicz, Ed., (Politechnika Wrocławska, Wrocław, 1994)
11. T. Ayano, K. Sakata, Spec. Publ. **192**, 141–158 (2000)
12. K.S. Al-Jabri, A.H. Al-Saidy, R. Taha, Constr. Build. Mater. **25**(2), 933–938 (2011)
13. I. Alp, H. Deveci, H. Süngün, J. Hazard. Mater. **159**(2–3), 390–395 (2008)
14. K.S. Al-jabri, M. Hisada, A.H. Al-saidy, S.K. Al-oraimi, Constr. Build. Mater. **23**, 2132–2140 (2009)
15. P.S. Ambily, C. Umarani, K. Ravisankar, P. Ranjan, B.H. Bharatkumar, N.R. Iyer, *Constr. Build. Mater.* **77**, 233–240 (Elsevier Ltd, 2015)
16. W. Wu, W. Zhang, G. Ma, Mater. Des. **31**, 6, 2878–2883 (2010)
17. S. Chandra, L. Berntsson, *Light. Aggreg. Concr.* (2002)
18. L. Domagała, Przegląd Bud. **9**, 38–44 (2016)
19. L. Domagała, M. Urban, Przegląd Bud. **4/2004**, 35–39 (2004)
20. L. Domagała, I. Hager, Cem. Wapno Beton **17**(3), 138–143 (2012)
21. A. Mueller, S. N. Sokolova, V. I. Vereshagin, Constr. Build. Mater. (2008)
22. C.L. Hwang, L.A.T. Bui, K.L. Lin, C.T. Lo, Cem. Concr. Compos. (2012)
23. L. Domagała, K. Kurzyniec, Energy Effic. Sustain. Build. Mater. Prod., 303–313 (2017)
24. W. Franus, G. Jozefaciuk, L. Bandura, M. Franus, *Minerals* (2017)
25. B. Gonzalez Corrochano, J. Alonso Azcarate, M. Rodas Gonzalez, B.G. Corrochano, M.R. Gonzalez, Int. J. Environ. Sci. Technol. (2011)
26. Y. Gao, L. Cheng, Z. Gao, S. Guo, Constr. Build. Mater. (2013)
27. K. Łuczaj, P. Urbańska, Mater. Bud. **1**(12), 44–47 (2015)
28. A. Sieciński, Badanie wybranych właściwości cieplnych betonu na kruszywie lekkim, (Politechnika Warszawska, 2017)
29. L. Domagała, *Procedia Eng.* **108**, 206–213 (2015)

Application of ceramic dust as a modifier reducing the extent of rheological deformations in airfield pavement concrete

Małgorzata Linek[1,*]

[1] Kielce University of Technology, Faculty of Civil Engineering and Architecture, Department of Transportation Engineering, Tysiąclecia Państwa Polskiego Street 7,25-314 Kielce, Poland

Abstract. The article presents material solution based on the application of ceramic dust as concrete mix component intended for airfield pavements. Material composition is intended for the application on the selected areas of pavement exposed to the influence of imposed thermal loads intensifying the stress strain state of concrete slabs. Due to the nature of loading of these parts it is necessary to reduce the extent of registered rheological deformations. Concrete containing dust additive is distinguished by more favourable porosity properties, more consistent cement matrix without visible discontinuities and with the formed different hydration products. Diversification of internal micro structure of cement concrete using the suggested dust has significant influence on the improvement of mechanical, physical and performance parameters. Also, assessment of the applied dust influence on the extent of the registered rheological deformations was presented. The analyses included concretes curing in standard conditions and concretes subject to thermal cycles representing the destructive influence of imposed loading. The obtained laboratory test results prove clearly the validity of the suggested solution. Reducing the extent of deformations is derivative of favourable changes observed in internal structure of concrete composite. Better formed contact areas provide the increased concrete parameters and consequently influence extending concrete durability.

1 Purpose and scope of the research

The purpose of the research was to define the influence of the applied ceramic dust on the change of parameters of hardened concrete exposed to the influence of the imposed thermal loads. Analyses included the determination of changes of extension of rheological deformations of concretes subjected to thermal cycles.

2 Research materials

* Corresponding author: linekm@tu.kielce.pl

The extension of the observed shrinkage deformations is referred to reactivity and physical and mechanical properties of aggregate. Concerning the materials used for the research purposes, the shrinkage was determined during drying up thereof absorbability. Based on the obtained results, these materials were included in the alkaline non-reactive group. Properties of the applied aggregate (table 1) influence the opportunity of reducing the size of defects within contact area. Contact area is formed by cement matrix of thickness range of 10-50 μm within the vicinity of aggregate grain. Contact areas are distinguished by the increased porosity which enables faster penetration of water and aggressive substances. They are areas where potential damages of internal structure of concrete can occur.

Table 1. Properties of the applied aggregate.

aggregate	granite grit	fine aggregate	ceramic dust
shrinkage during drying [%]	< 0.075	< 0.012	< 0.025
absorbability [%]	< 1.0	< 1.0	< 1.0
alkaline reactivity	non-reactive	non-reactive	non-reactive

Due to the fact that selection of concrete mixture components has significant influence on the size of shrinkage deformation, coincident initial assumptions for mixtures CC-1 and CC-2 design were assumed. Aggregate composition is decisive in case of shrinkage size. The designed aggregate mixes (table 2) of series CC-1 and CC-2 complied with the requirements of [1] and were considered as good grain size distribution of airfield pavements. The experiment assumed to perform two series of concrete. The first series was reference concrete (CC-1), in accordance with the requirements of [1]. Series CC-2 included concrete with partial fine aggregate substitute in the form of ceramic dust (0/2mm). Ceramic dust included crystalline components: quarc, mullite, cristobalite, corundum, very small amounts of clay minerals and anorthite.

Table 2. Material composition of mixes of series CC-1 and CC-2.

Mixes of series	Unit of measurement	Granite grit			Fine aggregate	Ceramic dust
		16/32 mm	8/16 mm	2/8 mm		
CC-1	kg/m^3	570	160	670	415	-
CC-2	kg/m^3	570	160	670	370	45

Mixes CC-1 and CC-2 included cement CEM I 42,5 N in the amount of 377 kg/m^3. Applied cement had reduced alkalis contents, its specific surface was 3431cm^2/g. Admixtures in the form of air entrainment agent (1.7 kg/m^3) and plasticizing admixture (0.7 kg/m^3) were used in these mixes. Maximum value of w/c ratio according to standard requirements [1] is 0.4; w/c ratio of the designed mixtures was assumed for 0.37. The lower coefficient value of w/c can be assumed that lower volume of capillary pores, which resulted in decreasing shrinkage.

Laboratory tests assumed the diversified number of thermal test cycles. A single test cycle included process of alternate heating and cooling - fig. 1. Recurrent influence of temperature stimulated the impact of aircraft on airfield pavements. The assumed heating and cooling time corresponded to the time during which the aircraft effected the most frequently used airfield pavement within the area of Poland. Velocity of gas stream emitted during the test was 30 m/s.

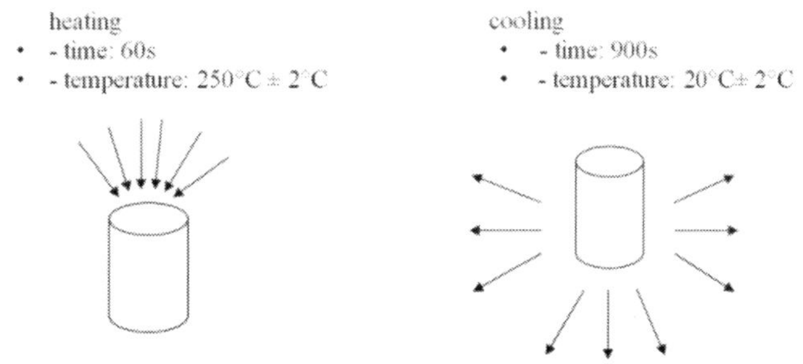

Fig. 1. Conditions assumed during the thermal research.

3 Research methods

During scientific research, the amount of the required samples was determined using student's T-distribution assuming the significance level of 0.05. The minimum essential number of samples ranged between 4 and 5, depending on the type of the conducted test. In case of such assumptions, 6 samples were selected, which, each time, were intended for the laboratory tests. In case of concrete series A, D, E, H and K the extent of (f_{ct}) tensile splitting strength was determined. The size of the test samples was 150x300 mm. Samples size and the loading method were consistent with the requirements of [1-3]. The samples were positioned centrically in the durometer which complied with the requirements of [4]. Along the bottom and upper edge of a sample, a pad was placed. These pads were made of fibreboard of density of 900 kg/m^3 and they were used for test purposes only once. Figure 2 presents the dimensions of the used pads. Load speed applied to the sample during the test was 0.05 MPa/s. In the course of the test, maximum load value was determined. The concrete sample is capable to transfer the load before it is damaged.

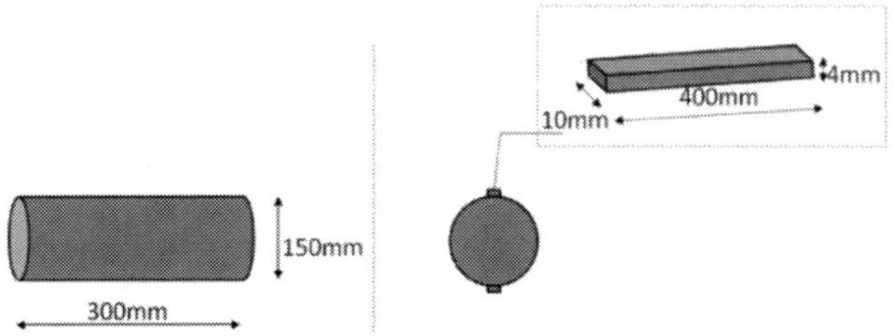

Fig. 2. Dimensions of samples and load diagram and dimensions of pads.

Concrete resistance (f_{ct}) was determined according to formula (1) in which F refers to maximum load applied to the sample [N], L – refers to the contact length [mm], while d – refers to the cross section of the sample [mm].

$$f_{ct} = 2\,\mathrm{F}\ /\ \pi L d \tag{1}$$

It was identified with concrete absorbability. In case of concretes of A series the ability of water absorption from the environment was determined. Samples of dimensions 150x150x150 mm intended for absorbability testing cured in water of 20°C for the first 28 days, and then they were to determine absorbability (n_w) in accordance with the guidelines of [1] according to (2); where G_1 – dry sample weight (sample dried at temperature 105°C), G_2 – weight of water saturated sample.

$$n_w = (G_2 - G_1)/G_1 \cdot 100 \tag{2}$$

It was assumed that defining the influence of the applied dust on the change of extension of shrinkage deformations will take place in case of five series. Each time, 6 samples of concrete CC-1 and CC-2 stored in the same environmental conditions were subject to the analysis. The first (A) series, reference series, included CC-1 and CC-2 concrete curing for the period of 28 days completely immersed in water of temperature 20 °C. Series D and H included CC-1 and CC-2 concrete stored from 28th day of curing in air-dry conditions. Series E and K included CC-1 and CC-2 concrete exposed to the influence of thermal cycles. The comparison of conditions during the test has been presented in Table 3.

Table 3. Test conditions for the research series (D, E, H and K).

series	Time	Temperature		Humidity	
	[s]	[°C]		[%]	
D	96000	20 ± 2		46-64	
E	96000	28 ± 2	20 ± 2	40-47	56-65
H	192000	20 ± 2		44-60	
K	192000	30 ± 2	20 ± 2	39-45	56-64

In order to determine single shrinkage of CC-1 and CC-2 concrete lots A, D, E, H and K six samples were used, in accordance with guidelines requirements [5], in the form of beams of cross section 150×150×600 mm. Changes of sample length in the course of their drying were measured by means of measuring instrument of reading capacity of more than 0.005 mm. Directly after removing each sample from a mould, locating distance *do* was determined. Socket extensometer and benchmarks stuck by means of quick-drying glue to the sample surface were used.

Assumed frequency of measurement series A performance included determination every 2 days for the period between the 1st and the 10th day, every 4 days for the period between the 7th and 28th day. Assumed frequency of measurement series D, E, H and K performance included determination every 2 days for the entire period.

Assumed research procedure complies with instruction [5]. In case of each analysed sample, after specified time, shrinkage deformation was determined ε_s, according to (3) formula, in which d_0 initial-basic distance (directly after unforming) between measuring points and d_t is distance determined after anticipated drying time:

$$\varepsilon_s = (d_0 - d_t) / d_0 \tag{3}$$

$$\varepsilon_{cs} = \varepsilon_{cd} + \varepsilon_{ca} \tag{4}$$

The alternative method to specify shrinkage deformations of concrete CC-1 and CC-2 in A series included the determination of assumed deformation values in accordance with the guidelines of PN-EN 1992-1[6] and [7]. As the value of total shrinkage deformation (ε_{cs}) the total extent of autogenic shrinkage (ε_{cd}) and the shrinkage caused as a result of drying (ε_{ca}) was assumed, according to the formula (4). The value of the assumed shrinkage caused by drying was determined according to (5) in compliance with the assumptions of standard [6].

$$\varepsilon_{cd(t)} = \beta_{ds(t,ts)} k_h \varepsilon_{cd,0} \tag{5}$$

$$\varepsilon_{ca(t)} = \beta_{as}(t) \varepsilon_{ca}(\infty) \tag{6}$$

The symbols in formula (5) coincide with symbols in [6], where: t refers to concrete age in the considered moment; ts refers to at the beginning of drying /expansion process; β_{ds} was determined according to the formula 3.10 of standard [6]; h_0 reliable cross section, h_0 = 75; u refers to perimeter of part of section exposed to drying process and k_h refers to the coefficient from the table 3.3 of standard [6] The value of autogenic shrinkage was determined according to (6) in compliance with the assumptions of standard [6]

Selected samples of CC-1 and CC-2 concrete subject to 100 and 200 thermal cycles were observed in scanning electron microscope. Samples preparation and interpretation of obtained results were in compliance with those described in subject literature [8]. Recent fracture were prepared from CC-1 and CC-2 concrete which were covered with carbon layer approx. 10nm thick. Preparation surface subject to observations by means of scanning electron microscope was not less than 1.0cm^2, and the scope of magnifying power was assumed between 200x to 100000x. During observations, special attention was paid to the structure of contact areas between aggregate grains and cement matrix, porosity characteristics and changes of cement matrix crystallization.

4 Results and discussions

Absorbability of concrete intended for airport pavements according to [1], which should not be higher than 5.0 % for class C30/37, 4.9 % for class C35/45, 4.9 % for class C40/50 and 4.8 % for class C45/55. According to the obtained laboratory tests results it was determined that concretes of series CC-1 and CC-2 complied with the requirements for the designed class C40/50. Moreover, CC-2 composed with the additive of dust is distinguished by the significantly lower absorbability. According to the analysis of the obtained test results of concrete series A it was proved that mechanical parameters of concretes CC-2 in all analysed cases are higher than parameters of concretes of CC-1. Significant influence (tab. 4) of the applied dust on the obtained tensile splitting strength. The applied dust is distinguished by the significant growth of tensile splitting strength after thermal cycles, which is associated with puzzolan properties. In a consequence of replacing cement with the suggested dust, the increase of average concrete tensile splitting strength in comparison to the reference concrete was proved. Increase in the strength was 31% after 200 thermal cycles. Reduction of concrete absorbability and increasing tensile splitting strength refers to the increased concrete durability.

Table 4. Determined parameters for CC-1 and CC-2 concrete.

parameters / concrete	N [%]			f_{ct} [MPa]		
	X	S_d	p-value	X	S_d	p-value
CC-1	3.98	0.08	< 0.000001	6.29	0.05	< 0.000001
CC-2	3.19	0.03	< 0.000001	7.31	0.04	< 0.000001
CC-1^{100}	4.41	0.06	< 0.000001	6.76	0.08	< 0.000001
CC-2^{100}	3.32	0.04	< 0.000001	7.54	0.06	< 0.000001
CC-1^{200}	4.89	0.07	< 0.000001	5.67	0.12	< 0.000001
CC-2^{200}	3.53	0.04	< 0.000001	8.20	0.07	< 0.000001

It was observed that in case of series A, the determined theoretical deformation of CC-1 concrete proved diversification with respect to the values determined experimentally. The most significant differences were observed between the 10th and the 28th day of curing. Values of the determined deformations were lower than the values measured experimentally – Fig. 3a. Theoretical deformations determined after the assumed research period was 0.0271‰ while the one measured experimentally was 0.0420%.

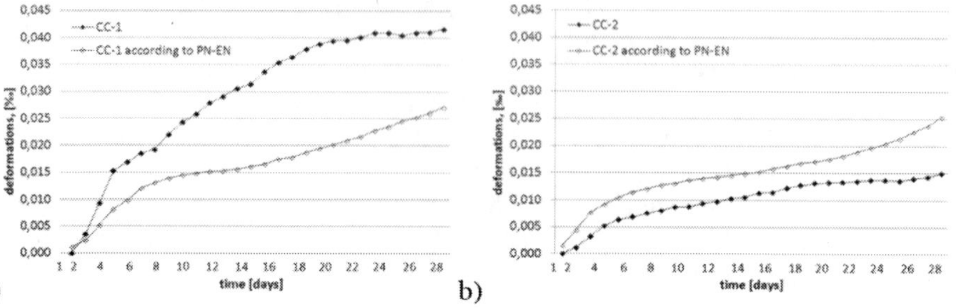

a) b)

Fig. 3. Deformations of concrete CC-1 (a) and CC-2 (b).

It was proved that in case of series A, the determined – theoretical deformations of CC-2 concrete were more extensive than the deformations obtained as a result of laboratory tests – Fig 3b. The length of curing period influences the extent of the registered deformations. In case of laboratory tests, deformations stabilized after 20 days of curing. Application of ceramic dust reduces CC-2 concrete susceptibility to external factors. Changes of environment humidity to a lesser extent influence the range of the registered rheological deformations. Determined theoretical deformation after the assumed research period was 0.0254‰, while the one measured experimentally was 0.0149%.

Table 5. Increasing shrinkage deformations of CC-1 and CC-2 concrete in various series (where: T indicates days, ε_p indicates initial deformation [‰], and ε_k indicates final deformation [‰]).

series	t	Concrete CC-1		Concrete CC-2	
	[days]	$\varepsilon_p \div \varepsilon_k$	$\frac{\varepsilon_k - \varepsilon_p}{t}$	$\varepsilon_p \div \varepsilon_k$	$\frac{\varepsilon_k - \varepsilon_p}{t}$
A	28	0.0012 ÷ 0.0417	1.45×10^{-3}	0.0004 ÷ 0.0150	5.21×10^{-4}
D	26.6	0.0417 ÷ 0.0453	1.35×10^{-4}	0.0150 ÷ 0.0169	7.14×10^{-5}
E	26.6	0.0417 ÷ 0.1547	4.11×10^{-3}	0.0150 ÷ 0.0352	7.59×10^{-4}
H	53.2	0.0417 ÷ 0.0858	8.29×10^{-4}	0.0150 ÷ 0.0190	7.52×10^{-5}
K	53.2	0.0417 ÷ 0.3951	6.64×10^{-3}	0.0150 ÷ 0.0555	7.61×10^{-4}

According to the obtained results from the shrinkage deformations analysis it was proved that dust additive used in the composition of CC-2 concrete has significant impact on the extent of reduced shrinkage (Table 5). Regardless of environmental conditions and the time of thermal cycles influence, CC-2 concrete is distinguished by reduced deformations. It was proven that CC-1^{A} concrete cured for 28 days in water with a temperature of 20 °C is distinguished by a deformation of 0.0417‰. In case of CC-2^{A} concrete this value was less than 0.0150‰ (Table 5 – A series). Deformations of CC-1^{A} concrete amount to 0.00145‰ and they are greater by order of magnitude than deformation of CC-2^A concrete. It was proven that change in the curing conditions (series D, E, H and K) also influences the phenomenon of increased shrinkage deformations. Samples of CC-1 and CC-2 concrete stored for 100 and 200 cycles in air-dry conditions (average temperature 20 °C and the humidity range 44-64%) are distinguished by average deformation lower than in case of thermal cycles. In each of these cases, deformations of CC-2 concrete are significantly lower than in case of CC-1 concrete. Deformations of CC-1^D concrete amount to 0.000135‰ and they are greater by order of magnitude than deformation of CC-2^D concrete. Similar tendency is also observed after extending the period with subsequent 100 thermal cycles. The average deformation of CC-1^H concrete is 0.000829‰, while in CC-2^H concrete it is only 0.0000752‰. The favourable phenomenon is also the fact that with CC-2 concrete, the growth of deformations after 200 cycles is lower than after 100 cycles. It was proven that CC-2 concrete exposed to thermal cycles (E and K series) is distinguished by significantly less shrinkage deformations. It should be emphasized that deformations observed in case of CC-2 concrete, in each of the discussed cases (cycles from 1 to 200), reached significantly lower values than in case of CC-1 concrete. Deformations of CC-1^E concrete amount to 0.00411‰ and they are greater by order of magnitude than deformation of CC-2^E concrete. Deformations of CC-2^E concrete amount to 0.000759‰. Similar tendency is also observed after extending the period with subsequent 100 thermal cycles. The average deformation of CC-1^K concrete is 0.00664‰, while in CC-2^K concrete it is only 0.000761‰. It is evidenced by the fact of the favourable influence of the applied dust on the reduction of deformations within a longer period of time.

The reasons for reducing deformations in CC-2 concrete with regard to CC-1 concrete should be found in favourable changes of internal structure of cement composite. According to SEM observations, it was determined that micro damages of cement matrix of concrete CC-2^{100} and CC-2^{200} series occur to a lesser extent in comparison to series CC-1^{100} and CC-1^{200} . Occasionally the occurring micro cracks of cement matrix of CC-2 concrete are distinguished by widths not exceeding 6μm in case of concrete CC-2^{100} and 8μm in case of concrete CC-2^{200}. In case of these concretes, there are continuous contact areas between cement matrix and aggregate grains (Fig. 4b). This is particularly advantageous phenomenon, due to the fact that contact areas are distinguished by the increased porosity and in case of the occurrence of additional micro cracks contributing to the increased penetration of water or water and chemical substances, can clearly reduce the time of structure durability. Discontinuity of contact areas can be observed in concretes of series CC-1^{100} and CC-1^{200} – Fig. 4a. Micro cracks in these areas are up to 10μm in CC-1^{100} concrete and up to 15μm in case of CC-1^{200}concrete, accordingly. In case of cement matrix of CC-1^{100} concrete, micro cracks of the width of up to 14μm and in matrix of CC-1^{200} concrete up to 18μm were observed. Observations by means of SEM proved that crystallization of ettryngite and fine-grained hydrated calcium silicate of C-S-H type prevails in cement matrix of CC-1^{200} concrete. There are also fractured aggregate grains which contribute to weakening of internal structure of concrete and consequently to the reduction of its resistance to rheological deformations.

a) 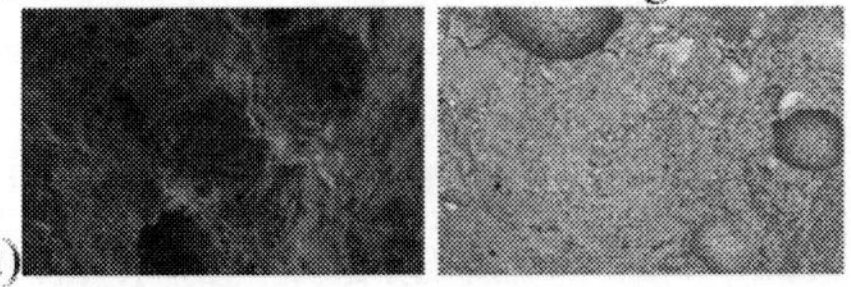 b) 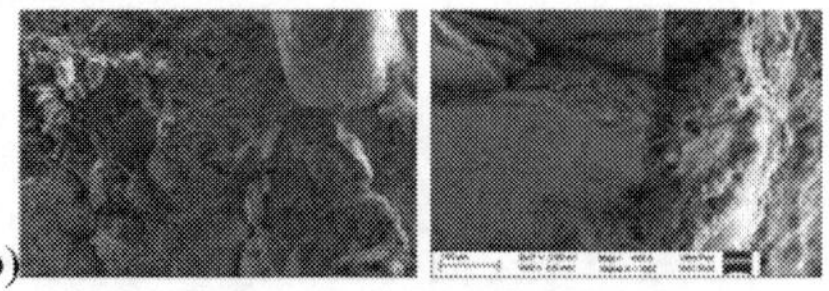

Fig. 4. Microstructure diversification (a) concrete CC-1^{200} and (b) concrete CC-2^{200}.

As a result of microstructure tests the influence of high temperature was proved which was generated on the surface of samples. Microstructure observations proved changes on the surface area. This was distinguished by the accumulation of carbonates in the amount strictly dependent on the assumed number of heating cycles. Within the surface area, increased porosity, with respect to layers located lower, was observed. During observations of CC-1 concrete series (Fig. 5a), numerous micro cracks were found, which were distinguished by significant widths and were not observed within surface layer of CC-2 concrete series (Fig. 5b). The phenomena observed in microstructure of concrete exposed to the influence of high temperature – series CC-2 (among others no micro cracks of surface area, continuous contact areas between cement matrix and aggregate grains and more favourable porosity characteristics) confirmed more favourable influence of modifier on parameters of hardened concrete. More favourably formed microstructure of CC-2 concrete series associated with better mechanical parameters (among others increased splitting resistance) and increased shrinkage deformations resistance proves the validity of using dust in the mix composition intended for airfield pavements.

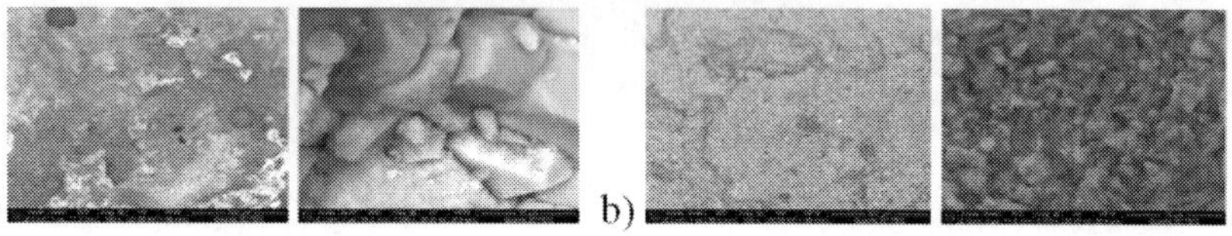

Fig. 5. The zone exposed to thermal cycles for concrete CC-1^{200} (a) and CC-2^{200} (b).

5 Conclusions

According to the conducted laboratory tests, the following conclusions have been reached:
- application of dust in the concrete mix composition has favourable influence on the reduction of the occurred micro damages of internal structure and porosity characteristics. The occurred hydration products and the size of individual crystals undergo change.
- the ceramic dust used in concrete mix contribute to the reduction of concrete absorbability by 0.79% after 28 days, 1.09% after 100 cycles and 1.36% after 200 thermal cycles.
- the ceramic dust used in concrete mix contribute to the increase of concrete tensile splitting strength after standard curing period by 14%.
- the ceramic dust used in concrete mix contribute to increase of concrete resistance exposed to the influence of 100 and 200 thermal cycles accordingly by 10% and 31%.
- the application of dust in the mix significantly influences the reduction of observed rheorogical deformations of hardened concrete by an order of magnitude.

References

1. NO 17-A204 (2015)
2. PN-EN 12390-2 (2011)
3. PN-EN 12390-6 (2011)
4. PN-EN 12390-4 (2011)
5. ITB Instruction No 194/98 (ITB, Warszawa, 1998)
6. PN-EN 1992-1-1 (2008)
7. W. Raczkiewicz, M. Bacharz, K. Bacharz, Adv. Mat. Sc. **15**(2), 22-29 (2015)
8. M. Linek, IOP Conf. Series: Mat. Sc. Eng. **245,** 032032 (2017)

MATEC Web of Conferences **163**, 03008 (2018) https://doi.org/10.1051/matecconf/201816303008
MATBUD'2018

Assessment of preliminary suitability of Fe- and Cr-rich waste material for the production of building ceramics

Ewelina Kłosek-Wawrzyn[1, *], and *Paweł* Murzyn[1]

[1]AGH University of Science and Technology, al. A. Mickiewicza 30, 30-059 Cracow, Poland

Abstract. This paper presents laboratory research on assessment of preliminary suitability of stainless steel waste material in building ceramics production. Application of the waste material in building ceramics production is innovative and might be environmentally friendly way to utilize it and to improve/modify the properties of the ceramic products. The waste was used as an additive to kaolinite-illite and montmorillonite-illite clay type. Selected properties of ceramics materials such as water adsorption, firing shrinkage, open porosity, apparent density, frost resistance, compressive strength and leaching of chromium and iron were determined. Improvement of the mechanical properties of the fired material and its colour intensification were observed. The studies confirmed the possibility of the utilization of this kind of waste material in production of building ceramics.

1 Introduction

Stainless steel is an iron-chromium alloy with minimum chromium amount of 11 %, which is their key alloying element. The addition of chromium above 11% creates a durable protective layer on the steel surface - a passive layer that provides anti-corrosive protection to the surface. To effectively prevent corrosion, the chromium contained in the steel must be fully dissolved in the solid solution and must not be in the form of chromium carbides [1].

Stainless steel mechanical processing is the final stage of the manufacturing process of many steel objects. It generates Fe- and Cr-rich wastes difficult to recycle. Use of the waste material in building ceramics production might be environmentally friendly way to utilize it and to improve the properties of the ceramic products. This topic has not been directly discussed in scientific literature yet, but it seems to have a great application potential. This paper presents laboratory research on assessment of preliminary suitability of stainless steel waste material in building ceramics production.

Until now attempts have been made to solidificate stainless steelmaking dust with clay in 1:1 ratio [2]. It was found that gained product can be used for the production of bricks or disposal and landfill. Research on use of stainless steel waste as pigment for ceramics was conducted [3-4]. Articles [5-6] presents the results of research on the use of this type

* Corresponding author: eklosek@agh.edu.pl

of pigment in the production of black ceramic tiles. Discussed work concerns the use of waste in the form of dust. Waste used in this work has a stainless steel fiber-like form mixed with fine grained mineral matter.

2 Materials and methods

2.1 Materials

Ceramic masses for building ceramics were prepared from two type of clay sediments popular in Poland: Triassic sediment signed as "P" and Miocene sea-sediment signed as "Z". As an additive Fe- and Cr-rich stainless steel waste was used.

Clays used in the research was milled to granulation below 2 mm. Fe- and Cr-rich steel waste was used without preparation. It was added to clay in an amount 30% by weight (Table 1).

Table 1. Dry mixes composition

Mass symbol	The content of raw clay [% mas.]	The content of waste [% mas.]
P	100 clay P	-
P30	70 clay P	30
Z	100 clay Z	-
Z30	70 clay Z	30

Water was added to the dry mixes until plastic consistency of masses was reached. The samples were hand-moulded. After drying, they were fired with the rate of temperature increase 100°C/h to 950°C with one hour temperature curing in 600°C and maximum temperature.

2.2 Methods

Chemical analysis of clay sediments and steel waste were performed using WDXRF method (WDXRF Axios maX PANalytical, with 4 kW Rh lamp). Mineral composition of clays was carried out by XRD method (Seifert-FPM XRD7 analyser). According to the method described in [7] untreated, burnt in 560°C and treated by ethylene glycol clays were examined (2θ measuring range was 4-30°). Thermal characteristics of ceramic masses components were characterized by simultaneous thermal analysis using NETZSCH STA 449 F3 Jupiter® and by evolved gases analysis using mass spectrometer QMS 403 Aëolos (heating rate 10°C/min, atmosphere: synthetic air, dynamic flow 40 ml/min). For complete characterization of steel waste its microstructure was observed using a scanning electron microscope FEI Nova 200 Nanos with EDS attachment.

Masses were characterised by mixing water content, drying shrinkage and volume density of dried samples.

Final materials were examined by their color, firing shrinkage, loss of weight upon firing, water adsorption, open porosity, apparent density, frost resistance, compressive strength according to method described in [7-9]. Leachability of chromium and iron were examined according to the standard PN-EN 12457-2 [10] and determined by inductively coupled plasma mass spectrometry ICP-MS method. The leachates with 1:10 dilution were analysed using a Perkin Elmer ELAN 6100 spectrometer. Analysis of the phase composition of final materials was carried out by XRD method (Phillips PW-1040 analyser, in the 2θ measuring range 5-60°).

3 Results and discussion

3.1 Characteristic of raw materials

Chemical compositions of the mass components are presented in Table 1. Clay P is red Triassic sediment from the deposit in Silesia (Poland). Its colour is from high hematite content which is reflected in high Fe_2O_3 content (Table 2). It can be use in clinker and porous ceramics (bricks, hollow bricks) production [7].

Clay Z is Miocene sea-sediment from the deposit from southern Poland. High content of CaO is a confirmation of calcareous characteristic of clay raw material. The chemical composition is closely related to the mineralogical composition, which is accurately described in the further part of the work.

Table 2. Chemical compositions of the mass components.

<table>
<tr><th rowspan="2">Oxide</th><th colspan="9">Content [% mas.]</th></tr>
<tr><th colspan="3">Clay P</th><th colspan="3">Clay Z</th><th colspan="3">Waste</th></tr>
<tr><td>SiO_2</td><td colspan="3">58.14</td><td colspan="3">52.1</td><td colspan="3">1.65</td></tr>
<tr><td>Al_2O_3</td><td colspan="3">17.99</td><td colspan="3">16.4</td><td colspan="3">7.48</td></tr>
<tr><td>Fe_2O_3</td><td colspan="3">9.81</td><td colspan="3">4.5</td><td colspan="3">55.20</td></tr>
<tr><td>TiO_2</td><td colspan="3">1.07</td><td colspan="3">0.61</td><td colspan="3">0.01</td></tr>
<tr><td>CaO</td><td>0.43</td><td rowspan="2">1.51</td><td rowspan="4">1.82</td><td>9.15</td><td rowspan="2">11.35</td><td rowspan="4">15.32</td><td>0.28</td><td rowspan="2">16.35</td><td rowspan="4">16.54</td></tr>
<tr><td>MgO</td><td>1.08</td><td>2.20</td><td>16.07</td></tr>
<tr><td>K_2O</td><td>0.20</td><td rowspan="2">0.31</td><td>3.50</td><td rowspan="2">3.97</td><td>0.02</td><td rowspan="2">0.19</td></tr>
<tr><td>Na_2O</td><td>0.11</td><td>0.47</td><td>0.17</td></tr>
<tr><td>SO_3</td><td colspan="3">0.07</td><td colspan="3">0.42</td><td colspan="3">0.07</td></tr>
<tr><td>Cr_2O_3</td><td colspan="3">-</td><td colspan="3">-</td><td colspan="3">11.39</td></tr>
<tr><td>Cl^-</td><td colspan="3">-</td><td colspan="3">-</td><td colspan="3">1.92</td></tr>
<tr><td>MoO_3</td><td colspan="3">-</td><td colspan="3">-</td><td colspan="3">0.14</td></tr>
<tr><td>MnO</td><td colspan="3">-</td><td colspan="3">-</td><td colspan="3">0.21</td></tr>
<tr><td>L.O.I. (in 1000°C)</td><td colspan="3">8.50</td><td colspan="3">10.18</td><td colspan="3">5.17 (in 465 °C)</td></tr>
<tr><td>total</td><td colspan="3">97.39</td><td colspan="3">99.53</td><td colspan="3">99.78</td></tr>
</table>

The main chemical components of the waste are: Fe_2O_3 and Cr_2O_3 from stainless parts. Alkaline oxides, Al_2O_3, SiO_2 and Cl originate from abrasive parts, such as corundum and mechanical processing fleece with talc [11-12]. Loss on ignition is given before Fe oxidation.

XRD analysis of clay sediments are given in Fig. 1-2. They show that the P-clay is a raw material of a kaolinite-illite character. It also contains small amounts of chlorite and β-quartz. The remaining minerals in the sample were identified by simultaneous thermal analysis.

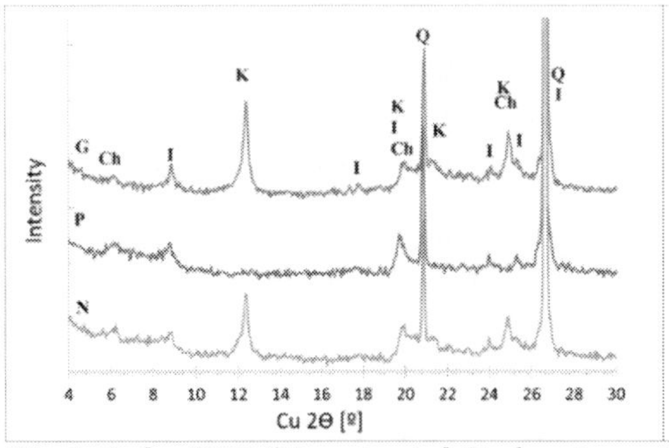

Fig. 1. P-clay XRD (N – untreated, P – burnt in 560°C for 2 hours, G – treated by ethylene glycol, *cts – Counts Per Step; Ch – chlorite; I – illite; K – kaolinite; Q – quartz)

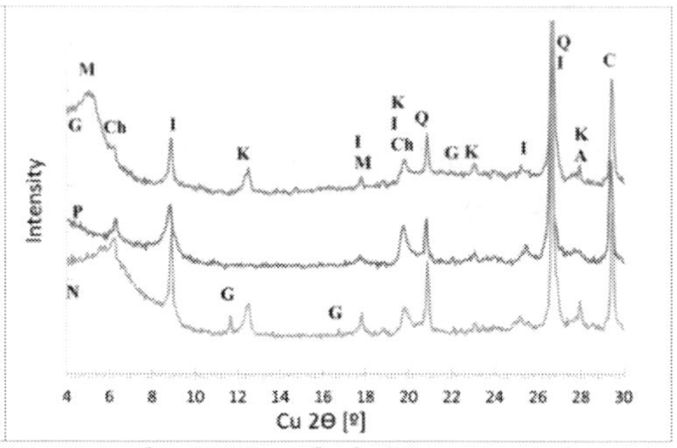

Fig. 2. Z-clay XRD analysis (N – untreated, P – burnt in 560°C for 2 hours, G – treated by ethylene glycol, *cts – Counts Per Step; M – montmorillonite ; Ch – chlorite; I – illite; K – kaolinite; Q – quartz, C - calcite)

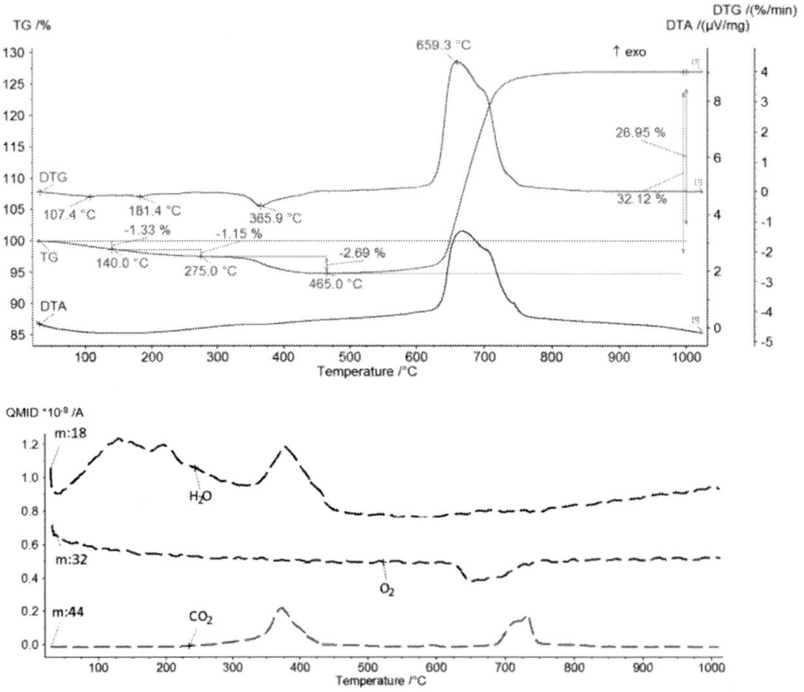

Fig. 3. DTA/EGA analysis of waste

Clay-Z is a montmorillonite-illite-kaolinite sediment. It contains also chlorite and calcite minerals. Calcite is characteristic component of calcareous clay raw materials. It consists of smaller amount of β-quartz than P-clay.

Fig. 3 presents simultaneous thermal analysis with separated evolved gases analysis of steel waste. Small double peak in ion current for m/z = 18 is present up to 275°C. It is due to water release from the material. In the range of 275 to 465°C simultaneously increase in water and carbon dioxide ion current combined with a weight loss is observed. This is probably related to the combustion of organic matter. Increase in carbon dioxide ion current is observed also at around 700°C. It may be caused by decomposition of carbonates. The most important thermal effect on the DTA curve is observed from around 600 to 800°C. It is exothermal oxidation of Fe from steel. It is associated with about 27% mass increase on TG curve and decrease in ion current on m/z-32 EGA curve which is due the decrease of oxygen concentration in the atmosphere near of the sample. During oxidation of Fe, also volume of the oxidized material increases.

Fig. 4. SEM micrograph of waste W.

Fig. 4 presents SEM micrograph of Fe- and Cr-rich waste. It has the form of steel fibres with remains of abrasive and polishing materials. Fibres have different length and width and they are intertwined with each other.

3.2 Properties of masses

Table 3 presents properties of the masses after drying. Water absorption of both masses after the waste addition slightly increased (respectively by about 5% for P-clay and 12% for clay-Z). Despite this, drying shrinkage of P-clay with waste addition is smaller than P-clay without addition. It is due to fibrous form of the waste. Bulk density of masses based on P-clay with the waste addition is higher than bulk density of P-clay without its content. Drying shrinkage and bulk density after drying of masses based on Z-clay are slightly different.

Table 3. The properties of the masses after drying

Mass symbol	W [% mas.]	W SD [% mas.]	Ss [%]	Ss SD [%]	ϱs [g/cm^3]	ϱs SD [g/cm^3]
P	20.22	0.05	8.65	0.21	1.97	0.02
P30	21.14	0.25	7.65	0.20	2.09	0.02
Z	24.60	0.05	10.18	0.17	1.87	0.01
Z30	27.48	0.10	10.36	0.26	1.89	0.01

W - mixing water content, Ss – drying shrinkage, ϱs – bulk density after drying,

3.3 Properties of materials after firing

Fig. 5 presents colours of samples after firing. In both cases addition of the waste intensifies colour of clay matrix. According to [13] colour change is occurring as result of the valence variation of Fe, new phases and the outward diffusion of Fe. From red in case P-clay sample colour changes to brown in case of masses with waste. From light orange in case Z-clay sample colour changes to gray-brown in case of masses with waste. According to XRD results (Fig. 6) it is due to higher hematite and magnetite content in samples with the waste. Reddish colour of P-samples is probably triggered by the higher amount of Fe_2O_3 in P-clay (Table 2) and hematite amount in P-material.

Firing temperature	P	P30	Z	Z30
950°C				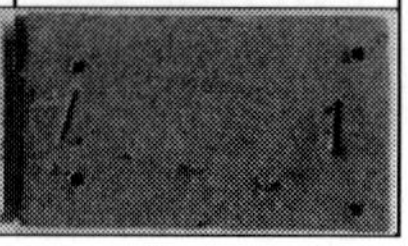

Fig. 5. Colours of samples after firing.

Table 4 presents properties of the materials after firing. In both cases high temperature expansion was observed for masses with the waste addition. It is probably due to iron oxidation during firing (Fig. 3). By the same phenomenon, masses of samples increase causing a decrease in firing weight loss. Water absorption and open porosity increases by the waste addition, but increase is higher for material made of Z-clay.

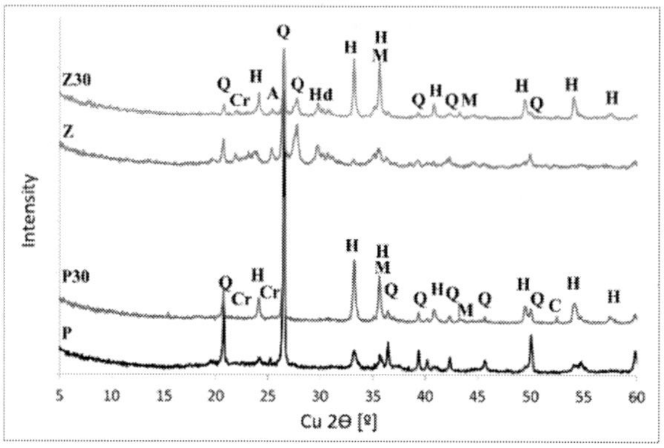

Fig. 6. XRD results of samples after firing (Q – quartz, Cr – crystobalite, H – hematite, M – magnetite, Hd – hedenbergite, A – anhydrite, C – corundum).

Despite the increase in porosity the apparent density increases with the waste addition. It is due to its higher real density than densities of clays. Fe- and Cr-rich waste stainless steel waste addition causes increase in compressive strength (by 11% and 31% respectively for P- and Z-material).

Each material is fully frost resistant according to methods shown in PN-B-12012:2007 [9] standard intended for unprotected type clay masonry units. According the standard PN-EN 771-1:2011+A1:2015 [8], this means that they can be used for masonry materials unprotected against external factors. The factor limiting this use is the high water absorption values (>6%). Therefore, they should be considered as suitable for use as materials for masonry protected against external factors.

Table 4. The properties of the materials after firing

Mass	Sf [%]	SD Sf	Fwl [%]	SD Fwl [%]	Aw [%]	SD Aw [%]	Po [%]	SD Po [%]	ϱp [g/cm³]	SD ϱp [g/cm³]	FR	Rsc [MPa]	SD Rsc [MPa]
P	1.11	0.08	6.99	0.00	13.27	0.07	27.22	0.09	1.97	0.01	Y	49.95	0.97
P30	-0.01	0.10	-2.35	0.03	13.84	0.24	29.21	0.18	2.20	0.01	Y	55.31	0.90
Z	2.07	0.11	12.12	0.02	14.11	0.09	25,53	0.12	1.81	0.01	Y	51.04	0.83
Z30	-0.28	0.19	1.64	0.08	20.08	0.07	37.72	0.09	1.90	0.01	Y	66.93	0.30

Sf – firing shrinkage, Fwl – firing waste loss, Aw – water absorption, Po – open porosity, ϱp – apparent density, FR – frost resistance, Y – full frost resistance, Rsc – compressive strength, SD – standard deviation

Table 5. Leaching tests results

Element\symbol	P0	P3	Z0	Z3
Cr	0.0004	0.0008	0.0009	0.0044
Fe	0.008	0.048	0.003	0.003

Table 5 presents leaching tests results. According to [14] acceptable value of total chromium leachability is 70 mg/l. There are no requirements for iron leaching. Higher chromium leachability was observed for material from Z-clay with stainless steel addition. Higher iron leachability was observed for material from P-clay with stainless steel addition. This issue requires a more detailed analysis.

4 Conclusions

In this paper the possibility use of Fe- and Cr-rich waste material from stainless steel elements mechanical processing in the production of building ceramics was analysed. As the basic constituent two types of clays was used: Triassic sediment signed as "P" and Miocene sea-sediment signed as "Z". The waste was added to clay in an amount of 30% by mass. Firing process temperature was 950°C.

Research of final ceramic materials confirmed suitability of use of the waste in the production of dark-coloured building ceramic products, especially as materials for masonry protected against external atmospheric factors.

The research work was financed by AGH-University of Science and Technology in Cracow, Poland, Faculty of Materials Science and Ceramics, Department of Building Materials Technology activity (No 11.11.160.184).

References

1. P. J. Cunat, *Praca ze stalą nierdzewną. Materiały i zastosowania, zeszyt 2.* (Euro Inox, Luxemburg, 2013)

2. M. Tang, J. Peng, B. Peng, D. Yu, Tang C. Trans. Nonferrous Met. Soc. China **18**, 1, (2008)
3. G. Costa, M. J. Ribeiro, T. Trindade, J. A. Labrincha, Boletin de La Sociedad Espanola de Ceramica Y Vidrio **46**, 1 (2007)
4. G. Costa, V. P. Della, M. J. Ribeiro, A. P. N. Oliveira, G. Monrós, J.A. Labrincha, Dyes Pigm. 77, 1 (2008)
5. R. Zhu, G. Ma, Y. Cai, Y. Chen, T. Yang, B. Duan, Z. Xue, J. Air Waste Manag. Assoc. **66**, 4 (2016)
6. X. Zhang, G. Ma, Y. Jin, P. Cheng., Ceram. Int. **40** (2014)
7. E. Kłosek-Wawrzyn, T*he effect of calcium carbonate on the sintering process of ceramic masses and the properties of ceramic building materials* (Akademia Górniczo-Hutnicza im. St. Staszica w Krakowie 2016), (in Polish)
8. PN-EN 771-1:2011+A1 (2015)
9. PN-B-12012 (2007)
10. PN-EN 12457-2 (2006)
11. B. Van Hecke, *Mechaniczne wykończenia powierzchni dekoracyjnych ze stali nierdzewnej. Materiały i zastosowania, zeszyt 6.* (Euro Inox, Luxemburg, 2005)
12. R. Crookes, Roger, *Wytrawianie i pasywacja stali nierdzewnej. Materiały i zastosowania, zeszyt 4.* (Euro Inox, Luxemburg, 2004)
13. H. Yurdakul, S. Turan, E. Ozel, Dyes Pigm. **91**, 2 (2011)
14. Polish Law Dz. U. poz. 38, (2013)

Hydration degree development of blast furnace slag blended cement pastes using thermogravimetric Analysis (TGA)

Walid Deboucha[1,*], *Nordine* Leklou[1], and *Abdelhafid* Khelidj[1]

[1] *LUNAM University, University of Nantes - IUT Saint-Nazaire, GeM, CNRS UMR 6183, Research Institute in Civil Engineering and Mechanics, France.*

Abstract. The hydration degree assessment of slag blended cement pastes with three different replacement levels (10, 20 and 40%) and two finenesses (400 and 500 m^2/kg) was investigated. The hydration degree was evaluated by using a modified TG method suggested by Deboucha et al., 2017. This method builds on the activity coefficient concept, which makes it possible to assume the amount of mineral additives that contribute to hydration reactions as well as cement. Results showed that the efficiency of blast furnace slag in developing hydration degree was conditioned by its level of replacement and fineness.

1 Introduction

Mineral additives are alternative materials, which are widely used as partial cement replacement with the attention of reducing the environmental impact of the cement production [1, 2, 3]. In addition to the hydration reactions of clinker phases, the presence of mineral additives have an influence on the hydration reactions of these phases due to their chemical and physical effects [4-5]. Hence, seeking ways to understand the impact of mineral additives on hydration reactions needs to be investigated.

The impact of slag on the hydration reactions of blended cement pastes is the focus of the present study. Two techniques were used to study the hydration of these systems. An isothermal calorimeter was used to quantify the influence of slag addition on hydration reactions at early ages. When we reached advanced times, thermogravimetric analysis (TGA) was used. Based on results obtained with this technique, a modified TG method suggested in previous article [6] was applied to quantify the impact of slag on hydration reactions of blended cement pastes, by measuring their hydration degree development. This method relies on the activity coefficient K concept, which allows to assume the amount of mineral additives that contribute to hydration reactions and then inferring the hydration degree.

[*] Corresponding author: walid.deboucha@univ-nantes.fr

 https://doi.org/10.1051/matecconf/201816304001

2 Experimental program

2.1 Materials and mix design

Ordinary Portland cement type I 52.5 N conforming to NF EN 197-1 [7] was used in this study. Slag (BFS) with two different degree of fineness (400 and 500 m^2/kg) was used as partial cement replacement. Table 1 presents the chemical compositions of cement and slag, which were provided in previous study [8]. Fig.1 provides the particle size distribution of cement and blast furnace slag using Laser diffraction (CILAS 1180L apparatus).

Table 1. Chemical composition and physical characteristics of materials [8].

Chemical composition (%)	OPC	BFS
SiO2	20.07	40.00
Al2O3	4.86	7.35
Fe2O3	3.07	2.28
CaO	64.25	42.69
MgO	0.95	5.19
So3	3.55	1.24
K2O	1.00	0.74
Na2O	0.18	0.04
LOI	1.09	0.25
Physical characteristics		
Specific gravity	3.14	2.98
Blaine (m^2/kg)	440	400/500

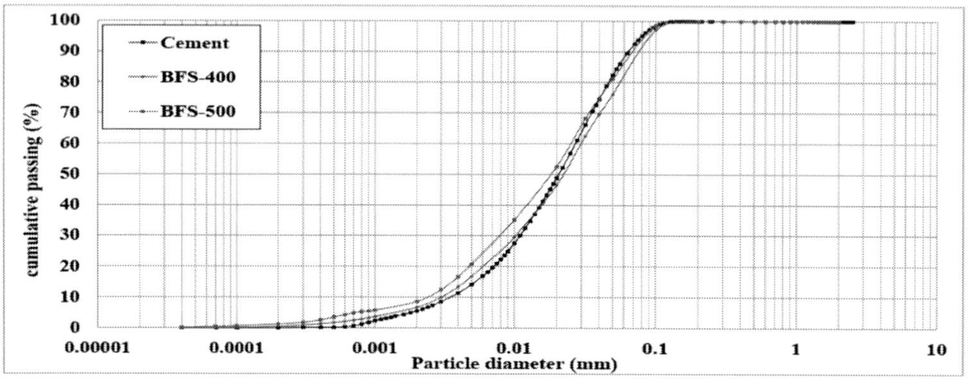

Fig. 1. Particle size distribution of slag and cement [8].

Cement pastes were prepared according to the procedure for mortar preparation, given by European standard EN NF 196-1 [9]. The Portland cement was replaced by slag with three replacement levels (10, 20 and 40 %). the water to binder ratio was selected at 0.35. the mix proportions of the combination of cement and slag are shown in Table 2.

Table 2. Mix proportions

mix	Cement (%)	BFS (%)	Slag Fineness (m^2/kg)	w/b (%)
PR	100	/	/	0.35
P10BFS400	90	10	400	0.35
P20BFS400	80	20	400	0.35
P40BFS400	60	40	400	0.35
P10BFS500	90	10	500	0.35

P20BFS500	80	20	500	0.35
P40BFS500	60	40	500	0.35

2.2 Methods

2.2.1 Isothermal calorimeter

The measurement of the released heat from the hydration reactions is widely used to quantify the hydration kinetics at early and very early ages. In this study, a multi-channel isothermal calorimeter TAM AIR device was used. After the end of mixing, two samples with weight of approximately 6g were extracted from the mixture and placed into standard plastic vials. The plastic vials were then loaded into isothermal calorimeter. The hydration heat was measured continuously for 88 hours at a constant temperature of 20°C.

2.2.2 Thermogravimetric analysis

The thermogravimetric analysis tests were carried out using a METTLER TOLEDO TGA/DSC1 with a balance accurate to 0.1 μg. The dynamic heating ramp varied between 25 and 1025 °C at a rate of 5°C/min. Tests were carried out under nitrogen atmosphere on 20-30 mg samples cured at 20 °C. The degree of hydration was determined from the thermogravimetric analysis data using a modified TG method. Equation (1) describes the methodology suggested for determining the chemically bound water (W_B).

$$WB= Ldh+Ldx+0.4(Ldc\text{-}Ldca)\text{-}(mc\times LOICC+mA\times LOIAC)+m_d \quad (1)$$

where Ldh, Ldx, Ldc, Ldca, LOICC, LOIAC, m_c, m_A and m_d are parameters defined in ref. [6].

The hydration degree for cement pastes was estimated according to Equation (2).

$$\alpha= W_B/\ W_{B\infty}\times\ L_{eq} \quad (2)$$

where Leq= m_C,+ K×m_A and K is the activity coefficient (the activity coefficient for slag, given by NF P 18-305 Standard is 0.9).

3 Results and discussion

3.1 Cumulative heat released

The heat released due to the chemical reactions can give an overview of the impact of mineral additives on hydration kinetics at early and very early ages. The cumulative heat released was measured continuously for 88 h at 20 °C using a multi-channel isothermal calorimeter as described earlier. It was observed from Fig.2 and 3 that the cumulative heat released for all cement pastes prepared with slag was lower than that of pure cement paste. The cumulative heat released decreased with the increase of slag content. Indeed, after 88 h of hydration, the cumulative heat released of cement paste containing 40 % of slag with a fineness degree of 400 m^2/kg was almost 23% lower than that of cement paste containing 10 % of slag with a fineness degree of 400 m^2/kg. This phenomenon is directly due to the dilution effect [10] and the latent hydration processes of slag [8-11].

With the increase of the fineness degree of slag from 400 to 500 m^2/kg, the cumulative heat released was slightly increased. For example, for cement paste prepared with 40 % of slag, the increase of the fineness degree of slag from 400 to 500 m^2/kg increased the cumulative heat released by about 4%. The increase of fineness degree of mineral additive in fact can accelerate and stimulate the cement particles hydration since; the physical effect of mineral additives in cementitious systems is due to its surface area [4-12]. In this study, the increase in the cumulative heat released was found to be very low when the fineness degree of slag increased from 400 to 500 m^2/kg. This can be explained by the convergence of fineness for slag and cement.

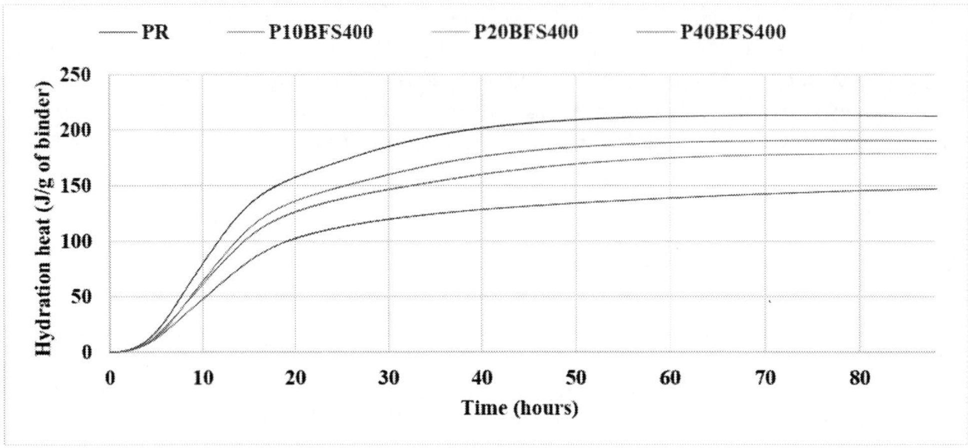

Fig. 2. Cumulative heat released for cement pastes containing BFS with fineness degree of 400 m^2/kg.

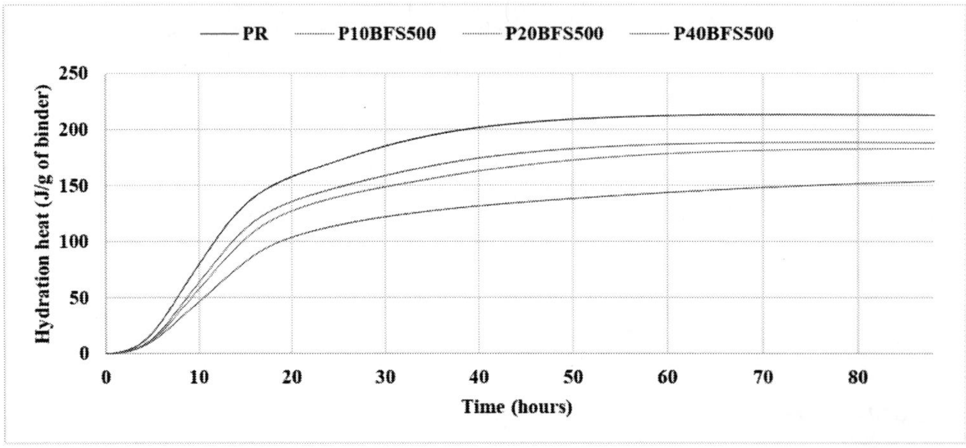

Fig. 3. Cumulative heat released for cement pastes containing BFS with fineness degree of 500 m^2/kg.

3.2 Degree of hydration

The degree of hydration of blended cement pastes containing slag with two finenesses degree of 400 and 500 m^2/kg was determined using a modified TG method. The mass loss between 105 and 1000 °C was considered to determine the chemically bound water as recommended by Scrivener et al. [13]. The hydration degree was calculated according to Equation (2).

The hydration degree of the pure cement paste and cement pastes containing slag with a fineness degree of 400 m^2/kg up to 90 days is shown in Fig.4. The pure cement paste showed higher degree of hydration at all tested ages. The contribution of slag with a fineness degree of 400 m^2/kg to the hydration development was not quite significant due to the lower surface area of this latter (400 m^2/kg) compared to the surface area of Portland cement (440 m^2/kg).

The presence of slag with a fineness degree of 500 m^2/kg in cementitious materials meanwhile, had a greater effect on the hydration reactions compared to the slag with a fineness degree of 400 m^2/kg. In fact, at all tested ages, the cement pastes containing slag with a fineness degree of 500 m^2/kg showed a greater hydration degree as compared to the others cement pastes containing slag with a fineness degree of 400 m^2/kg at the same replacement level. For instance, the 90 days hydration degree of P10BFS500 was higher by about 6 % than that of P10BFS500. Similarly, the 90 days hydration degree of P40BFS500 was higher by about 5 % than that of P40BFS400.

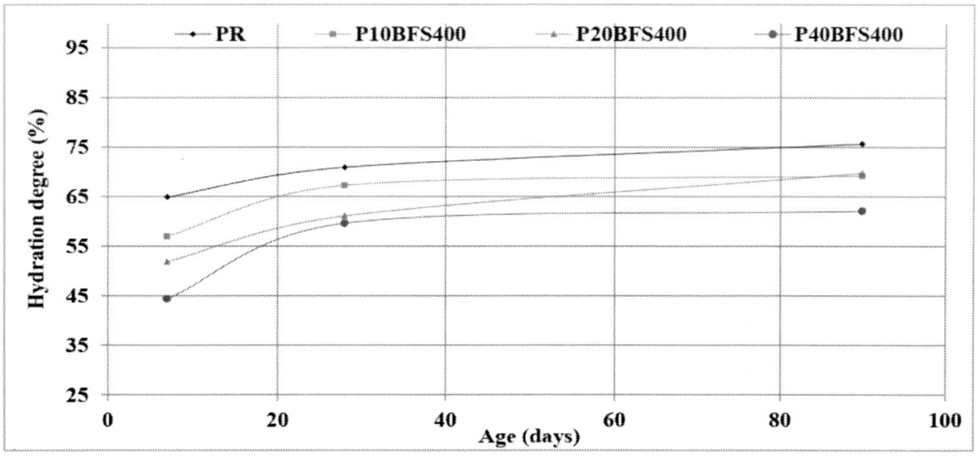

Fig.4. Hydration degree evolution for cement pastes containing BFS with fineness degree of 400 m^2/ kg.

From results presented in Fig.5, it was observed also that the difference in the hydration degree between the pure Portland cement paste and cement pastes prepared with slag decreased with increasing curing time and even, in the case of cement paste made with 10 % of slag at a fineness degree of 500 m^2/kg, the 90 days hydration degree was similar to that of pure Portland cement paste. This enhancement in hydration degree could be attributed to the greater surface area of 500 m^2/kg for slag compared to the 440 m^2 for Portland cement where the slag particles can act as nucleation site for cement hydrates. Indeed, as reported for example by Berodier and Scrivener [13] and Cyr et al. [14], the heterogeneous nucleation effect is dependent on several factors such as the replacement level and the fineness of mineral additive. Berodier and Scrivener [4] found that a similar surface area for both cement and mineral additive is not efficient in the heterogeneous nucleation process. However, this effect is more remarkable for fine mineral additive particles.

The hydration degree of cement pastes containing slag agree well with the hydration heat evolution where both hydration degree and hydration heat released for cement pastes prepared with slag proportionally decreased with the increase in replacement level. This finding supports the hydration degree values accuracy calculated according to the modified TG method.

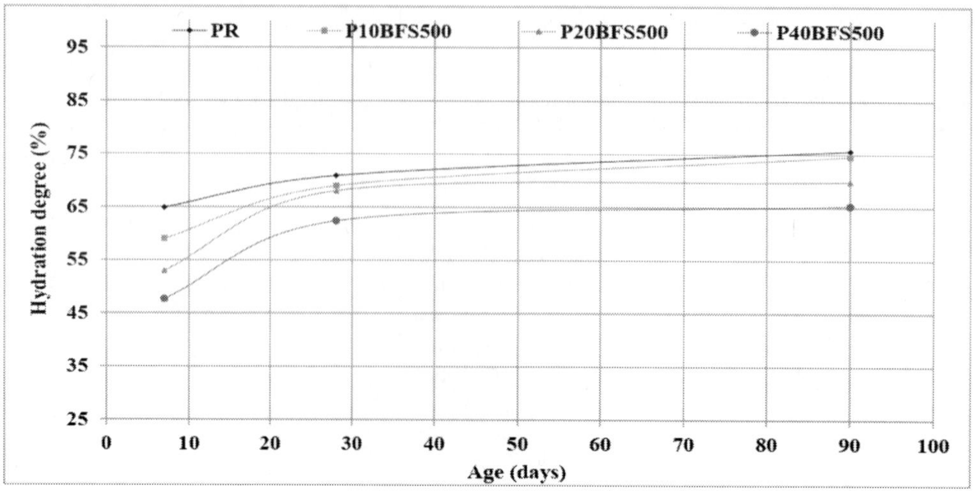

Fig.5. Hydration degree evolution for cement pastes containing BFS with fineness degree of 500 m^2/kg.

4 Conclusions

In this study, the impact of slag addition on hydration reactions in binary blended cement pastes was investigated. The hydration degree of these systems was estimated using modified TG method. The hydration degree values showed that the level of replacement and the fineness degree of slag conditioned the hydration degree of slag blended cement pastes. It was found that the increase in slag content caused a decrease in the hydration degree at all testing ages, due to dilution effect. It was observed meanwhile, that the increase in the fineness degree of slag stimulated the hydration reactions, which resulted in enhanced hydration degree.

References

1. D.M. Roy and G.M. Idorn, Amer. Concrt. Instit. J. **79**, 444–457(1982)
2. W. Deboucha, N. leklou, A. Khelidj, M.N. Oudjit, Energ Proc. **139**, 689- 695 (2017)
3. S.E. Chidiac and D.K Panesar (2008), Cem. Concr. Comp. **30**, 63–71 (2008)
4. E. Berodier and K. Scrivener, Amer. Ceram. Soc. J. **97**, 3764–3773 (2014)
5. M. Cyr, P. Lawrence, E. Ringot, Cem. Concr. Res. **36**, 264-227(2006)
6. W. Deboucha, N. leklou, A. Khelidj, M.N. Oudjit, Constr. Build. Mater. **146**, 687–701 (2017)
7. NF EN 197–1 (2000)
8. W. Deboucha, N. Leklou, A. Khelidj, KSCE J. Civ. Eng, 1-7 (2018)
9. EN NF 206–1 (2004)
10. H. Beushausen, M. Alexander, Y. Ballim, Constr. Build. Mater. **29**, 533–540 (2012)
11. W. Deboucha, M. N. Oudjit, A. Bouzid, L. Belagraa, Proc. Eng. **108**, 254 – 261 (2015)
12. P. Lawrence, M. Cyr, Cem. Concr. Res. **33**, 1939–1947 (2003)
13. K.L. Scrivener, B. Lothenbach, N. De Belie, E. Gruyaert, J. Skibsted, R. Snellings, A. Vollpracht, Mater. Struct. **48**, 835–862 (2015)
14. M. Cyr, P. Lawrence, E. Ringot, Cem. Concr. Res, **35**, 719–730 (2005).

The impact of lithium nitrate on the physical and mechanical properties of Portland cement

Justyna Zapała-Sławeta[1,*], *Zdzisława* Owsiak[1]

[1]Kielce University of Technology, al. Tysiąclecia Państwa Polskiego 7, 25-314 Kielce, Poland

Abstract. The effectiveness of lithium nitrate as a chemical additive which reduces the negative effects of alkali aggregate reaction was subject to research by scientists in many centres around the world. The literature data on the impact of lithium nitrate on the physical and mechanical properties of cements are rare. Without a precise definition of the impact of lithium nitrate on the cement properties, it is extremely hard to determine its real advantages in practical usage. In this paper, studies were undertaken to assess the impact of $LiNO_3$ on the properties of pastes and mortars with Portland cement. The rate of hydration of the cement with lithium additive was examined by isothermal calorimetry, measurements of setting time and phase composition of cement pastes in the initial stages of hydration. The influence of the admixture on the compressive strength development of mortars after 2, 7 and 28 days of hardening was also researched. Results indicate that lithium nitrate accelerates the early hydration of Portland cement, affecting the precipitation of hydration products. The compressive strength of mortars with lithium admixture decrease after 28 days, although 2 an 7-day strength were greater than the control mortars.

1 Introduction

The problem of the alkali aggregate reaction was first noticed in the United States of America in the 1940s, when Stanton provided evidence for the existence of internal corrosion processes, leading to concrete expansion and cracking [1]. Detailed analyses of the alkali – reactive aggregates reaction, conducted by many researchers worldwide indicated that the chemical interaction in concrete may occur between alkali and different types of aggregates. In 1957 Swenson demonstrated the occurrence of a reaction between alkali and carbonate aggregate, indicating that it is a different type of a reaction between aggregate and sodium and potassium hydroxides than the one discovered by Stanton [2]. The effects of the alkali aggregate reaction were observed in elements such as concrete sidewalks, concrete pavements, bridges, dams and other structures exposed to humidity over an extended period of time [3].

The negative effects of alkali-silica reaction in concrete can be reduce by some supplementary cementing materials which leads to reduction the amount of alkalis available for reaction with aggregate [4-7]. Another way of diminishing deleterious expansion due to

[*] Corresponding author: jzapala@tu.kielce.pl

alkali aggregate reaction is the addition of lithium nitrate to the concrete mix, which was first proposed by McCoy and Cadwell in 1960s [8]. As a result of the research programmes conducted later on by the U.S Department of Transportation, Federal Highway Administration Research and Technology, it has been proven that lithium nitrate was used as a compound efficiently limiting internal corrosion of the concrete in airport pavements in Atlanta as well as bridges and highways among others in Mexico, Texas and Pennsylvania [9].

Choosing the right amount of lithium nitrate, which would limit concrete expansion caused by the alkali aggregate reaction, depends on the amount of the alkali in the cement as well as the type, the amount and the location of reactive minerals in the aggregate [10,11]. Lithium nitrate is introduced into the concrete mix together with mixing water and the amount of the lithium ions introduced corresponds to the molar ratio of Li to [Na+K]. The optimum amount of lithium ions, expressed by means of the molar ratio of the total amount of alkali in the cement oscillates around the value of 0.72 – 1.0 [12,13]. For economic reasons, the evaluation of the efficacy of the use of lithium nitrate in the smallest possible amounts is pivotal. However, the possibility to use greater amounts of lithium nitrate should be considered for the sake of additional or a more long-lasting protection of concrete exposed to internal corrosion. Apart from determining the efficiency of $LiNO_3$ with regards to limiting the linear expansion of concrete, it is crucial to research the effect adding the lithium additive has on the physical and mechanical properties of cement. The research concerning the impact of lithium salts on the properties of the cement pastes and mortars was scarce. The results of these analyses are inconclusive. Some researchers indicate among others that the cement strength decreases and the setting time is reduced as a result of a reaction with the lithium compound, whereas others show no record of a significant effect $LiNO_3$ might have on the above mentioned properties [14].

In this paper, studies were undertaken to assess the impact of lithium nitrate on the properties of Portland cement pastes and mortars. What was evaluated was the effect of $LiNO_3$ on i) cement hydration, by means of determining the changes in the kinetics of reaction ii) the strength of the Portland cement mortars.

2 Materials and methods

In order to efficiently prevent the alkali aggregate reaction using lithium nitrate it is crucial to add the correct amount of this salt. The molar ratio of the lithium ions to the contents of alkali in the cement, which amounts to 0.74 and is the so called recommended dosage, is considered to be sufficient in most cases. In this paper, the Li/[Na+K] molar ratio, amounting to 0.74, was assumed to be the 100% of the $LiNO_3$ dosage. This ratio may increase for aggregates with greater reactivity. Due to the above, 100% and 200% of the recommended dosage of the lithium nitrate was used during the tests. For comparison purposes, tests were executed on cement without the lithium additive. The tests were executed using CEM I 42,5R Portland cement pastes and mortars, the composition of which is presented in table 1.

Table 1. The chemical composition of the CEM I 42,5R [%].

Material	SiO_2	Al_2O_3	Fe_2O_3	CaO	MgO	SO_3	K_2O	Na_2O	TiO_2	Na_2O_e
Cement	18.47	5.40	2.80	60.77	1.42	3.34	0.93	0.18	0.45	0.79

The heat of hydration of the cement without the lithium additive as well as of the cement containing 100% and 200% of the $LiNO_3$ dosage was measured using the BT 2.15 differential scanning calorimeter, while working in semi adiabatic conditions. The samples for the calorimetric tests were prepared in the form of pastes with the w/c coefficient of 0.5 and an appropriate amount of $LiNO_3$. At the same time, cement phase compositions were being determined, using the X-ray analysis (XRD). What was determined was the effect of the lithium additive on the setting times of cement pastes in the Vicat needle, in accordance with the method described in the PN-EN 196-3:2006 standard [15]. The compressive strength was determined using mortar beams measuring 4x4x16 cm, while the w/c ratio in the samples without the lithium additive and with the $LiNO_3$ was constant and equal to 0.46. The compressive strength of the mortars was determined after 2, 7 and 28 days of hardening.

3 Results

3.1 Heat of hydration

The calorimetry test results of the cement pastes being analysed are presented in Figure 1 in the form of a heat release rate versus time curve.

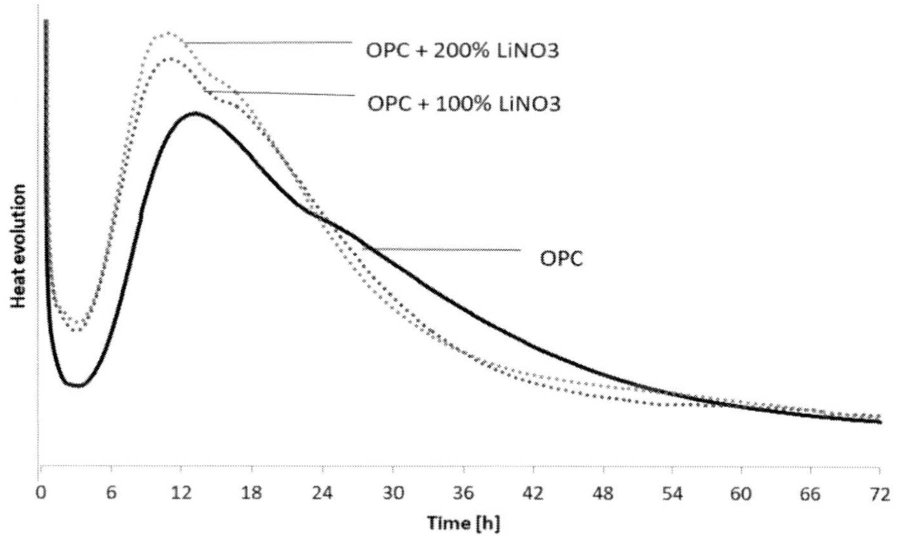

Fig. 1. The calorimetric curves of Portland cement with lithium additive

When comparing the course of the calorimetric curves, an increase in the heat release rate was noticed in the induction period in the pastes with the lithium admixture. It may be stated, that in $LiNO_3$ pastes the induction period dissipates and the cement paste does not regain its initial temperature. The analysis of a further fragment of the curve, which shows the so-called reaction acceleration effect connected to the quick crystallization of the products is moved in time towards the beginning of hydration. As it is shown in the subject matter literature, a greater number of alkali leads to an increase in the reactivity of the C_3S and C_3A phase [3]. The energy transfer effect in the cement pastes with $LiNO_3$ indicating the tricalcium silicate hydration phase is moved in time by approximately 3 h. There were greater differences noted in the process of the hydration of C_3A occurring with $LiNO_3$ present. The energy transfer effect connected with the hydration of C_3A in the cement pastes with $LiNO_3$, which is visible in the form of a characteristic bulge, is moved significantly in time with the movement in time ranging from 7.5h (100% $LiNO_3$) to 8.5h (200% $LiNO_3$). Additionally, it needs to be noted that mutual movements of the energy transfer effect connected with the hydration of C_3S relative to the same energy transfer effect originating from the C_3A phase decrease with the increase of the amount of $LiNO_3$ being introduced.

3.2 Phase composion

The results of the X-ray analyses executed within the initial periods of hydration (up to 72h) confirm the acceleration of the hydration processes of the C_3A phase with the lithium additive present. Lithium nitrate increases the solubility of gypsum, leading to the decreases in the peak intensity, with the angle equal to 2ϴ = 11.6°(Fig. 2). At the same time, the intensity of the ettringite peak increase (9.1° Theta). Gypsum originating peak was only found in the cement sample without the lithium additive after 12 h.

Portlandite peaks, which are the evidence of the progress of the calcium silicates hydration also prove to be more intense in the pastes with the lithium additive. Additionally, with the increase of the amount of $LiNO_3$ the hydration of the C_3S phase progresses as well, which has also been noticed by Bentz [16]. Lithium ions may play a role in the hydration processes similar to that of the remaining alkali ions. The acceleration of the hydration process of phase C_3S is probably the result of the nucleation processes occurring in the hydration products containing calcium, which occur faster as a result of the decrease in the solubility of calcium compounds in the pore solution. Moreover, as it is indicated by Kurdowski after Jawed and Skalny, all the salts causing oversaturation of a solution relative to scarcely soluble calcium compounds, contribute towards keeping the concentration of calcium low in the liquid phase, and therefore to increasing the solubility of the initial phase C_3S [17]. These statements are in accordance with the results of the X-ray analyses of the course of hydration processes. It needs to be remembered that the addition of an electrolyte to the cement paste is connected both with the cation and the anion impact on the electrical double layer. The ions which accelerate hydration, some of which are lithium ions and nitrate ions, cause the double layer to become thinner and lower the zeta potential, which facilitates the coagulation of the colloidal phase C-S-H [18].

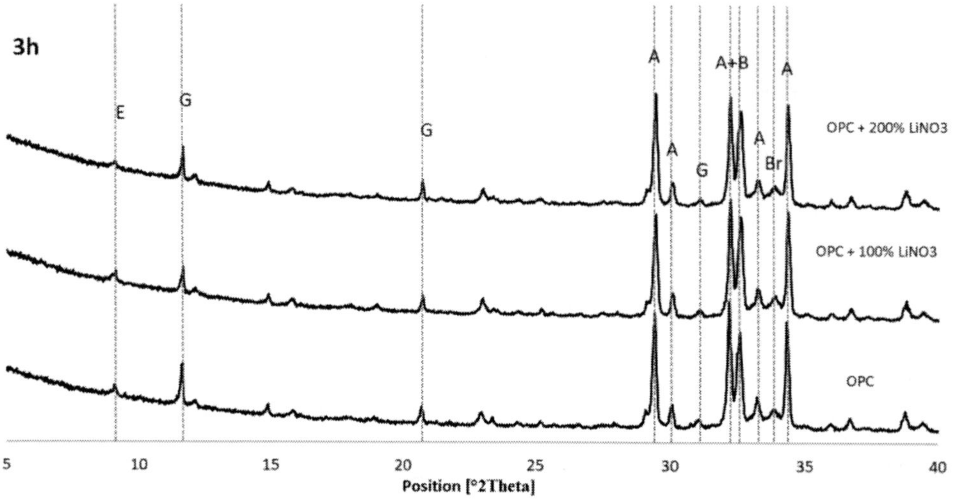

Fig. 2. X-ray patterns of Portland cement with 100% and 200% of the $LiNO_3$, dosage, hydrating for 3 h. E - ettringite, G - gypsum, A - Alite, B - Belite, Br – Brownmillerite.

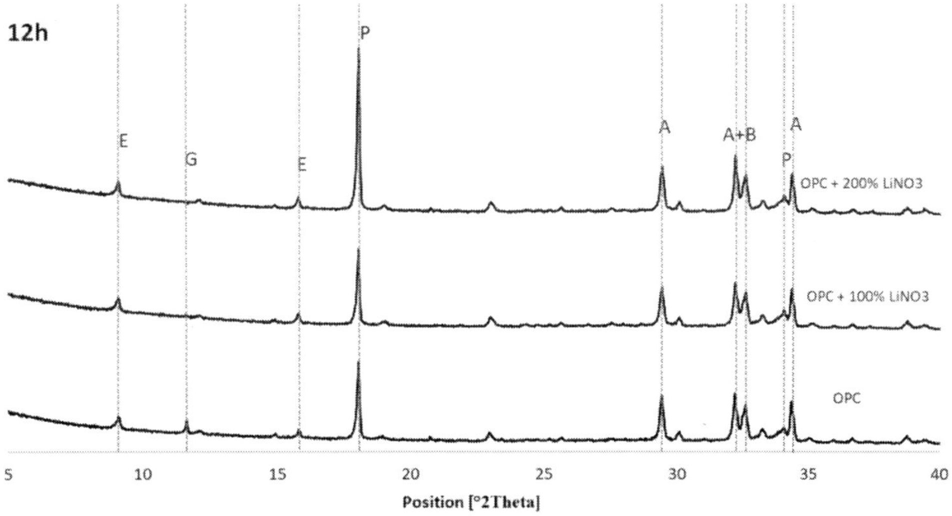

Fig. 3. X-ray patterns of Portland cement with 100% and 200% of the $LiNO_3$, dosage, hydrating for 12 h. E - ettringite, G - gypsum, P - Portlandite, A - Alite, B – Belite.

After 24 hours of cement hydration, the amount of portlandite is also the greatest in the sample with the lithium additive, which is indicated by the increased intensity of peaks with 18.0 and 34.1°Theta (Fig. 4). The results of the X-ray analyses are in good correlation with the results of the calorimetric analysis of the cement pastes. After 72h, with the increase of the amount of lithium nitrate, the intensity of the portlandite peak increases insignificantly, compared to the paste without the additive (Fig. 5).

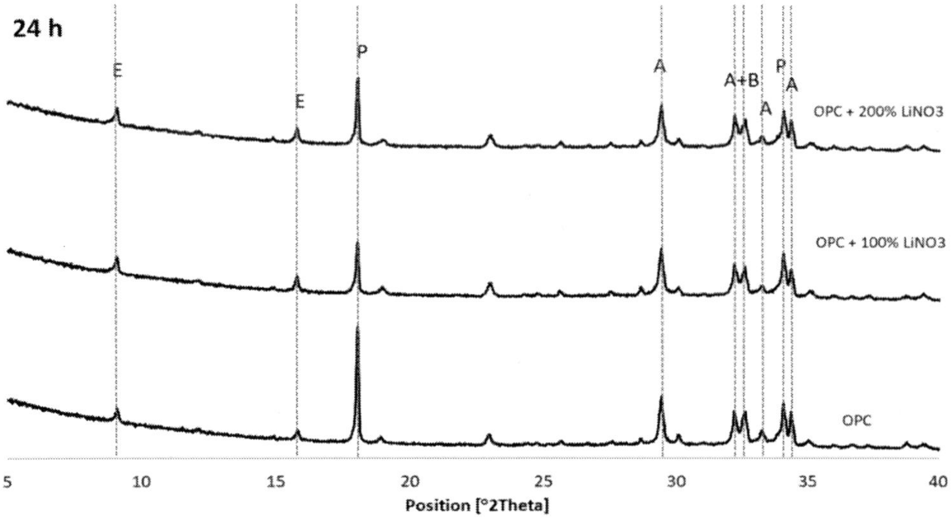

Fig. 4. X-ray patterns of Portland cement with 100% and 200% of the $LiNO_3$, dosage, hydrating for 24 h. E - ettringite, G - gypsum, P - Portlandite, A - Alite, B – Belite.

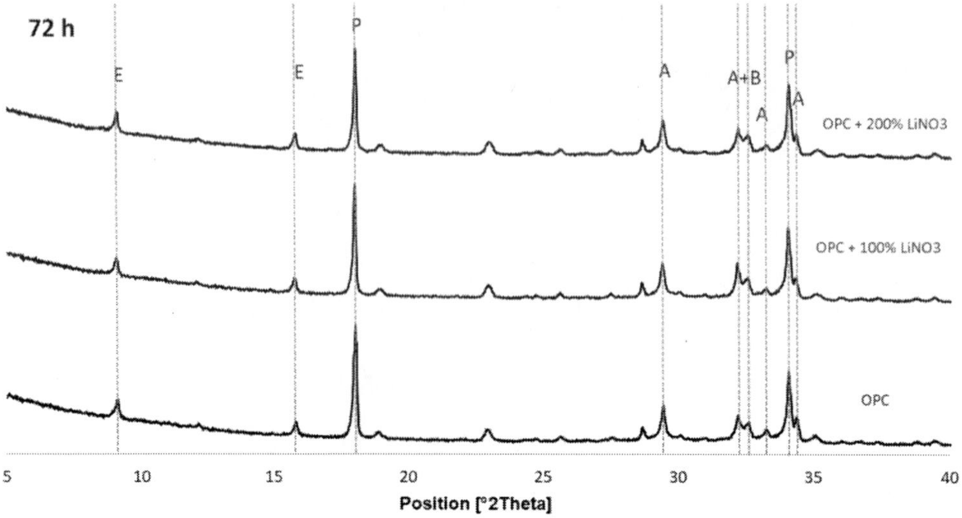

Fig. 5 X-ray patterns of Portland cement with 100% and 200% of the $LiNO_3$ dosage, hydrating for 72h. E - ettringite, G - gypsum, P - Portlandite, A - Alite, B – Belite

3.3 Setting time and compressive strength

Figure 6 shows the test results for the setting time of cement pastes with different contents of lithium nitrate. At the same time, it was found that adding the lithium additive causes the decrease of the cement's water demand. Lithium nitrate causes the beginning of cement setting to shorten. The $LiNO_3$ additive amounting to 100% and 200% causes the loss of plasticity time in the cement paste to accelerate by 9% and 15% accordingly. The completion of the setting of the pastes with a lithium additive is also shortened (by 35-50 minutes) relative to the sample without the lithium compound introduced.

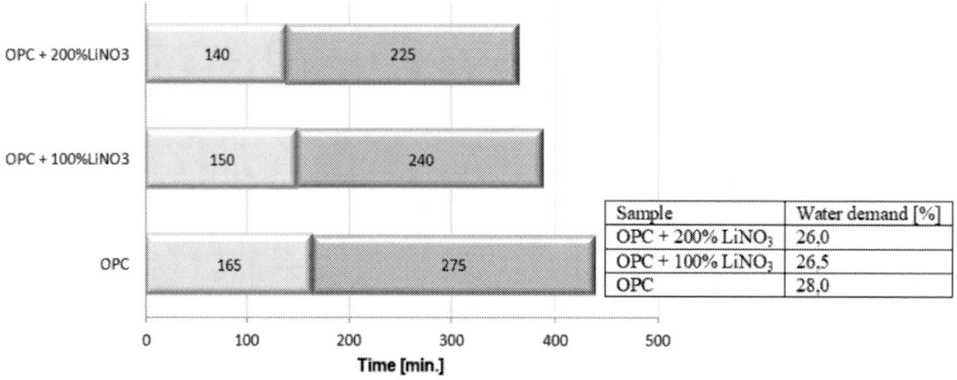

Sample	Water demand [%]
OPC + 200% $LiNO_3$	26,0
OPC + 100% $LiNO_3$	26,5
OPC	28,0

Fig. 6 The setting time and the water demand of the Portland cement with different amounts of the lithium nitrate additive

Figure 7 shows the test results of the compressive strength of cement mortars with different contents of lithium nitrate after 2, 7 and 28 days of hardening.

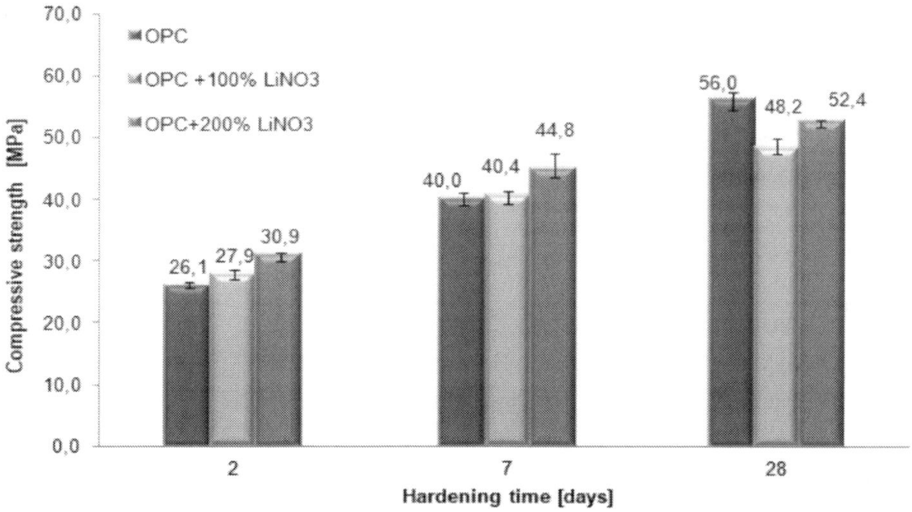

Fig. 7. Compressive strength of cement mortars with the $LiNO_3$ additive.

Adding the lithium additive in the amount of 100% of the recommended dosage of $LiNO_3$ causes the compressive strength to increase after 2 and 7 days of hardening. The greatest increase in the compressive strength was found in mortars with 200% of the recommended dosage of the chemical additive. The compressive strength of the mortars with the $LiNO_3$ additive after 28 days of hardening is lowered. In the case being analysed by the author the compressive strength is lowered by 14% and 6% accordingly for the dosages of 100% and 200% of lithium nitrate.

4. Conclusions

The conducted tests have shown that lithium nitrate influences the hydration processes of the Portland cement. The calorimetric tests of the cement pastes with 100% and 200% of the recommended dosage of $LiNO_3$, where the dosage of 100% corresponds to the molar ratio of Li to [Na+K] = 0.74, and is used the most frequently to limit the alkali aggregate reactions, are evidence for the acceleration of the hydration of phases C_3A and C_3S. The test results of the phase composition indicate more intense reflexes of ettringite in the samples with the lithium additive introduced and greater intensities of the portlandite peaks accordingly. It is possible that the acceleration in the C_3S hydration is a result of the decrease in the solubility of the products of hydration containing calcium, which accelerates the nucleation process, whereas quicker hydration of the C_3A is a result of the increase in the solubility of this phase influenced by the lithium compound. The beginning and the end of cement setting time are both shortened, however an insignificant interval increase was found between the beginning and the end, as more $LiNO_3$ was introduced. Relative to the reference paste, the beginning of the setting is shortened by 9% -15%, whereas the end is shortened by 13% - 18%. The acceleration of the beginning and the end of the cement setting for 100% and 200% of the recommended $LiNO_3$ was also observed by Millard and Kurties, in particular in low-alkali cement [3,14]. For high-alkali cement the differences relative to the reference sample were slight. Lithium nitrate affects the mortar strengths. The test results remain in accordance with the pattern showing that setting and hardening accelerators at the same time cause a decrease in durability after 28 days. The tests performed by Wang, who analysed the effect of LiOH on the mechanical properties of mortars, indicate that a decrease in the durability of mortars with a lithium additive and a high-alkali cement and an increase in the durability of low-alkali mortars is possible [19]. Millard and Kurties also made note of the negative effect of bigger doses of lithium on compressive strength [3]. These findings are particularly crucial due to the necessity of increasing the amount of $LiNO_3$ in order to protect reactive aggregate grains, advocated by many researchers.

References

1. T.E. Stanton, Proc. ASCE **66**, 1781 (1940)
2. E.G. Swenson, ASTM Bull. **226,** 48 (1957)
3. M.J. Millard, K.E. Kurtis, Cem. Concr. Res. **38**, 500 (2008)
4. M. Thomas, Cem. Concr. Res. **41**(12), 1224 (2011)
5. Z. Owsiak, P. Czapik, Cement Lime Concrete **3**, 6 (2014)
6. Z. Owsiak, P. Czapik, Bull. Pol. Ac.: Tech. **63 (1)**, 4 (2015)
7. J. Zapała-Sławeta, J. Mater Struct. **50**, 217 (2017)
8. W.J. McCoy, A.G. Caldwell, ACI Mat. J. **22(9)**, 693 (1951)
9. K.J. Folliard, M.D. Thomas, K.E. Kurtis, FHWA Report: FHWA-RD-03-047, July 2003
10. Z. Owsiak, J. Zapała-Sławeta, Cement Lime Concrete **20**, 25 (2015)
11. J. Zapała-Sławeta, Z. Owsiak, Bull. Pol. Ac.:Tech. **65 (6)**, 773 (2017)
12. X. Feng, *Effects and mechanism of lithium nitrate on controlling alkali-silica reaction*, (PhD Thesis, University of New Brunswick, Canada, 2008)
13. Z. Owsiak, J. Zapała-Sławeta, Ceram.- Silik. **57 (2)**, 138 (2013)
14. M. Millard, K. Kurtis, *Lithium Admixtures ($LiNO_3$) And Properties of Early Age Concrete* (An IPRF Research Report Innovative Pavement Research Foundation Airport Concrete Pavement Technology Program ,Skokie, USA, 2006)
15. PN – EN 196 – 3 (2006)
16. D.P. Bentz, ADV CEM RES **18**, 65 (2006)
17. W. Kurdowski, *The Chemistry of Cement and Concrete* (SPC, Kraków, 2010)
18. R. Kondo, M. Daimon, E. Sakai, H. Oshiyama, J. Appl. Chem. Biotechnol. **27**, 191 (1977)
19. W.Ch. Wang, Constr. Build. Mater. **50**, 727 (2014)

Effect of boron-containing aggregates on setting and hardening of Portland cement mortars

Aneta Antolik[1,*], *Michał A.* Glinicki[1], and *Marek* Gawlicki[2]

[1]Institute of Fundamental Technological Research, Polish Academy of Sciences, Pawińskiego 5B, 02-106 Warsaw, Poland
[2]Institute of Ceramics and Building Materials, Department of Process and Environmental Engineering, Oświęcimska 21, 45-641 Opole, Poland

Abstract. Multicomponent cement-based composites are known as versatile structural materials for enhanced radiation shielding. The use of selected elements, like boron, cadmium, or rare earth elements, provides an increased neutron shielding capacity. Because of profusion, reasonable costs and large cross-section for neutron capture, boron containing minerals are suggested as aggregates for radiation shielding concrete. Despite many advantages, boron additives may act as cement setting retarders. Uncontrolled setting and hardening is not acceptable in radiation shielding concrete technology. In this work we present results from isothermal calorimetry measurements on cement mortars with boron-containing aggregates. Four types of boron aggregates were used in the studies: colemanite, ulexite, borax and boron carbide. Based on calorimetric curves, the beginning of setting time was determined. Additionally early mortar strength was investigated and linear relationship between the heat generated in the isothermal calorimeter and the early compressive strength has been observed. The use of isothermal calorimetry allowed us to estimate the limits for the content of boron compounds to be used cement mortar.

1 Introduction

Due to the continuous development of nuclear power and the use of neutron radiation, new solutions for shielding are needed. Radiological protection for personnel and environment requires capturing of highly penetrating neutron radiation. Particular elements, like hydrogen, boron, cadmium, or gadolinium, are known to exhibit outstanding neutron shielding capacity [1-3]. Such elements, preferably in a form of abundant minerals, could be used as fillers or aggregates in radiation shielding concrete [4, 5]. Because of availability, reasonable costs and large cross-section for neutron capture, boron containing minerals are suggested [6] as beneficial aggregates for radiation shielding concrete.

Boron has a lot of advantages in radiological protection, e.g. high thermal neutron absorption cross section, does not emit highly penetrating gamma radiation. Boron naturally occurs as isotopes ^{10}B (20%) and ^{11}B (80%), the first of them is more important in the shielding. ^{10}B due to its thermal neutron high absorption capacity results in

* Corresponding author: aantolik@ippt.pan.pl

the emmission of easily absorbed weak gamma radiation. The cost of boron-containing minerals is relatively low. However boron compounds are damaged by the absorption of neutrons ($^{10}_{5}B + ^{1}_{0}n \rightarrow ^{7}_{3}Li + ^{4}_{2}He$). Moreover, boron compounds are expected to interfere with the setting of Portland cement [7, 8].

The objective of this investigation is to study the effects of various boron-bearing compounds on cement setting and hardening in order to evaluate the limits for possible use in radiation shielding cement mortars.

2 Materials and laboratory test methods

2.1 Materials

Four types of boron-containing aggregates with varied content of boron were tested: colemanite (0.5 – 2 mm, B = 15.78 wt%), ulexite (0.5 – 2 mm, B = 13.34 wt%), sodium tetraborate pentahydrate (borax, <1.6 mm, B = 14.85 wt%) and boron carbide (90 – 125 μm, 94% of purity, B = 73.56 wt%).

Boron-containing aggregate was used as a partial replacement for quartz sand in Portland cement mortar (CEM I 42.5 R). The boron aggregate content was selected in such a way as to obtain the required amount of boron in relation to the cement mass. Mortar samples were prepared with constant water to cement ratio w/c = 0.5 and aggregate to cement ratio a/c = 0.75.

2.2 Testing methods

Isothermal calorimetry

For the determination of rate of heat generation and total heat generated during cement hydration process an isothermal calorimetry method was used. The benefits and limitations of calorimetric methods are known from numerous studies, e.g. [9,10]. Tests were carried out on Calmetrix I-Cal 2000 HPC isothermal calorimeter (23°C, 72 h). On Fig. 1 the calorimetric measurement stand has been presented. Mortar samples are placed in channels, where the difference between rate of heat generated by the mortar and the reference sample is measured. As a reference sample, material in which heat is no generated is chosen (in this case aluminium samples).

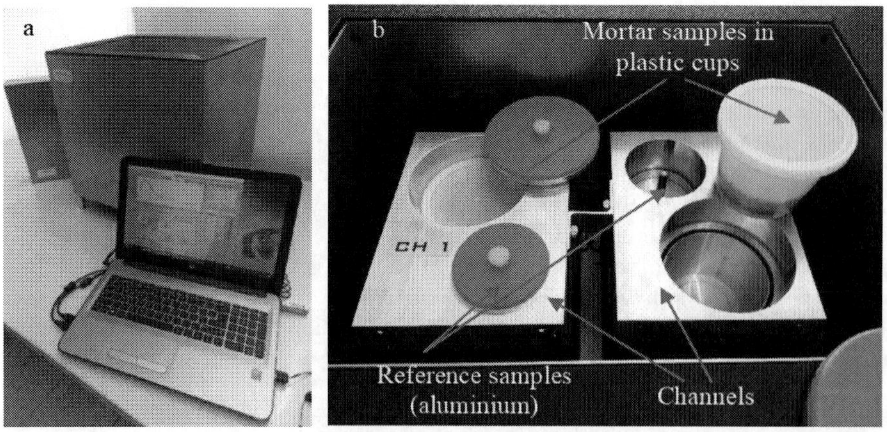

Fig 1. a) Isothermal calorimeter Calmetrix I-Cal2000 HPC, b) interior of the calorimeter chamber.

Compressive strength tests

According to the standard EN 196-1, compressive strength tests were carried out. Tests were performed on the 40 x 40 x 160 prisms after 72 h and after 28 days of hardening. Three tests were made for each mortar.

3 Results and discussion

Results from isothermal calorimeter tests are shown on Fig. 2 a-d. On graphs a-c the rate of heat generation for four different boron-containing aggregates is presented with the same percentage content of boron (2.5, 5.0, 10.0 wt% of B) in relation to the cement mass.

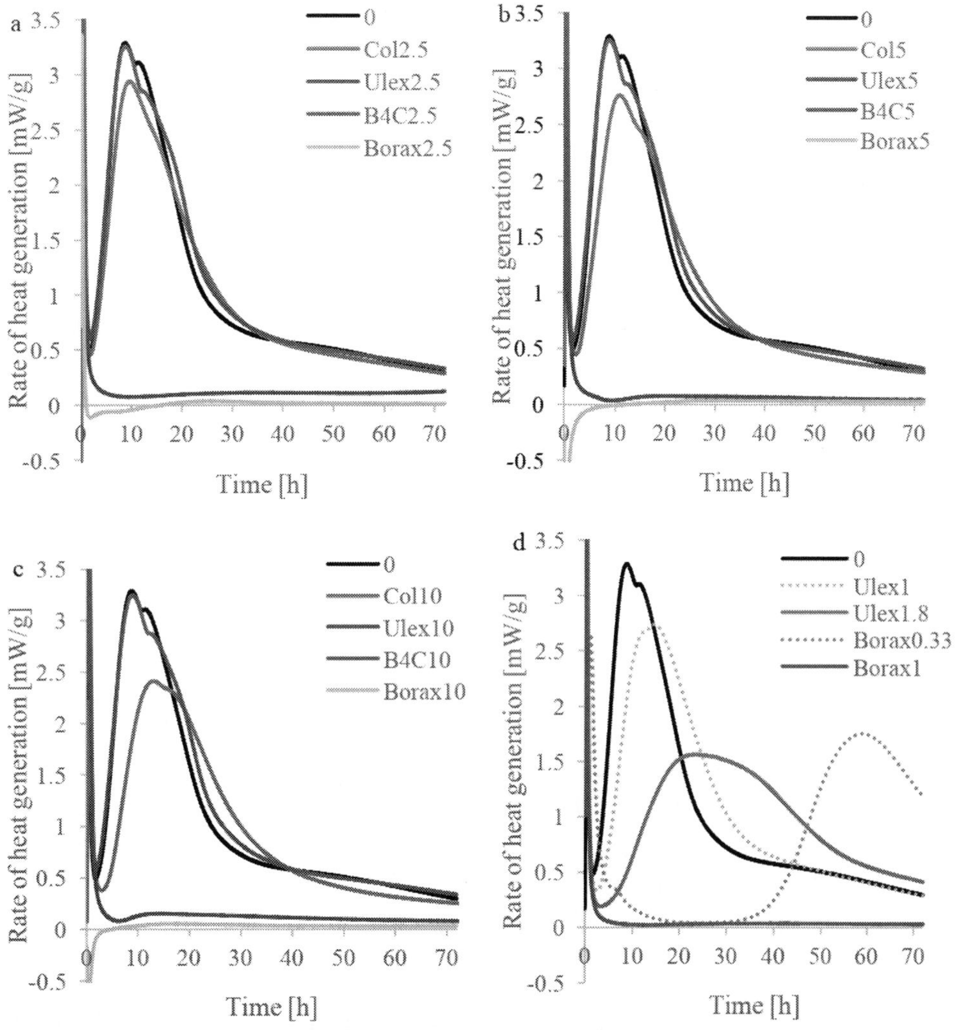

Fig. 2. The rate of heat generated per 1 g of cement during hydration of cement in mortar with different boron-containing aggregates (colemanite, ulexite, boron carbide, borax): (a) 2.5% (b) 5% (c) 10% of boron per cement mass, (d) 1 and 1.8% of boron per cement mass in the form of ulexite and 0.33 and 1% of boron in the form of borax, 0 - mortar without addition of boron-containing aggregate.

In the case of colemanite, it was noticed that along with the increase of aggregate content in the mortar, the maximum is lowered and the calorimetric curve shifts towards longer times.

however, the hydration and hardening process with high boron content (10%) is still quite good. Boron carbide is much better presented, whose calorimetric curve runs almost as in the reference mortar (without boron additives). Two more minerals: borax and ulexite, affect very negatively on the process of hydration. In graphs with boron contents from 2.5 to 10%, the calorimetric curves are flat, mortars showed no setting. In the case of borax (above 5% of B), heat is initially taken instead of being generated (endothermic process). Due to the strongly retarding effect of ulexite and borax, additionally tests were carried out with lower contents of these minerals, Fig. 2 d. Slight delay of the hydration process is observable at 1% B in the form of ulexite, 1.8% B causes stronger changes while the addition of 0.33% B in the form of borax causes a delay of the process to almost 40 hours.

Using calorimetric curves, the initial time of setting for particular mortars was determined. Two methods proposed by Ge et al. [11] and Bobrowicz [12] and presented in [13] were used. The first method is based on the first derivative of the function of the rate of heat generation in time (its maximum), the second method is based on the point of intersection of tangent to the minimum and tangent to the point of inflection. Results are presented on Fig. 3.

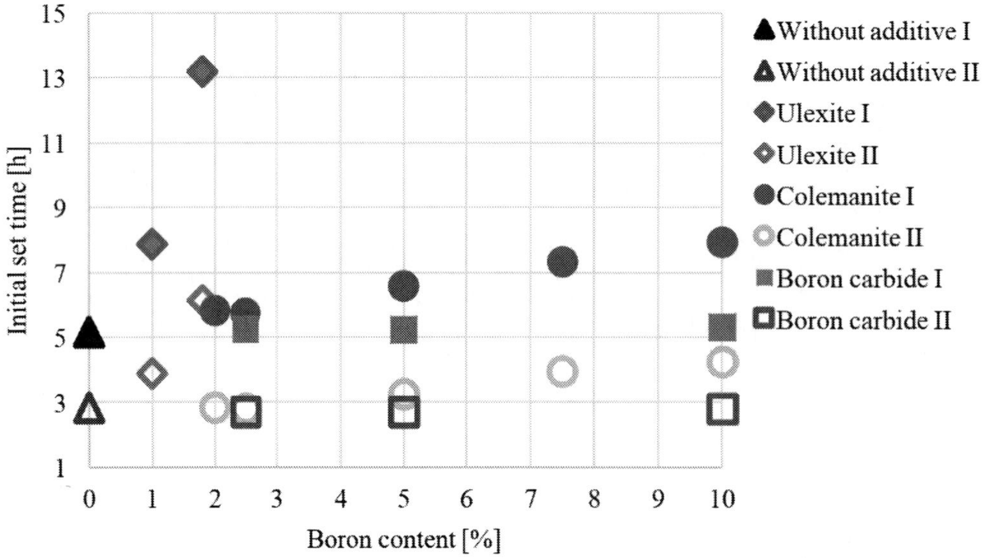

Fig. 3. Initial setting time determined by two methods: I (the first derivative) and II (intersection of tangents) as a % of boron content in relation to cement mass in mortars.

The setting times determined by these two methods differ almost twice between each other. However, qualitatively the results are similar. The second method (intersection of tangents) seems to be more suitable for these materials, taking into account Vicat measurements of the initial setting time of the reference mortar which is 2.8 h. Along with the increase of boron content in the mortar, the initial setting time rises. Only the addition of boron carbide does not change the initial setting time. With the addition of 0.33% of B in the form of borax, the initial setting time is 34.5 h by method I and 39 h by method II. To increase the readability of the results, in Fig. 4, the initial setting time of mortar with borax was intentionally omitted.

For the compressive strength tests, samples with the optimal boron content for each aggregate were selected (Ulex1.8, Kol5, B4C10, Borax0.33). The best strengths after 72 h of hardening were found in boron carbide (28.4 ± 1.2 MPa) and colemanite (20.8 ± 0.2 MPa) samples compared to the reference mortar (25.4 ± 0.8 MPa). Due to the long delay of the initial setting time of mortar with ulexite (10.3 ± 0.6 MPa) and borax (1.1 ± 0.6 MPa), they had much lower strength. 72h compressive strength constituted about 60% of 28-day strength in the case of reference mortar, and mortars with boron carbide and colemanite. For mortars with ulexite and borax the increase was much bigger due to low early strength. Regardless of the aggregate used, the 28-day strength of mortars with boron-bearing additives was lower than the reference mortar.

In Fig. 4 the relationship between the heat of hydration measured by isothermal calorimetry and the compressive strength obtained from tests according to EN 196-1 is shown. There is a clear increase in compressive strength of mortar with the increase in generated heat. The points are arranged along a straight line, so it can be considered that the calorimetric measurements allow for predicting the early strength of mortar with fairly good accuracy.

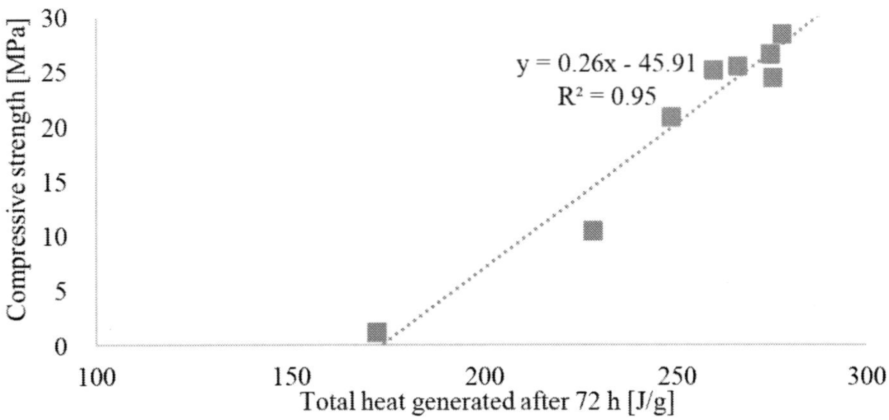

Fig. 4. Compressive strength of mortar with boron containing additives after 72h of hardening as a function of hydration heat of cement released within 72 hours, determined on the basis of isothermal calorimetry.

The applicability of isothermal calorimetry for studying effects of boron-bearing aggregates in mortar has been confirmed. Considering numerous challenges associated with the design of concrete for radiation shielding structures [14] the method is supposed to be effectively supporting the selection of components of concrete to avoid its uncontrolled early hardening. This is needed to ensure the homogeneity of concrete not to weaken its radiation shielding capacity.

4 Conclusions

The investigation has shown that the addition of boron-containing aggregates into cement mortars significantly changes their functional properties.

The largest delaying effect on the initial setting time of cement had borax and ulexite, in greater quantities, they inhibited cement hydration process. Colemanite slightly delayed cement setting, while boron carbide did not cause any effects.

The linear relationship between heat generated after 72 h of hydration determined by isothermal calorimetry and compressive strength after 72 h of hardening was observed.

The isothermal calorimetry may be an appropriate tool for assessing the compatibility of aggregates with a cement matrix.

The financial support from the National Centre for Research and Development in Poland is gratefully acknowledged (Project PBSII/A2/15/2014 "Durability and efficiency of concrete shields against ionizing radiation in nuclear power structures").

References

1. R.G. Jaeger, E.P. Blizard, A.B. Chilton, M. Grotenhuis, A. Hönig , T.A. Jaeger, H.H. Eisenlohr (Eds.), *Engineering Compendium on Radiation Shielding, Volume 2: Shielding Materials* (Springer-Verlag Berlin Heidelberg, 1975)
2. M.F. Kaplan, *Concrete radiation shielding: nuclear physics, concrete properties, design and construction* (Longman Scientific & Technical, Harlow, England, 1989)
3. M. Mazgaj, T. Piotrowski, Brittle Matrix Composites **11**, 361-370 (2015)
4. A.M. Brandt, D. Jóźwiak-Niedźwiedzka, Cement Wapno Beton **18**, 1-22 (2013)
5. M.A. Glinicki, Longterm performance of concrete in shielding structures of nuclear power plants (in Polish) (IPPT PAN, Warszawa 2015)
6. M.H. Kharita, S. Yousef, M. Alnassar, Prog. Nucl. Energy. **53**, 207-211 (2011)
7. L. Wadsö, C.P. Cooper-Jensen, P.M. Bentley, Phys. Rev. Accel. Beams. **20** 1-8 (2017)
8. W. Kurdowski, *Cement and Concrete Chemistry* (Springer, 2014)
9. M. Gawlicki, W. Nocuń-Wczelik, Ł. Bąk, J. Therm. Anal. Calorim. **100**, 571–576 (2010)
10. W. Nocuń-Wczelik, P. Czapik, Constr. Build. Mater. **38**, 980-986 (2013)
11. Z. Ge, K. Wang, P.J. Sandberg, J.M. Ruiz, J. Adv. Concr. Technol. **7**, 355-366 (2009)
12. J. Bobrowicz, Cement Wapno Beton **20**, 401-410 (2015)
13. M.A. Glinicki, A. Antolik, M. Gawlicki, Constr. Build. Mater. **164**, 731-738 (2018)
14. K. Kurtis, Y. Xi, M.A. Glinicki, J. Provis, E. Giannini, T. Fu, Concr. Int. **39**, 29-35 (2017)

Influence of temperature on the properties of non-air-entrained and air-entrained cement mortars modified with different plasticisers and superplasticisers types

Beata Łaźniewska-Piekarczyk[1,*]

[1]Sielsian University of Technology, Faculty of Civil Engineering, Department of Building Materials and Process Engineering, Akademicka 5, 44-100 Gliwice, Poland.

Abstract. The paper presents the results of the research on the impact of low and high temperature on the air content and consistency of non-air-entrained and air-entrained and plasticized or superplasticized mortars according to PN-EN 480-1, which were made with CEM I cement. In case of the mortar with CEM I, various types of plasticizers and superplasticizers as well as aeration admixtures differing in chemical base were used. The analysis of the results of the research indicates that the effect of temperature influences the consistency and air-content of the mortar depending on the result of the type of plasticizing and superplasticizing admixture.

1 Introduction

The most commonly used chemical additives in building practice are admixtures modifying rheological properties of concrete mix, plasticizing (plasticisers) and significantly plasticizing (superplasticizers). These additives are different because of the various action on concrete mix. Therefore, the proper selection of these admixtures plays an extremely important role in the concrete technology [2, 3]. Admixture effectiveness, with respect to both fresh and hardened concrete, depends on variables related to the cement, aggregates, and admixtures themselves, and also depends on external factors such as air temperature. A very important issue related to the use of plasticizing admixtures, and especially superplasticizing admixtures in building practice, is the compatibility of another admixtures with cements and additives and influence of temperature on the effectiveness of their cooperation. In case of frost-resistant and hydrotechnical concretes, air-entraining admixtures are used in addition to plasticizing and superplasticizing admixtures. Air-entraining admixtures are surface-active substances consisting of a long hydrocarbon chain terminated with a hydrophilic group. Chemical bases of air-entraining admixtures are fatty acid salts, alkaline salts of wood resins and alkaline salts of sulphates and sulphonates of organic compounds. In the presence of the air-entraining admixture, evenly spaced air bubbles of 20 to 200 μm are formed during the concrete mixing. The air bubbles created a break in the

* Corresponding author: beata.lazniewska@polsl.pl

continuity of capillaries, because of which the capillarity of the water decreases and frost resistance of concrete increases. Research conducted by [4, 5] has shown that it is very important to use different plasticizing and air-entraining at the same time to be compatible with each other. That is because, the use of superplasticizing admixtures with air-entraining admixtures can affect the degree of aeration of concrete mix, and superplasticizers themselves, added in larger quantities, can aerate the mixture to a large extent [9]. The generalization and, more importantly, the prediction of the influence of temperature on the properties of mixtures on cementitious binders, especially modified with many simultaneous admixtures present in its volume, is therefore a difficult issue and requires experimental verification.

The aim of the studies analysed in the article is to evaluate the interaction of various superplasticizers, plasticizers and air-entraining admixtures at different temperatures, both reduced and elevated, due to the properties of-fresh mortar.

2 Research methodology

2.1 Properties of materials and mortar compositions

The tests were performed on reference mortar acc. to PN-EN 480-1:2014 - Admixtures for concrete, mortar and grout. Test methods. Reference concrete and reference mortar for testing. The mortar composition is shown in Table 1. Basic information on the properties of cement, air-entraining admixture, plasticizers and superplasticizers is presented in Table 2 and Table 3.

Table 1. Mortar composition

Component of mortar		Amount in grams
Cement CEM I 42.5 R		450.0
Water	w/c = 0.50	225.0
Standard sand		1350.0

Table 2. Mechanical, physical and chemical properties of CEM I 42.5 R

Properties of cement	Unit	Average value	Standard
2 Days Strength	MPa	28.0	≥ 20.0
28 Days Strength		57.2	≥ 42.5 ≤ 62.5
Start of binding	min	170	≥ 60
Volume stability acc. to Le Chatelier	mm	0.4	≤ 10
Specific surface	cm^2/g	3681	no requirements
Loss on calcination	%	2.87	≤ 5.0
Content of insoluble residue	%	0.79	≤ 5.0
Sulphate content (as SO_3)	%	2.50	≤ 4.0
Chloride content Cl^-	%	0.059	≤ 0.10
$Na_2Oeq.$	-	0.63	no requirements

The choice of the type of superplasticizer from the generic group, especially from PCE and PCP groups, was made based on the preceding preliminary tests, during which superplasticizers were eliminated, which significantly increased the air content in non-aerated mortar. In turn, the dosage of plasticizing and superplasticizing admixtures was selected so that non-air entrained mortar at 20°C had a similar slump flow diameter of approximately 21 cm.

However, the amount of aeration admixtures in the air-entrained mortar was determined based on the air content test in aerated concrete acc. to PN-EN 480-1 in the range of 5-7%. The results of this test, for the same amount of aeration admixture with respect to the cement mass, mortar of the same standard, is aerated in 10%. The dosage of admixtures with respect to the cement mass is given in Table 3.

Table 3. Characteristics of superplasticizing, plasticizing and air-entraining admixtures.

Admixture type		Main chemical base	Dosage of admixture
Superplasticizer	PCE	Polycarboxylate ether	0.220
Superplasticizer	PCP	Polycarboxylate	0.220
Superplasticizer	N	Modified naphthalene	0.110
Superplasticizer	A	Acrylate-polyethylene glycol	0.220
Superplasticizer	F	Modified phosphonates	0.220
Plasticizer	NSF	Naphthalene sulfonate	0.440
Plasticizer	MSF	Sulfonated melamine formaldehyde	0.440
Plasticizer	MLS	Maleated lignosulfonate	0.440
Air-entraining	AEA	Synthetic surface-active compounds	0.015
Reference mortar	-	-	-

2.2 The scope and method of determining the properties of mortars

Air-entrained mortars for three ranges of their temperatures, 10, 20 and 30°C were subjected to consistency and air-content tests. The study of the air content in mortars was performed using the method acc. to PN- EN 1015-7:1999: Methods of test for mortar for masonry. Determination of air content of fresh mortar. The mortar consistency test was carried out in accordance with the standard PN-EN 1015-3:1999: Methods of test for mortar for masonry. Determination of consistence of fresh mortar (by flow table).

3 Analysis of test results

The influence of temperature on the consistency of non-air-entrained mortars was analysed in the publication of Ghafoori and Diawar [10], Schmidt [11], who found that above 20 °C, the flow of non-air-entrained mixes decreases with increasing temperature. The test results are consistent with those presented in Fig. 1. There are also different reports on the behaviour of cement mortar with PCE plasticizers at various temperatures [12].

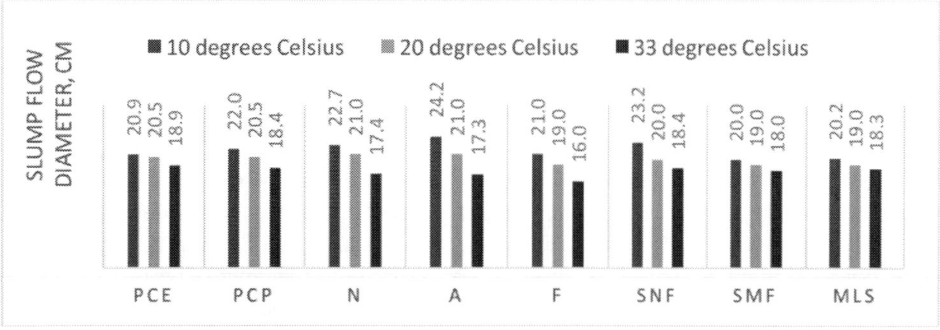

Fig. 1 Comparison of the influence of temperature on the flow diameter of non-air-entrained mortars with various types of superplasticizing and plasticizing admixtures.

Results of the research [13] indicate that polycarboxylate-based superplasticizers should be used at a dosage close to the saturation in order to ensure adequate rheological behaviour at high temperature and prolonged mixing time, while melamine sulfonate- and naphthalene sulfonate-based superplasticizers should be used at dosages beyond the saturation level. This was the reason for the doubling of the amount of plasticizers due to superplasticizers in analysed studies (Table 3).

In order to know the effect of temperature influence on superplasticizers, first you need to know their behaviour at different temperatures, especially considering their spatial structure. Water-reducing admixtures stand out among the most used in concrete. They can be composed of various active ingredients, including lignosulfonate, naphthalene, and polycarboxylate. Such admixtures cause cement grains to disperse, and thus to have a plasticizing effect—through mechanisms of electrostatic repulsion and steric hindrance [1, 2]. Electrostatic repulsion, which explains how naphthalene and lignosulfonates work, is brought about by the adsorption of an (electrostatically negative) sulfonic group onto the surface of the cement grain, giving it a negative electrostatic charge. The predominance of negative charges on the surface of the grains leads to repulsion between particles, thus releasing the water that is held between them. In case of steric hindrance, the mechanism that describes how polycarboxylates work is that the particles adsorbed onto the cement grain have long side chains, which have greater repulsive capacity because they serve as physical impediments that prevent grains from bonding. Admixture effectiveness, depends on temperature. Test results [15] also show that the increase in the ambient temperature causes a decrease in mortar flow diameter, and the amount of reduction in the flow depends on the superplasticizer's contribution to the cement mass and type of the superplasticizer or the plasticizer. The publication [14] indicates that under the influence of temperature increase, or the presence of ions from salts easily dissociable in water, there are breaks in hydrogen bonds connecting the polyether chain with water (Fig. 2). As a result, the chain becomes entangled and the steric block gradually decreases until it disappears completely.

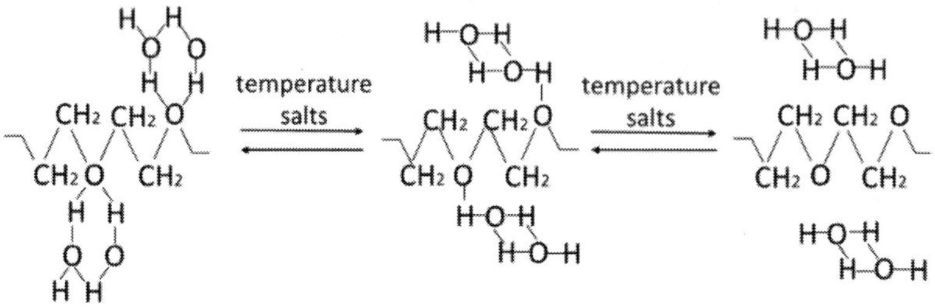

Fig. 2. Dehydration of polyether chains under the influence of temperature [14].

The results indicate that the increase in temperature leads to increase in saturation dosage; lignosulfonate had the most pronounced retarding effect, followed by polycarboxylate, and finally, naphthalene, and considering the weather conditions in the area where the study was carried, the final finding would be that the naphthalene-based admixture had the best performance [15]. Test results analysed in the paper showed that SNF mortar, among ~~the~~ other plasticizers, has the largest flow diameter at low temperature (Fig. 1), like cement paste with acrylates (A). In turn, the greatest reduction of flow at high temperature was demonstrated by the mortar with phosphonate (F). And the smallest mortar flow reduction at high temperatures corresponds to PCE and lignosulphonate (MLS) mortar. According to publication [12], there may even be the increase in the flow diameter of self-contaminated concrete from PCE at a higher temperature relative to the lower one.

In case of non-aerated mortars, the increase in temperature causes their reduction of the flow diameter (Fig. 1), and this in turn causes the increase in its aeration (Figs 3 and 4).

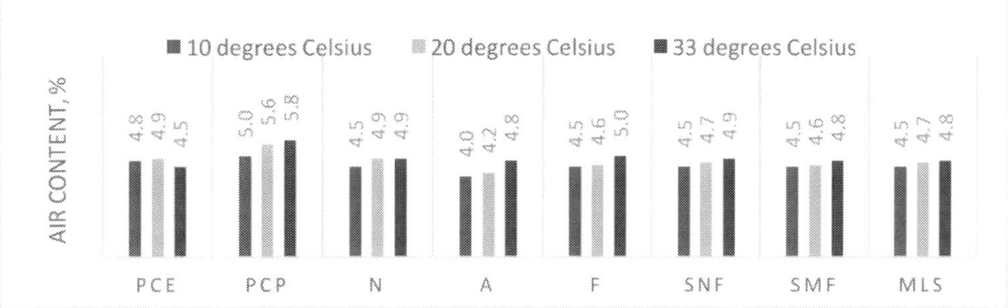

Fig. 3. Comparison of the influence of temperature on the air-content of non-air-entrained mortars with various types of superplasticizing and plasticizing admixtures.

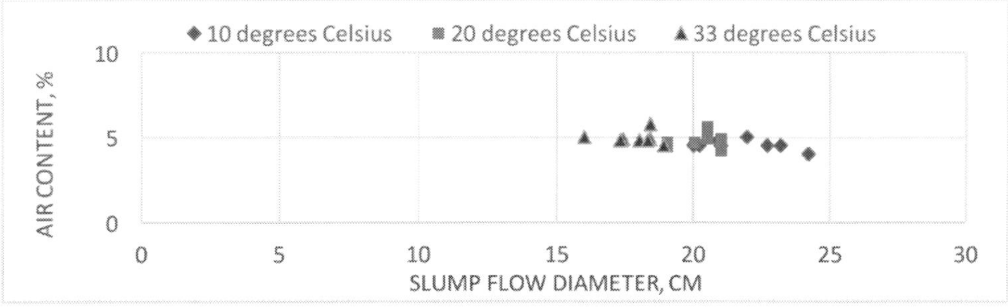

Fig. 4. Comparison of the relationship between the air-content in non-air-entrained mortars of different temperature and their flow diameter.

The influence of temperature on the consistency of air-entrained mortars, with various plasticizing or plasticizing admixtures, was not previously analysed in literature, in such scope as in the article. As shown by the results of the tests shown in Figs. 5 and 6 the impact of the superplasticizer and plasticizer type on the properties of air-entrained cement mortar at various temperatures is also important. The analysis of the test results indicates that the temperature rise always causes a reduction in the flow of the aerated and liquefied mortar, regardless of the type of admixture (Fig. 5). The type of plasticizing and superplasticizing admixture also strongly disturbs the relationship between the air content in mortar and its consistency in different temperatures (Fig. 7).

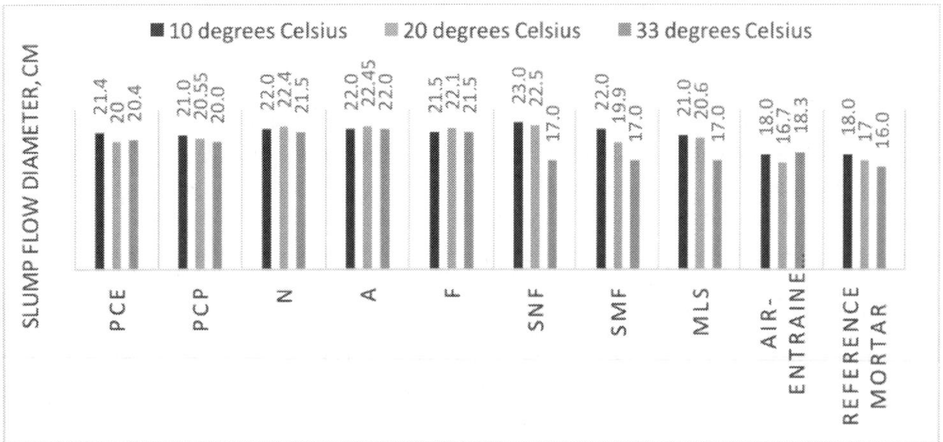

Fig. 5. Comparison of the influence of temperature on the flow diameter of air-entrained mortars with various types of superplasticizing and plasticizing admixtures.

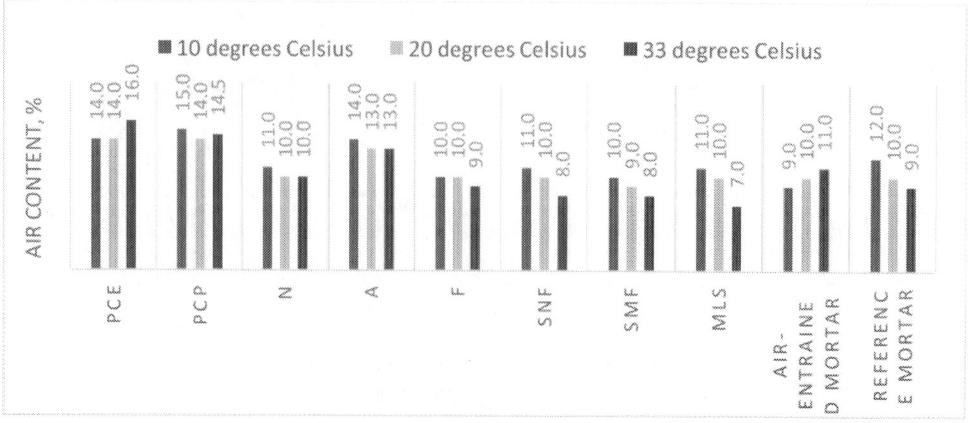

Fig. 6. Comparison of the influence of temperature on the air-content of air-entrained mortars with various types of superplasticizing and plasticizing admixtures.

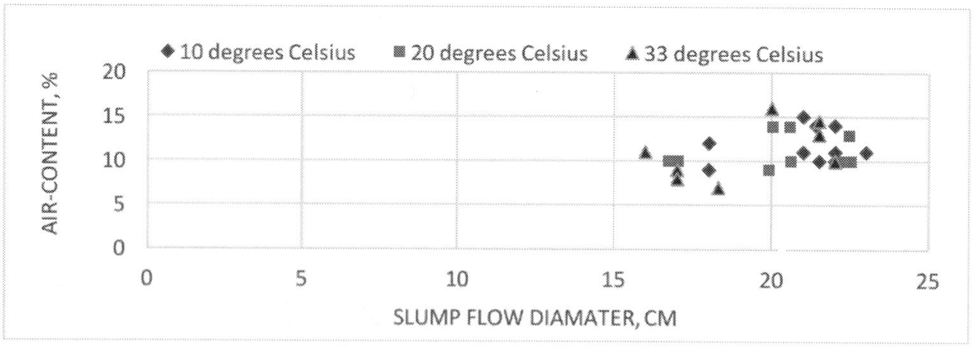

Fig. 7. Comparison of the relationship between the air-content in air-entrained mortars of different temperature and their flow diameter.

However, the air content and diameter of the flow of air-entrained mortars, with the increase of temperature, depending on the type of superplasticizing admixture, slightly increases or decreases, or does not change at all (Fig. 6). However, in case of superplasticizers, the reduction in the diameter and air-content of the mortar is less as the temperature increases than in case of plasticizers (Fig. 5). In addition, the size of the change in the flow and air-content of the mortar in the temperature range of 20-30 deg. C and 10-20 deg. C depends on the type of plasticizer and superplasticizer, which is consistent with the results of the studies [15].

Interestingly, in case of some PCE and PCP superplasticisers, the flow diameter and air content in the mortar can increase as the temperature increases. Similarly, in case of aerated mortar with lignosulphonate (MLS) the increase in temperature causes the increase in the air content, but no longer in the diameter of the flow. The reason for this is the strong influence of lignosulfonate plasticizer on the surface tension of water, as well as some PCE and PCP superplasticisers. The mechanisms of SP action include dispersion of cement particles by reduction in surface tension of mixing water and ~~a~~ decrease in frictional resistance because of the line-up of linear polymers along the concrete flow direction and lubrication properties produced by low molecular weight polymers [8]. The presence of functional groups (oxygen in form of etheric group (-O-), hydroxyl group (-OH) and carboxyl group) produce water surface tension decrease, producing flocculation of associated molecules and increase in moisture of not only grains of cement but also the whole mineral framework (Fig. 8).

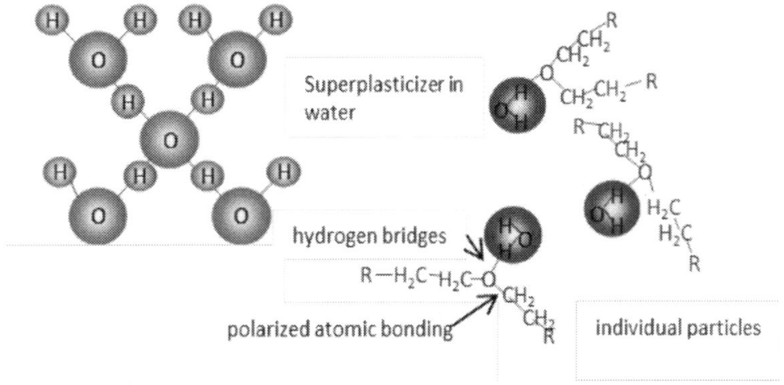

Fig. 8. Water conglomerates deflocculating under the influence of superplasticizer as the effect of surface tension diminution [16].

In case of high temperature of mortar, PCE and MLS admixtures reduced the flow diameters of the non-aerated mortar the least and caused the increase in the air-content of aerated mortar in case of PCE. The surface tension of aqueous solutions of some PCP and PCE superplasticizers is very low, almost like the admixtures. What's more, when such a superplasticizer is added to the air-entrained mortar, the surface tension of water decreases even more [17], and the increase of water temperature causes an even greater reduction of this value. The degree of reduction in surface tension corresponds to the air-entraining admixture [17]. Hence, the effect of increasing the flow of air-entrained mortars containing PCE and PCP and with lignosulfonate as the temperature increases. The surface tension, due to intermolecular interactions, is highly dependent on temperature. The dependence of the surface tension of the liquid on temperature is described by Eötvös (1886) equation [0]:

$$\sigma\left(\frac{M}{d}\right)^{\frac{2}{3}} = K(T_k - T - \delta) \tag{1}$$

where:
σ - surface tension of the solution,
M - is the molar mass,
D - density of liquid (hence M / d - is a volume of one mole),
T_k – critical temperature,
δ - is a correction of about 6 deg (it turns out that the disappearance of surface tension occurs at a temperature slightly lower than T_k).

4 Conclusions

The effect of the temperature influence on the behaviour of air-entrained mortar, modified with superplasticizing or plasticizing admixtures, is variously directed, depending on the type of admixture and associated mechanism of the influence of admixture on cement paste. In the analysed area of the test results, it is concluded that:

- Mortar with SNF, among other investigated plasticizers SMF and MLS, has the largest flow diameter at low temperature, like cement paste with acrylates. In turn, the greatest reduction of flow at high temperature was demonstrated by the mortar with phosphonate. And the smallest mortar flow reduction at high temperatures corresponds to PCE and lignosulphonate mortar.
- The increase in temperature of the mortar slightly increases the flow diameter of the air-entrained mortar, i.e. for some superplasticisers, i.e. highly reducing the surface tension

of water, or reducing the mortar flow diameter, for other superplasticisers and all plasticizers.

- The increase in temperature of air-entrained and liquefied mortar causes various changes in its air-content: no change, or slightly increased air-content of the mortar (in case of superplasticizers with a greater impact on surface tension of water), and finally lowering (in case of other superplasticisers and all plasticizers, except for lignosulphonate where the air content increases).

References

1. P.C. Aitcin, Cement Wapno Beton **5** (2006)
2. L. Kucharska, Cement Wapno Beton **2** (2000)
3. W. Kurdowski, Cement Wapno Beton **5**, 296-305 (2010)
4. S. Hanehara, K. Yamada, Cem. Concr. Res **29** (1999)
5. S. Erdogdu, Cement Wapno Beton **30** (2000)
9. B. Łaźniewska-Piekarczyk, Cement Wapno Beton **3**, 133–145 (2009)
10. N. Ghafori, H. Diawara, Constr. Build. Mater. **24**, 946-955 (2010)
11. W. Schmidt, *Design concepts for the robustness improvement of self-compacting concrete – Effects of admixtures and mixture components on the rheology and early hydration at varying temperatures* (Eindhoven: Technische Universiteit Eindhoven, 2014).
12. G. Cygan, J. Gołaszewski, M. Drewniok, The effect of temperature on the properties the self-compacting concrete, ACE **LXII**(3), 25-32 (2016)
13. S. Al Martini, Ph.D., M. Nehdi, JMCE **21**(8) (2009)
14. S. Grzeszczyk, M. Sudoł, Cement Wapno Beton **6**, pp. 325-331 (2003).
15. W.R.L. Silva, L.R. Prudêncio Jr., A.L. Oliveira, G. Damo, E. Tochetto, ACE **2010,** 10 (2010)
16. T. Rudnicki, Autostrady **4**, 22-25 (2004)
17. J. Szwabowski, B. Łaźniewska-Piekarczyk, Cement Wapno Beton **4**, 205–215 (2008)
18. K. Pigoń, Z. Róziewicz, *Physical chemistry* (PWN, Warszawa, 6, 2018)

Effect of redispersible polymer powder on setting time of thin-bed mortars

Marcin Kulesza[1,*], *Dawid* Dębski[1], and *Jadwiga* Fangrat[2]

[1]Atlas sp. z o. o., Kilińskiego 2, Łódź, 91-421, Poland
[2]Instytut Techniki Budowlanej, Filtrowa 1, Warszawa, 00-611, Poland

Abstract. The aim of work is to describe and to complement the existing knowledge on the effect of redispersible polymer powder on the setting time of cement mortar as measured by means of ultrasonic apparatus. The mortars were simple mixtures of filler in the form of quartz sand, Portland cement CEM I 42.5 R as binder and methyl hydroxyethyl cellulose. The mortar additives were two commercially available redispersible polymer powders in an amount of 0-5% of total weight. The chosen polymers had different chemical structure: styrene-acrylic copolymer (SA), vinyl acetate-ethylene copolymer (ET), and minimum film forming temperature (MFFT). The amount of water was constant and equal to 20% of the total weight of the mortar. Conclusions: - even the small amount of redispersible polymer powder added to the cement mortar has a significant influence on mortar's initial and final setting time and this effect strongly depends on the percentage of the additive, - the chemical structure of the studied redispersible polymer powders did not influence the behaviour of the whole chemical system, - the further studies on technical properties, such as mechanical strength and permeability are planned in order to find more information on the structure of hardened mortar.

1 Background

The consequence of development in architecture and construction engineering is the constant increase of technical requirements for construction products, including thin-bed mortars based on cement. The increase of the requirements regards mainly technical parameters such as mechanical strength or durability for which the expectations are presented in the standards. Fulfilment of these high requirements by building mortars, which are simple mixtures of cement binder and mineral filler, is impossible for many applications. Therefore, the modification by addition of chemical additives and polymer binders is necessary. The basic chemical additives, significantly effecting the achievement of technical parameters of cement mortars, are modified natural polymers, mainly cellulose ethers and synthetic polymer binders frequently in form of redispersible polymer powders (RDP).

* Corresponding author: marcin.kulesza@atlas.com.pl

The effect of above mentioned chemical additives on technical parameters of mortars, mechanism of their operation, and their mutual interaction and compatibility is the subject of research since the half of XX century.

Despite numerous studies published on this subject, significant research effort and development, there are still gaps in knowledge, particularly the mechanisms of processes that occur in the modified cement mortars are not precisely described. Application of chemical additives is based rather on macroscopic observations and empirical evaluation of their influence on the technical parameters. For this reason, each analysis of the effect of single chemical additive on the performance of building mortars is a valuable complementation of knowledge and has high application potential.

Recently, many studies were made on influence of cellulose ethers on the properties of cement mortars. The effect of cellulose ethers on increase of water retention, improvement of consistence, increase of chemical resistance, and improvement of some mechanical properties (for instance adhesion) was observed [1-3]. At the same time, the presence of cellulose ethers in the cement mortar significantly increases its setting time [2, 4]. There are several published hypothesis describing this phenomena. Among explanations it is proposed, that this delay is caused by decreasing mobility of ions in the mortar containing cellulose ethers, which affects the hydrolysis process and precipitation of structural constituents of the solution [5].

1.1 The research problem

Another additives widely used in building industry are polymers playing mainly the role of complementary binder. The most frequently used polymers are ethylene-vinyl acetate copolymer (EVA), styrene-acrylic copolymer (SA), vinyl acetate-ethylene copolymer (ET), styrene-butadiene copolymer (SB), and also some homopolymers, and more frequently terpolymers. Usually, they are used in the form of redispersible polymer powders (RDP), which are water polymeric dispersions subjected to the process of spray-drying. This process is reversible – as a result of repeated contact with water, redispersible polymer powders form the system having properties very similar to the primary dispersion. There is evidence that, similarly to cellulose ethers, the redispersible polymer powders influence the properties of building mortars having also the long-term effect. Usually, the presence of polymer binder causes positive effects i.e. increase of water retention, adhesion and flexural strength, decrease of permeability, as well as improvement of workability. The exemplary negative effect of RDP additive is the decrease of compressive strength probably due to the increased porosity of the mortar [6-7].

The present work aims to describe and to complement the existing knowledge on the effect of redispersible polymer powder on the setting time of cement mortars as measured by means of ultrasonic apparatus by example of two chosen commercially available polymeric powders.

2 Materials and methods

The experiments were performed in the Research and Development Laboratory of Atlas Company. The research programme included the measurement of setting time of representative cement mortar modified with two commercially available redispersible polymer powders as additives. The data for the modified mortars were collected and analysed with comparison to the adequate data obtained for the non-modified mortar.

One of the components of studied mortar was methyl hydroxyethyl cellulose, which may cause delay in the process of cement setting, but at the same time it is necessary in the system to ensure the proper water retention and workability of the mortar. Therefore, the influence of 0.3% MHEC on the setting time of a mixture of quartz sand and cement alone, was also determined.

Ultrasonic method of the setting time determination was used in the study. This method, based on the analysis of the velocity of ultrasonic wave propagation in hardening mortar, was chosen as it exhibits multiple advantages over typical setting time determination (with Vicat's apparatus). The measurement of setting time by ultrasonic method can be conducted without the destruction of the sample, what allows for performing the analysis in a continuous manner. High reproducibility, accuracy and resolution are key advantages of ultrasonic assay comparing it with other methods of setting time determination. Thus, even the subtle changes in initial and final setting time of the modified mortar can be observed.

2.1 Materials

The studied mortars were simple mixtures of filler in the form of quartz sand (Table 1), Portland cement CEM I 42.5 R (Table 2) as binder and methyl hydroxyethyl cellulose in order to obtain basic workability and water retention. The mortar additives were two commercially available redispersible polymer powders in an amount of 0-5% of total weight. The chosen polymers had different chemical structure: styrene-acrylic copolymer (SA) and vinyl acetate-ethylene copolymer (ET) and minimum film forming temperature (MFFT, 0 °C and +5 °C for ET and SA polymer powder respectively). Details of all studied mortars are presented in the Table 3. The amount of water was constant and equal to 20% of the total weight of the mortar.

Table 1. The properties of studied quartz sand according to PN-85/H-11001.

Main fraction	0.20/0.315/0.16
Average grain size D_{50}	0.25 mm
Sintering temperature	1550 °C
Humidity	< 0.5%
Chemical analysis	SiO_2 > 99% Fe_2O_3 < 0.07% carbonates < 0.08%

Table 2. The properties of studied cement according to EN 197-1.

Initial setting time	213 min
Final setting time	276 min
Early strength (N2)	31.6 MPa
Standard strength (N28)	58.7 MPa
Volume stability (LeChatelier)	0.75 mm
Surface area acc. Blaine	3786 cm^2/g
Loss on ignition	2.91%
Insoluble residue	0.79%
Content of sulphates	3.00%
Chloride content	0.065%

Table 3. The composition of studied mortars.

Composition (in %)	00	0	1ET/ 1SA	2ET/ 2SA	3ET/ 3SA	4ET/ 4SA	5ET/ 5SA
CEM I 42.5R	30	30	30	30	30	30	30
Quartz sand 0.1-0.5 mm	70	69.7	68.7	67.7	66.7	65.7	64.7
Methyl hydroxyethyl cellulose (MHEC)	0	0.3	0.3	0.3	0.3	0.3	0.3
Redispersible polymer powder ET or SA	0	0	1	2	3	4	5

2.2 Testing method

In the course of experiments two series of setting time measurements of mortars were performed by means of the apparatus for measuring the velocity of ultrasonic wave propagation for:

(1) the primary mortar (mixture of sand and cement)
(2) the reference mortar (mixture of sand and cement and MHEC)
(3) the modified mortars (samples mortars - reference mortars with addition of different type and amount of redispersible polymer powder)

The kinetic traces of mortar setting were obtained by constant measurement of velocity of ultrasonic wave propagation between the transmitter and emitter of the UltraSonic Multiplex Tester IP-8 (Ultratest GmbH, Germany). The mortar was placed in a 40 mm diameter cylinder connected to the apparatus. The initial and final setting time of the mortar was identified as the soundwave velocity reached 100 ± 1 m/s and 400 ± 1 m/s respectively.

3 Results

Initial setting time should be sufficiently long in order to ensure the possibility of work with the material during its application. On the other hand, the final setting time, and the strength growth should be as short as possible not to delay the subsequent technological phases.

3.1 Effect of methyl hydroxyethyl cellulose and RDP on setting time of cement mortar

The comparison of kinetic traces for the „00" and „0" samples unequivocally shows the influence of methyl hydroxyethyl cellulose on retardation of setting of cement mortars (Figure 1). Observed retardation and its level are in agreement in the observations by other research teams. Addition of both the MHEC (0.3%) and RDP (1% of SA or ET powder) to the mortar results in further retardation of the setting. However the effect of retardation of RDP, regardless the increased dosage is significantly lower than of the MHEC.

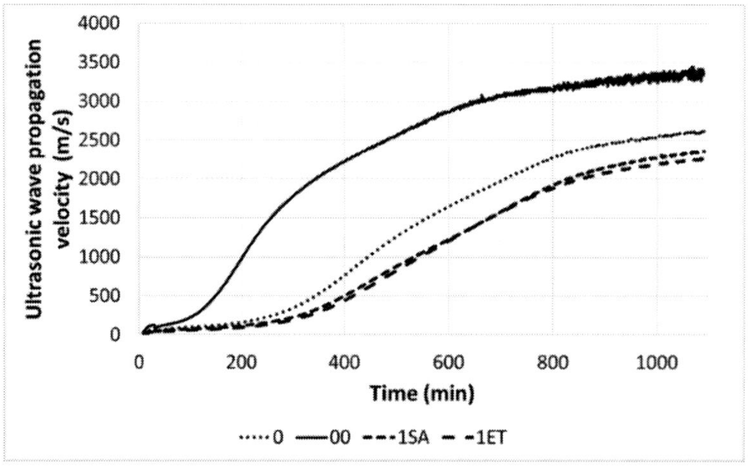

Fig. 1. Comparison of the ultrasound wave velocity during hardening of tested mortars: 00 - without any additives, 0 – containing 0.3% MHEC, 1SA – containing 0.3% MHEC and 1% SA polymer powder, 1ET – containing 0.3% MHEC and 1% ET polymer powder.

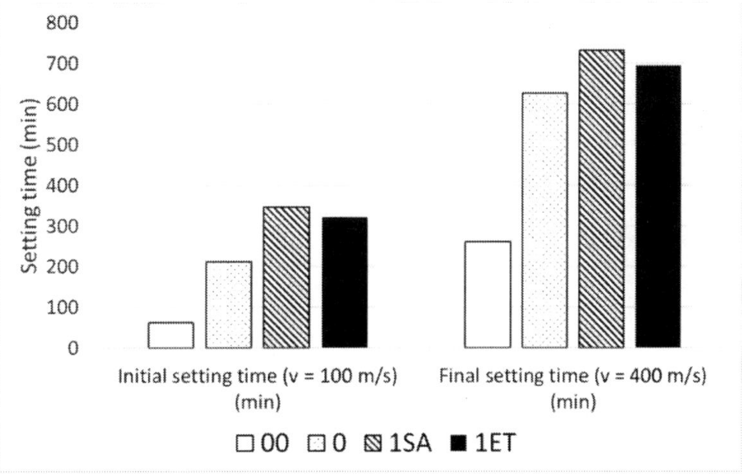

Fig. 2. Initial and final setting times of the tested mortars. The times were determined as periods after which the propagation velocity of ultrasound wave has reached the indicated value.

The comparison of initial and final setting time of the tested mortars has been shown in Figure 2. The presence of the MHEC alone caused approx. 3.4-fold prolongation of initial setting time respectively. The presence of 1% of styrene-acrylic RDP caused additional retardation of setting time which was equal to 1.6-fold of the reference mortar. RDP based on vinyl acetate caused the similar effect (1.5-fold retardation). The retardation is also visible for the final setting time – the addition of MHEC resulted in approx. 2.4-fold retardation of setting time; including the 1% of styrene-acrylic RDP in the sample lead to 1.2-fold further prolongation of the final setting time compared to 1.1-fold increase of setting time for 1% of vinyl acetate polymer powder. Although both MHEC and RDP affected the setting of cement mortar, they have more significance in modification of initial than final setting time.

3.2 Effect of RDP concentration on initial and final setting time of cement mortar

The setting time of a series of cement mortar samples containing different RDP levels (from 1-5%) was also determined in the same experimental setup. The experimental results of measured velocity of ultrasonic wave propagation have been presented graphically on Figure 3. The retardation effect of the redispersible polymer powder depends strongly upon the concentration of the RDP in the cement mortar. This effect is in agreement with observations of other research teams which performed similar experiments but with different methodology and apparatus [8-10]. The similar observation was made for both styrene-acrylic and vinyl acetate-ethylene copolymers used in study.

The analysis of the initial and final setting time retardation, presented as a ratio of the setting time of the sample with RDP (t) and reference sample (t_0) has been shown in the Figure 4 and Table 4. In both experiments the retardation effect of the RDP depended linearly upon the concentration of polymer powder. The presence of polymer in the mortar had more distinguishable influence on the initial setting time, than on the final setting time. The influence on the initial setting time varied from 1.6-fold increase for 1% of styrene-acrylic RDP to almost 2.2-fold increase for 5% of the same RDP. Simultaneously, the final setting time was increased only 1.2 times when 1% of SA RDP was present, while the sample containing 5% of RDP had the setting time increased 1.3 times. Vinyl acetate-ethylene copolymer has shown nearly same retardation effect. Similar results were observable in the experiment described in the section 3.1 with MHEC addition.

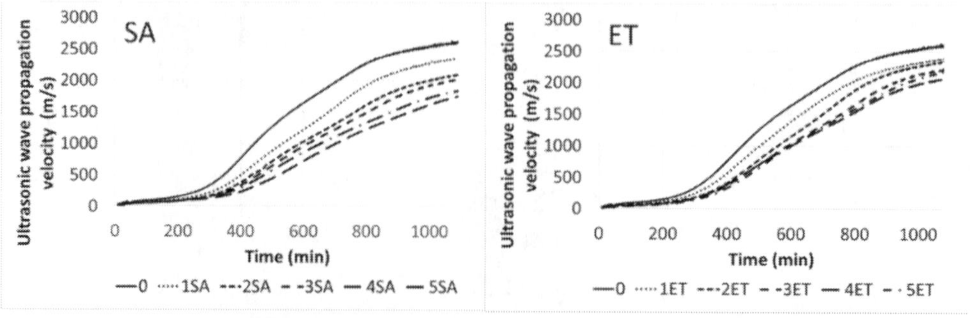

Fig. 3. The velocity of ultrasonic wave propagation as a function of type and amount of redispersible polymer powder in the mortar.

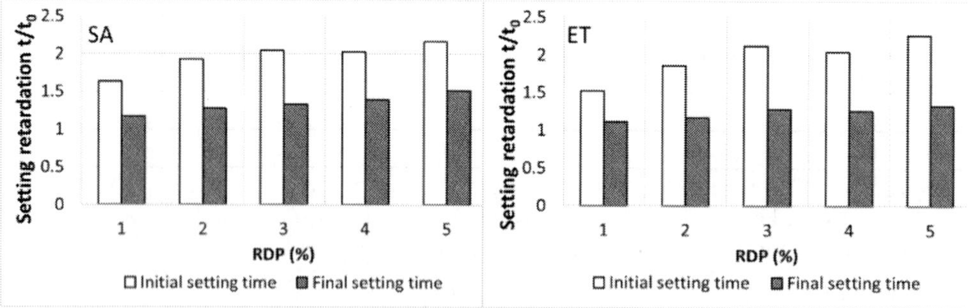

Fig. 4. Effect of delay of initial and final setting time as a function of the content of redispersible polymer powder in the mortar.

Table 4. The setting time retardation (t/t_0) caused by the addition of RDPs used in the study

Polymer content (in %)	Setting time retardation (t/t_0)			
	Initial setting time		Final setting time	
	SA	ET	SA	ET
1	1.5	1.6	1.2	1.1
2	1.9	1.9	1.3	1.2
3	2.1	2.0	1.3	1.3
4	2.0	2.0	1.4	1.3
5	2.3	2.2	1.5	1.3

4 Conclusions

1) Despite numerous studies published so far on the effect of redispersible polymer powders on the setting time of cement mortars, there are still gaps in the existing knowledge.

2) Even the small amount of redispersible polymer powder added to the cement mortar has a significant influence on its initial and final setting time and this effect strongly depends on the percentage of the additive.

3) The chemical structure of the studied redispersible polymer powders does not influence the behaviour of the whole chemical system.

4) The further studies on technical properties, such as mechanical strength and permeability are planned in order to find more information on the structure of hardened mortar.

References

1. A. Beeldens, D. Van Gemert, H. Schorn, Y. Ohama, L. Czarnecki, Mater. Struct. **38**, 601 (2005)
2. A.M. Betioli, J. Hoppe Filho, M.A. Cincotto, P.J.P. Gleize, R.G. Pileggi, Constr. Build. Mater. **23**, 3332 (2009)
3. K.H. Khayat, Cem. Conc. Compos. **20**, 171 (1998)
4. X. Kong, S. Emmerling, J. Pakusch, M. Rueckel, J. Nieberle, Cem. Conc. Res. **75**, 23 (2015)
5. Ł. Kotwica, J. Małolepszy, Cem. Wapno Beton **29**, 282 (2009)
6. Y. Ohama, Cem. Conc. Compos. **20**, 189 (1998)
7. L. Patural, P. Marchal, A. Govin, P. Grosseau, B. Ruot, O. Devès, Cem. Conc. Res. **41**, 46 (2011)
8. J. Pourchez, P. Grosseau, R. Guyonnet, B. Ruot, Cem. Conc. Res. **36**, 1777 (2006)
9. J. Schulze, O. Killermann, Cem. Conc. Res. **31**, 357 (2001)
10. D.A. Silva, P.J.M. Monteiro, Cem. Conc. Res. **36**, 1501 (2006)
11. N.K. Singh, P.C. Mishra, V.K. Singh, K.K. Narang, Cem. Conc. Res. **33**, 1319 (2003)
12. G. Bundyra-Oracz, W. Kurdowski, Mater. Construcc. **61**, 227 (2011)

Hydrophobisation of mortars containing waste polyurethane foam

Danuta Barnat-Hunek[1], *Sylwia* Duda[2,*], *Monika* Garbacz[2], and *Grzegorz* Łagód[2]

[1]Faculty of Civil Engineering and Architecture, Lublin University of Technology, str. Nadbystrzycka 40, Lublin 20-618, Poland
[2]Faculty of Environmental Engineering, Lublin University of Technology, str. Nadbystrzycka 40a, Lublin 20-618, Poland

Abstract. This paper investigates the physical and mechanical properties of mortars modified with waste polyurethane foam (WPF) and the influence of hydrophobising agents on the physical characteristics. The sand was substituted with WPF (2-4 mm grain size) in the mortars which consisted of 5, 10 and 15% of foam, respectively. The contact angle of lightweight mortars was determined (θ_w) in the function of time, prior to and after the frost resistance test. The surface free energy characterizing the wettability and adhesion of mortars under normal conditions and following frost erosion was calculated with Neumann's method based on the obtained data. The structure of mortars, adhesion of lightweight aggregate to cement paste and the structure of thin hydrophobic film were demonstrated by means of scanning electron microscopy. The mortars subjected to hydrophobisation process revealed a slight mass change caused by freezing and thawing processes: 0.1% for methyl silicone resin, 2.1% for alkyl-alkoxy-silane, and 9.2% for the samples which were not hydrophobized. On average, the contact angle of the standard mortars was 3 times lower than the one of hydrophobic material. The best results illustrating the efficiency of hydrophobisation were obtained for methyl silicone resin.

1 Introduction

Due to a growing problem of construction waste utilization, engineers are seeking possibilities of their management. One of them is recycling of energy saving waste in construction materials. A proof of the effectiveness of this solution is more common application of heat insulating mortars with the use of waste [1-3]. An example of artificial aggregate achieved from material recycling can be waste polyurethane foam (WPF) used in lightweight mortars [4-6] and lightweight concrete [7,8]. Materials containing this type of lightweight aggregate have smaller mass and strength but bigger absorbability and thermal insulation [4-8]. The application of waste polyurethane foam in heat insulating mortars considerably reduces heat transmission coefficient by the mortar. However, waste polyurethane foam can be a reason for increased wettability of mortars, hence after they have been prepared, a surface hydrophobisation should be considered. The purpose of our further research [3, 9-11] was the evaluation of the possibility

* Corresponding author: sylwia.m.duda@gmail.com

of applying hydrophobising preparations based on silica-organic compounds in the protection of lightweight mortars and concrete with construction waste, as well as water and wastewater management.

The authors would like to call attention not only to the problem of construction waste recycling, including polyurethane foam (WPF), but also to their low resistance to water and frost. The purpose of these studies is to determine physical reference properties of hydrophobised insulating mortars containing recycled foam, because the literature lacks information on this subject. Owing to its unique, pitted structure after having been grated, the foam has good sorption properties and can cumulate the sufficient amounts of ice and salt. The authors have confirmed in their research that the application of different hydrophobising agents causes a significant reduction of wettability and the increase of frost resistance of lightweight mortars.

2 Materials and methods

2.1 Materials

The content of 3 mineral mixtures of lightweight mortars with WPF has been prepared. In the mortars the sand was partially substituted with the foam of 0-4 mm granulation. WPF came from waste obtained during the construction of a single family house. The waste was grated in a way to achieve grains of size not bigger than 4 mm. The samples were marked as follows: M5, M10, M15 – mortars containing 5, 10, 15 % WPF respectively. Table 1 presents the percentage content of mortars with addition of utilised foam in proportion by capacity.

Table 1. Composition of mortars with polyurethane foam, proportion by capacity, vol. %.

Components	M5	M10	M15
Portland cement CEM I 32.5 R	25	25	25
Polyurethane foam 0/4 mm	5	10	15
Quartz sand 0/4 mm	15	10	5
Hydrated lime	25	25	25
Water	30	30	30
Vinyl acetate copolymer	0.15	0.15	0.15

The hydrated lime parameters were in line with the requirements of PN-EN 459-1:2015-06 [26] and the value of its apparent density was 390 $kg{\cdot}m^{-3}$. The composition of the lime, in terms of oxide, was as follows: CaO – 95.5%, CO_2 – 2.1%, MgO – 0.5%, SO_3 – 0.1%, free water – 1.5%. The Portland cement CEM I 32.5R was developed according to the Polish standards PN EN 197-1:2012 [27] and PN-B-19707:2013-10 [28]. CEM I 32.5R that was used had the following parameters: a specific surface of 3985 $cm^2{\cdot}g^{-1}$, density of 3.05 $g{\cdot}cm^3$, the initial setting time of 248 min, compressive strength after 2 days – 17.6 MPa and after 28 days – 43.2 MPa, loss on ignition of 5.0% by weight cement, volume stability < 10 mm.

It should be pointed out that the foam is a very compressible material in comparison to other components of mortar such as cement, lime, water and sand. Its density can thus vary depending on exerted pressure. Absorptivity (vol. %) of WPF was determined after it has been immersed in water for 24 hours. Compressive strength on three cubic rigid WPF samples (50x50x50) was measured on a Controls 50 kN universal testing machine at 20 °C. The test was conducted to deformation of 10% and the speed of crosshead movement was 5 mm/min. The composition of WPF was determined based on EDS. Table 2 presents the main physical properties of WPS on the basis of authors' research.

Table 2. Composition and properties of waste polyurethane foam.

Density (kg/m^3)	Absorp-tivity (%)	Total porosity (%)	Compressive strength (MPa)	C	O	N	H	Ca	others
				weight (%)					
22	15.1	95	0.170	64.55	20	7.2	6.4	1.0	1.03

The following hydrophobic agents were used in the research:

- The methyl silicone resin i.e. undiluted bicarbonate agent which can be used to hydrophobize porous materials and cement-lime plasters. The parameters of this agent are as follows: viscosity of 2.846 $Pa{\cdot}s{\cdot}10^{-3}$, density of 0.82 g/cm^3, surface tension of 24.30 $N/m{\cdot}10^{-3}$ and the active ingredient concentration is 11% [3,15].
- Micromolecular alkyl-alkoxy-silane with aliphatic hydrocarbons as carriers, $R_n{-}Si{-}(OR')_{4-n}$, where R_n is alkyl-alkoxy-silane derivative which was used as a hydrophobising preparation for porous building materials of mineral origin, sand-lime bricks, mineral renders. It has the following parameters: the viscosity of 1.79 $Pa{\cdot}s{\cdot}10^{-3}$ at 20°C, the density of 0.80 g/cm^3 and siloxane content reaching approximately 7 % by mass [3, 15].

2.2 Methods

The porosity, density and bulk density were determined in line with EN 1015-10:2001 [16] standard. The absorptivity of mortars was investigated after 1 and 14 days, according to BS 1881-122, 2011 standard [17]. Before the testing, six samples of 40×40×160 mm were prepared and dried. In order to determine the thermal conductivity coefficient λ, plate apparatus FOX 314, TA Instruments, New Castle, USA with 300×300×50 mm plates was used. Three plates were prepared for each mortar. The samples were subjected to drying at the temperature of 20°C and 0°C for a heating and cooling plate, respectively, in order to obtain a constant weight. Water vapour permeability of mortars was determined on the basis of EN ISO 7783:2012 regulation [18]. Cylindrical samples 2 cm in thickness and 10 cm diameter were made of mortars. Three layers of hydrophobising preparation were applied with a brush on the surface of one of the samples. The measurements were carried out three times in constant temperature of 23°C and relative humidity, 93% on one side of the sample and 50% on the opposite side. The duration of the test was 8 – 9 days depending on diffusion through the surface of the material. The analysis of compressive strength was conducted in accordance with EN 1015-11:2001 standard [19], following sample curing for the duration of 28 days. The direct method, established in EN 12012:2007 [20] standards was implemented in the calculation of frost resistance. The dimensions of the utilized samples were 40×40×160 mm. All samples were subject to 50 cycles of freezing and thawing processes. However, the resistance has not been determined after the frost resistance test, but the samples were dried to constant mass, and mass loss was determined. The surface free energy (SFE) was established by measuring the contact angle of the considered mortars [3, 9, 11]. For that purpose, a test stand consisting of a goniometer and a camera for capturing the images of 2 mm^3 drops placed on the sample surface with a micropipette [3, 9] was used [15]. Adopting the Neumann model required the application of distilled water. The test involved placing 6 drops on each sample. The contact angle was analyzed before and after the frost resistance test. The SFE was calculated with the Neumann model using the following equation [15, 21]:

$$\cos\theta_w = \left[e^{-0.000125(\gamma_s-\gamma_w)^2} 2\sqrt{\frac{\gamma_s}{\gamma_w}} - 1 \right] \tag{1}$$

where: γ_w – SFE of water, θ_w – water contact angle. The value γ_w = 72.8 mJ/m^2 was assumed for distilled water [9, 15].

To determine the structure of mortars and the interfacial transition zone between the paste and aggregates, a scanning electron microscopy SEM was used. Observations were conducted by means of Quanta FEG 250 microscope, by FEI, Hillsboro, USA.

The surface hydrophobisation was performed by applying the preparation with a brush for three times. Next, all the samples were seasoned for the period of 7 days in laboratory conditions in order to start hydrolytic polycondensation.

The considered samples included:
M5, M10, M15 – control mortars without a hydrophobising layer;
M5.R, M10.R, M15.R - mortars with a methyl silicone resin solution;
M5.A M10.A, M15.A - mortars with alkyl-alkoxy-siloxane.

3 Results and discussion

The mechanical and physical properties of the considered heat-insulating mortars without a hydrophobising agent surface are shown in Table 3.

Table 3. Properties of heat-insulating mortars with WPF.

Descriptive statistics	Apparent density (kg/m^3)	Density (kg/m^3)	Open porosity (%)	Total porosity (%)	Thermal conductivity (W/m^{2}K)	Compressive strength (MPa)
M5						
Mean	1.47	2.41	25.1	38.07	0.441	19.4
SD	0.2	0.23	0.24	0.5	0.09	0.9
M10						
Mean	1.41	2.39	28.02	39.0	0.395	17.3
SD	0.1	0.3	0.75	0.75	0.05	0.41
M15						
Mean	1.38	2.38	30.1	41.1	0.302	16.7
SD	0.32	0.87	1.32	1.54	0.02	0.95

*Standard deviation

The achieved properties are characteristic of lightweight mortars. The value of heat transfer coefficient λ enables to classify the mortar as a heat-insulating one. This mortar demonstrates high porosity (38.07-41.1%), which is related to the increase of absorptivity in relation to the traditional mortars. However, despite high porosity and low density, good compression strength (16.7-19.4MPa) was also obtained. The increased quantity by 10 % of WPF addition causes the reduction of bulk density M15 by 6 % in comparison to M5. This amount of WPF addition in M15 has also resulted in the abatement of its compression strength by 14% and thermal conductivity coefficient by 31.5%, which raises its insulating properties in relation to M5. This is connected with the increase of open porosity by about 17%. Numerous studies have demonstrated that the application of lightweight aggregate often caused a substantial decline of mechanical properties of materials [2, 6, 9, 22]. However, in the analysed mortars this decline was not considerable as an admixture of vinyl acetate copolymer was applied. Polymers are commonly used in the modification of mortars and concrete due to their significant role in general enhancement of the efficiency of cement based materials, which was described by Wang et. al [23] and Wiliński et. al [24].

The results of research on the effectiveness of hydrophobisation of mortars with WPF in terms of absorptivity, contact angle, water vapour permeability and frost resistance has been presented in Table 4.

Table 4. Influence of hydrophobisation on the properties of heat-insulating mortars.

Mortars	Absorptivity 1 day (%)	Absorptivity 14 days (%)	Water vapour permeability (10^{-12} kg/m·s·Pa)	Contact angle (°)	Contact angle after frost test (°)	Mass loss of samples (%)
M5	4.4	24.1	5.7	27.5	17.6	6.2
M5.R	0.1	10.3	2.3	102.5	96.1	1.0
M5.A	0.4	15.7	4.3	100.3	82.0	1.2
M10	6.6	27.3	9.6	21.7	15.7	10.0
M10.R	0.1	11.1	3.0	91.5	69.6	2.2
M10.A	0.3	16.4	4.1	87.2	61.3	3.4
M15	8.1	29.7	12.3	15.3	7.3	14.7
M15.R	0.2	15.4	5.2	90.1	60.5	2.6
M15.A	0.5	18.2	7.3	80.1	54.9	3.8

The value of absorptivity of reference mortar after 1 day ranged between 4.4 and 8.1%. After hydrophobisation, absorptivity was at the level of merely 0.1 and 0.5%. However, after 14 days of the test, the differences between the reference samples and hydrophobised ones decreased considerably and amounted to 29.7% for M15 and 24.1% for M5. Another parameter that confirms high wettability of the material is contact angle before hydrophobisation which was 3.7, 4.2 and 5.9 times higher for M5, M10 and M15 mortars, respectively. This indicates reduced adhesive properties and hydrophobicity of porous mortar surface. Silica-organic compounds in aliphatic hydrocarbons modified the surface properties of the mortars with WPF. The decrease of water vapour permeability of mortars with WPF after the hydrophobisation has been observed, which especially in the case of methyl-silicon resin (R) reduces the material ability to diffusion cumulated inside the moisture even from 58% to 69% for M15 and M10, respectively. However, mortars with lightweight aggregate have higher permeability than the ones made of normal materials, which was presented in the papers of Sugiyama et al. [25], (1.7–1.9 times higher). This increase of permeability in reference mortars can be caused by the fact that WPF is a highly porous material meaning it has relatively high diameter of capillaries, which makes the water vapour flow easier [26]. Additionally, the foam is highly netted which means the presence of ducts or capillaries that are placed in the whole mortar and by connecting with each other they make a thin net enabling vapour to easily flow through.

Despite the decrease of water vapour permeability, agent R was the best at protecting the surface of mortars from humidity and frost corrosion. The frost resistance of the mortars can be formulated with only moderate confidence level, because the analysis only involved the mass loss of samples following the frost resistance test and the changes in SFE. No studies reflecting the changes in compression strength after freezing and thawing cycles were conducted. However, on the basis of the performed research, the authors obtained preliminary, but sufficient information on the changes in wettability and frost resistance of the material.

Methyl-silicon resin (R) increased the frost resistance by 77-84%, while alkyl-alkoxy-siloxane (A) raised it by 66-80% in relation to the reference samples. The layer of polisiloxane resin protected the surface against the infiltration of water, and consequently, against the destructiveness of ice [3,9-11].

When the surface of the mortar has to be resistant to corrosive environment, the preparation with the lowest possible content of SFE should be used. Figure 1 shows the alteration of SFE due to hydrophobisation of mortars, as well as SFE before and after the frost resistance test.

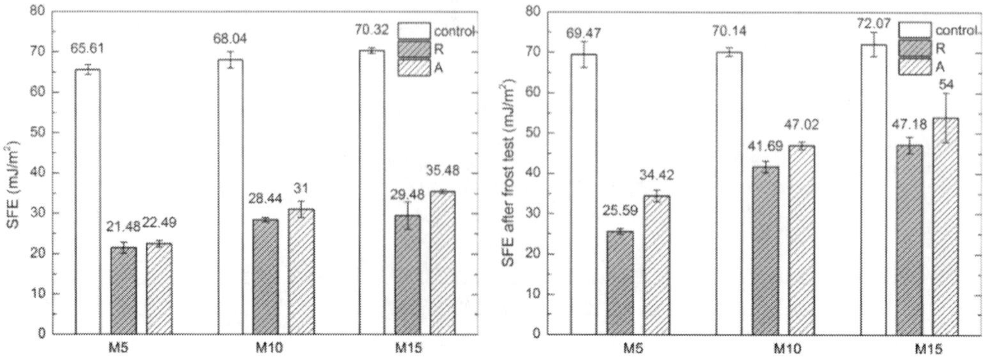

Fig. 1. SFE of mortars with WPF before and after the frost resistance test.

The M5.R mortar after hydrophobisation is characterized by the lowest SFE value which is about 3 times lower than the value of the referential mortar. Other mortars have been effectively protected to a similar extent (1.7-2.2 times). Lower SFE results in lower adhesive properties, higher resistance of the material to the infiltration of water and corrosive compounds, as well as higher resistance to freezing and thawing cycles, as evident from frost resistance tests. Calculations of SFE after frost resistance tests demonstrated a decrease in the efficiency of hydrophobisation. The contact angle of reference mortars following 50 cycles of freezing and thawing was smaller by 36% and 52% for M5 and M15 mortars, respectively. SFE raised by about 16% for M5-15.R and 35% for M5-15, respectively, which signifies weaker properties of the layer of polisiloxane resin.

Fig. 2. Structure of M15 mortar: (a) 500x, (b) 10000x.

The macromolecular resin (R) hermetically covered the structure of mortars with WPF resulting in a slightly better efficiency of hydrophobisation than the agent A with smaller molecules. In the case of absorbent, more absorptive mortars with bigger amount of WPF, the number of preparation layers that are applied with a brush should be increased or a low-pressured spray should be applied in order to enhance the effectiveness of the agent A with smaller molecules.

Figure 2 presents the structure of M15 mortar with the highest content of WPF. In reality, mechanical properties can be explained with the presence of WPF, as well as the quality of interfacial zone between cement paste and aggregate, which significantly affects its strength. Cement paste is dense while WPF aggregate is highly porous (Fig. 2a). A very good adhesion of an aggregate grain to coarse pellets of grated WPF has been noticed (Fig. 2a,b). In the contact zone there are no cracks or scratches. Figure 3b shows crystallized C-S-H phase.

All of these properties are the reason for a very good strength of cement paste, which has been confirmed by the analysis of mechanical properties. On the other hand, it has been observed that interfacial transition zone is slightly higher in the case of recycled aggregates than in the case of sand.

This is related to high water absorption by porous WPF aggregate, which has also been described by Rodríguez et al.[27] in his research.

4 Conclusions

The following main conclusions can be drawn:

- The increase of the amount of WPF has led to about 16% higher porosity of cement paste which makes the exchange of humidity with the ambient environment easier, water vapour permeability (up to 54%) and reduces thermal conductivity coefficient by about 35%.
- The increase of the amount of foam by 10% caused a significant increase in weight loss after freezing and thawing cycles by as much as 51% while the reduction of compressive strength was 16%.
- Contact angle after hydrophobisation was 3.7-5.9 times lower than for control mortars, which proves the effectiveness of hydrophobisation.
- Hydrophobisation by methyl-silicon resin (R) increased frost protection by 77-84%, while hydrophobisation by alkyl-alkoxy-siloxane (A) raised it by 66-80% in relation to the reference samples.
- The weight loss after freezing and thawing cycles is closely dependent on water absorptivity and hydrophobisation of the surface explicated by means of SFE.
- The highest hydrophobisation efficiency of lightweight mortars with WPF was obtained by applying methyl silicone resin in an organic solvent.

These results reinforce the idea of using waste polyurethane foam in the production of lightweight insulating aggregates with the application of hydrophobisation as an effective protection against humidity and frost.

References

1. H. Ez-zaki, B. El Gharbi, A. Diouri, Constr. Build. Mater. **159** (2018)
2. C. Giosuè, M. Pierpaoli, A. Mobili, M.L. Ruello, F. Tittarelli, Materials **10** (2017)
3. D. Barnat-Hunek, G. Łagód, B. Klimek, Thermophysics **2017**, 040004 (2017)
4. P. Mounanga, W. Gbongbon, P. Poullain, P. Turcry, Cement Concr. Compos. **30** (2008)
5. S. Kamonkunanon, Int. J Civil Arch. Struct. Constr. Eng. **8** (2014)
6. J. Gadea, A. Rodríguez, P.L. Campos, J. Garabito, V. Calderón, Cement Concr. Compos. **32** (2010)
7. A. B. Fraj, M. Kismi, P. Mounanga, Constr. Build. Mater. **24** (2010)
8. R. Boser, T. Ragsdale, C. Duvel, J Ind. Tech. **18** (2002)
9. D. Barnat-Hunek, G. Łagód, R. Siddique, Constr. Build. Mater. **155** (2017)

10. Z. Suchorab, D. Barnat-Hunek, P. Smarzewski, Z. Pavlik, R. Černy, Ecol. Chem. Eng. S **21** (2014)
11. D. Barnat-Hunek, P. Smarzewski, G. Łagód, Z. Suchorab, Ecol. Chem. Eng. S **22** (2015)
12. PN-EN 459-1 (2015-06)
13. PN-EN 197-1 (2012)
14. PN-B-19707 (2013-10)
15. D. Barnat-Hunek, P. Smarzewski, Constr. Build. Mater. **102** (2016)
16. PN-EN 1015-10 (2001)
17. BS 1881-122 (2011)
18. EN ISO 7783 (2012)
19. EN 1015-11 (2001)
20. EN 12012 (2007)
21. D. Li, A.W. Neumann, Langmuir **9** (1993)
22. M.L. Torres, P.A. Garcia-Ruiz, Cem. Concr. Compos. **31**(2) (2009)
23. R. Wang, J. Li, T. Zhang, L. Czarnecki, Bull. Polish Acad. Sci. Tech. Scis. **64** (2016)
24. D. Wiliński, P. Łukowski, G. Rokicki, Polimery **61** (2016)
25. T. Sugiyama, T.W. Bremner, Y. Tsuji, Cem. Concr. Res. **26** (1996)
26. E.P. Kearsley, P.J. Wainwright, Cem. Concr. Res. **31** (2001)
27. A. Rodríguez Sáiz, J.M. Manso Villalaín, A. Aragón Torre, J.J. González, Resour. Conserv. Recy. **53** (2009)

Frost resistance of concretes containing ground granulated blast-furnace slag

Paweł Łukowski[1,*], *Ali* Salih[2] , and *Joanna J.* Sokołowska[1]

[1]Warsaw University of Technology, Faculty of Civil Engineering, al. Armii Ludowej 16, Warsaw 00-637, Poland
[2]Roadstone Ltd., A CRH Group Company, Fortunestown, Tallaght, Dublin 24, Ireland

Abstract. The paper deals with the influence of addition of ground granulated blast-furnace slag (GGBS) on the frost resistance of concrete. GGBS is a valuable modifier of concrete, having the latent hydraulic properties and particularly improving the chemical resistance of concrete. However, the performance of concretes with blast-furnace slag under freezing and thawing action is still not explained fully and remains a subject to discussion. The authors have investigated the concretes containing various amounts of GGBS and the portland cement CEM I, with various values of water to binder ratio, with and without the use of air-entraining admixture. The results of research show that the addition of blast-furnace slag causes some worsening of the frost resistance of concrete. The extent of this worsening depends on the water to binder ratio and the aeration of the concrete. However, even under the least favourable conditions, the concretes with GGBS addition have met the requirements of frost resistance after 200 cycles of freezing and thawing, given in the Standard PN-B-06265.

1 Introduction

Ground granulated blast-furnace slag (GGBS) is one of the concrete additions. According to PN-EN 206:2016 [1], GGBS is type II addition, together with fly ash and silica fume. It belongs to the group of latent hydraulic materials; PN-EN 206 recommends *k* value of 0.6 for GGBS used with cements CEM I and CEM II/A. The blast-furnace slag is also an important constituent of many common cements according to PN-EN 197-1 [2].

Ground granulated blast-furnace slag is formed by rapid cooling of the liquid slag, which is produced during smelting of the iron ore in the blast furnace. After grinding, it forms fine powder; the grinding process, increasing the specific area of the material, is of favour for its hydraulic activity. The slag is, therefore, the waste material from the production of the iron. According to Neville [3], there is about 300 kg of the slag produced together with 1000 kg of the pig iron.

The main components of the blast-furnace slag are calcium oxide CaO (30-50%), silicon dioxide SiO_2 (28-40%), aluminum oxide Al_2O_3 (6-24%) and magnesium oxide MgO (1-18%). The slag shows hydraulic ability after activation by such substances like lime (also

* Corresponding author: P.Lukowski@il.pw.edu.pl

this produced during Portland cement hydration), sodium hydroxide, sodium carbonate, water glass, gypsum, and others.

The use of the ground granulated blast-furnace slag as the concrete addition meets the requirements of the strategy of sustainable development in construction and is justified from the economical point of view. As reported by Deja [4], the mineral additions can have a positive influence on the durability of the binders; this effect can also be attributed to GGBS [5-8]. However, the technical properties of cement concretes containing ground granulated blast-furnace slag are not fully recognized yet and their durability under frost action still creates the particular uncertainty.

The aim of the presented research was to recognize the effect of ground granulated blast-furnace slag addition on the durability of concrete under freezing and thawing action. The blast-furnace slag was used as a partial replacement for cement (the maximum degree of substitution was 70% of the cement mass). The variable factors were water to binder ratio in the concrete mix and use (or not) of the air-entraining admixtures.

2 Materials and methods

2.1 Materials used and mix proportions

Portland cement CEM I 42,5R was used for manufacturing the concretes for testing. The chemical composition of the cement as well as the slag, determined according to PN-EN 196-2 [9], are presented in the Table 1.

Table 1. Chemical composition of the cement and slag used in research.

Component	Content, mass %	
	CEM I 42.5R	GGBS
CaO	63.30	42.16
SiO_2	19.39	35.89
Al_2O_3	5.05	10.86
Fe_2O_3	3.38	0.61
MgO	1.51	7.13
TiO_2	0.20	0.76
Alkalis (Na_2O_{eq})	0.66	0.62
Sulfates (SO_3)	3.42	0.11
Chlorides	0.04	0.02
Sulfides	–	0.50
Loss on ignition	2.53	1.16
Unsoluble residue	0.54	0.18

Density of the ground granulated blast-furnace slag (determined by the use of LeChatelier method according to PN-B-06714-02 [10]) was 2860 kg/m^3, which belongs to the range of the values typical for this material. The SEM observations showed that the morphology of the GGBS grains is also typical with domination of irregular shapes (Fig. 1).

Fig. 1. SEM images of GGBS grains.

The aggregate conforming to PN-EN 12620 [11] was used to make the tested concretes. The aggregate consists of natural sand and crushed hornfels aggregates with the fractions 2/4 mm, 4/8 mm and 8/16 mm. Tap water was used as the mixing water. The commercially available superplasticizer (HRWRA) and air-entraining agent were used as the admixtures for concrete.

The concrete mixes with various water to binder ratio (where binder means cement or the sum of cement and GGBS), i.e. 0.45, 0.40 and 0.37, are presented in the Table 2. The conistence S3 acc. to PN-EN 206 was assumed, i.e. the slump 100-150 mm acc. to PN-EN 12350-2 [12]. The content of a polycarboxylate superplasticizer was adjusted to obtain this consistence class. The content of the air-entraining agent (when used) was constant and equal to 0.25% of the cement mass.

Table 2. Mix compositions of the tested concretes

Components	Water to binder ratio											
	0.45				0.40				0.37			
	Contents per w 1 m^3 of the concrete mix, kg											
% of GGBS in binder mass	*0*	*30*	*50*	*70*	*0*	*30*	*50*	*70*	*0*	*30*	*50*	*70*
GGBS	0	102	170	238	0	108	180	252	0	111	185	259
Cement	340	238	170	102	360	252	180	108	370	259	185	111
Water	153				144				137			
Sand 0/2 mm	519				521				523			
Gravel 2/4 mm	389				390				392			
Gravel 4/8 mm	372				273				274			
Gravel 8/16 mm	758				761				765			
HRWRA	1.40				1.48				1.52			

2.2 Methods of testing

The air content in the concrete mixes was measured acc. to PN-EN 12350-7 [13].

The frost resistance of concretes was determined following PN-B-06265 [14] (the national supplement to EN 206). The method uses 100 mm cubic specimens; 6 of them is periodically freezed in air (4 h) and thawed in water (4 h), while the other 6 is stored for comparison. After 200 freezing and thawing cycles the changes of mass and compressive strength of the specimens were determined. Also, the visual inspection of the specimens was done. The tests were performed after 28 days of curing.

3 Results and discussion

An air-entraining agent was used in the constant amount 0.25% of the cement mass. This way the air content in the mixes was obtained as required by PN-EN 206 for the concretes exposed to frost (exposition classes XF2-XF4), i.e. ≥ 4%. Only slight influence of the slag on the level of air-entraining was observed (Fig. 2).

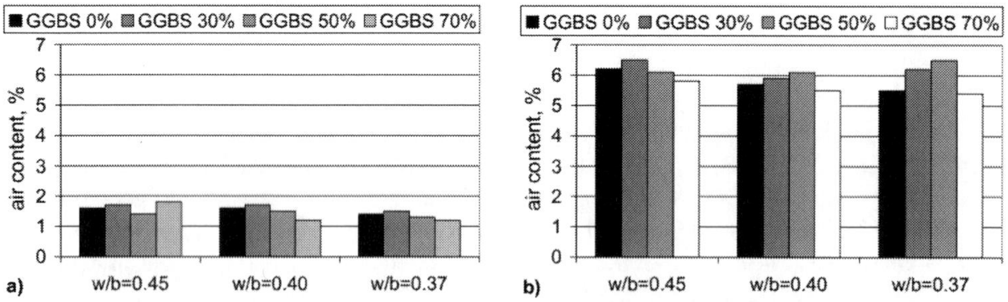

Fig. 2. Air content in the concrete mixes with various amounts of GGBS: a) without air-entraining agent, b) with air-entraining agent.

Figures 3-5 present the absolute loss of the compressive strength of the tested concretes after 200 cycles of freezing and thawing. The percentage loss of strength, which is the basis for evaluation of the frost resistance of concrete acc. to PN-B-06265, is presented in the Figures 6 and 7, while the loss of mass is presented in the Figure 8.

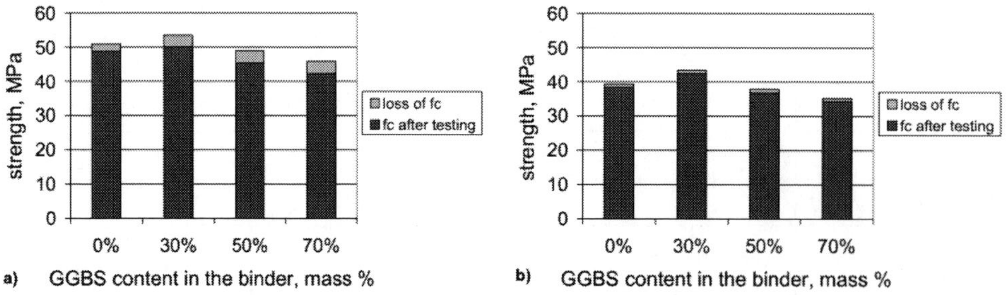

Fig. 3. Absolute loss of compressive strength of concretes with various amounts of GGBS; water to binder ratio 0.45, after 200 cycles of freezing and thawing: a) without air-entraining agent, b) with air-entraining agent.

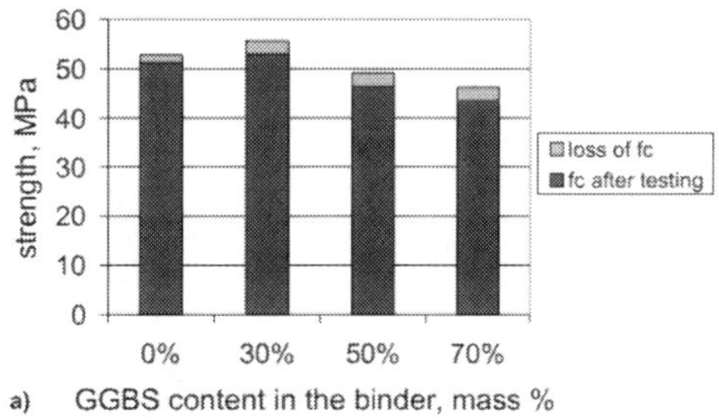

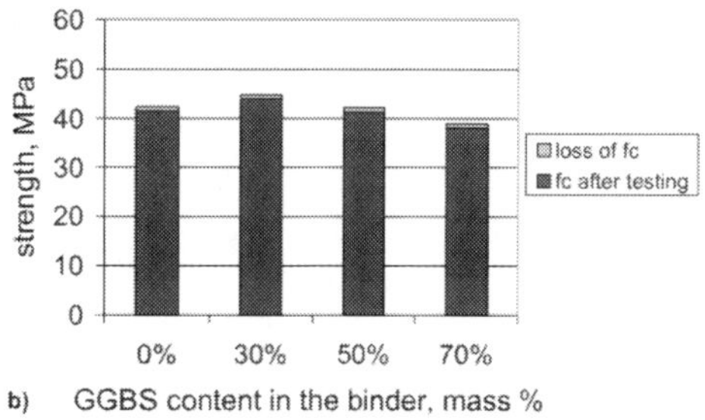

Fig. 4. Absolute loss of compressive strength of concretes with various amounts of GGBS; water to binder ratio 0.40, after 200 cycles of freezing and thawing: a) without air-entraining agent, b) with air-entraining agent.

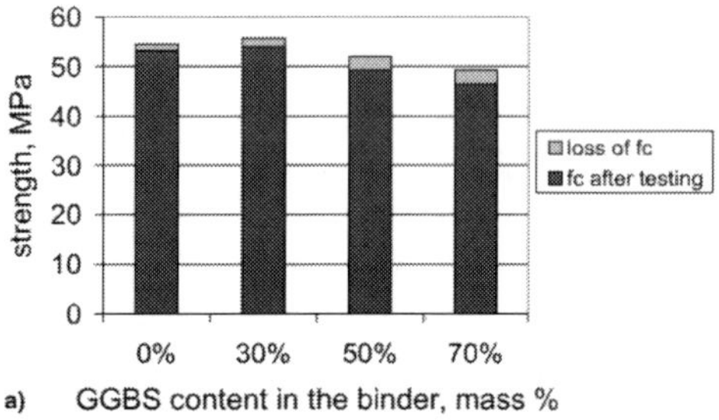

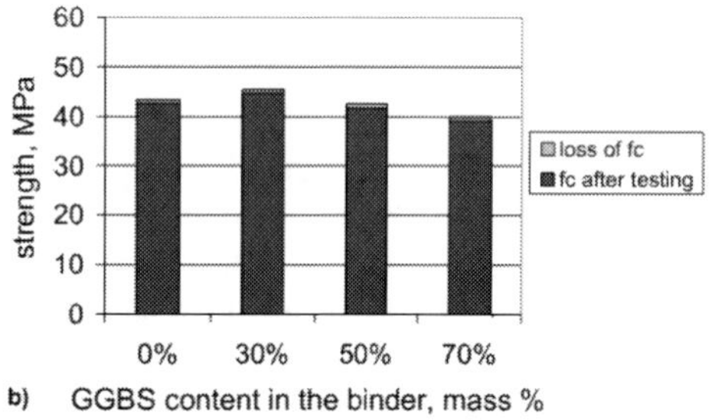

Fig. 5. Absolute loss of compressive strength of concretes with various amounts of GGBS; water to binder ratio 0.37, after 200 cycles of freezing and thawing: a) without air-entraining agent, b) with air-entraining agent.

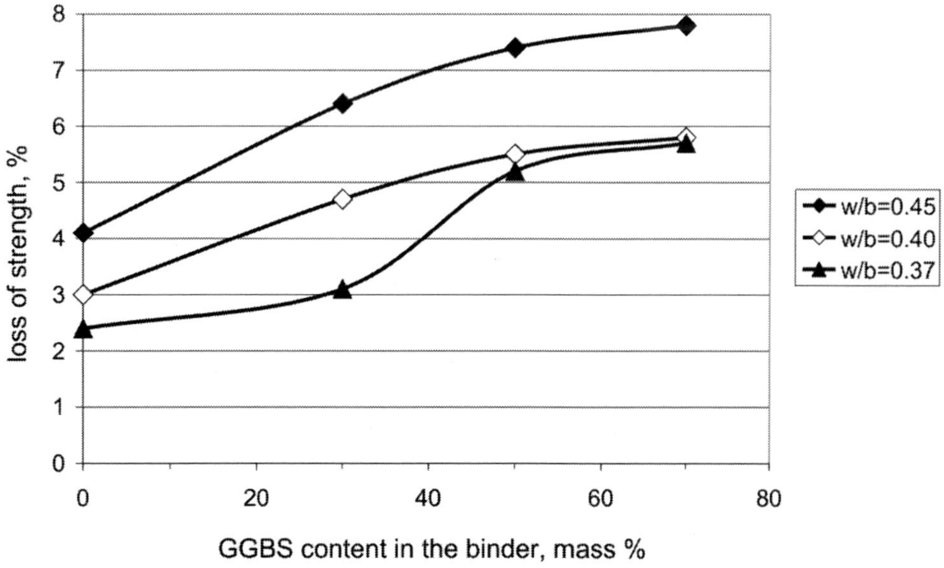

Fig. 6. Relative loss of compressive strength of concretes with various amounts of GGBS and water to binder ratio after 200 cycles of freezing and thawing: the concretes without air-entraining agent.

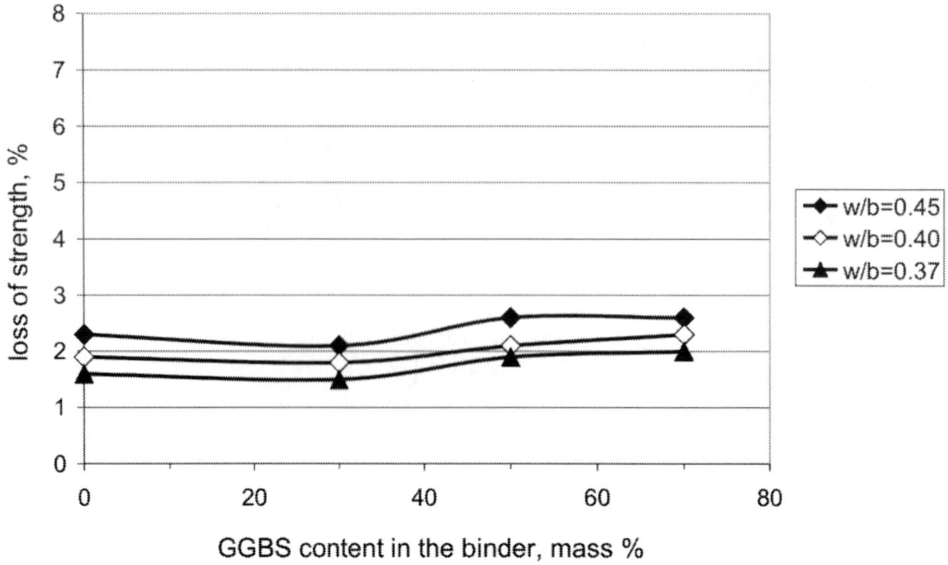

Fig. 7. Relative loss of compressive strength of concretes with various amounts of GGBS and water to binder ratio after 200 cycles of freezing and thawing: the concretes with air-entraining agent.

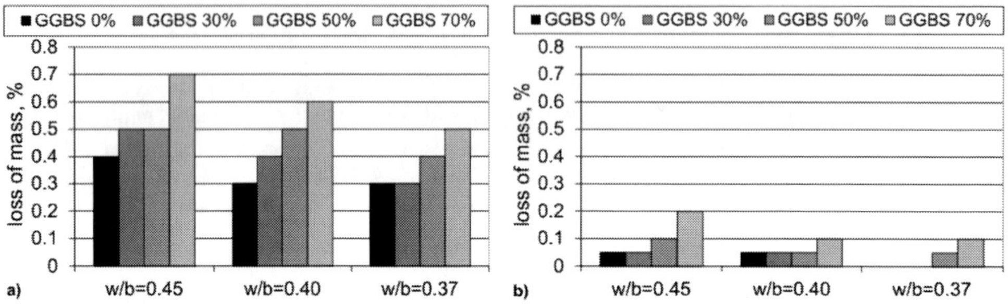

Fig. 8. Loss of mass of concretes with various amounts of GGBS and water to binder ratio; CEM I; after 200 cycles of freezing and thawing: a) without air-entraining agent, b) with air-entraining agent.

The last element of the evaluation of concrete resistance to frost action is visual inspection of the specimens during and after the given number of freezing and thawing cycles. The status of the example set of specimens is documented in the Figure 9.

Three criteria of evaluation of concrete frost resistance have been formulated in PN-B-06265:

- strength decrease of specimens after freezing and thawing considering the reference (not freezed) specimens should not exceed 20%,
- loss of mass of specimens after freezing and thawing should not exceed 5%,
- there is no cracking visible on the specimens after freezing and thawing.

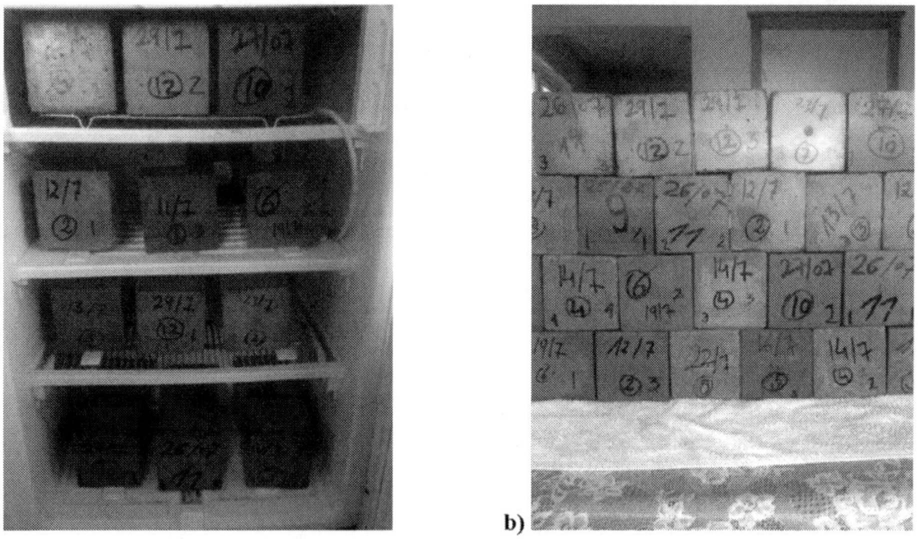

a) b)

Fig. 9. Tested concrete specimens: a) in the freezing chamber, b) after 200 cycles of freezing and thawing

Following these requirements we have observed that:

- the use (or not) of an air-entraining agent, the water to binder ratio and the addition of ground granulated blast-furnace slag all influence the degree of compressive strength loss of the concrete after 200 cycles of freezing and thawing. This decrease is, as expected, deeper in the case of the concretes without air-entraining agent and with higher value of water to binder ratio.

  The effect of GGBS on the strength loss of concrete after freezing and thawing depends on aeration of concrete (or the lack of it). When no air-entraining agent is used, the loss of strength grows with increasing content of slag. However, the compressive strength of such concretes after 200 cycles of freezing and thawing did not lower by more than 8%, even at 70% GGBS content in the binder.

  The influence of GGBS is different in the case of the aerated concrete. The maximum loss of strength after 200 cycles of freezing and thawing did not exceed 3% then; the degree of strength lowering only slightly depends on the slag content. Some minimum has been even observed for 30% GGBS content in the binder.

  Therefore, the determined losses of compressive strength are in any case far from the value allowed by PN-B-06265, i.e. 20%.
- The higher content of ground granulated blast-furnace slag caused bigger losses of mass of concretes after freezing and thawing. However, the losses were very slight in every case and did not exceed 0.7% for the non-aerated concretes and 0.2% for aerated concretes. The mass losses of the concretes containing GGBS after 200 cycles of freezing and thawing did not approach to the value allowed by PN-B-06265, i.e. 5%.
- The visual inspection of the specimens, both during and after the cycles of freezing and thawing, has revealed no cracks nor other damages of concrete.

Taking into consideration the aforementioned observations, we conclude that the addition of ground granulated blast-furnace slag causes some worsening of the frost resistance of concrete, if no air-entraining agent is used. However, the criteria of frost resistance formulated in PN-B-06265 are still met even for large 70% content of GGBS in the binder. The proper aeration protects the concrete containing blast-furnace slag against the consequences of the cyclic freezing and thawing in the same degree as the concrete without addition of GGBS.

4 Sum up and conclusions

The subject of this paper was the evaluation of the influence of ground granulated blast-furnace slag (GGBS) on the frost resistance of concrete. The use of blast-furnace slag as a mineral addition of type II for concrete, with replacement of part of the cement with GGBS, is favourable from the ecological point of view, allowing to rational utilization of the industrial waste or by-products. Moreover, it makes possible to decrease the consumption of Portland cement, production of which causes significant emission of carbon dioxide. All this is in accordance with the strategy of sustainable development in construction. The concretes containing the cheaper than cement GGBS are also economically attractive.

The particular subject of the study was to discuss possibility of using concrete containing ground granulated blast-furnace slag in the moderate climate. In such conditions, the material in the structure is usually exposed to cyclic freezing and thawing action. The results of the investigation allowed to learn that the addition of the blast-furnace slag causes some worsening of the frost resistance of concrete; the degree of this worsening depends on the water to binder ratio and the aeration of concrete. However, even under the least favourable conditions, the concretes with GGBS addition have met the requirements of the frost resistance after 200 cycles of freezing and thawing, given in PN-B-06265.

Taking the above into consideration, we conclude that the properly designed, manufactured and cured concretes, containing even large amount of the ground granulated blast-furnace slag, can be used in the conditions of threat of the frost aggression.

The authors would like to thank Faculty of Civil Engineering of Warsaw University of Technology for the support of this research.

References

1. PN-EN 206 (2016)
2. PN-EN 197-1 (2012)
3. A.M. Neville, *Properties of Concrete* (5th ed. Pearson, 2012)
4. J. Deja, *10th International Congress on Chemistry of Cement 4*, 4iv015 (1997)
5. J. Deja, Ceramika – Ceramics **83** (2004)
6. J. Małolepszy, J. Deja, *4th CANMET/ACI International Conference on Durability of Concrete 1*, 189 (1997)
7. J. Deja, Cement Concrete Comp. **25**, 357 (2003)
8. P. Łukowski, A. Salih, Procedia Engineering **108**, 47 (2015)
9. PN-EN 196-2 (2013)
10. PN-B-06714-02 (1976)
11. PN-EN 12620+A1 (2010)
12. PN-EN 12350-2 (2011)
13. PN-EN 12350-7 (2011)
14. PN-B-06265 (2017) (draft)

The impact of air content on the durability of concrete under combined sulphate and freeze-thaw attack

Julia Marczewska[1,*], and *Wojciech* Piasta[1]

[1] Department of Civil Engineering and Architecture, Kielce University of Technology, al.1000-lecia PP 7, 25-314 Kielce, Poland

Abstract. The aim of the study was to determine the optimal air content in concretes subjected a combination of exposure conditions. Five series of concretes with different air contents were tested. The article reports the experimental results of the combined sulphate expansion and freeze-thaw attack in ordinary Portland cement concrete. Some of the concrete samples were immersed in a 5% sodium sulphate solution for nine months, and then frozen in air and thawed in water. The resistance of concrete to plain freezing and thawing was also tested. In order to compare the degree of concrete damage in both environments, long-term linear deformations (expansion) and compressive strength of concretes were investigated. To explain the complex damaging process the air-void structure in concretes was investigated. The changes in concrete microstructure after storing in sodium sulphate solution were observed under the scanning electron microscope (SEM). The prior sulphate attack significantly accelerated the deterioration of non-air-entrained concrete and air-entrained concrete with low content of air voids. Optimal air entrainment will protect concrete from the harmful effects of the interaction of sulphate attack and freeze-thaw damage.

1 Introduction

In water structures, bridges, tunnels and foundations, simultaneous or alternate occurrence of frost damage and sulphate corrosion is frequent.

According to Powers [1], water freezes gradually. As the temperature decreases, the ice content in the capillary pores increases, and because the freezing water increases its volume by about 9%, it is expelled from the capillaries. The amount of ice increases gradually with decreasing temperature and the length of time. The flow and thus the pressure increase with the increasing freezing rate and water quantity [2], causing swelling and then disruption in concrete.

In non-air entrained concrete, transport system for gas, water and solutions is provided by the w/c ratio-dependent capillary pores [3], the predominant factor affecting the frost durability of concrete. The deterioration rate of non-air entrained concrete depends to the

* Corresponding author: jmarczewska@tu.kielce.pl

highest extent on the degree of pore saturation with water, which is mainly related to the sorption and permeability of concrete. Concrete containing large diameter pores has the highest permeability. Nevertheless, freezing of water in small diameter pores is the source of the highest stresses [4].

The pore system includes pores always present in the hardened cement paste (gel, capillary pores and pores of entrapped air [5]) and the air pores incorporated deliberately to enhance the resistance of concrete freeze-thaw damage. Considering the air entrainment characteristics, the freeze-thaw resistance of concrete can be estimated on the basis of the pore size distribution, the spacing ratio and the content of pores with a diameter smaller than 300 μm [6]. In fact, the spectrum of pore sizes is continuous and the pores can occur in a wide range of sizes (from 0.02 to 1mm [2]); they can also coalesce under increasing air content or form larger systems in cement pastes when air entrainment has not been perfect [7]. It thus seems reasonable to examine whether air entrainment and different air contents have a significant effect on the freeze-thaw damage rate in the case of concretes previously subjected to sulphate attack [8]. It should be remembered that sulphate attack products will precipitate, changing the microstructure of the concrete. The physical effects of sulphate attack and freeze-thaw cycle-related deterioration were investigated but in the range of non-air-entrained concretes [9, 10].

In recent years, the theory of sulphate expansion due to the growth of ettringite crystals and the accompanying crystallization pressure has gained importance as the most probable hypothesis for concrete swelling under the influence of external sulphate attack [11]. Some researchers have long pointed out that during the attack of the sulphate solution, expansion is not necessarily dependent on the amount of ettringite formed [12]. According to Scrivener [13, 14], expansion is not related to the amount of ettringite but depends on the sites in which ettringite crystallizes. The expansion is induced by an increase in size of ettringite crystals formed as a result of Al $(OH)^{4-}$, Ca^{2+} and SO_4^{2-} and OH^- reaction in supersaturated solutions. The crystallization pressure can be transferred to the microstructure of the cement paste only where it meets the resistance, and this occurs in very small pores in the C-S-H phase at the monosulphate phase location. The precipitation of ettringite in very small pores restraining the growth of the crystals will cause mechanical pressure on the pore walls and the expansion of the paste microstructure, whereas the free, unrestrained growth of large ettringite crystals in large pores will not be the cause of expansion in the hardened cement paste. Scherer [11] shows that the crystallization pressure is inversely proportional to the size of pores in which the crystal grows. Thus, the crystals of ettringite growing in small pores of the order of tens of nanometres are probably the most significant factor in generating internal stresses leading to paste expansion during the attack.

There is, therefore, some similarity in the mechanisms of freeze-thaw damage and sulphate attack [15, 16]. For either of them to occur, the capillary pores must be filled with water or solution and, above all, higher stresses will be exerted by the growth of ice or ettringite crystals in small-diameter pores.

2 Experimental procedure

2.1 Aim and scope of the experiment

The aim of this study was to investigate and evaluate the effects of the different air content on durability of concretes exposed to complex deterioration conditions.

Non-air entrained and air entrained concrete mixes with an air content of 3-4%, 4-5%, 6-7% and 8-9% were prepared according to the assumption of the experimental design. The air content was measured on freshly mixed concrete by pressure method. The bulk density and consistency were additionally determined. An optical microscope was used to examine the structure of the air pores in the hardened concretes. The samples were also tested for absorption and compressive strength at 28 days of curing in water. After prior sulphate attack the frost durability of concretes was tested and the plain frost durability of concretes for comparison. Scanning electron microscopy (SEM) was used to explain changes in the microstructure and deterioration rate in the air-entrained concretes.

2.2 Materials

Five series of concrete with Portland cement were prepared according to chemical composition shown in Table 1. The aggregate composition was prepared in accordance with [17]. Coarse basalt aggregate (2-8, 8-12 mm) and natural quartz sand (0-2 mm) were used. Coarse aggregate sizes were limited by sieving grains over 12 mm due to non-standard specimen size. The aggregates volume fraction was the same in all mixes, with the following proportions:

- coarse basalt aggregate 2/8 - 33%
- coarse basalt aggregate 8/12 - 33%
- fine quartz aggregate 0/2 - 34%

Table 1. Chemical and phase composition of Portland cement [%].

Cement	Chemical composition								
	SiO_2	Al_2O_3	Fe_2O_3	CaO	$CaO_{(wolne)}$	MgO	SO_3	Cl^-	Na_2O_{eq}
CEM I	21.31	5.01	3.36	65.06	1.53	1.95	0.74	0.03	1.14
	Phase composition								
CEM I	C_3S		C_2S		C_3A		C_4AF		
	67.83		13.57		7.76		11.70		

Tap water was used as batch water in accordance with PN-EN 206 [18]. The water-cement ratio was 0.55. Synthetic surfactants were used as an air entraining agent with a density of about 1.0 g/cm3. Table 2 shows the composition of the concrete mixtures.
The amount of air entraining agent was varied to obtain the air contents compliant with the experimental design assumptions:
- 3.2% air content (series designated as γ1) – insufficiently air entrained concrete [18]
- 4.5% air content (γ2) – properly air entrained concrete (acc. to PN-EN 206)
- 6.5% air content (γ3) - properly air entrained concrete (acc. to PN-EN 206)
- 8.5% air content (γ4) – air entrainment higher than the recommended limit [18]

The non-air entrained mix had the air content of 1.6% (γ5 - concrete).

Table 2. Composition of concrete mixes per 1 m^3.

Component	Content [kg/m^3]				
	γ1	γ2	γ3	γ4	γ5
Coarse basalt aggregates 2-8	655.6	655.6	655.6	655.6	655.6
Coarse basalt aggregates 8-12	655.6	655.6	655.6	655.6	655.6
Quartz sand 0-2	695.2	695.2	695.2	695.2	695.2
Cement	334	334	334	334	334
Water	185*	185*	185*	185*	185*
Air entraining admixture	0.07	0.1	0.3	0.57	-

* - mass of water in the non-air entrained concrete; total mass of water with the air entraining agent in the air entrained concrete mixes.

Cubes with dimensions 100x100x100 mm were prepared from each mix for compressive strength testing and air voids system characterization. Beams with dimensions 60x60x350 mm were prepared for investigating longitudinal strains during freezing/thawing and sulphate attack. The samples were cured in water for 28 days in accordance with PN-EN 12390-2 [19]. Half of the samples were immersed in a 5% sodium sulphate solution for nine months, and then frozen in air and thawed in water. In order to minimize the influence of uncontrolled factors, the samples were prepared while maintaining identical manufacturing technology (identical order of added ingredients, mixing time, mixing intensity, temperature, contact time of cement with water before measurements).

3 Results

The properties of fresh concrete mixes, non-air entrained and air entrained with different air content are summarized in Table 3. Table 4 shows the average values from the results of compressive strength and water absorption tests after 28 days of curing in water.

Table 3. Result for the freshly mixed concrete mixes.

Concrete	Consistency, slump [mm]	Bulk density [kg/m^3]	Air content [%]
γ1	9	2493	3.2
γ2	15	2455	4.5
γ3	23	2413	6.5
γ4	130	2328	8.5
γ5	9	2516	1.6

The incorporation of additional air through the use of air entraining admixture reduced the bulk density of the mixtures designated as γ1, γ2, γ3 and γ4 by 0.9, 2.4, 4.1 and 7.5%, respectively. The air entrainment changed the consistency of the mixes, making them more fluid.

Table 4. Compressive strength and absorption at the age of 28 days.

Concrete	CEM I γ1	CEM I γ2	CEM I γ3	CEM I γ4	CEM I γ5
Absorption [%]	4.96	5.25	5.96	6.05	5.83
Compressive strength [MPa]	52.1	49.6	43.0	36.6	55.2

The results of the compressive strength test clearly illustrate the reduction in concrete strength after the use of air entraining admixture in proportion to the amount of admixture used. Absorption values increased with increasing air content in the air entrained concrete mixes.

To further examine the concretes, the pore system of the air entrained concrete mixes was evaluated (Table 5). The porosity of the hardened concrete was characterized using an automatic image analysis system and a computer program on properly prepared polished concrete surfaces. The photographs from the optical microscope (Figure 1a, 1b, 1c) showed evident differences in the number and distribution of air pores in the concrete samples. Examination of the pore system indicated an increase in the total air content in the hardened concrete, in the specific surface area of the pores and in the micropore content with increasing amount of admixture used and, consequently, with the increasing air content in the fresh concrete mix. At the same time the spacing factor decreased. The designed air content was achieved, as confirmed by both hardened concrete and concrete mix.

Table 5. Air pore system in the hardened concrete.

Concrete	Air pore system in the hardened concrete				Air content in fresh mixes
	A [%]	α [mm^{-1}]	$\bar{L}$ [mm]	A300 [%]	Zp [%]
γ1	3.11	15.25	0.332	1.04	3.2
γ2	4.19	18.18	0.299	1.28	4.5
γ3	6.66	28.91	0.152	3.71	6.5
γ4	9.16	34.75	0.090	5.41	8.5

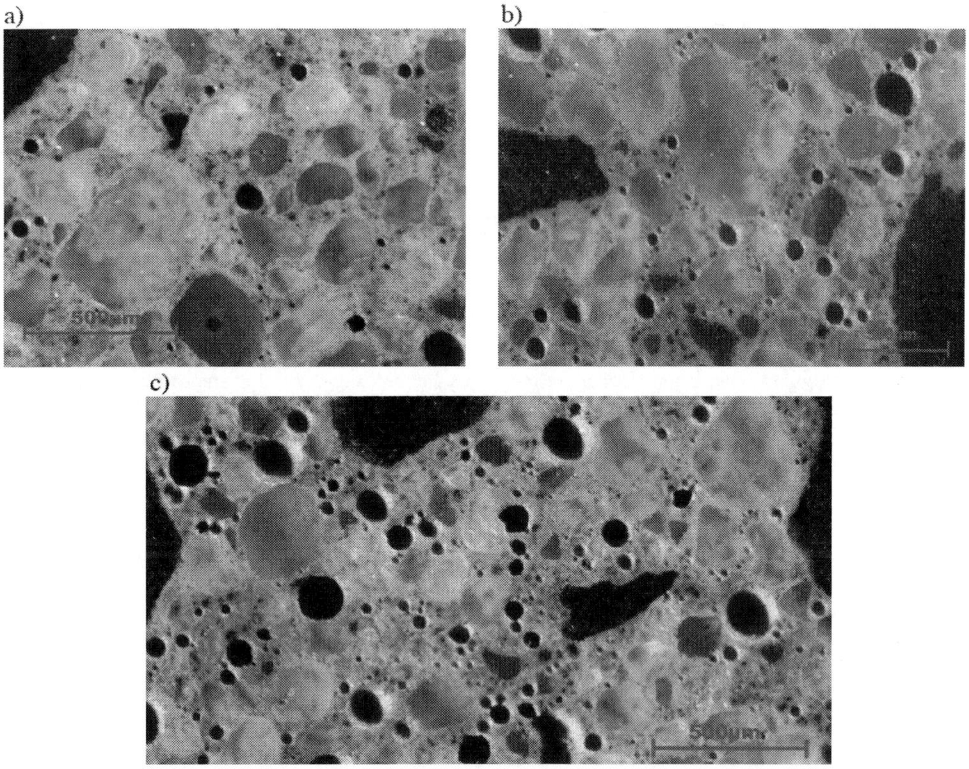

Fig. 1. View of the distribution of real air-pore structures on the surface of concrete samples a) γ2: A = 4.19% $\bar{L}$ = 0.299; b) γ3: A = 6.66% $\bar{L}$ = 0.152; c) γ4: A = 9.16% $\bar{L}$ = 0.090.

Irregular pores (indicative of entrapped air and capillary pores) were very rarely observed under optical microscope (Fig 1c). This can be associated with higher permeability of concretes for liquid and gaseous substances, which is related to the lowered resistance of concretes to chemical attack and cycles of freezing and thawing.

After performing the basic tests of the hardened concretes, the samples were divided into two groups. The first group included the samples subjected to 200 freeze-thaw cycles after nine months immersion in a solution of sodium sulphate were. The non-air entrained concrete showed a fast progressing increase in deformations as early as at 20 freeze-thaw cycles (Fig. 2). The final measurement of the non-air entrained samples after 200 cycles of freezing and thawing revealed the greatest deformation (over 2.5 ‰) of all the samples. The concrete with an air content of 3.11% reached low deformation - only 1.1‰ at the end of the test. The air entrained concretes of air content 4.19% and 6.66% had a linear deformation of about 0.6‰ after about 9 months immersion in the sodium sulphate solution, while during 200 freeze-thaw cycles this value showed a negligible increase and at the end of the test they showed the smallest deformation of 0.8‰. The concrete with the highest air content (γ4) had the final linear deformation (2.1‰) under complex deterioration conditions (Fig. 2) similar to that of the non-air entrained concrete.

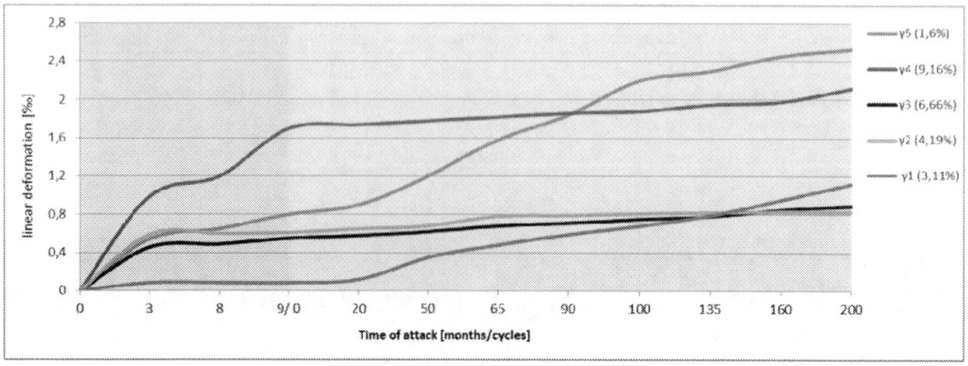

Fig. 2. Deformation of concretes subjected to Na_2SO_4 solution and then freeze-thaw cycles.

The other concrete samples were only subjected to cyclic freezing and thawing process. After 300 cycles, the non-air entrained concrete samples and those with the highest air content showed the highest and similar values of linear deformation. It is interesting that the concrete with 9.16% air content showed the highest linear deformation (1.4‰), higher than that of the non-air entrained concrete (Fig. 3). Throughout the test period, i.e., after 300 freeze-thaw cycles, the linear deformations in concretes with 4.19 and 6.66% air content did not exceed 0.2‰, while the deformation in concrete containing 3.11% of air reached 0.4‰ in the final stage.

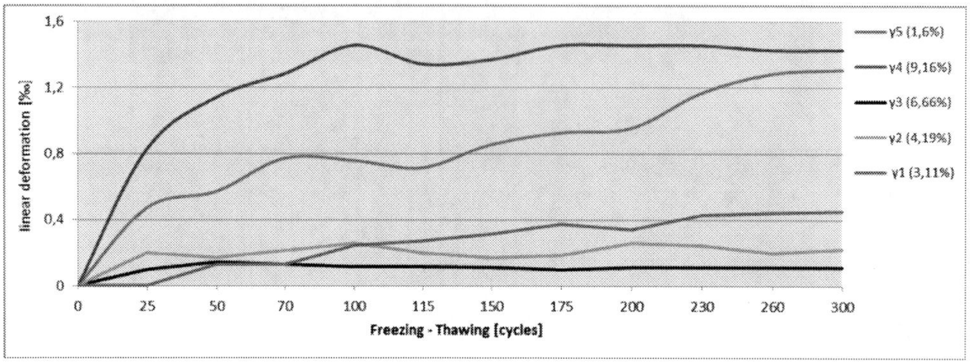

Fig. 3. Linear deformation of concretes after freeze-thaw cycles.

Compressive strength is given as a percentage of the 28-day strength for each concrete. All concretes subjected to the combined attack showed from 8 to 25% higher compressive strength measured after 9 months immersion in Na_2SO_4 solution in relation to the 28-day strength (Fig. 4). After this period, a drop in the strength during 200 freezing cycles was observed. In contrast, the concretes subjected only to freezing and thawing showed nothing but increase in the strength at the end of the test.

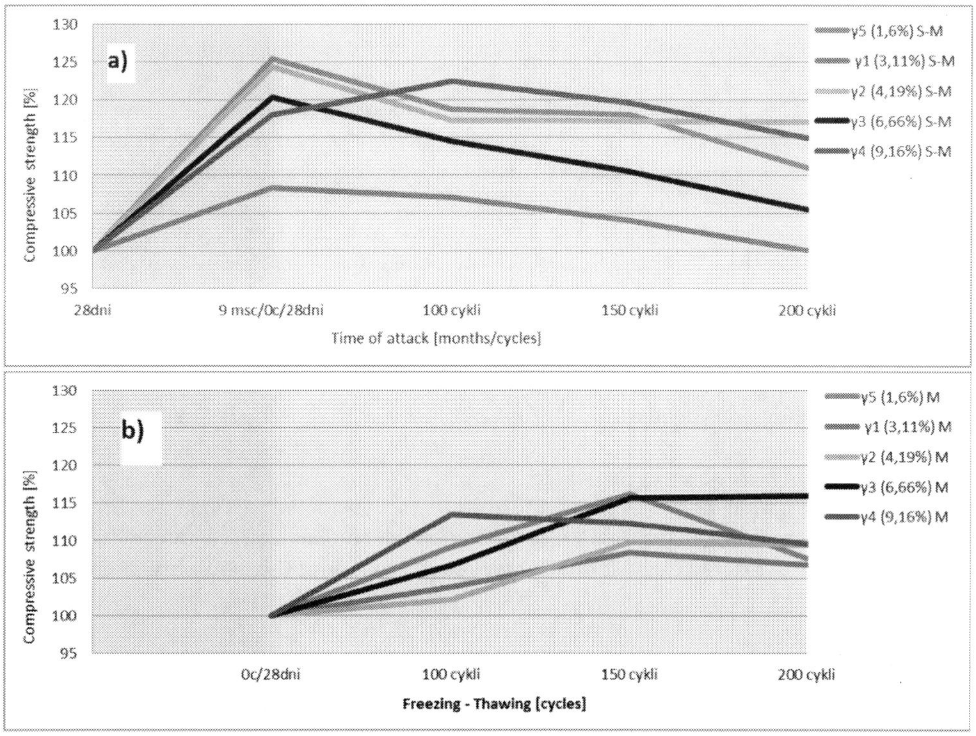

Fig. 4. Strength for the concretes: a) after 9 months of sulphate attack and after freeze-thaw cycles, b) frozen and thawed, expressed as percentage at 28 days for each concrete mix.

The air pores of the concretes (Fig. 5) were filled with ettringite crystals formed in the zone of active sulphate corrosion, i.e., at a distance of up to 1.5 mm from the surface of the concrete sample. The degree of pore filling, however, was dependent on the amount of air entrainment.

a) b)

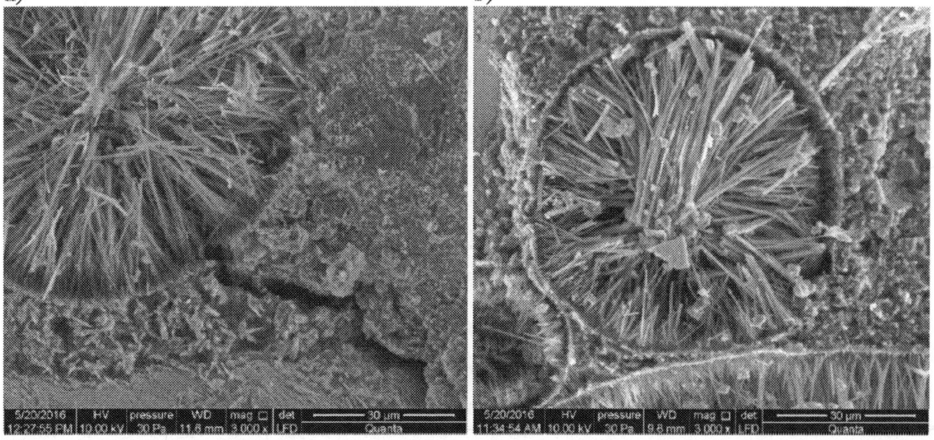

Fig. 5. SEM images of the microstructure of concretes after 9 months of immersion in the sodium sulphate solution before being subjected to freeze-thaw cycles a) concrete with a 3.11% air content; b) concrete with a 9.16% air content.

4 Discussion and conclusions

The best frost resistance evaluated in experimental studies was found in concretes with an air content of 4.19% and 6.66%. Prior sulphate attack reduced their freeze-thaw resistance only slightly. The performance of other air entrained concretes (3.11% and 9.16% air content) deteriorated considerably and was similar to that of non-air entrained concretes. In order to explain that behaviour, SEM was employed for microstructural observations of concrete, with particular attention paid to the air voids. Figures 5a) and 5b) show the air voids filled with ettringite crystals. The largest degree of filling was found in the pores with a diameter smaller than 150 μm. It can therefore be assumed that the lower resistance of air entrained concretes previously subjected to sulphate attack was caused by sulphate attack products filling a portion of the pore volume. The resistance to freezing and thawing of non-air-entrained concretes, which were previously affected by the Na_2SO_4 solution, declined noticeably. The properly air entrained concretes (4.19% and 6.66%) did not show major deformations during the initial sulphate attack and subsequent freeze-thaw cycles. The combined attack, however, led to a slightly accelerated strength reduction as compared to the effects of freeze-thaw cycles alone.

The incorporation of 4.19 and 6.66% air was considered optimal for maintaining durability under freeze-thaw damage and combined sulphate attack-freeze-thaw cycles conditions. The use of too high air content was found to reduce the durability of concrete.

References

1. T. Powers, Highway Research Board, PCA Bulletin **29**, 184-211 (1949)
2. W. Kurdowski, *Chemia cementu i betonu,* (PWN, Warszawa, 2010)
3. I. Soroka, *Portland Cement Paste and Concrete,* 88 (Macmillan, London, UK, 1979)
4. Z. Rusin, *Technologia betonów mrozoodpornych* (Polski Cement, Kraków 2002)
5. P.K. Mehta, P.J. Monteiro, *Concrete structure, properties and materials*, (Prentice-Hall, Inc., Englewood Cliff, New Jersey, Second Edition, 1993)
6. P.W. Brown, D. Shi, J. Skalny, *Material Science of Concrete II,* 83–110 (1991)
7. J. Wawrzeńczyk, W. Kozak, Constr. Build. Mat. **112**, 360-365 (2016)
8. W. Piasta, J. Marczewska, M. Jaworska, Procedia Eng. **108**, 5562 (2015)
9. L. Jiang, D. Niu, L. Yuan, Q. Fei, Cold Reg. Sci. Technol. **112**, 112-117 (2015)
10. D. Wang, X. Zhou, Y. Meng, Z. Chen, Constr. Build. Mater. **147**, 398-406 (2017)
11. G.W. Scherer, Cem. Concr. Res. **29**(8), 1347 - 1358 (1999)
12. P.W. Brown, H.F.W. Taylor, *Sulfate Attack Mechanisms* (Am. Ceram. Soc. 1999)
13. C. Yu, W. Sun, K. Scrivener, Cem. Concr. Res. **43**, 105–111 (2013)
14. W. Kunther, B. Lothenbach, K. Scrivener, Cem. Concr. Res. **46**, 23–29 (2013)
15. D. Niu, L. Jiang, Q. Fei, Jour. Wuhan Univ. Tech. Mater. Sci. Ed. **28**(6), 1172-1176 (2013)
16. H. Yang, X. Shen, M. Rao, X. Li, X. Wang, Adv. Mater. Sci. Eng **2015** (2015)
17. PN-EN 12620+A1 (2010)
18. PN-EN 206 (2014-04)
19. PN-EN 12390-2 (2011)

Chemical corrosion of external stairs – case study

Przemysław Czapik[1,*], and *Zdzisława* Owsiak[1]

[1]Department of Building Engineering Technologies and Organization, Kielce University of Technology, Al. Tysiąclecia Państwa Polskiego 7, 25-314 Kielce, Poland

Abstract. On the basis of examinations of the efflorescences formed on the concrete surface, an attempt was made to analyze the sources of concrete corrosion without entering inside the construction. The concrete stairs revealed the symptoms of leaching, as a result of alkali-aggregate reactions developing beneath the surface. As a result of this corrosion process and the carbonation propagating from the concrete surface, the carbonate efflorescences were found. Their phase composition was determined by X-ray diffraction. In order to identify whether the efflorescences were the results of the alkali-silica reaction or alkali-carbonate reaction, the microstructure was investigated using the scanning electron microscope together with energy dispersive spectroscopy.

1 Introduction

The construction of the concrete entrance stairs to the public utility building built between 2009 and 2013 is the object of the research reported in this work. The stairs were built of concrete containing dolomite aggregate and quartz sand. Granite tiles were laid on the stairs and stair risers, while the side walls of the structure were painted with the paint for concrete. After several years, numerous deposits of efflorescence appeared on the side walls of the stairway structure in the form of massive, layered encrustations (Fig. 1 a and b) or, appearing near cracks, caused severe surface scaling (Fig. 1c). The observed phenomena are a proof of concrete corrosion [1-5]. To identify the corrosion type, the samples from the eroded surface were collected for examination, without disturbing the structure of stairs [1, 6]. The phase composition and microstructure of samples were investigated.

* Corresponding author: p.czapik@tu.kielce.pl

Fig. 1. a), b) Layered massive encrustations on the side surfaces of stair structures, c) surface scaling near horizontal cracks.

2 Materials and Methods

The samples taken from the massive encrustations shown in Fig. 1a) and b) were designated as samples A and B, respectively, and samples taken from the flakes formed around the cracks were designated as C, (Fig. 1c). The collected material was ground to the grain size below 0.063 mm and averaged.

The phase composition of the samples was identified by X-ray diffraction in an Empyrean diffractometer equipped with a Cu lamp and a X'Celerator silicon detector. The analyses were carried out in the range from 5 to 60 $^{o}\theta$, and the ICDD PDF-2 database was used to interpret the diffractograms. The phase analysis was then supplemented with a microstructure study performed in the scanning electron microscope Quanta FEG 250, equipped with an X-ray microanalyzer. The microstructure testing was done on powdered, unsputtered samples deposited on the adhesive carbon tape glued to the microscope base. The tests were carried out under low vacuum conditions (pressure = 30 Pa) using a 5 kV electron beam.

3 Results and discussion

3.1 Phase analysis

Figure 2 shows the XRD patterns of the samples. The X-ray analysis of samples A and B shows that they consist mainly of calcite (Fig. 2a, b). The calcium carbonate formed as a result of carbonation of the calcium hydroxide exuding onto the stair side surfaces [6]. This phenomenon can be observed on concrete structure elements, e.g., bridges, viaducts [4], especially when the proper drainage of water has been neglected. The formation of efflorescence does not significantly reduce the durability of the concrete, but only affects the aesthetics of the structure.

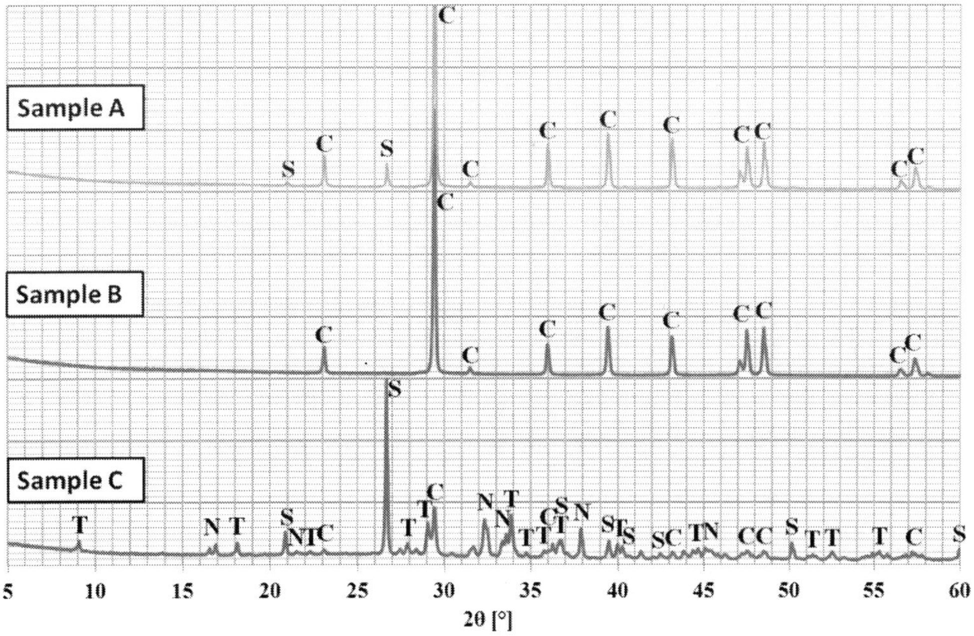

Fig. 2. XRD patterns of samples taken from the stair side surfaces: C - calcite S - quartz, N - thermonatrite, T – trona.

The phase composition of sample C (Fig. 2c) differs greatly from that of samples A and B. The sample C contains calcite, quartz and hydrated sodium carbonates – trona ($Na_2CO_3 \cdot NaHCO_3 \cdot 2H_2O$) and thermonatrite ($Na_2CO_3 \cdot H_2O$). Hydrated sodium carbonates may be formed by the carbonation of the alkali-aggregate reaction products. There are two types of alkali-aggregate reactions that pose a risk to concrete: alkali-silica reaction and alkali-carbonate reaction. Trona, a crystalline product, forms when sodium ions are responsible for alkali-silica reaction [7, 8]. The source of sodium ions was mainly de-icing salt used to remove ice from the surface of the stair. According to Katayama [9], trona could also result from the carbonation of sodium hydroxide exuding onto the concrete surface.

Thermonatrite can form during dedolomization during alkali-carbonate reaction according to the equation (1) [9]:

$$CaMg(CO_3)_2 + 2NaOH \xrightarrow{H_2O} Mg(OH)_2 + CaCO_3 + Na_2CO_3 \cdot H_2O \qquad (1)$$

or from sodium carbonate (trona) at a temperature over 57°C [10], which may have been present on the south side of the stairs during the summer months. Thermonatrite could also form as a result of the carbonation of sodium hydroxide with silica reaction products [11].

Calcite may have been formed both during the carbonation of portlandite and in the dedolomization process.

In contrast to the samples A and B, sample C exhibits a high background on the XRD pattern.

3.2 Microstructure

Microstructural analysis of samples A and B confirms the results of the X-ray phase composition studies. The efflorescences consist of calcium carbonates less or better developed (Fig. 3). The well-formed calcite crystals (Fig 3a) are dominant, as identified from the diffractograms. Spherical agglomerations of metastable vaterite (Fig 3b) occur less frequently. The metastable vaterite, which results from the carbonation of concrete, transforms into calcite over time [12].

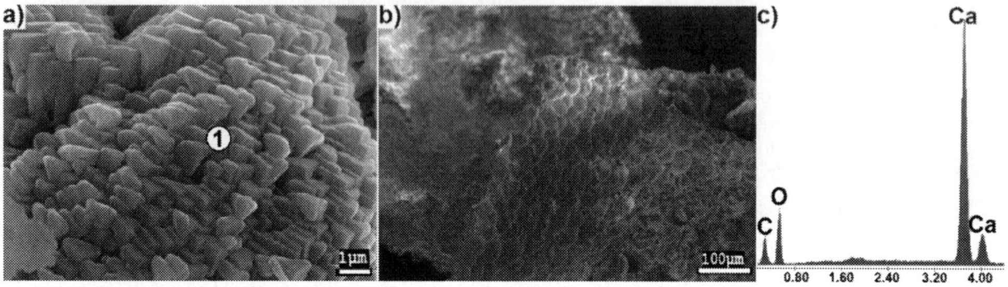

Fig. 3. Microstructure of sample B: a) calcite crystals (mag. 10 000×); b) spherical agglomeration of vaterite (mag. 3 500 ×); c) EDS at point 1.

Needle- and bar-like crystals are the main components of sample C (Fig. 4a). With high content of sodium and an increased content of carbon (Fig. 4b), they may be the hydrated sodium carbonates identified in the X-ray analysis [13]. The presence of silicon in the samples containing sodium carbonates can be explained by the occurrence of residual uncarbonated silica gel. A significant proportion of potassium is also observed in some of the carbonates (Figs 5b, 5e).

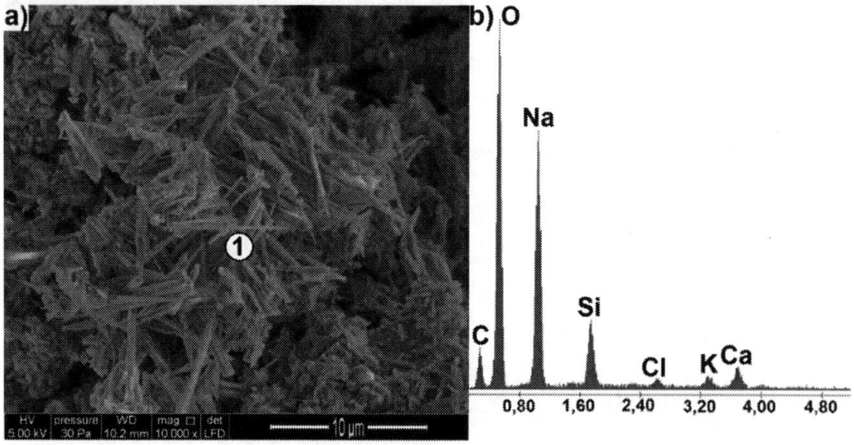

Fig. 4. Microstructure of sample C: a) hydrated sodium carbonates; b) EDS at point 1.

A dense amorphous phase was observed in the corroded concrete fragments (Fig. 4a, c), the presence of which was suggested by a high background exhibited on the XRD pattern. The X-ray microanalysis reveals a significant silicon content typical for the alkaline silica gel formed as a result of the alkali-silica reaction with sodium and potassium [14, 15, 16, 17]. The presence of chlorine and even single crystals of sodium chloride are also detected (Fig. 5f).

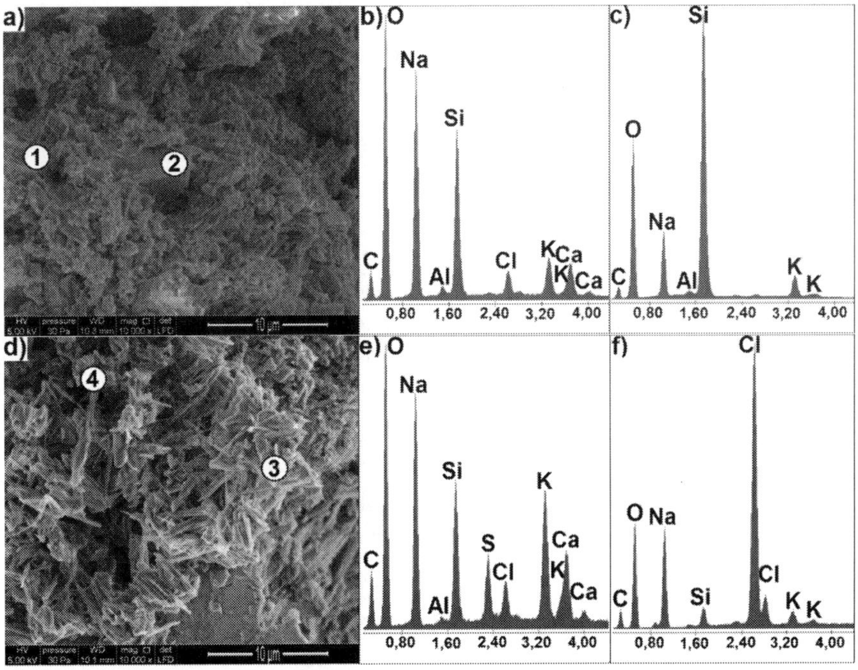

Fig. 5. a), d) Microstructure of sample C with the EDS at points 1 (b), 2 (c), 3 (e) and 4 (f) respectively.

Sample C contains single portlandite crystals coated with hydrated sodium carbonate (Fig. 6). The presence of portlandite may suggest that the calcite detected during the X-ray analysis could be formed partially as a result of the transformation of thermonatrite [9]:

Fig. 6. Crystal of portlandite covered in hydrated sodium carbonates.

$$Ca(OH)_2 + Na_2CO_3 \cdot H_2O \xrightarrow{H_2O} CaCO_3 + 2NaOH + H_2O \tag{2}$$

4 Summary

As it results from the data presented above, the encrustations and surface scaling on the surface of the stairs is the consequence of calcium hydroxide washing out and subsequent formation alkali-aggregate reaction products. Calcium hydroxide and alkali-aggregate reaction products transported to the surface of concrete are subjected to the carbonation process, resulting in the formation of calcite, trona and thermonatrite. These phases constitute the "building material" of encrustations and scaling present on the concrete surface.

The X-ray analysis revealed the presence of hydrated sodium carbonates, formed both during the carbonation of sodium hydroxide and alkali-silica reaction products. The amorphous, compact silicate gel of sodium silicate observed as a part of the SEM image suggests the occurrence of alkali-silica reaction [2]. The study of cores taken from the stairs allowed determining the contribution of particular types of alkali-aggregate reactions in the processes of concrete corrosion.

The reaction products of alkali-aggregate are rich in sodium. The X-ray microanalysis indicated the presence of chlorine and even sodium chloride crystals in the samples. Thus, the sodium ions derived from sodium chloride can promote the alkali-aggregate reaction. The source of sodium chloride is the de-icing salt used to remove ice from the stairs [18]. The granite tiles used on the surface of the stairs are made of granite of high water absorption, and the simultaneous lack of insulating layers on the surface of the concrete stairs causes the migration of de-icing salt solutions into the concrete. Calcium hydroxide, sodium hydroxide and hydrated sodium silicates emerge on the side surfaces of the stairs. Carbonation of these products on the side surfaces of the concrete stairs produces hydrated sodium carbonates, whose presence was confirmed in the study.

The occurrence of the alkali-aggregate reaction could be prevented by using mineral additives or chemical admixtures during the production of concrete [1, 14, 19, 20, 21]. They would increase the tightness of concrete and counteract the alkali-aggregate reaction by binding the sodium and potassium ions in the C-S-H phase [1]. The use of a binder with mineral additives can also counteract the leaching. In the pozzolan reaction relatively easily soluble calcium hydroxide can be replaced by the C-S-H phase.

As a result of this observed corrosion processes the concrete structure becomes loosened and weakened to such an extent that the granite tiles can fall off the stairs.

References

1. W. Kurdowski, *Chemia Cementu i Betonu* (Kraków 2010)
2. M.D.A. Thomas, B. Fournier, K.J. Folliard, Y.A. Resendez, Report No. FHWA-HIF-12-022 (2011)
3. Z. Owsiak, J. Zapała-Sławeta, P. Czapik, Bull. Pol. Ac.: Tech. **63**, 23-30 (2015)
4. W. Piasta, Eng. Fail. Anal. **79**, 606-614 (2017)
5. M.D.A. Thomas, B. Fournier, K.J. Folliard, Report No. FHWA-HIF-13-019 (2013)
6. W. Grzmil, Z. Owsiak, Cement Wapno Beton **18**, 137-144 (2013)
7. C. Balachandran, J.F. Muñoz, T. Arnold, Cem. Con. Res. **92**, 66-74 (2017)
8. C.E. Tambelli, J.F. Schneider, N.P. Hasparyk, P.J.M. Monteiro, J. Non-Cryst. Solids **352**, 3429-3436 (2006)
9. T. Katayama, Mater. Charact. **53**, 85-104 (2004)
10. K.J. Cho, T.C. Keener, S.-J. Khang, Powder technol. **184**, 58-63 (2008)

11. L. Baingam, T. Nawa, E. Iwatsuki, T. Awamura, Constr. Build. Mater. **95**, 820-831 (2015)
12. S. Liu, Z. Dou, S. Zhang, H. Zhang H., X. Guan, C. Feng J. Zhang, Constr. Build. Mater **150**, 591-594 (2017)
13. R.S. Gärtner, G.-J. Witkamp, Hydrometallurgy **88**, 75-91 (2007)
14. J. Zapała-Sławeta, Z. Owsiak, Bull. Pol. Ac.: Tech **65**, 773-778 (2017)
15. P. Czapik, Z. Owsiak, Cement Wapno Beton **21**, 79-85 (2016)
16. Z. Owsiak, J. Zapała-Sławeta, Ceramics – Silikáty **57**, 138-145 (2013)
17. Z. Owsiak, Ceramics – Silikáty **47**, 108-115 (2003)
18. A. Shayan, A. Xu, G. Chirgwin, H. Morris, Cem. Concr. Res. **40**, 563-568 (2010)
19. Z. Owsiak, J. Zapała-Sławeta, Cement Wapno Beton **20**, 25-31 (2015)
20. Z. Owsiak, P. Czapik, Cement Wapno Beton **19**, 152-157 (2014)
21. Z. Owsiak, P. Czapik, Cement Wapno Beton **18**, 310-320 (2013)

Calcarreous Fly Ash Content Signifacence To Self-Terminating Carbonation Model Of Concrete

Paweł Woliński[1,*], *Piotr* Woyciechowski[1] , and *Grzegorz* Adamczewski[1]

[1]Warsaw University of Technology, Faculty of Civil Engineering, Al. Armii Ludowej 16, 00-637 Warsaw, Poland

Abstract. Calcareous Ely Ash is a combustion by-product of Brown coal. Its technical properties depend on the combustion conditions – in the paper calcareous fly ash obtained from the traditional grate furnaces is considered. The presented paper shows the results of accelerated carbonation tests according to the draft of CEN / TS 12390-12 standard. The obtained models have self-terminating character. Concretes of water to cement ratio (w / c) from 0,35 to 0,55 and Ely Ash to cement ratio (p / c) from 0,2 to 0,5 were investigated.

1 Introduction

Due to a wide range of modern concrete applications, this material is exposed to many different types of interactions. The durability of concrete depends on its resistance to the impact of a given environment and it can be effectively shaped in particular at the material and technological stage [1]. Appropriate selection of concrete components quality and its composition, as well as performance conditions, such as the method of casting, compacting and, in particular, curing have a decisive impact on the durability of the final product. Due to the economic and ecological reasons it is justified to introduce waste components into the concrete, eg ashes. One of these types of additives is the low-calcium fly ash, which is not explicitly normalized in Poland with regard to concrete use. Fly ash limestone (W) is defined as "a very fine dust, having hydraulic and / or pozzolanic properties. It consists essentially of reactive calcium oxide (CaO), reactive silicon dioxide (SiO2) and aluminum oxide (Al_2O_3). The residue contains iron oxide (Fe_2O_3) and other compounds" [2]. PN-EN 206: 2014-04 [3] standard defines many different aggressive environments described in exposure classes. Among them, particular importance undoubtedly should be attributed to the impact of atmospheric CO_2 - resulting in the phenomenon of carbonation of concrete [4,5]. This phenomenon is accompanied by a gradual lowering of the pH of the concrete progressing slowly from the outside of the element towards the core. The concrete layer in which this process took place, i.e. where the pH value fell below a certain limit, is considered to be a carbonated zone. The rate of carbonation depends on many factors: the ratio of water to cement (w / c), CO_2 exposure time, concrete moisture, free content of lime, i.e. reaction capable, in cement stone, etc. Steel reinforcement in concrete environment is protected against corrosion if concrete it is not contaminated with aggressive substances in relation to

* Corresponding author: g.adamczewski@il.pw.edu.pl

steel, and its pH is high enough to ensure durability of so-called a passive film on the reinforcement surface. The passive film on the surface of the steel is formed thanks to the alkaline environment inside the concrete and is stable up to the pH level around 10 - 11. The pH of the concrete may, however, change during time. When concrete is hardened the excess mixing water can allow carbon dioxide from the atmosphere surrounding the building element to diffuse into the existing fine pores of the concrete. The reaction of carbon dioxide with calcium hydroxide leads to the formation of calcium carbonate, $CaCO_3$. When the carbonation front reaches the surface of the steel reinforcement, the passive layer protecting against corrosion disappears. When there is enough water and oxygen, the steel corrosion process begins.

2 Potential of using fly ash in cement composites

In Poland, fly ash is formed primarily in the Bełchatów Power Plant, but even a product from this one source has a high variability of chemical and physical properties. Despite the above, addition of this fly ash to cement in an amount up to 30% provides high strength after 90 days of hardening, including a significant increase between 28 and 90 days. It is believed that after treatment, fly ash is a valuable mineral supplement for the production of cement [9] and is therefore currently used in this field.

The PN EN 197-1 [2] standard distinguishes four types of cements in which fly ash designated in the cement composition with the letter W is used:

1) Portland multi-component ash cement (CEM II / A-W and CEM II / B-W);
2) multi-component mixed Portland cement (CEM II / A-M and CEM II / B-M);
3) pozzolanic cement (CEM IV / A and CEM IV / B);
4) multi-component cement (CEM V / A and CEM V / B).

They may contain various, depending on the cement type, amount of W lime ash (from 6 to 35%, and in some cases even up to 55%) [3].

An important direction for the use of fly ash in the production of building materials may also be the production of light aggregates. Although numerous production technologies for such aggregates have been developed, their production has not yet been developed on a larger scale. It is probably influenced by their relatively high price and too weak product promotion [4].

Other applications of calcareous fly ash include:

1) production of dry mortars,
2) strengthening of native soil and construction of road embankments,
3) hygienization of sewage sludge,
4) completion and reclamation of areas adversely affected by mining activities.

As the addition in concrete, fly ash is not standardized in Europe, which is due to the fact that it is still not a material well researched in the context of such use. Accordingly, the possible production of concrete with fly ash limestone should be carried carefully and with attention given to the issues of concrete rheology and in particular the problems associated with the deterioration of fluidity mix, loss of consistency, as well as the compatibility issues in a binder-additive-admixture system.

The potential limitation of the use of ash as a component of concrete may also be associated with unfavorable aspects of its impact on durability. The influence of fly ash on the durability of concrete limestone is an issue not widely discussed in the literature, in comparison with the impact of the siliceous fly ashes and opinions on this matter are sometimes divergent. Addition of this type of fly ash, which belongs according to ASTM classification [5] to a class C causes that the concrete has a very low permeability to water and solutions containing chlorine compounds, if used as a complement of the finest aggregate fraction and not as a replacement of cement [6]. Similarly, the introduction of lime to fly ash cement composition (ie. As a substitute for part of the clinker) increases the resistance of mortars permeability to chloride ions [7]. According to Dąbrowska [7], this resistance increases with increasing amounts of lime fly ash in concrete composition, extension of the period of curing and the application of lime fly ash activated mechanically (milling significantly increases the activity of the ash).

When replacing part of the cement with fly ash, the compressive strength of the concrete decreases, especially in early hardening period. During the hardening (28 days and longer) concrete with the addition of fly ash achieves strengths similar to or even higher than concrete without ash content [6].

The concrete in which a significant part of the cement was replaced by fly ash may have low resistance to frost. The main reason for the reduction in frost resistance is the slow course of pozzolanic reaction and the low rate of strength increase in concrete. Proper curing of cement-ash concretes aged for a longer period may reduce this decrease [6].

Cement with the addition of calcareous fly ash increases the susceptibility of mortars to carbonation, especially at its higher content in the cement composition. The depth of carbonation is lower for mortars with extended moisture curing in the initial hardening period [7].

Dąbrowska [7] has also investigated the influence of calcareous fly ash on the sulfate corrosion resistance solutions such as Na_2SO_4, $MgSO_4$, (NH_4) $2SO_4$ and the mixture of Na_2SO_4 and $MgSO_4$ and in artificial sea water. An adverse influence of calcareous ash was found in the case of contents up to approx. 30% of cement. Only 50% addition of fly ash to the cement composition allows to limit the expansion and decreases in compressive strength of mortars in each of the aggressive solutions, compared to Portland reference mortars CEM I. Introduction to the cement composition of fly ash in combination with another main component cement (ie granulated blast-furnace slag or silica fly ash) allows to increase cement resistance to sulfate corrosion and permeability of chloride ions. There is a noticeable synergic effect between these mineral components, especially at their higher content in the cement composition.

3 Modification of concrete with fly ash and a carbonation of concrete

Evaluation of the carbonation of concretes containing fly ash should take into account the microstructure of the hardened cement paste, since fly ash binds a part of $Ca(OH)_2$ as a result of pozzolanic reaction. As a result, the content of calcium hydroxide, which is capable of reacting with atmospheric carbon dioxide, is lower, thus the carbonation depth may be higher, since the same amount of CO_2 is sufficient to carbonate a larger area of concrete, according to Neville [8] on the basis of Bier's research [9]. These studies show that the depth of carbonation is higher when the content of Ca $(OH)_2$ in the cement paste is lower. It follows that the presence of fly ash can accelerate the process of carbonation and move the carbonation front into the concrete. However, there is also an opposite effect, associated with the formation of a more compact microstructure of the hardened paste with ashes, which reduces the diffusivity and the limits the rate of carbonation. It can therefore be concluded that the effect of fly ash on the course of carbonation consists of two opposite effects: • accelerating - associated with the deficiency of $Ca(OH)_2$ used for the pozzolanic reaction, whose carbon dioxide diffuses is "searching" deeper in the concrete; • inhibitory - associated with the structure sealing by fine grain ash and additional pozzolanic reaction products. The results of the published studies do not resolve unequivocally [10], which effect is dominating. It depends on the interaction of ingredients and external factors, however, one of the most important factors is curing. Proper concrete curing is essential for pozzolanic reactions to take place with the formation of a tight structure. It was found that badly cured ash concrete is very rapidly carbonating and poor curing of concretes with mineral additives after a year of exposure in the urban atmosphere resulted in a very large carbonation depth, reaching around 20 mm. Properly cured B30 (C25/30) concretes, made using cement with 30% admixture of fly ash, showed only about 1% greater depth of carbonation in comparison with concretes of the same class from Portland cement [11].

The method of introducing ashes into the concrete mix, ie whether the additive is introduced as a substitute for part of the cement or as an increase in the amount of binder, is also important. In the first case, the effect of the $Ca(OH)_2$ deficit increases, in the second - the role of the sealing effect. There are not too many data in the publications on carbonation testing of concretes with lime ashes. The high content of free lime in the ash may suggest that the initial availability of $Ca(OH)_2$ for the carbonation reaction will be high, especially that the free lime in these ashes is particularly reactive [6]. One can therefore expect a large initial rate of carbonation. On the other hand, the intense course of the pozzolanic reaction will deplete the available calcium hydroxide, while filling the porosity structure with the products of this reaction, which should reduce the CO_2 diffusion capacity and inhibit the course of carbonation.

4 Formulation of finite carbonation model for concretes with calcareous fly ash

To formulate the model, an element of the equation (1 / sqrt (t)) was used, which determines the hyperbolic character of the carbonation function. Woyciechowski [12] presented the general model of carbonation in the form of:

$$h=a-b*(1/\mathrm{sqrt}(t)) \quad (1)$$

where: h - carbonation depth a, b - material and technological factors t - exposure time

In the general form of the model (1) it was assumed that the carbonation depth h is a function of the inverse of the element from the time of carbonation:

$$h=f(1/\mathrm{sqrt}(t)) \quad (2)$$

The model was developed for each of the examined compositions in a detailed form (1).

The free expression of the function (a) for each case (composition) is expressed in a complex form (3), so that its dependence on the constants w/c and p/c in the given composition is visible.

$$a=A+B*(w/c)+C*(p/c) \quad (3)$$

The research program assumed, for technological and practical reasons, values of the w/c ratio from 0.35 to 0.55 and p/c from 0.20 to 0.50. The calculated values of the above coefficients for the experiment plan were determined for five intermediate input values x1 and x2, determined for conventional values xi of 0, ± 1, ± 1,414 (Table 2). The research program was developed in the form of a two-factor, partial-rotal-quasi-uniformal plan with a double repetition at the central point. The plan was considered optimal due to three criteria of selection, ie the criterion of implementation, informativeness and efficiency.

Table 1. List of real and code variables of the series of the experiment plan

Nr. Of series	Coded values		Real variables	
	x1	x2	x1^ (w/c)	x2 ^(p/c)
1	-1	-1	0.379	0.24
2	1	1	0.521	0.46
3	-1.414	0	0.35	0.35
4	1.414	0	0.55	0.35
5	0	-1.414	0.45	0.20
6	0	1.414	0.45	0.50
7	0	0	0.45	0.35
8	-1	1	0.379	0.46
9	1	-1	0.521	0.24
10	0	0	0.45	0.35

5. Results

The Portland cement CEM I 32.5 R (compliant with PN-EN 197-1: 2012 [2]), tap water in accordance with PN-EN 1008: 2004 [13], and aggregate of 0/2 mm fraction (river sand) and gravel 2/8 mm, 8/16 mm, in accordance with PN-EN 12620 + A1: 2010 [14] were used for the tests. Calcareous ash used for research came from the Bełchatów power plant. The compositions of concrete mixes are given in Table 2.

Accelerated test of resistance to carbonation of concrete was carried out according to the procedure given in the draft standard EN 12390-12 [15]. The test is performed on a minimum of 2 samples with dimensions 100mm × 100mm × 500mm. The result of the test is the average of 20 measuring points. The samples were matured after curing not less than 42 days from the date of forming - including one day covered in the mould, 27 days in water and 14 days in laboratory conditions (RH 50 ÷ 65 %, T = 18 ÷ 250C), dried to constant weight. Carbonation took place in a hermetic carbonation chamber in an environment that meets the following conditions: forced air circulation with CO_2 concentration = (4 ± 0.5)%, T temperature (20 ± 2) 0^C, RH = (55 ± 5)%. The basic exposure time is 70 days. In the case of the presented tests, the study was carried out after 28, 56, 70 and 90 days of exposure. To determine the depth of carbonation, a phenolphthalein solution (1 g phenolphthalein dissolved in 70 g ethanol and then diluted with 30 g distilled water) was used. The solution was sprayed on a fresh cross-section of the sample and the colour change to violet after about 60 minutes was observed.

Table 2. Quantification of components (kg) per 1 m3 of concrete mixes

Component / Nr. of composition	Cement, kg	Water, kg	Aggregate, kg	Fly ash, kg	w/c	p/c	Superplasticizer % of cement mass	
							A	B
1	361.2	137	1872.2	88.1	0.379	0.244	1.50	2.49
2	299.6	156	1817.6	136.6	0.521	0.456	1.03	0
3	338.6	118.5	1904.4	118.5	0.35	0.35	1.50	2.20
4	317.3	174.5	1784.6	111.0	0.55	0.35	0.12	0
5	363.3	163.5	1816.6	72.6	0.45	0.2	1.44	0

6	297.8	134	1861.1	148.8	0.45	0.5	1.50	1.80
7	327.3	147.3	1841.2	114.5	0.45	0.35	1.50	1.44
8	312.4	118.5	1895.5	142.5	0.379	0.456	1.50	2.55
9	343.8	179	1781.6	83.5	0.521	0.244	0.48	0
10	327.3	147.3	1841.2	114.5	0.45	0.35	1.50	1.44

In total, under accelerated carbonation conditions, 9 concrete compositions were analyzed mathematically and with different w / c ratios and p / c ratios determined according to the experiment plan, obtaining a very good agreement of results with a hyperbolic model in the form of

$$h=A+B*(w/c)+C*(p/c)+b/sqrt(t) \qquad (4)$$

The correlation coefficients of hyperbolic models were developed on the basis of averages of measurements indicate a good match of the adopted hyperbolic model to the results of laboratory tests obtained under accelerated carbonation conditions (R = 0.85-0.99) regardless of the accepted range of material and technological variables (tab. 3). The adequacy of the actual course of the phenomenon and models obtained using the measurement averages is confirmed by the fact that in all measurement series the variability index of results did not exceed 10%. A wide range of material and technological variables adopted in the conducted experiments and high statistical compliance of experimental results with the tested general carbonation model and small dispersion of measurement results in the series of measurements allow to accept as true thesis that the process of carbonation is finite in the concrete space, and to the mathematical description changes in the range of carbonation can be used in the carbonation model, which is a hyperbola, the asymptote of which is the maximum range of carbonation. The basis for the adoption of the hyperbolic model is the recognition of the carbonation process to be finite in time as a result of the self-limiting nature of the phenomenon, due to the gradual decrease of the CO_2 diffusion rate and the rate of carbonation aspiring asymptotically to zero. The maximum carbonation depth h_{max} is the asymptote of the hyperbolic function with the equation:

$$h_{max} = \lim_{t\to\infty} h\left(\frac{w}{c};\frac{p}{c};t\right) = a + b * \left(\frac{w}{c}\right) + c * \left(\frac{p}{c}\right) \qquad (5)$$

Table 3. List of models of carbonation in accelerated conditions

Nr. Of compositio n	Model of carbonation, *h, t* – depth and time of carbonation	Ordinate of the asymptote, mm	Coefficient of correlation, R
1 w/c=0.379 p/c=0.24	*h*=(7.500938)+(2.907884)*0.379+(1.878495) *0.24+(-27.1548)/sqrt(*t*)	9.04	0.96
2 w/c=0.521 p/c=0.46	*h*=(9.74987)+(5.126820)*0.521+(4.538526)* 0.46+(-54.3055)/sqrt(*t*)	14.49	0.93
3 w/c=0.35		2.25	0.62

p/c=0.35	h=(1.769975)+(0.686727)*0.35+(0.686727)* 0.35+(-8.35332)/sqrt(t)		
4 w/c=0.55 p/c=0.35	h=(12.31813)+(6.822073)*0.55+(4.378277)* 0.35+(-65.9656)/sqrt(t)	17.59	0.96
5 w/c=0.45 p/c=0.2	h=(10.88112)+(4.954678)*0.45+(2.258276)* 0.20+(-50.5864)/sqrt(t)	13.56	0.94
6 w/c= 0.45 p/c= 0.5	h=(9.372020)+(4.274658)*0.45+(4.738282)* 0.50+(-50.3489)/sqrt(t)	13.66	0.92
7 w/c=0.45 p/c=0.35	h=(7.701497)+(3.522767)*0.45+(2.762448)* 0.35+(-37.2595)/sqrt(t)	10.25	0.85
8 w/c=0.379 p/c=0.46	h=(2.424346)+(0.983052)*0.379+(1.171441) *0.46+(-11.5269)/sqrt(t)	3.33	0.99
9 w/c=0.521 p/c=0.24	h=(11.88171)+(6.240330)*0.521+(2.929116) *0.24+(-60.4545)/sqrt(t)	15.83	0.94

6 Conclusions

The study has shown that the calcareous ash used in the research as a substitute for aggregates in the amount of 20 to 50% of cement allowed to obtain concrete with very low carbonation depth (from 7 to 9 mm after 70 days in 4% CO_2 concentration), further indicate that with constant cement content and w / c value, there is an optimal ash content, leading to a lower susceptibility to carbonation. The dynamics of the carbonation depth development changes in the case of an increase in the ash content - the intensive progress of the process is noticed until the 56th day of carbonation, if the ash content is high.

References

1. M. Thomas, Portland Cement **IS548** (2007)
2. PN-EN 197-1 (2012)
3. Z. Giergiczny, *Popiół lotny w składzie cementu i betonu* (WPŚ, Gliwice, 2013)
4. K. Galos, A. Uliasz-Bocheńczyk, Gosp. Sur. Min.**21**(1), 23-42 (2005)
5. ASTM C618-15 (2015)
6. Z. Giergiczny, *Rola popiołów lotnych wapniowych i krzemionkowych w kształtowaniu właściwości współczesnych spoiw budowlanych i tworzyw cementowych*, 325, (Politechnika Krakowska, Kraków, 2006)

7. M. Dąbrowska, *Wpływ PLW na odporność korozyjną kompozytów wykonanych z cementu portlandzkiego wieloskładnikowego* (Rozpr. Dokt., Politechnika Śląska, 2014)
8. A.M. Neville, *Właściwości betonu* (Stowarzyszenie Producentów Cementu, Kraków, 2012)
9. Th. A. Bier, Materials Research Society Symposium **85**, 123-34 (1987)
10. J.A. Bogas, S. Real, B. Ferrer, Cem. Concr. Comp. **71**, 110-121 (2016)
11. *Zastosowanie popiołów lotnych z kotłów fluidalnych w betonach konstrukcyjnych*, (IPPT PAN, Warszawa, 2010)
12. P. Woyciechowski, Prace naukowe budownictwo **157**, (Oficyna wydawnicza PW, Warszawa, 2013)
13. PN-EN 1008 (2004)
14. PN-EN 12620 +A1 (2010)
15. EN 12390-12 (2010)

Influence of cement type on carbonation of concrete mixtures

Christos Tassos[1,*], *Kosmas* Sideris[1], *Alexandros* Chatzopoulos[1], *Nikolaos* Pistofidis[2], and *Emmanouil* Chaniotakis[2]

[1]Laboratory of Building Materials, Democritus University of Thrace, Xanthi, 67100 Greece
[2]TITAN Cement Company, Kamari Plant, Elefsina 19200 Greece

Abstract. This research aims to investigate the influence of cement type to carbonation. For this purpose mixtures of four different cement mortars and eight different concretes that have been prepared with four different cements were left exposed for one year in open air in northern Greece. Results indicate that the type of cement influences the carbonation rate. Concrete mixtures produced according to the definitions of EN 206 standard perform lower service life against carbonation induced corrosion if the choice of the cement type is not carefully examined.

1 Introduction

The durability of concrete constructions is an important issue these days with huge economic, social and environmental impact. This problem was to be solved by the new regulations which use restrictions on composition according to the environmental category, in which the construction is exposed.

The exposure category of constructions is defined with clarity in EN 206-1, but the severity of environmental exposure depends on the local climate conditions. In recent times, efforts have been made to correlate the severity of exposure against carbon dioxide with meteorological data, mainly annual fluctuations in the relative humidity of the atmosphere. However, intense fluctuations in humidity and high average annual temperatures accelerate the carbonation. Another important factor is the type of the cement used. According to EN 206, in the case of exposure to carbonation, there are no predictions for the use of different types of cement.

This research aims to investigate the influence of cement type to carbonation. For this purpose mixtures of four different cement mortars and eight different concretes that have been prepared with four different cements were left exposed for one year in open air in northern Greece. Results indicate that the type of cement influences the carbonation rate. Concrete mixtures produced according to the definitions of EN 206 standard perform lower service life against carbonation induced corrosion if the choice of the cement type is not carefully examined.

* Corresponding author: ctassos@civil.duth.gr

2 Experimental program

Four different cement mortars and eight different concretes were produced. The cement mortars produced according to the European Standard EN 196-1 [1]. The concrete mixtures belonged in the strength classes C25/30 and C30/37 according to EN206-1 [2]. All concretes were prepared using four different cements, CEM I 42.5R, CEM II/A-M(P-LL) 42.5N, CEM II/B-M(W-P-LL) 32.5N and CEM II/(B-LL) 32.5N. In the first four concrete mixtures, CEM I 42.5R or CEM II 42.5 (A-M)N cement was exclusively used for the two strength classes mentioned above. In the remaining four, a cement combination of CEM II/B-M(W-P-LL) 32.5N or CEM II (B-LL) 32.5 N was used along with CEM CEM II/A-M(P-LL)42.5N. Coarse aggregates consisted of crushed granite with a maximum size of 32mm. The fine aggregate used, were crashed and natural silicate sand. High range water reducing carboxylic ether polymer admixture was added at different dosages to achieve the desired workability. The chemical analysis of the cements used is shown in Table 1.

Table 1. Chemical analysis of used cements.

	CEM I 42.5 R	CEM II/A-M (P-LL) 42.5N	CEM II/B-M (W-P-LL) 32.5N	CEM II (B-LL) 32.5N
SiO_2	20.86	23.85	28	14.4
Al_2O_3	5.38	5.22	9	5.1
Fe_2O_3	2.73	4.13	5.5	2.4
CaO	60.88	58.2	48	62.5
MgO	2.31	3.2	2.2	1.6
SO_3	3.13	3.3	3.2	3.2
K_2O	1.93	0.68	1.5	0.1
Na_2O	-	0.32	0.5	0.2
TiO_2	-	0.24	-	-
P_2O_5	-	0.06	-	-
SrO	-	0.03	-	-
Cr_2O_3	-	0.02	-	-
ZnO	-	0.01	-	-
MnO	-	-	-	-
Specific weight	3.1	3.1	3.15	3.05

Standard type of sand was used in cement mortar mixes, while gravel (2-16 mm) and rubble (8-32 mm) were used in the concrete mixes, resulted from crushing of natural rock in a crusher. The proportions as well as the properties of the fresh mixtures are presented for all concrete and cement mortars prepared in Tables 2 and 3.

Table 2. Mix design characteristics of concrete mixtures prepared.

Mix design	C 25/30 CEM I 42.5	C 30/37 CEM I 42.5	C 25/30 CEM II 42.5	C 30/37 CEM II 42.5	C 25/30 Mixed	C 30/37 Mixed	C25/30 Mixed	C30/37 Mixed
kg/m^3	C1	C2	C3	C4	C5	C6	C7	C8
CEM I 42.5	300	350	0	0	0	0	0	0

CEM II 42.5	0	0	300	350	150	200	150	200
CEM II 32.5	0	0	0	0	150	150	0	0
CEM II 32.5 (B-LL)N	0	0	0	0	0	0	150	150
River Sand	400	400	400	400	400	400	400	400
Crashed Sand	500	500	500	500	500	500	500	500
Gravel	270	270	270	270	270	270	270	270
Rubble	630	630	630	630	630	630	630	630
Water	180	175	180	175	180	175	180	175
Sup/er	1.17	1.4	1.17	1.4	1.17	1.4	1.17	1.4
Retarder	0.56	0.69	0.56	0.69	0.56	0.69	0.56	0.69
W/C	0.6	0.5	0.6	0.5	0.6	0.5	0.6	0.5
Slump (cm)	16	17	22	15	18.5	21	15	15

Table 3. Mix design characteristics of cement mortars prepared.

Mix design	CEM I 42.5R	CEM II/A-M(P-LL) 42.5N	CEM II/B-M (W-P-LL) 32.5N	CEM (B-LL) II 32.5N
gr/1,7lt	M1	M2	M3	M4
CEM I 42.5	450	0	0	0
CEM II 42.5	0	450	0	0
CEM II 32.5	0	0	450	0
CEM II 32.5 (B-LL)	0	0	0	450
Sand	1350	1350	1350	1350
Water	225	225	225	225
W/C	0.5	0.5	0.5	0.5
flow (cm)	13.5	13.5	14	14

The cement mortar specimens were prisms 40x40x160mm, while concrete specimens prepared were 150mm edge cubes and cylinders 60x100mm. The cylindrical and the prism specimens were used to measure carbonation depth. These specimens were maintained in a curing chamber (T = 20 °C, RH> 98%) for 3 days and then placed in a laboratory environment until the age of 28 days. After this age, a set of specimens were placed into the accelerating carbonation chamber (T = 20 °C, relative humidity = 55%, CO_2 = 1%), while the rest of them were exposed to the outside open air space of the laboratory. This environment corresponds to the exposure class XC4 according to EN 206. All carbonation specimens remained in the exposure environment – accelerated carbonation chamber or open air environment- until the test age. Specimens used for compressive strength measurements were cured in the curing room until the age of the test.

3 Experimental results and discussion

3.1 Compressive strength

The compressive strength was measured for all concrete and cement mortar mixtures at ages of 2, 7 and 28 days. These values are shown in Tables 4 and 5 and in Figures 1 and 2.

Table 4. Compressive strength (MPa) of concretes prepared.

(MPa)	C1	C2	C3	C4	C5	C6	C7	C8
fc_2	26.4	30	23.9	24	18.1	22.7	15.3	22.7
fc_7	36.1	50.4	32.7	39.7	24.3	34.5	22.4	32.0
fc_{28}	46.1	58.7	39.7	47.6	35.0	43.3	32.5	40.2

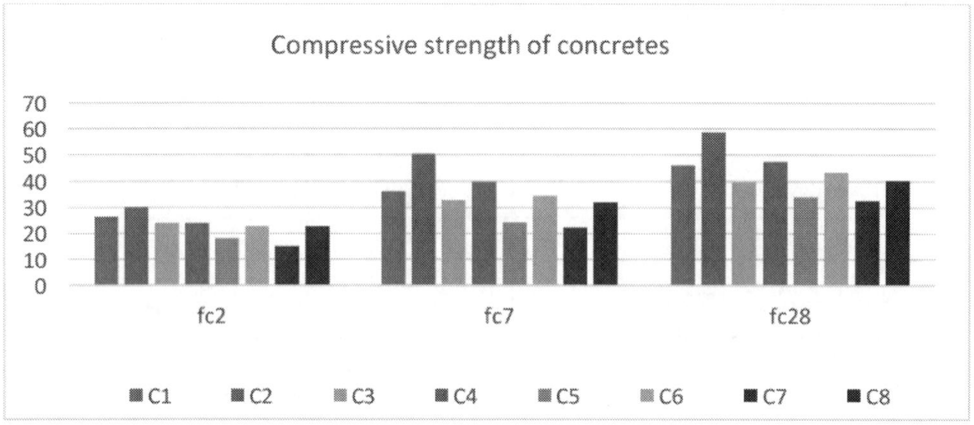

Fig. 1. Compressive strength of concretes (MPa).

Table 5. Compressive strength (MPa) of cement mortars.

(MPa)	M1	M2	M3	M4
fc_2	25.7	18.8	11.8	11.1
fc_7	38.4	30.0	21.9	20.6
fc_{14}	52.6	40.6	30.0	28.1
fc_{28}	54.4	48.1	42.3	41.3

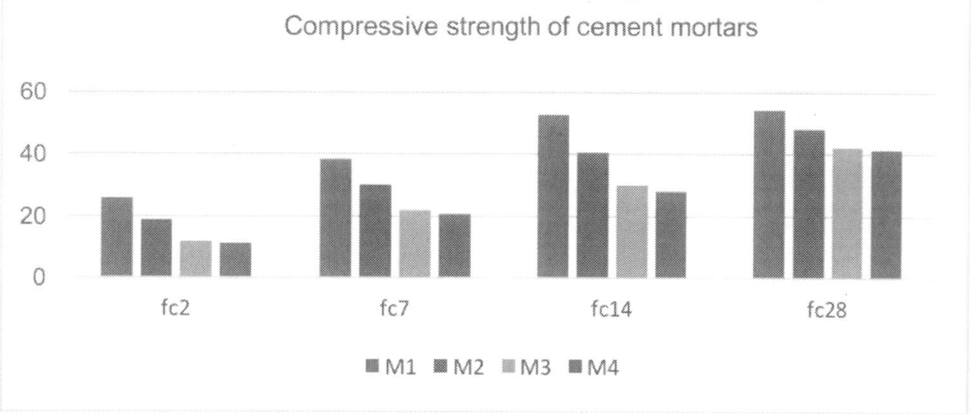

Fig. 2. Compressive strength of cement mortars (MPa).

It is well shown that all mixtures prepared develop the compressive strength needed in order to fulfill the strength requirements of their class according to EN206. Mixture C7 was the exception to this rule. The use of CEM II/B-M (W-P-LL) 32.5N and CEM II(B-LL)32.5N cements in combination with high w/c ratio resulted to lower compressive strength values at the age of 28 days.

3.2 Carbonation depth

The carbonation depth was measured by spraying the freshly broken surfaces of the samples with a phenolphthalein [4] indicator according to the procedure described in Standard EN 14630 [3]. The carbonation depth of all mixtures is shown in Tables 6 to 9.

Table 6. Accelerated carbonation depth of concretes prepared.

Carbonation (mm)	C1	C2	C3	C4	C5	C6	C7	C8
28 days	3.5	3.0	4.3	3.1	8.1	4.5	10.4	6.5
56 days	5.3	4.6	7.1	4.8	12.6	7.4	16.3	9.9

Table 7. Accelerated carbonation depth of cement mortars.

Carbonation (mm)	M1	M2	M3	M4
28 days	2.5	2.3	5.8	6.5
56 days	3.5	3	8.1	9.4
90 days	4.3	3.8	10.3	11.8
120 days	4.8	4.3	11.7	13.9
180 days	5.8	5.3	14.0	17.0

Table 8. Natural carbonation depth of concretes prepared.

Carbonation (mm)	C1	C2	C3	C4	C5	C6
6 months	3.0	2.5	3.3	2.8	5.9	3.3
9 months	4.2	3.4	4.5	3.8	7.9	4.3
12 months	5.2	4.1	5.5	4.9	9.5	5.3

Table 9. Natural carbonation depth of cement mortars.

Carbonation (mm)	M1	M2	M3	M4
3 months	0.5	0.6	1.5	1.9
6 months	0.9	0.8	2.5	3.0
9 months	1.0	0.9	2.9	3.5
12 months	1.1	1.0	3.3	3.9

The influence of cement type on the carbonation resistance of mixtures is clearly displayed on Tables 7 and 8. Both mortars and concretes prepared with type II cements found to be more vulnerable against carbonation. This vulnerability increased in mixtures C5 and C6 produced with combinations of CEM II 45.2 and CEM II32.5 cements. The higher pozzolanic content of these mixtures resulted to higher carbonation depths measured. The carbonation depth measured on mixture C5 produced with higher w/c ratio was almost twice the one performed by mixtures C1 and C3 (Table 6). Increased vulnerability of blended cement concretes produced with higher w/c ratios is in accordance of findings of other researches [5, 6].

Carbonation rate was further identified by calculating the carbonation coefficient k of concretes. These values were derived from linear regression between the values measured until the age of one year. The coefficients k calculated for all concrete and mortar mixtures prepared are demonstrated on Tables 10, 11 and figures 3 and 4 respectively.

Service life against carbonation was assessed by adopting the following formula:

$$x = k \cdot t^{0.5} \tag{1}$$

where k is the carbonation coefficient expressed in (mm/year$^{0.5}$) and t is the hydration time expressed in years. [7].

Table 10. Coefficient k of concretes.

	C1	C2	C3	C4	C5	C6
k	5.0138	4.0007	5.3382	4.6463	9.2897	5.1654

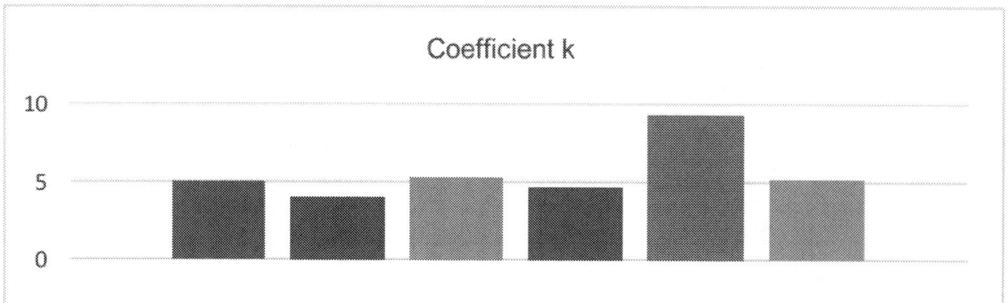

Fig.3. Coefficient k of concretes.

Table 11. Coefficient k of cement mortars.

	M1	M2	M3	M4
k	1.1474	1.0093	3.3707	3.9964

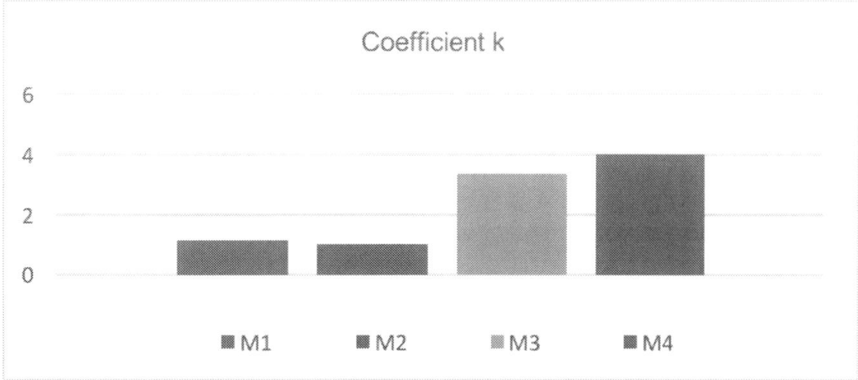

Fig. 4. Coefficient k of cement mortars.

The following scenario is about a construction exposed to natural carbonation [8]. According to EN206-1 [2] the reinforcement cover ($c_{min,dur}$) is 25 mm for concrete structures exposed to XC2 class (mixtures C1, C3 and C5) and 35mm for mixtures exposed to XC4 class (mixtures C2, C4 and C6),. Table 12 and Figure 5 present the time needed for carbonation to reach the reinforcement in each of the 6 concretes.

Table 12. Service life (initiation period) of concrete mixtures prepared (years).

Life Time	C1	C2	C3	C4	C5	C6
years	25	77	22	57	7	46

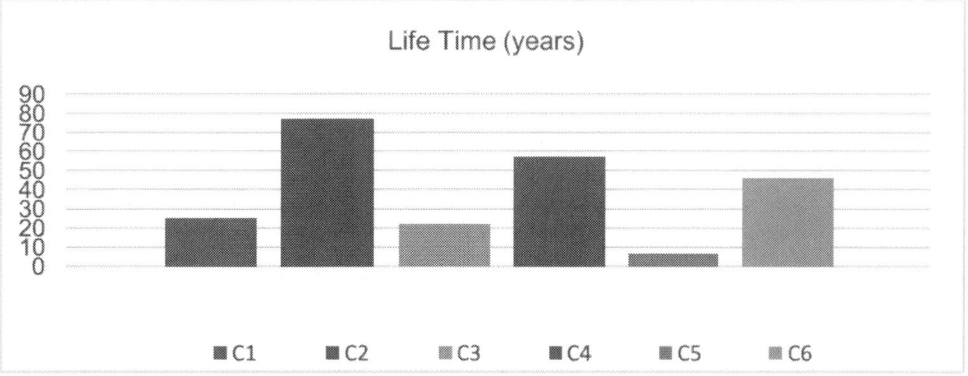

Fig. 5. Service life against carbonation (Years).

4 Conclusions

Carbonation rate of concrete mixtures is significantly influenced by the type and the strength category of the cement used. Concrete mixtures produced with combinations of CEM II/A-M(P-LL) 42.5N and CEM II/B-M(W-P-LL) 32.5N cements performed increased carbonation depth as compared with mixtures produced with CEM I or CEM II/A-M(P-LL) 42.5N only. Service life or reinforced concrete structures produced with these mixtures are therefore decreased down to 60% for C30/37 mixtures and 28% for C25/30 mixtures with high w/c ratio.

The required 50 years of service life is achieved only when a mixture with high cement content and low w/c ratio is used. However even in this case the choice of cement is crucial since the combination of type II cements of different strength grades is decreasing the service life close to the limit value of 50 years.

Use of lower grade concretes produced with higher w/c ratio and lower cement content does not allow the concrete structure to achieve the specified service life irrespective of the cement type adopted.

References

1. EN 197-1 (2000)
2. EN 206 (2011)
3. EN 14630 (2006)
4. K. Sisomphon, L. FrankeCem. Con. Res. **37**, 1647 (2007)
5. A Leemann, P. Nygaard, J. Kaufmann, R. Loser, Cem. Con. Comp **62** 33 (2015)
6. R. Neves, F. Branco, J. de Brito, Cem. Con. Comp. **41** 9 (2013)
7. M. Tarek Uddin, Md. Nazrul Islam, S. Kumar Sutradhar, M. Habibur Rahman Chowdhury, A. Hasnat, J. M. Khatib, *Third International Conference on Sustainable Construction Materials and Technologies* **1** (Kyoto, Japan, 2013)
8. J. Khunthongkeaw, S. Tangtermsirikul, T. Leelawat, Con. Build. Mat. **20**, 744 (2006)

Analysis of protective features of concrete in precast prestressed floor slabs (HC type) against chloride penetration

Zofia Szweda[1,*]

[1]Department of Building Structures, Silesian University of Technology, Akademicka 5, 44-100 Gliwice, Poland

Abstract This paper describes protective features of concrete used in precast prestressed floor slabs against chloride penetration. HC-500 hollow core slabs are designed for constructing ceilings in skeletal buildings with reinforced concrete. Material and structural protection used in slabs determines the acceptable range of using HC slabs, taking into account durability of floors during their normal exploitation. These slabs can be used in buildings where environmental conditions affect them according to the following exposure classes: X0, XC1, XC2, XC3 (according to standards: EN 206-1: 2003 [1], EN 1992-1-1:2004 [2]). Values of diffusion coefficient for concrete collected directly from precast prestressed HC-500 slabs, were determined during short (24h and 48h) tests accelerated by the electric field. In those tests we used a thermodynamic migration model and a reverse equation of chloride migration in concrete. Taking into consideration values of diffusion coefficient and diffusion equation, the time required for chloride concentration to reach critical value for reinforcement and the thickness of reinforcement cover were established under the condition that the structure durability was subjected to the exposure class XD3 (chapter 4 of the standard PN 1992-1-1[2]).

1 Introduction

Wide-span floor systems can be used to construct one-space interior with many functional qualities. Such floor systems facilitate very quick assembly of a construction, without using shuttering and supports that prevent creeping. Consequently, we obtain significant economic benefits and become independent of atmospheric conditions [3]. Nowadays, a variety of hollow core slabs produced in Poland is growing. The described HC-500 slabs were precast prestressed elements for floors in buildings of different type and construction. According to [4], HC slabs should be mainly used in frame constructions with reinforced concrete or steel structure while maintaining the fundamental static nature of a slab as a freely supported beam. HC-500 slabs covering the maximum span of 21 m are recommended for warehouses, industrial and commercial buildings, and service buildings. The slabs were made of C50/60 concrete, w/c = 0.3, with Portland cement CEM II 52.5 R. Material and structural shielding

* Corresponding author: zofia.szweda@polsl.pl

of the slabs: concrete class, cement type and content in concrete, cover thickness of prestressing tendons; determines the permissible application of HC slabs regarding floor durability during standard operation. Such slabs can be exclusively used in facilities where environmental conditions (corrosion inducing agent) directly affect the slabs and correspond to the following exposure classes: X0, XC1, XC2 i XC3 (according to standards: EN 1992-1-1:2004 [2], EN 206-1: 2003 [1]).

EN 1992-1-1:2004 [2] standard specifies the permissible critical concentration of chloride ions (C_K = 0.1 or 0.2% by cement weight) in prestressed constructions. Due to safety reasons, this value is lower than in case of reinforced concrete structure. British standard BS 8110: Part 1: 1997 [5] limits the maximum concentration of chlorides from all composition materials to the level of 0.1 ÷ 0.4% by cement content depending on the used type of reinforcement, cement and concrete. American standard ACI 318M-02/318RM02 [6] specifies the maximum content of chloride ions in concrete as 0.06% of cement weight in prestressed concrete elements, where concrete is in direct contact with prestressed steel [7].

The purpose of this paper was to determine values of chloride diffusion coefficient for concrete directly drilled from precast prestressed concrete slabs HC-500. To determine the coefficient, we performed short (24 and 48-hour) tests accelerated by the electric field and applied a thermodynamic migration model with the inverse problem of the equation for chloride migration in concrete [8]. Considering values of determined coefficients and the solved equation for diffusion, we defined time after which chloride concentration could reach its critical value on the reinforcement surface. We also specified the cover thickness for a structure with the given durability (50 years), structural class S4 and exposure class XD 3 - cyclic wet and dry environment, parts of bridges exposed to spray containing chlorides, pavements, car park slabs (in accordance with Chapter 4, PN 1992-1-1 [2]).

2 Experimentally determined and calculated diffusion coefficient

The tests were performed on elements drilled directly from the top surface of precast prestressed floor slabs HC-500, using a diamond drilling machine with a diameter of 8 cm (Fig. 1). The slabs were produced from the concrete mix specified in Table 1 [4].

Table 1. Components of concrete mix and average compressive strength of concrete f_{ctm}.

Concrete	Components	Mass [kg/m^3]	w/c	f_{cm} [MPa]
C50/60	Cement Cem II A-S 52.5	550	0.31	65
	Crushed BASALT	1304		
	Sand	580		
	Superplasticizer	6.24		
	Water	149		

The tests were performed on six cylindrical specimens with a diameter of 80 mm and height of 50 mm, whose side surfaces were coated with resin to ensure unidirectional flow of chloride ions. Containers with 3% NaCl solution were attached firmly to the top surface of cylindrical specimens. Stainless steel cathodes with a size adjusted to cross section of cylindrical elements, were placed inside the containers. The tests were simultaneously performed on three specimens connected in parallel to form the electric circuit (Fig. 1a). They were placed on a wet sponge below which a platinum-plated titanium mesh anode was. The specimens were soaked in distilled water for 72 hours to increase electrical conductivity of concrete. Then, a set of specimens was connected to direct current supply with voltage U = 18 V. The tests on chloride migration were performed for two times t_1 = 24 and t_2 = 48 hours. NaCl solution was replaced after 24 hours. During the whole period of tests, the temperature of solution was constant, at about 20°C. When the migration process was completed, the specimens were left under ambient conditions for 72 hours. Concrete powder was collected from each specimen, from10 consecutive depth increments of thickness g = 2 mm using "Profile Grinding Kit" from German Instruments AS company. We aggregated powder concrete from three similar specimens and added distilled water in a 1:1 weight ratio. That dense solution was stirred every few hours per 24 hours, and then passed through a medium-sized filter. Washing out was repeated twice. The experimental solutions were subjected to chemical analysis to determine concentration c^1 [mg/dm^3] of chloride ions in the liquid by argentometric titration in accordance with the Mohr's method (standard ISO 9297) [9] – Table 2.

Water in a 1:1 weight ratio was used to wash out powder concrete twice. Water weight by volume γ_w = 1 kg/dm^3 was used to estimate volume V_r of the experimental liquid, and then to determine mass density ϱ^1 of chloride ions [8].

$$\rho^1 = \frac{m^1}{V_c} = \frac{2c^1\gamma_c}{\gamma_w} \tag{1}$$

Table 2 presents calculated results for mass density ϱ^1 of chloride ions migrating in concrete specimens under the electric field.

Table 2. Concentration c^1 in the experimental solution and mass density ρ^1 in concrete – migration of ions Cl^- under the electric field.

Interval [mm]	Coordinate [mm]	Cl^- Concentration in solution c^1 [mg/dm^3]		Density of mass Cl^- in concrete ρ^1 [kg/m^3]	
		t_1 = 24 h	t_2 = 48 h	t_1 = 24 h	t_2 = 48 h
0 ÷ 2	1	306.60	577.6	1.61	3.03
2 ÷ 4	3	114.10	57.8	0.60	0.30
4 ÷ 6	5	40.30	29	0.21	0.15
6 ÷ 8	7	24.60	14.3	0.13	0.07
8 ÷ 10	9	18.60	14.1	0.10	0.07
10 ÷ 12	11	13.10	12.1	0.07	0.06
12 ÷ 14	13	16.00	15.4	0.08	0.08
14 ÷ 16	15	5.30	4.9	0.03	0.03
16 ÷ 18	17	1.80	8	0.01	0.04

Chloride ions migrate in concrete through liquid present in pores containing dissolved components of hydrated cement. We used the thermodynamic model of chloride ingress in concrete described in the paper [8], to analyse ion flow in pore solution. We identified a theoretical representative of volume X - Fig. 1c, and then defined concrete structure with water-filled pore space as an inert component $\alpha = 0$ which was not directly involved in the process. The following anions $Cl^- - \alpha = 1$, $OH^- - \alpha = 2$ and cations $Na^+ - \alpha = 3$, $K^+ - \alpha = 4$ and $Ca^{2+} - \alpha = 5$ were the components involved into this process.

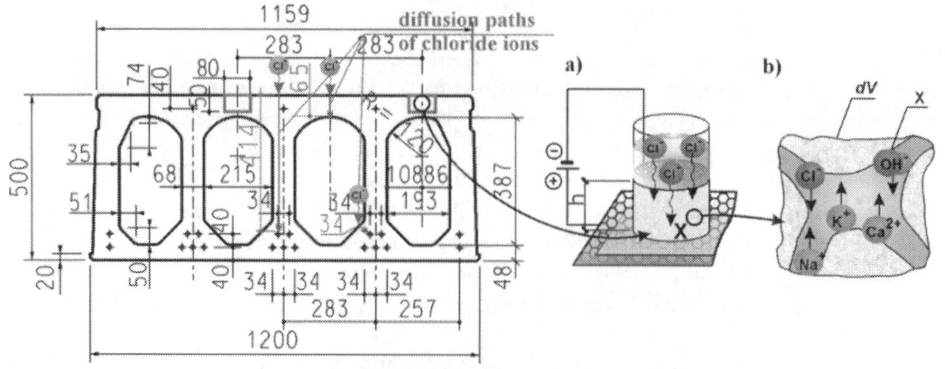

Fig. 1. Cross-section of HC500-19/R120 precast prestressed floor slabs and paths of chloride ion diffusion to prestressing tendons. The model of ion ingress in concrete: a) electrical scheme, b) the process components in the representative volume element.

The sample was parametrised with the coordinate system, where the top surface in contact with chloride solution corresponded to the plane x = 0, whereas the plane x = h specified the location of its bottom part. To determine experimentally the distribution of chloride ion concentrations $\rho^1(x)$, the powder material was collected from the depth increments of a specimen. The places, for which concentrations were determined experimentally, were located within the distance a from the top surface of the specimen. Concentration distributions were determined for t_1 and $t_2 > t_1$ during the time period of migration. Considering voltage U between the electrodes, we assumed potential $\varphi(0) = U$ at point x = 0 and $\varphi(h) = 0$ at bottom of the specimen(coordinate x = h). The potential distribution along the specimen height was approximated by the linear function.

As a result conversions, we obtained an expression determining the coefficient diffusion of chloride ions $D^1 = 1/Q$, that is, the converse diffusion resistance for the whole tested zone of the concrete within the range a

$$D^1 = \frac{\bar{j}^1(a)a\Delta t}{\frac{z^1 FUg}{RTh}\left[\bar{\rho}_1^1 + \bar{\rho}_2^1 + \ldots + \bar{\rho}_n^1\right]\Delta t - B} \qquad B = \int_0^a Q_x\left[\rho^1(x,t+\Delta t) - \rho^1(x,t)\right]dx \qquad (2)$$

where $\bar{j}^1(a)$ is the value of chloride ions flowing through the plane within the distance x = a (Fig. 3), $\bar{\rho}_1^1$, $\bar{\rho}_2^1$, and $\bar{\rho}_n^1$ densities of Cl^-at midpoints of consecutive intervals [0, g], [g, 2g], …, [(n-1)g, a] averaged over time Δt . The first term of the dominator determines the steady state of chloride ion flow, whereas the second one - B refers to the non-steady state [8].

On the basis of measured mass distribution ρ^1 of chloride ions migrating in concrete under the electric field, we obtained the reliable value of diffusion coefficient from the relation (2). Calculated results are presented in Table 3.

Table 3. Calculated results for the diffusion coefficient of ions Cl^-.

Zone	t [h]	$D_s^1 \cdot 10^{12}$ [m²/s]	$D_{ns}^1 \cdot 10^{12}$ [m²/s] Non-stationary influence 10%	20%	30%	40%	50%
I	24	0.05	0.06	0.07	0.08	0.09	0.11
	48	0.14	0.16	0.19	0.20	0.24	0.29
II	24	0.3	0.33	0.39	0.42	0.51	0.61
	48	0.2	0.22	0.26	0.28	0.35	0.41
III	24	0.87	0.96	1.13	1.22	1.48	1.74
	48	0.39	0.43	0.51	0.55	0.66	0.78
IV	24	1.17	1.28	1.52	1.63	1.98	2.33
	48	0.47	0.52	0.62	0.66	0.80	0.95
V	24	1.47	1.61	1.91	2.05	2.49	2.93
	48	0.52	0.57	0.68	0.73	0.89	1.04
VI	24	1.68	1.85	2.18	2.35	2.85	3.36
	48	0.5	0.55	0.66	0.71	0.86	1.01
VII	24	1.85	2.03	2.40	2.59	3.14	3.70
	48	0.5	0.57	0.68	0.73	0.89	1.04
VIII	24	2.48	2.72	3.22	3.47	4.21	4.95
	48	0.44	0.48	0.57	0.61	0.74	0.87
$\overline{D}^1$		**0.72**	**0.79**	**0.94**	**1.01**	**1.22**	**1.44**

The obtained value of diffusion coefficient $\overline{D}_s^1 = 0{,}72 \cdot 10^{-12}$ m²/s n for the steady state of the tested concrete classified the tested concrete to the group with very low chloride permeability [8, 10, 11, 12]. The parameters of the tested concrete were superior to the requirements of the standard EN 206-1 [1] in terms of minimum values of concrete durability, cement content and w/c ratio required for the analysed exposure class XD3.

3 Determination of protective features of concrete used in precast prestressed floor slabs (HC type) against chloride penetration

Protective properties of concrete cover were determined by calculating an increase of chloride ion concentration from experimentally determined diffusion coefficient for concrete that had been directly collected from precast prestressed floor slabs HC500-19/R120. For calculations we used concrete cover c_{nom} = 35 mm for tendons Ø9.3 made of steel Y1860, typical for those hollow-core slabs. The tendons formed the top reinforcement protecting concrete on the top surface of prefabricated units against cracking in the initial stage, directly after compression, and they protected long slabs against breaking down while they were lifted. Additionally, we determined the cover c_{nom} = 55 mm according to Table 4.5N, EN 1992-1-1 Eurocode 2 [2] assuming the structure durability (50 years) for structural class S4 and exposure class XD3 - cyclic wet and dry environment, parts of bridges exposed to spray containing chlorides, pavements, car park slabs. Diffusion coefficients of chlorides in steady state conditions was used for calculations because, according to the tests on concrete performed by the author [8, 10, 12], reliable values of diffusion coefficient could be obtained after six-month diffusion under steady state conditions. According to EN 1992-1-1 Eurocode

2 [2], prestressing tendons in the top part of a slab were prone to corrosion induced by chloride ions reaching the critical concentration was C_K = 0.2% by cement mass. We also determined the shortest path of chloride ion ingress in prestressing tendons with a diameter Ø12,5 (Fig. 1) located in the bottom part of a slab. That path was a sum of thickness of a top slab in the structural-floor tile - 65 mm and a tendon cover at the inner side of the hollow core - 34 mm (Czarnecki, Wojciechowski 2014[7]. Corrosion risk for tendons in the bottom part of main reinforcement was assumed to be induced by critical concentration of chloride ions equal to 0.1% according to EN 1992-1-1 Eurocode 2 [2].

Changes in chloride concentration between reinforcement and concrete cover x = 35, x = 55 and x = 99 mm were calculated according to the known solution for diffusion equation [7, 12, 13]

$$\rho^1_{cal}(x = 35,55;t) = \rho^1_{0,cal}\left(1 - erf\frac{x}{2\sqrt{D^1_{ns}t}}\right) \tag{3}$$

where $\rho^1_{0,cal}$ is the calculated mass density of chlorides at the the element boundary, erf – the Gauss error function, t- time.

Mass density of chloride ions $\rho^1(x = 35,55,99;t)_{cal}$ was referred to constant density of cement mass ρ_{cem} in concrete by determining relative concentration of chlorides

$$C^1(t) = \frac{\rho^1(x=(35;55;99),t)_{cal}}{\rho_{cem}} 100\% \tag{4}$$

Chloride concentration equal to $C^1_0 = 0{,}8\%$ by cement weight was assumed at concrete edge. Changes in chloride concentration over time t were computer determined. These changes are illustrated in Fig. 2.

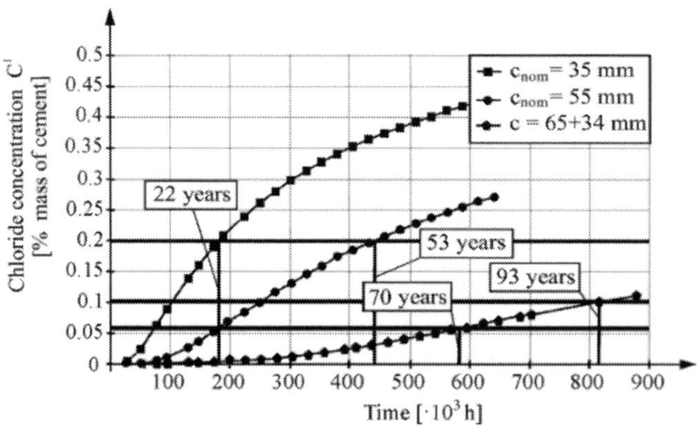

Fig. 2. Time-dependent changes of chloride concentration in concrete-steel contact zone for three different thickness values (35, 55 and 99 mm) of concrete coatings of steel prestressing tendons.

They referred to the interface between prestressing tendons in the top part of concrete and concrete cover - 35 mm and 55 mm, respectively, and for the shortest determined path of chloride ingress in lower edges of prestressing tendons in the main reinforcement – 99 mm. Time, after which prestressing tendons were prone to corrosion, was specified by comparing chloride concentration to the critical value. Corrosion risk for top tendons will appear after

22 years of the structure operation providing that cover thickness is 35 mm. The minimum thickness of cover for top tendons for structural class S4 and exposure class XD3, that is, the thickness of 55 mm would protect upper tendons against corrosion for 53 years. Assuming that chloride ions penetrate bottom tendons from the top surface of floor slab, and the limit value of chloride concentration is $C_K = 0.1$, corrosion of tendons is possible after 93 years of operation. For the limit value of chloride concentration $C_K = 0.06$, bottom tendons will corrode after 70 years (Fig. 2).

4 Conclusions

On the basis of the accelerated tests performed on chloride ion migration into cylindrical elements drilled directly from the top surface of precast prestressed floor slabs HC500-19/R120, we determined the diffusion coefficient of chloride ions in concrete used to produce the analysed slabs. Changes in chloride concentration over time were determined for the shortest ingress path of chloride ions at the interface between top prestressing tendons and concrete cover used in such slabs, for the cover of minimum thickness recommended for XD3 class, and on the surface of bottom prestressing tendons. We determined time periods, after which those tendons could corrode assuming that penetration of chloride ions would be observed only at the surface of the top floor slab.

It may be difficult to draw firm conclusions about the predicted durability of analysed slabs on the basis of the obtained results. Due to the complex work of top and bottom tendons in the prestressed structure, the detailed static analysis should be performed. Its scope should include the possible scenario of tendons unbonded to concrete because of corrosion, and the resulting effect on bearing capacity of bottom tendons. The analysis of slab bearing capacity depending on thickness of used concrete cover is also required [14].

Critical value of Cl^- concentration is not clearly specified and depends on numerous factors, such as: a type of cement, C_3A content in cement, a type of concrete mix, w/c ratio, a type of steel surface. Therefore, tests on the development of corrosion risk are needed. They should involve the following electrochemical techniques: Electrochemical Impedance Spectroscopy (EIS) and Linear Polarization Resistance (LPR), and should indicate concentration of chloride ions on the steel surface of prestressing tendons that induces corrosion in the specimens drilled directly from the tested floor slabs [15]. Values of diffusion and migration coefficients for elements drilled from the described slabs should be also determined on the basis of binding European standards [16].

References

1. PN-EN 206-1
2. PN-EN 1992-1-1
3. A. Ajdukiewicz, J. Mames, *Konstrukcje z betonu sprężonego* (PC, Kraków, 2004)
4. G. Troszczyński, *Wytyczne projektowania stropów z płyt sprężonych „HC". Poradnik dla projektantów* (Steelco, 2013)
5. BS 8110: Part 1: 1997 (1998)
6. ACI 318M-02/318RM02 (2002)
7. L. Czarnecki, P. Woyciechowski, *XIII Konferencja Naukowo-Techniczna*, 102-124 (WPRB, 2014)
8. Z. Szweda, A. Zybura, Procedia Eng. **57**, 1121–1130 (2013)
9. PN-ISO 9297

10. Z. Szweda, *Badania oporu wnikania chlorków do betonu na cementach portlandzkim zwykłym i niskoalkalicznym*, 148-153(OpK, 2016)
11. Z. Szweda, Roczniki Inżynierii Budowlanej, 79-86, (2016)
12. Z. Szweda, T. Ponikiewski, J. Katzer, J. of Clean. Prod. **156**, 569-576, (2017)
13. Z. Szweda, A. Śliwka, *12th Central European Congress on Concrete Engineering CCC 2017*, 118-126 (2017)
14. Z. Szweda, A. Śliwka, Procedia Eng. **193**, 417-422 (2017)
15. T. Jaśniok, M. Jaśniok, Procedia Eng. **108**, 316-323 (2015)
16. Z. Szweda, *Wybrane zagadnienia teoretyczne i doświadczalne w badaniach materiałów i konstrukcji budowlanych*, 451-463 (Gliwice, 2017)

The content of chlorides in blast-furnace slag cement as a factor affecting the diffusion of chloride ions in concrete

Justyna Kuziak[1,*], *Piotr* Woyciechowski[1], *Rafał* Kobyłka[2], and *Aldona* Wcisło[2]

[1] Warsaw University of Technology, Al. Armii Ludowej 16, Warsaw 00-637, Poland

[2] Institute of Agrophysics, Polish Academy of Sciences, Doświadczalna 4, Lublin 20-290, Poland

Abstract. Blast-furnace slag cement (CEM III) has a wide range of acceptable Cl^- content according to EN 197-1: 2012. This makes possible to use alternative fuels for the production of cement, which can increase the chlorides content in the clinker. However, it raises several new research problems, among them the problem of increase of the risk of Cl^- penetration intensity due to the higher concentration of chlorides in cement, presented in this paper. Studies have shown that after a short hardening time (28 days) the rate of diffusion of chloride ions increases slightly with the increase in the content of chlorides in the cement CEM III, but after full stabilization of concrete properties (1 year), the content of chlorides in cement CEM III practically does not affect the diffusion rate of chloride ions in concrete.

1 Introduction

Chloride ions are one of the two main causes of corrosion of steel reinforcement in concrete. Because of the reason above, existing standards contain limitations of chloride ions content in concrete as well as its components. According to EN 197-1: 2012 [1], the acceptable content of chloride ions in cement is 0.1%. The exception is blast-furnace slag cement (CEM III), which may contain more than 0.1% chlorides. The total Cl^- content in concrete in Europe is limited by EN 206: 2013 [2] and is in range from 0.1 to 1.0 depending on the type of reinforcement.

The higher allowed content of Cl^- in CEM III makes possible to use alternative fuels for the production of cement, which can increase the chlorides content in the clinker. However, it raises several new research problems, e.g.:

1) does the higher concentration of chlorides in cement affect the protective ability of concrete cover for reinforcement ?

2) does the higher concentration of chlorides in cement increase the risk of Cl^- penetration intensity?

The first problem was investigated in the Kuziak et al. [3], where influence of chloride ions content in cement type CEM III, in range 0.085-0.5%, on steel reinforcement was

* Corresponding author: jkuziak@il.pw.edu.pl

presented. Authors found no effect of chlorides content in the CEM III, over a range of tested values, on the corrosion of steel. Noncarbonated concrete cover, even at the highest chlorides content (0.5% in cement), protected reinforcement from corrosion.

Research focused on the second problem - influence of chloride ions content in CEM III on the diffusion of chloride ions in concrete made with this cement - is presented in this paper.

The effective coefficient of chloride ions diffusion (D_{eff}) is a measure of the rate of chlorides penetration into concrete. It depends primary on the composition of the cement (the ability to bind chloride ions by the cement matrix), the water/cement ratio and the age of the concrete. Concrete containing cement with ground granulated blast-furnace slag is characterized by low D_{eff} and value of D_{eff} decreases with the age of concrete [4]. To authors' knowledge there are no publication dealing with the effect of initial content of Cl^- in cement on chlorides diffusion in concrete.

2 Materials and methods

The chloride ions diffusion research was performed for concrete specimens with water/cement ratio 0.44 made of cements type CEM III A 42.4 with chlorides content of 0.1, 0.2 and 0.5%. The tests were carried out after 28 days and after a year of concrete hardening. Samples were stored before tests in a climate chamber (temperature = 20±2°C, RH > 95%).

Chlorides diffusion test for specimens after 28 days of hardening were carried out in a two-chamber vessel. These chambers, with volume of approx. 1.2 dm^3, were separated by concrete membrane (120x120x25 mm) (Fig. 1). One of the chambers was filled with a solution of NaCl with concentration 3% and the other one with distilled water. Chlorides diffused from the first chamber into the concrete. The concrete surface in contact with the solutions was ca. 81 cm^2.

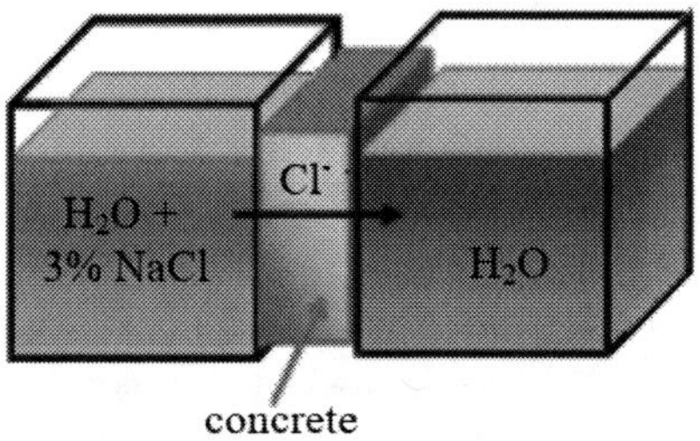

Fig. 1. The experimental set-up for the chlorides diffusion tests.

The concrete samples (120x120x50 mm) that hardened for a year were placed in a vessel filled with solution of NaCl with concentration 3%, so that the samples were completely immersed in the solution.

After 29-32 days concrete specimens were removed and concrete samples from various depths for determination of chloride ions were taken using the Profile Grinder Kit. Thicknesses of the taken concrete layers were 1 mm. Chloride ions determination was carried out using the Volhard's method. The determination of chlorides content in reference concrete (concrete not in contact with the NaCl solution) was also carried out.

The effective diffusion coefficient of chloride ions was determined by fitting the obtained values of chloride ions content in concrete to equation (1).

$$c = c_0[1 - \mathrm{erf}\left(\frac{x}{2\sqrt{D_{eff}t}}\right)] \quad (1)$$

where:

c – chlorides content in concrete at depth x,
c_0 – chlorides content in the near-surface layer of concrete,
erf – error function,
x – distance from the concrete surface,
t – time of chlorides diffusion in concrete.

To include the initial chlorides content in concrete (resulting from the chlorides content in all concrete components), calculation were also made for chlorides content after correction for the initial content (after subtraction the chlorides content in the reference samples that were not placed in the NaCl solution).

3 Results

The obtained results of the initial content of chloride ions in reference concrete and the content of chloride ions in concrete, to which chloride ions penetrated, are presented in Table 1-3 and Fig. 2. The chlorides penetration was lower in concrete tested after one year of hardening than in concrete tested after 28 days of hardening, which indicates the densification of the concrete microstructure over the time. In the case of samples that hardened for 28 days, a slight increase in the D_{eff} (faster penetration of chlorides into the concrete) was observed (Table 4, Fig. 3) with an increase in the chlorides content in cement from which the concrete samples were made. Chlorides penetrating the concrete can be bound by hydrated calcium aluminates present in hardened cement matrix of concrete (2) [5].

$$3CaO{\cdot}Al_2O_3{\cdot}6H_2O + 2Cl^- + Ca^{2+} + 4H_2O \rightarrow 3CaO{\cdot}Al_2O_3{\cdot}CaCl_2{\cdot}10H_2O \quad (2)$$

The slower penetration of chlorides into the concrete with a lower initial chlorides content may be associated with a greater ability to bind chlorides by hardened cement paste with a low chlorides content.

Also, the correction of the results by the initial chlorides content did not change the observed trend, however the increase of D_{eff} was slower than in the case of calculations without correction (Table 4, Fig. 3).

After a year of hardening of the concrete, reduction in D_{eff} was observed (Table 4, Fig. 4) - about two times in the case of samples made of cement containing 0.1% of chlorides and approx. 4-6 times in concrete with cement with a higher initial chlorides content.

Slower diffusion of chloride ions in older concrete is related to the concrete microstructure densification in time due to the hydration reactions of the binder components. This effect is particularly visible in the case of concrete with cement CEM III, which is characterized by a relatively low degree of hydration after 28 days. The dependence of D_{eff} values on time describes the equation (3) [6]:

$$D_{eff} = D_{eff\,0}\,(t_0/t)^m \quad (3)$$

where $D_{eff\,0}$ is the value of the effective diffusion coefficient determined on the basis of measurements at time t_0, and m is a parameter characterizing the change in the diffusion coefficient over time. The value of m depends mainly on the type of cement and water/cement ratio. Based on D_{eff} values obtained after 28 days and one year of concrete hardening, the parameter m is 0.3 for concrete made of cement with a chlorides content of 0.1% and 0.6-0.7 for concrete made of cement with a higher initial chlorides content. The observed effect is consistent with literature data, according to which the D_{eff} of chlorides in concrete with CEM III shows a significant decrease in time, and the obtained values of *m* are from 0.27 to 0.8 [6, 7, 8]. In the case of concretes made of Portland cement, the changes in the diffusion coefficient over time are much lower (according to literature data *m* is 0.1-0.25). This is related to the slower and longer hydration of cement type CEM III, which results in a slower increase in the strength of the concrete, as well as in the extension period of the concrete microstructure densification.

Table 1. The initial content of chloride ions in reference concrete to which chlorides did not penetrate.

Concrete	Concrete with CEM III with Cl^- content = 0.1%	Concrete with CEM III with Cl^- content = 0.2%	Concrete with CEM III with Cl^- content = 0.5%
%Cl^- in concrete	0.02	0.03	0.07

Table 2. The content of chloride ions in concrete after the diffusion test of chloride ions - after a 28 days of concrete hardening.

Depth, mm	%Cl^- in concrete with CEM III with Cl^- content = 0.1% (after 32 days of Cl^- penetration)	%Cl^- in concrete with CEM III with Cl^- content = 0.2% (after 32 days of Cl^- penetration)	%Cl^- in concrete with CEM III with Cl^- content = 0.1% (after 30 days of Cl^- penetration)
0-1	0.63±0.04	0.60±0.19	0.61±0.02
1-2	0.44±0.02	0.59±0.14	0.48±0.01
2-3	0.26±0.03	0.37±0.08	0.35±0.01
3-4	0.17±0.02	0.25±0.04	0.26±0.01
4-5	0.11±0.00	0.17±0.02	0.24±0.02
5-6	0.06±0.01	0.12±0.04	0.19±0.03
6-7	0.05±0.02	0.08±0.02	0.18±0.02
7-8			0.16±0.00

Table 3. The content of chloride ions in concrete after the diffusion test of chloride ions - after a year of concrete hardening.

Depth, mm	%Cl^- in concrete with CEM III with Cl^- content = 0.1% (after 29 days of Cl^- penetration)	%Cl^- in concrete with CEM III with Cl^- content = 0.2% (after 30 days of Cl^- penetration)	%Cl^- in concrete with CEM III with Cl^- content = 0.1% (after 30 days of Cl^- penetration)
0-1	0.47±0.15	0.60±0.10	0.42±0.20
1-2	0.26±0.13	0.27±0.01	0.25±0.17
2-3	0.13±0.10	0.15±0.04	0.14±0.08
3-4	0.07±0.05	0.07±0.00	0.10±0.04
4-5	0.06±0.04	0.04±0.01	0.08±0.01
5-6	0.03±0.01	0.03±0.00	0.07±0.01
6-7	0.02±0.01	0.03±0.01	0.08±0.01
7-8	0.02±0.00	0.02±0.00	0.07±0.00
8-9	0.02±0.00	0.02±0.01	0.08±0.00
9-10	0.01±0.00	0.02±0.00	0.08±0.00

Table 4. The dependence of the effective diffusion coefficient of chlorides in concrete on the initial content of chlorides in the cement.

Age of concrete	%Cl^- in CEM III	0.1	0.2	0.5
28 days	D_{eff}, 10^{-12} m^2/s	**1.7**±0.4	**3.2**±0.7	**5.4**±1.7
	$D_{eff\ corrected}$[1)], 10^{-12} m^2/s	**1.5**±0.4	**2.6**±0.6	**3.5**±1.0
1 year	D_{eff}, 10^{-12} m^2/s	**0.9**±0.4	**0.7**±0.1	**1.3**±0.6
	$D_{eff\ corrected}$[1)], 10^{-12} m^2/s	**0.8**±0.4	**0.5**±0.1	**0.6**±0.4

[1)] D_{eff} calculation based on chlorides content in concrete after correction for initial chlorides content.

After the correction of results for the initial content of chlorides in the concrete, similar D_{eff} values were obtained in all tested samples, which indicates no significant influence of the initial content of chloride ions in the cement type CEM III on the diffusion rate of chloride ions in concrete. Therefore, the factor determining the rate of penetration of chloride ions into concrete made of CEM III after a year of hardening is the age of concrete and the associated concrete microstructure densification.

A)

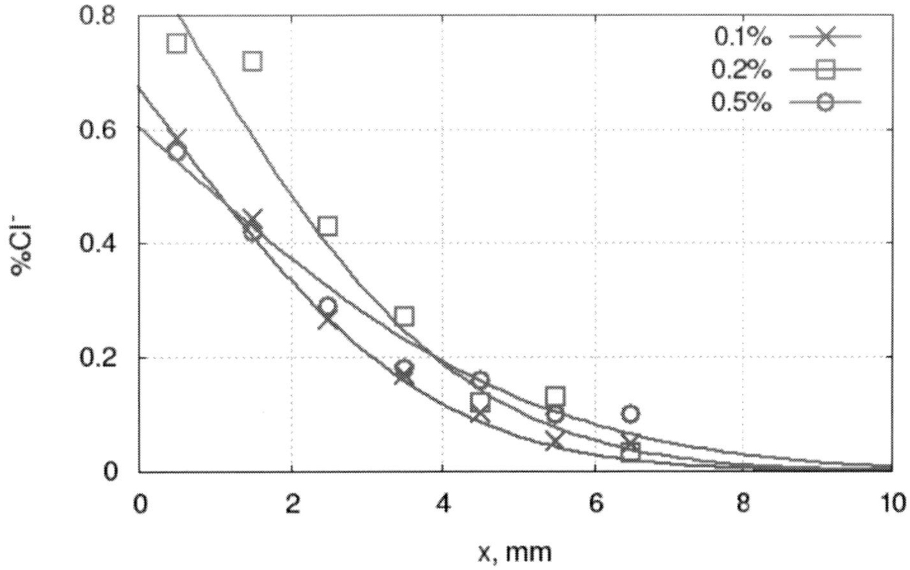

B)

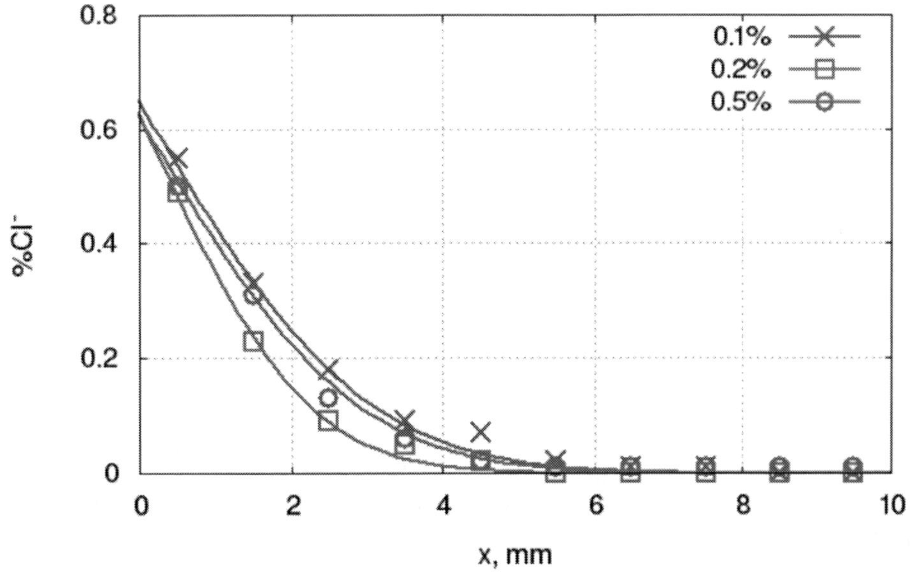

Fig. 2. Dependence of chlorides content in concrete made of cement CEM III containing 0.1, 0.2 and 0.5% chlorides, on the concrete depth (points) fitted to equation (1) (lines). A) after 28 days, B) after one year of hardening of concrete. Results after correction for initial chlorides content in concrete.

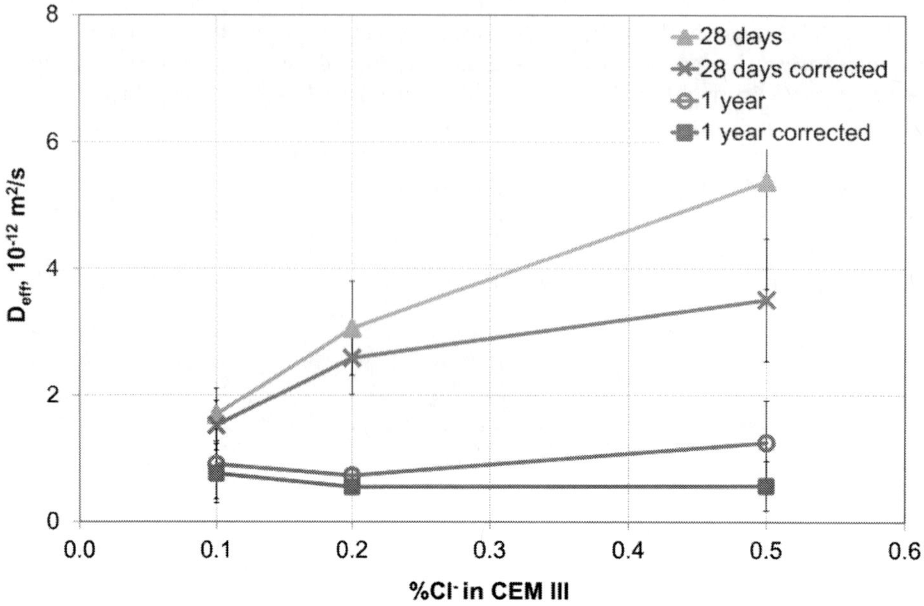

Fig. 3. Change of the effective diffusion coefficient of chlorides in concrete with cement CEM III with chlorides content 0.1, 0.2 and 0.5%.

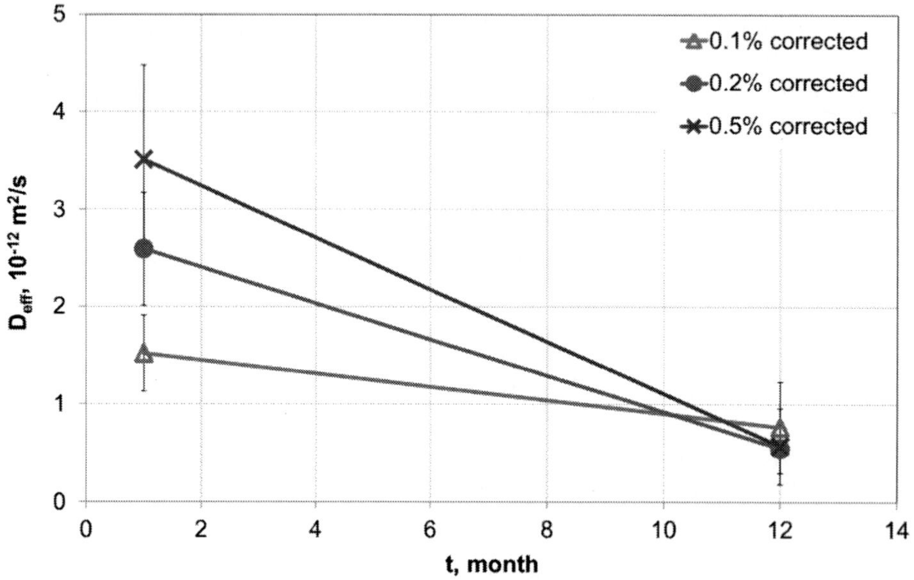

Fig. 4. Time dependence of the effective diffusion coefficient of chlorides in concrete made of cement type CEM III containing 0.1, 0.2 and 0.5% Cl^-. Results after correction for initial chlorides content in concrete.

4 Conclusions

On the basis of conducted investigations, the following conclusions were found out:

1. After a short hardening time (28 days) the rate of diffusion of chloride ions in concrete made of cement type CEM III increases slightly with the increase in the content of chlorides in the cement, but after full stabilization of concrete properties (1 year), the content of chlorides in cement CEM III practically does not affect the diffusion rate of chloride ions in concrete.

2. The age of concrete affects the diffusion rate of chloride ions. With the elongation of the hardening time from one month to one year, D_{eff} decreased - from 2 to 6 times depending on the content of Cl^- in cement.

3. The higher Cl^- content in concrete results in stronger influence of hardening time on D_{eff} value, but when Cl^- content is higher than 0.2% - differences between D_{eff} values are small. Calculated values of parameter m are as follows: 0.25 for Cl^- content equal to 0.1% and 0.6-0.7 for Cl^- content as high as 0.2-0.5%.

References

1. EN 197-1 (2012)
2. EN 206 (2013)
3. J. Kuziak, P. Woyciechowski, A. Wcisło, Corr. Prot. **59**, 169 (2016)
4. K.Y. Yeau, E.K. Kim, Cem. Concr. Res. **35** ,1391 (2005)
5. W. Kurdowski, *Cement and concrete chemistry* (Springer Science & Business, 2014)
6. D.H. Chisholm, N.P. Lee, *20th Biennial Conference of the Concrete Institute of Australia*, 96 (2001)
7. J.E. Carlsen, *2nd International RILEM Workshop on Testing and Modelling the Chloride Ingress into Concrete*, 161 (2000)
8. P.B. Bamforth, Mag. of Concr. Res. **51**, 87 (1999)

The influence of metahalloysite addition on tobermorite formation studied by X-ray powder diffraction and scanning electron microscope

Anna Skawińska[1,*]

[1]Institute of Ceramics and Building Materials, Division of Glass and Building Materials in Cracow, Cementowa Str. 8, Cracow 31-983, Poland

Abstract. This paper presents the results of the studies carried out in the model systems and concerning the tobermorite synthesis with an addition of metahalloysite. Quartz sand and quicklime were the main raw material constituents. The mixtures in the form of slurries underwent hydrothermal treatment with an addition of metahalloysite (5%, 10%, 15%, 20% and 30%) for 4 hours and 12 hours. The resultant composites were analysed for their phase composition using X-ray powder diffraction. The microstructure was examined using the Scanning Electron Microscope. Tobermorite was the principle reaction product. When 30% metahalloysite was added to the mixture containing CaO and SiO_2, the formation of katoite was found.

1 Introduction

Tobermorite is the principal phase formed in autoclaved building composites, the one which provides the final products with appropriate mechanical properties. Until now three tobermorite phases have been distinguished on the basis of their basal spacing: 14Å tobermorite- $Ca_5Si_6O_{16}(OH)_2{\cdot}7H_2O$, 11 Å tobermorite- $Ca_5Si_6O_{16}(OH)_2{\cdot}5H_2O$ and water free 9 Å tobermorite $Ca_5Si_6O_{16}(OH)_2$. The most important type is 11Å tobermorite, which is the major phase found in aerated autoclaved concretes. Understanding the mechanism of tobermorite formation in the autoclave process is very important for the production of AAC [1-5]. Huston et al. [6] stated, that tobermorite formation proceeds in three stages: formation of amorphous C-S-H phase, increase in semi-crystalline tobermorite content and its crystallization. Numerous studies shown, that reactive silica, aluminium, Na and K or sulphates compounds have significant influence on tobermorite synthesis [6-9].

The tobermorite was synthesized with an addition of metahalloysite. Metahalloysite was obtained by dehydroxylation of halloysite [10-11]. This reaction can be presented as the schematic formula

$$Al_2Si_2O_5(OH)_4 \rightarrow Al_2Si_2O_7 + 2\,H_2O \qquad (1)$$

Since tobermorite is important phase of autoclaved building composites, the influence of metahalloysite addition and autoclaving time on this phase formation in constant autoclaving

* Corresponding author: a.skawinska@icimb.pl

 https://doi.org/10.1051/matecconf/201816305008

conditions corresponding to saturated water vapour pressure at temperature of 180°C were examined.

2 Materials, methods and samples preparation

Raw materials used to obtain autoclaved composites included silica sand and quicklime. The silica sand chosen met the PN-EN 196-1 criteria. Grains with the size less than 0.5 mm were ground to reach the Blain fineness of 6560 cm^2/g. The quicklime was obtained by subjecting pure calcium carbonate to the heat of 1000°C for 2 h. Natural halloysite, used to obtain the metahalloysite, came from the Dunino mine, near Legnica. Metahalloysite was obtained by thermal treatment of halloysite at the temperature of 750°C for 2 hours. This mineral was used as an admixture 5%, 10%, 15%, 20% and 30% by weight of the joint weight of CaO and SiO_2.

Tobermorite was synthesized at the CaO/SiO_2 molar ratio of 0.86. The ratio of water w to the amount of solid constituents s was 2 (W/S=2). The constituents were placed in plastic containers and homogenized for 5 minutes. Metahalloysite was added towards the end of mixing the main components. The slurries were then poured into PTFE moulds and moved into a laboratory autoclave and held at 180°C for 4 and 12 hours at a saturated steam atmosphere. The temperature in the autoclave was elevated with the rate of 1.5°/min. The specimens were dried under vacuum to prevent them from being contaminated with CO_2.

The phase composition and the microstructure of the composites was examined using an X-ray diffraction method (EMPYREAN manufactured by PANalytical) and scanning electron microscope (Quanta 250 FEG manufactured by ThermoFisher Scientific) in the low vacuum mode. The steam atmosphere was 30 Pa. The microstructure of halloysite and metahalloysite are shown in Fig.1a and Fig.1b respectively. Details of raw material composition are given in Table 1.

a) b)

Fig. 1. The microstructure of a) halloysite b) metahalloysite.

Halloysite is consist of irregular forms and spatially dispersed tubes. Metahalloysite is amorphous and the particles with tubular morphology, typical for halloysie, are not found.

Table 1. Chemical composition of raw materials as determined by X-ray fluorescence.

Oxide [%]	Quicklime	Quartz	Metahalloysite
SiO_2	NS*	99.6	49.6
K_2O	NS*	NS*	0.1
TiO_2	NS*	NS*	2.0
Al_2O_3	NS*	NS*	41.5
Fe_2O_3	NS*	NS*	5.7
P_2O_5	NS*	NS*	0.8
CaO	99.8	NS*	0.3

NS*- not studied

3 Results

Sample designations and metahalloysite percentage share in mixes composed of quicklime and silica are given in Table 2.

Table 2. Designation of autoclaved mixes with metahalloysite.

Designation of samples	A1	MA1	MA2	MA3	MA4	MA5
Metahalloysite content [%]	0	5	10	15	20	30

Phase compositions of mixes composed of quartz sand and CaO with metahalloysite after 4h and 12h of autoclaving process are presented on Figs. 2-3.

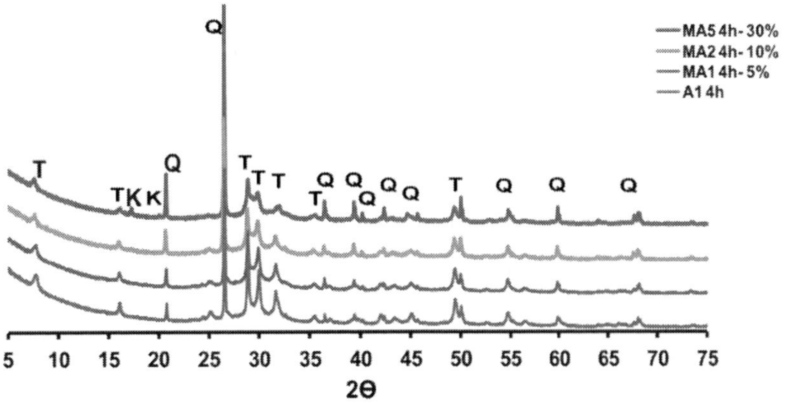

Fig. 2. X-ray diffraction patterns of mixes with metahalloysite addition, autoclaving time 4 hours. T-tobermorite, K- katoite, Q- quartz.

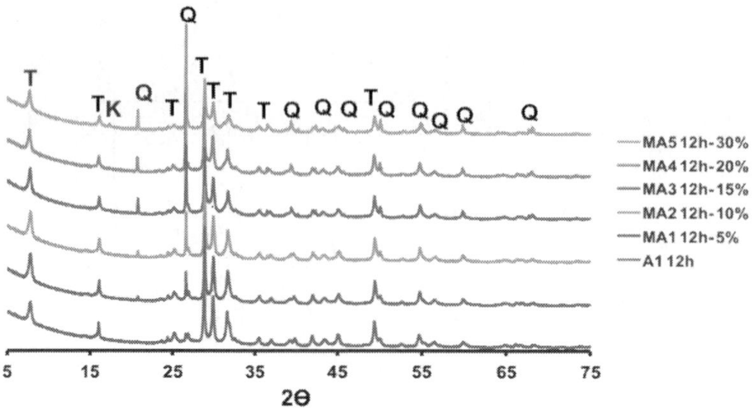

Fig. 3. X-ray diffraction patterns of mixes with metahalloysite addition, autoclaving time 12 hours. T- tobermorite, K- katoite, Q- quartz.

X-ray pattern analysis (Figs. 2,3) shown that primary synthesis product in samples with methalloysite is tobermorite. Except this phase, quartz is also present, but the intensity of this phase decreased with reaction time (after 12 hours). Katoite occurs in samples with 30% of metahalloysite after 4 hours. This phase appears when Al/(Al+Si) molar ratio exceeds 0.15. Hydrogarnet did not disappear after 12 hours. According to Klimesh et al. [12-16], katoite is indirect phase in tobermorite formation. Hydrogarnets durability decreases with increase in silica content and the extension of autoclaving time. Metahalloysite was not present in any sample. It means that this mineral completely reacted, and SiO_2 contained in metahalloysite is more reactive than quartz sand, therefore it quickly reacts with calcium ions.

Microscopic examinations confirmed the phase composition results from the XRD analysis. Figs. 4-5 show SEM selected images of the morphology of calcium silicate hydrates (C/S= 0.86) with metahalloysite. Observations under scanning electron microscope have shown, that time of synthesis and presence of mineral additions influence on products morphology. After 4 hours of autoclaving process, poorly crystallized calcium silicate hydrates are observed. Well-developed concentrations of tobermorite were observed in the samples without metahalloysite after 12 hours. In the sample containing 30% of metahalloysite, tobermorite is accompanied by hydrogarnets. The typical octahedral morphology of hydrogarnet occurring in twinned crystals is illustrated in Fig.5. Tobermorite occurs in the form of flattened laths.

a) b)

Fig. 4. Microstructure of reference sample after a) 4 hours b) 12 hours.

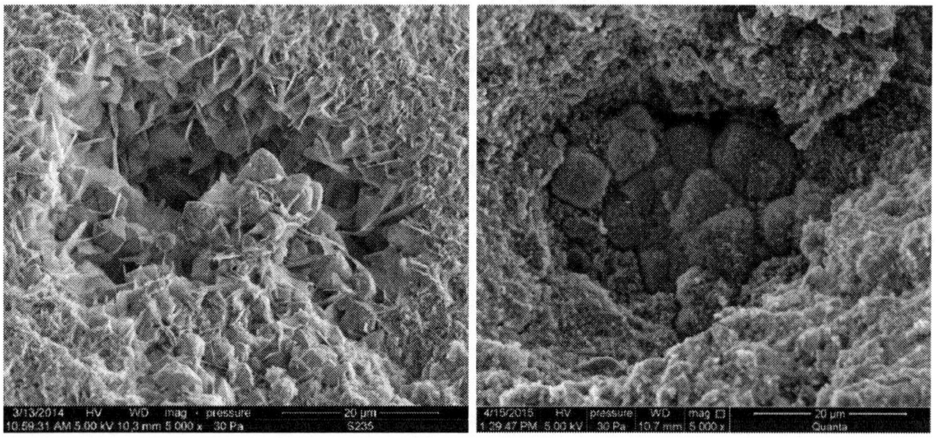

Fig. 5. Microstructure of sample with 30% of metahalloysite addition.

4 Analysis of the results

Tobermorite under hydothermal conditions can be synthesized with an addition of clay mineral such as metahalloysiye. Test results of samples with metahalloysite addition shown, that variable phase composition is obtained depending on the amount of used metahalloysite. Only tobermorite and quartz are observed in samples with metahalloysite addition in amount to 20%. Metahalloysite addition higher than 20% caused, except tobermorite, also katoite formation. Microstructure analysis under scanning electron microscope shown, that extension of hydrothermal treatment time favours the formation of calcium silicate hydrates with ordered structure. Hydrogarnets crystals in the form of rhombic octahedron were present in samples with 30% of metahalloysite addition.

References

1. S.A.S. El-Hemaly, T. Mitsuda, H.F.W. Taylor, Cem. Concr. Res. **7**, 429 (1977)
2. H.F.W. Taylor, *The Chemistry of Cements, Academic Press* (London, 1964)

3. K. Luke, H.F.W. Taylor, G.L. Kalousek, Cem. Concr. Res. **11**(2), 197-203 (1981)
4. H.F.W. Taylor, *Proc.4th Intern. Symp.Chem.Cem.,* 176-232 (Washington, DC, 1960)
5. H.F.W. Taylor, J.Am.Chem. Soc., 163-167 (1953)
6. J.H. Huston, R.S. Maxwell, S.A. Caroll, Geochem.Trans. **10**, 510-519 (2009)
7. A.J. Hamad, Inter. J.Mat. Sci.Eng. **2**(2), 152-157 (2014)
8. K. Matsui, J. Kikuma, M. Tsunashima, T. Ishikawa, S. Matsuno, A. Ogawa, M. Sato, Cem.Con. Res. **41**, 510–519 (2011)
9. E.V. Fomina, V.V. Strokova, World Applied Science Journal **25**(1), 48-54 (2013)
10. L. Stoch, *Minerały ilaste*, (Wydawnictwa Geologiczne, Warsaw, 1974 in Polish)
11. S. Guggenheim et al., Clays and Clay Minerals **54**, 761 (2006)
12. D.S. Klimesh, A. Ray, Cem. Con. Res. **28**(9), 1309-1316 (1998)
13. D.S. Klimesh, A. Ray, Cem.Con. Res. **28**(9), 317-1323 (1998)
14. D.S. Klimesh, A. Ray, Cem.Con. Res. **28**(8), 1109-1117 (1998)
15. D.S. Klimesh, A. Ray, Thermochimica Acta **334**, 115-122 (1999)
16. D.S. Klimesh, A. Ray, J.Thermal Analysis and Calorimetry **56**, 27-34 (1999)

Alkali-silica reaction and microstructure of concrete subjected to combined chemical and physical exposure conditions

Daria Jóźwiak-Niedźwiedzka[1,*], *Mariusz* Dąbrowski[1], *Karolina* Gibas[1], *Aneta* Antolik[1], and *Michał A.* Glinicki[1]

[1]Institute of Fundamental Technological Research, Polish Academy of Sciences, Pawińskiego 5b, 02-106 Warsaw, Poland

Abstract. Salt solutions are used to ensure safe driving conditions during winter. NaCl deicer is the most often used brine in Polish climatic zone. The chemical effects of this type of chloride-based deicer in wetting and drying (WD) and temperature cycles on concrete need to be better understood. This research was focus to study the microstructure of air-entrained pavement concrete after combined chemical (10% of NaCl) and physical (WD and 60°C) exposure conditions. The adopted WD and temperature regime was designed to verify the hypothesis that regularly alternating wetting and drying cycles with external alkali supply from deicer salt will provoke the Alkali-Silica Reaction (ASR). The aggregates varied their origin and mineralogical composition. The microscopic examination was carried out on concrete specimens using SEM with EDX. The microscopic analysis has shown that main reason for concrete deterioration during cyclic chemical and physical exposure conditions was both physical influence - WD cycles and the chemical influence – ASR (primarily, the fine aggregate which lead to form of alkali-silica gel). The expansive gel was shown to be capable of destroying the test specimens. Also differences in mineralogical composition of coarse aggregates influenced on the concrete prism expansion due to ASR.

1 Introduction

In central, eastern and northern Europe, and in the north America the climatic conditions in the autumn and winter season are particularly severe for concrete construction and concrete infrastructure. The successive wetting/drying (WD) and freezing/thawing (FT) cycles cause both external and internal concrete deterioration. In a winter season, the outdoor structures may (in certain cases) be exposed to over 100 cycles. The exposure conditions become even more severe in the urban and suburban areas located in cold regions as a result of heavy use of deicing salts. The safe driving conditions during winter months are of primary concern, hence large quantities of deicing chemicals are applied in order to clear the pavement and to allow for the normal traffic flow, [1]. Highway deicers often contain chlorides as freezing-

* Corresponding author: djozwiak@ippt.pan.pl

point depressants because of their cost-effectiveness. The sodium chloride (NaCl) is the most frequently used deicer, [2]. Chloride-based deicers may have detrimental effects on concrete infrastructure – which is often manifested by expansion, mass change, reduction of the dynamic modulus of elasticity, [3] and, most of all, strength, [4]. Generally, the results of several past studies [2, 3, 5-7] on the effects of deicers on concrete vary greatly, depending on such factors as the exposure conditions, concentrations of the deicers, and the test temperature.

Previous research studies have shown that both $MgCl_2$ and $CaCl_2$ deicers are known to accelerate the alkali-carbonate reaction in concrete containing reactive dolomite aggregates [8, 9]. Rajabipour et al. [10] prepared a list of known and still not fully understood phenomena related to the ASR. They stated, among others, that the ingress of NaCl into concrete, as a deicing salt, has been thought to exacerbate certain cases and laboratory studies of ASR, while having no effect on expansion in others.

For the performance testing of concrete in respect to ASR risk two methods were developed in Germany [11, 12]. Both methods comprise of cyclic exposure of concrete specimens to wetting-drying-soaking in NaCl solution while the changing the temperature withing the range from –20 to 60°C and from +20 to 60°C, respectively. The VDZ method [Muller] has been implenented at IPPT PAN for the ASR risk evaluation of domestic aggregates in concrete.

In this study the most frequently applied deicing solution - NaCl was used to evaluate the physico-chemical alterations in air-entrained pavement concretes. To assess the effect of salt on concrete microstructure, a high concentration of NaCl was assumed – 10%. The concretes were subjected to wetting/drying, which were designed to be reasonably realistic with respect to the actual field exposure conditions, and temperature up to 60°C was applied. The adopted WD and temperature regime was designed to verify the hypothesis that regularly alternating wetting and drying cycles with external alkali supply from deicer salt will provoke the Alkali-Silica Reaction (ASR). The microscopic examination was carried out to analyze the causes for this kind of behaviour. The microstructure of concretes specimens exposed to varied conditions was observed using Scanning Electron Microscope (SEM) with Energy Dispersive X-Ray Analysis (EDX).

2 Experimental program

2.1 Materials and mixture proportions

Portland cement CEM I 42.5R (Na_2O_{eq}=0.6%), natural siliceous sand with a maximum grain size of 2 mm (specific gravity 2.65 g/cm^3), tap water, a superplasticizer based on a modified policarboxylanes and air entraining admixture were used for concrete manufacturing. Six different coarse aggregates were used: amphibolite (A), basalt (B), granite (G), dolomite (D), limestone (L), and crushed from glacial deposit (Gr). The petrographic analysis of the above aggregates is presented in [13, 14]. Concrete mixes were designed with constant cement content (360 kg/m^3), constant water to cement ratio of 0.45, a constant slump of 100÷180 mm and the air content from 5.5% to 7.5% in the fresh mix. To achieve the target slump and air content the required amount of chemical admixtures was established experimentally.

Prismatic 75x75x285 mm specimens after casting, were protected from moisture loss and stored in the moulds at 20±2°C for 1 day. Then the prisms were demoulded and stored in a water for 27 days (temp. 20°C). After the 27 days of pre-conditioning, the prisms were subjected to 10 fourteen-day cycles: 5 days in 60°C, 2 days 20°C and 10% of NaCl solution, 6 days in 60°C and a minimum 98% R.H., 1 day 20°C and 98% R.H. The physical changes taking place in the prismatic specimens were monitored which included biweekly measurements of resonant frequencies, mass and length changes.

2.2 Test methods

The microstructural analysis was performed on the 25x42x10 mm specimens cut from the prisms after 10 cycles of alternating storage. Schematic illustration of the specimen prepared for the test and longitudinal section of beam with approximate location of SEM specimens is presented in Fig. 1. The small specimens from the beams were cut from their edges and from the middle. Above selected localization of the specimens was chosen as the most (Z) and least (S) influenced by NaCl deicer.

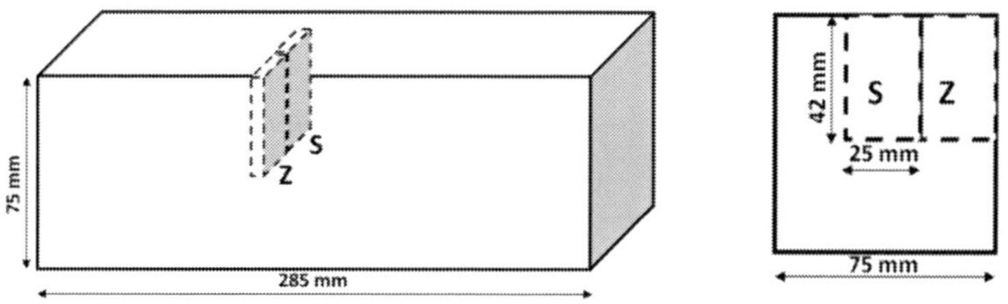

Fig. 1. Schematic of prismatic specimen used for experiment and the fragment of transverse section of the specimen showing approximate location of specimens for microstructural analysis.

After completion of exposure period, the specimens for microstructural investigation were obtained by slicing these specimens with a slow speed diamond saw. The samples were then dried in an oven at 50 °C for 3 days, vacuum-impregnated with a low-viscosity epoxy, lapped and polished using a special procedure for SEM specimens. Each specimen was prepared in such a way that the polished face which was to be examined was a cut surface and two of the edges of this face of interest were originally exposed to the deicer (localization Z) or one edge (localization S).

A strip of conductive tape was then attached to each polished sample, after which they were coated with a thin layer of carbon for about a minute. Each of the specimen was thoroughly examined using JEOL JSM-6380 LA SEM-EDX in the backscatter mode using an acceleration voltage of 15 kV.

3 Results and discussion

3.1 Visual inspection

The visual inspection revealed that almost all concrete specimens showed the crack network on the surface, also the single pop-out or the beginning of the surface scaling of the edges were observed for specimens with aggregate from glacial deposits (Gr), Fig. 1. It was also stated that concrete specimens with basalt aggregate (B) did not undergo any visible distress during the period of 140 days subjected to physical and chemical exposure conditions.

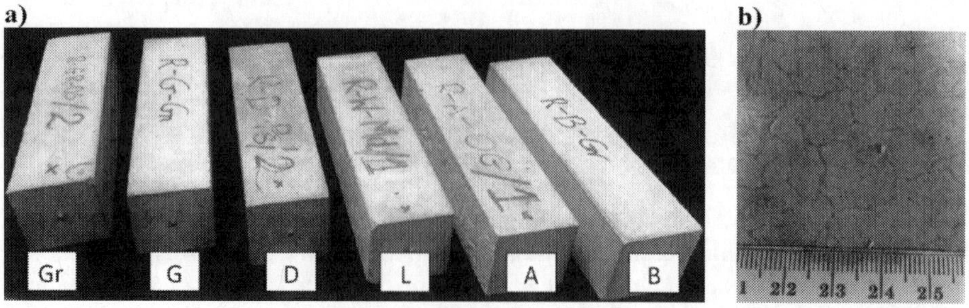

Fig. 2. Specimens subjected to 10 cycles of WD conditions, 10% of NaCl and 60°C: a) general view - one beam from each concrete, prismatic specimens 75x75x285 mm, b) surface view - with granite aggregate (G).

3.2 SEM analysis

When subjected to WD and temperature cycles and exposed to 10% of NaCl solution the concrete specimens with amphibolite aggregate (A), granite (G) and aggregate from glacial deposits (Gr) suffered more extensive surface deterioration than other concrete specimens. The cracks which were localized in the upper part of the analyzed specimen were mostly parallel to the surface and were empty. The lower part of the analyzed specimens, ~40 mm from the top, did not showed any spectacular cracking, Fig. 3.

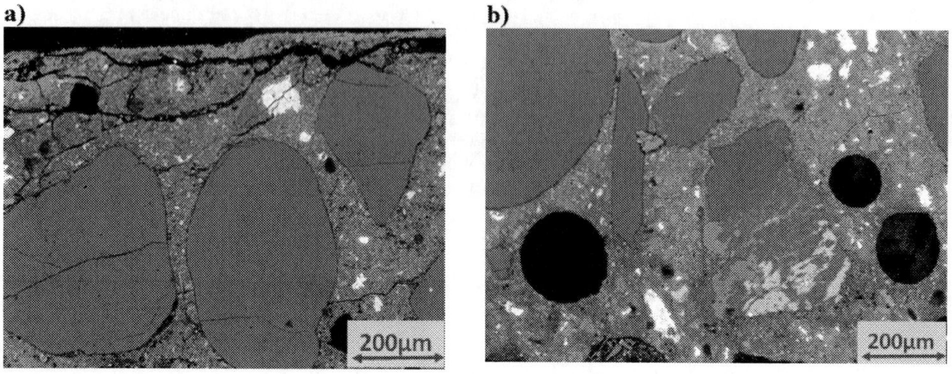

Fig. 3. Scanning electron micrograph for the a) top surface (up to 0.6 mm), and b) lower part, 40 mm of concretes with amphibolite aggregate (A); specimens localisation - internal (S), scale bar = 200 µm.

In all tested specimens changes in the matrix due to severe exposure conditions were visible. The cracks in the matrix were present in all analyzed concrete specimens, regardless of whether it was an external (Z) or an internal (S) specimen. The most severe matrix degradation has been observed in concrete with glacial deposit aggregate (Gr). Figure 4 shows the macro and micro cracking in concrete made with (Gr) aggregate. Cracks were present in the matrix and the debonding of aggregate was visible in all the concrete specimens, but the most extensive debonding was present in cover zone.

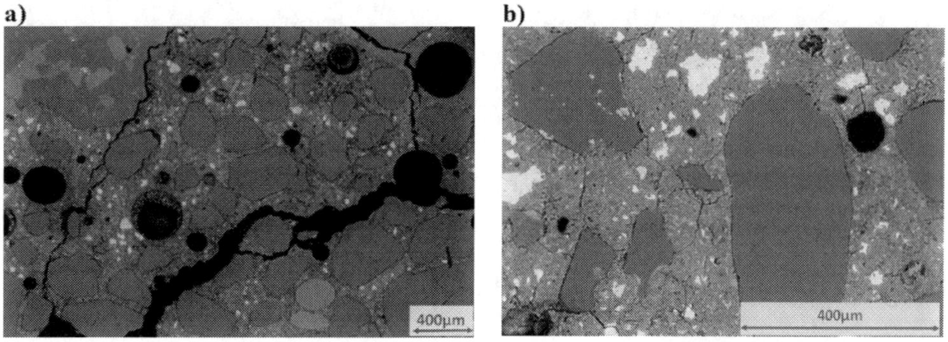

Fig. 4. Scanning electron micrograph of the cracked concrete with glacial deposit aggregate (Gr), a) larger cracks, b) smaller cracks, scale bar = 400 µm; specimens localisation - internal (S).

In both kinds of analyzed specimens, from the edge of the beams (Z) and from the middle of the beams (S), the ettringite and Friedel's salt were present in the cover zone, Fig. 5. Ettringite needles were totally or partially filling the air-voids and Friedel's salt was localized in the cement matrix.

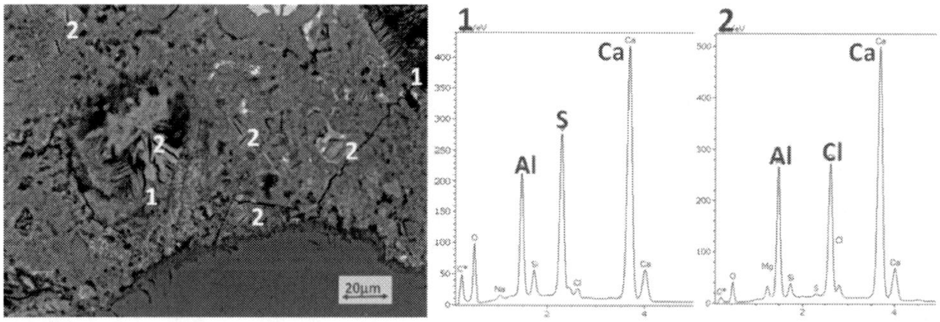

Fig. 5. Scanning electron micrograph of the cement matrix with visible a) ettringite deposit (1), b) Friedel's salt (2); concrete with glacial deposit aggregate (Gr), scale bar=20 µm; specimens localisation – external (Z).

Chloride penetration profiles for concrete specimens in 10% of NaCl solution were comparable for all tested specimens. The microstructural changes for all concretes were observed in terms of formation of Friedel's salt in the top and bottom (~40 mm, from the surface) regions of the specimens exposed to NaCl salt. The lowest deposition of Friedel's salt has been observed in concrete specimens which were taken from the middle of the tested beams. Additionally Friedel's salt was present only in a cement matrix, the air-voids contain only a slight signs of this product. The salt was homogenously distributed in all cement matrix in cover zone. Fig. 6 presents an example of mapping of aluminium and chloride content in concrete with amphibolite aggregate (A). The brighter dots of Al and Cl are overlapping, which corresponds to Friedel's salt.

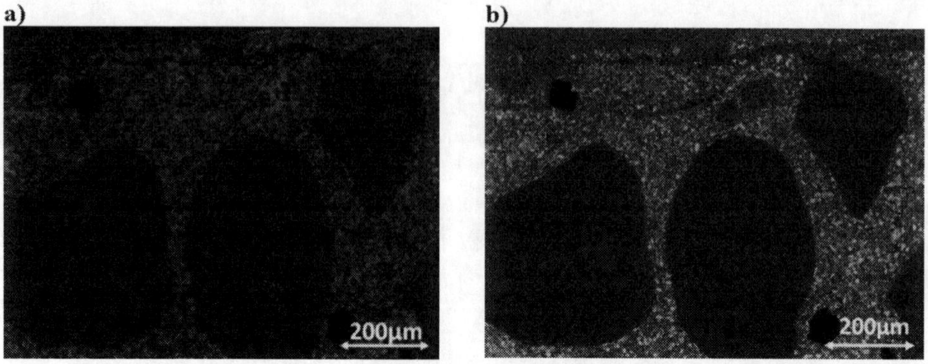

Fig. 6. Mapping taken on the top surface (up to 0.6 mm) of concretes with amphibolite aggregate (A): a) aluminium content, b) chloride content; specimens localisation - internal (S), scale bar=200 μm.

When the symptoms of ASR were visible in the analyzed specimens, they occurred in places that were furthest from the outer edges of the beams. The greatest occurrence of the alkali-silica gel was in the bottom of the specimens labeled (S). The analyzed specimens differed in vulnerability to ASR occurrence, depending of the origin of coarse aggregate. The most severe degradation caused by ASR was visible in concrete with aggregate from glacial deposit (Gr), amphibolite (A) and with granite aggregate (G). The alkali-silica gel was filling cracks and air-voids in tested concrete. Due to the considerable variety of rock and mineral types which were found in the analysed crushed gravel from glacial deposits 13, the highest potential to ASR occurrence was expected. The example of the Si-Ca-Na-K gel in concrete with (Gr) aggregate is showed in the Fig. 7. In this particular example the sandstone with silica (chalcedonite) cementation was responsible for the ASR phenomenon.

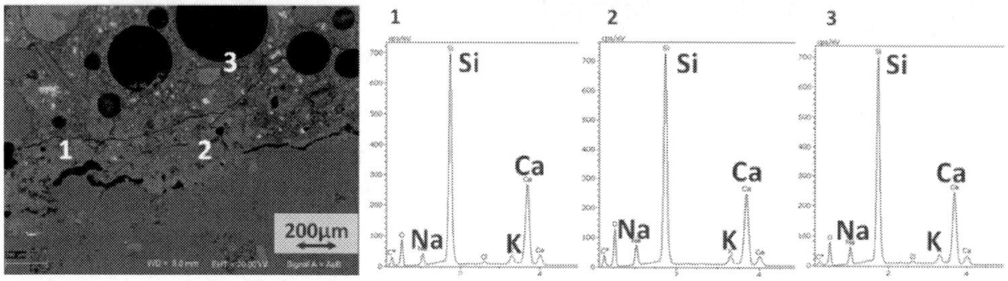

Fig. 7. Scanning electron micrograph of the cracked concrete with glacial deposit aggregate (Gr), the Si-Ca-Na-K gel is shown; scale bar=200 μm; specimens localisation - internal (S).

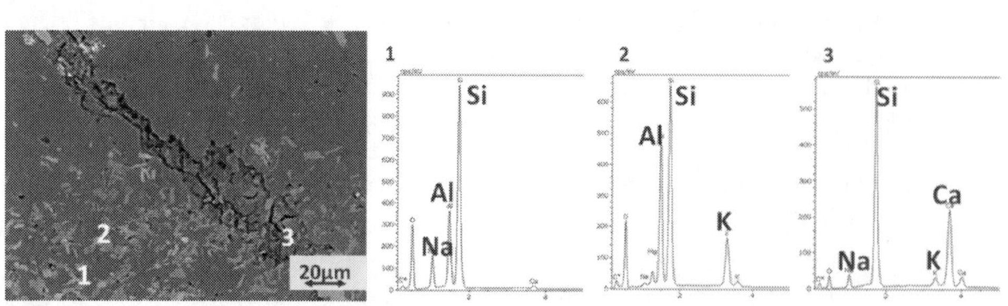

Fig. 8. Scanning electron micrograph of the cracked concrete with amphibolite aggregate (A); 1- albite, 2-K-feldspar, 3-the Si-Ca-Na-K gel; scale bar=200 μm; specimens localisation - internal (S).

Concretes with amphibolite and granite aggregate showed also significant signs of ASR. The SEM-EDX analysis revealed that the cracks in above aggregates were filled by Si-Ca-K-Na gel, Fig. 8. Minerals like K-feldspar and albite were found in coarse aggregate, and they suggest to provoke ASR. The cracks in an aggregates started in the center of grains and the sodium from deicer would start the reaction at the grain boundary, like in Fig. 10.

In coarse aggregate, dolomite (D) and limestone (L), and in siliceous sand visible signs of ASR were present – Si-Ca-Na gel in cracks, Figs. 9, 10. In the Fig. 9 the fibrous crystals are visible beneath dolomite grain. They are filling the main crack, which propagates through whole grain. Fig. 10 presents a cracked sand particle with visible ASR gel, which comes through the sand grain into the air-void from one side, and into the cement matrix from other side.

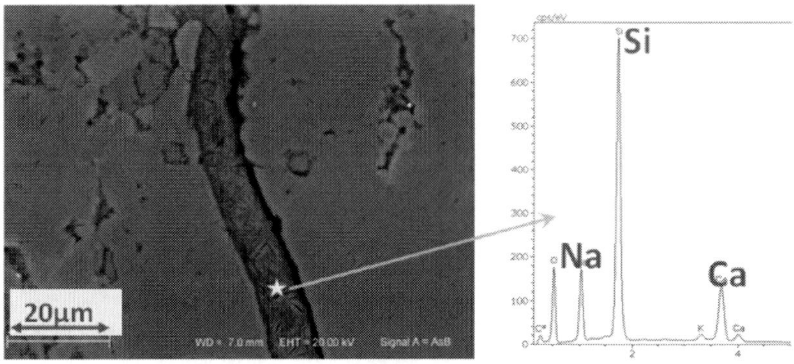

Fig. 9. Scanning electron micrograph of the cracked dolomite grain (D), the Si-Ca-Na gel is shown; scale bar=200 µm; specimens localisation - internal (S).

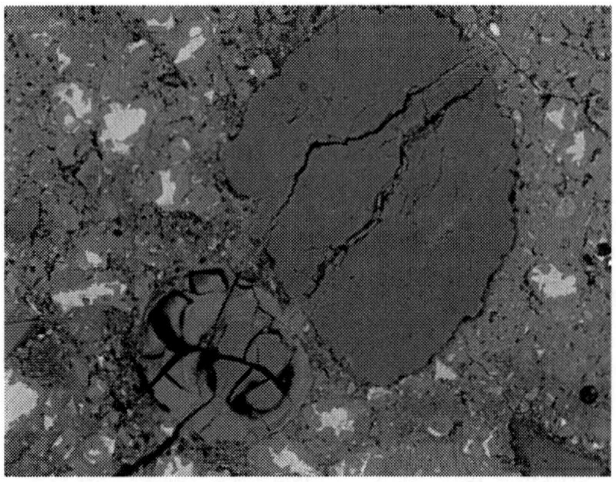

Fig. 10. Scanning electron micrograph of the sand grain with alkali-silica gel in cracks and in the air-void, scale bar = 100 µm.

4 Discussion

The observed damage due to ASR was various depending on the type of aggregate. The alkali-silica gel passed through whole grains of siliceous sand and then into the cracks and air-voids, Fig. 10. The highest damage as a result of ASR was found in the multimineral coarse aggregates. These aggregates included minerals which consisted of potassium and sodium, like albite or K-feldspar, Fig. 8. In the specimens with amphibolite aggregate (A), granite (G) and aggregate from glacial deposits (Gr) the significant signs of ASR were present. Owsiak et al. [15] showed that the most reactive gravel components in their research was glauconite sandstone with clay–limestone binder and chalcedony. The SEM-EDX analysis revealed that the cracks in above aggregates were filled by Si-Ca-K-Na gel, Fig. 8. The cracks in an aggregates started in the center of grains and the sodium from deicer would start the reaction at the grain boundary. It is in agreement with earlier assumptions made by Bérubé et al. [16], that ASR damage in thin structures is reduced by leaching. In thick structures, the effect of NaCl is limited by its penetration depth. The ASR takes place in the inner, unaffected region as usual, i.e. is not affected by the external alkalis. In this case the formation of a swelling alkali silica gel inside the concrete leads to inhomogeneous expansion which induces cracks, [16].

In all analyzed specimens the significant changes in the concrete microstructure after WD and temperature cycles and also 10% of NaCl solution were visible. The presence of ettringite, Friedel's salt and alkali-silica gel were observed, however with varying intensity.

Microstructural changes for concrete specimens were observed in terms of formation of Friedel's salt. The presence of Friedel's salt in the matrix resulting to the NaCl application is in accordance with other researchers findings, [16-19]. Sutter et al. [20] confirmed the formation of Friedel's salt in voids in analyzed concrete specimens exposed to NaCl solution and examined by petrographic methods. Sibbick and Page, [18] analyzed the effect of NaCl exposure on ASR in concrete with regard to the alkali content of the concrete and the C_3A content of the cement. They found that the ingress of NaCl deicer increased the NaOH concentration of the pore solution owing to the reaction of chloride with the C_3A hydration products and portlandite.

Bérubé et al. [16], showed that the high sodium concentration in the near-surface layer of the concrete might not cause severe ASR owing to a decreased OH^- concentration in this area. Heising et al. [17] stated that the chloride binding on hydrated cement phases and the corresponding release of OH^- ions does not significantly promote ASR. The OH^- concentration, and therefore pH, of the concrete pore solution is mainly lowered by potassium leaching, but the OH^- concentration is also lowered by the formation of Friedel's salt. Both studies [17] and [16] concerned long-term soaking of concrete specimens in NaCl solutions at various concentrations. The effect of cyclic wetting and drying, which significantly affects the durability of the concrete, has not been considered. Wang et al. [6] investigated the impact of various deicing chemicals and exposure conditions on concrete materials. It was showed that sodium concentration of the NaCl immersed specimen in 6% solution was high in the near-surface layer of concrete (over 5000 and 7000 ppm at 20 and 60 WD cycles, respectively), but it dropped considerably with specimen depth. They stated that NaCl deicing chemical might penetrate into the concrete no more than 26 mm in depth.

The results from the conducted research according to the ASR are in agreement with above statements. The alkali-silica gel has been found in the central part of the analyzed beam specimens. The gel was present in the matrix and in the fine and coarse aggregate. The gel consisted of both, sodium and potassium ions. According to [17], the Si concentration in the pore solution increases in the presence of dissolved NaCl. So the ASR damage should be intensified. Higher silicon solubility was explained by the higher total alkali concentration which increased surface silicate solubility, the formation of an aqueous complex $NaHSiO_3^0$ and a higher ionic strength. These effects promoted the sensitivity of silicate minerals to ASR, the formation of alkali silica gel and finally ASR damage, [17]. Authors [17] analysed only alkali reactive aggregate, therefore, they have achieved significant damage caused by ASR.

5 Conclusions

The wetting and drying and elevated temperature cycles with application of NaCl deicer has been shown to promote ASR and to affect concrete both physically and chemically. The damage due to the alkali-silica reaction was present in all tested specimens, but in places farthest from the edge of concrete beams. The highest degradation was in concretes with amphibolite aggregate (A), granite aggregate (G) and aggregate from glacial deposits (Gr). These aggregates are characterized by minerals containing sodium and potassium, like albite and K-feldspar. The limestone and dolomite aggregates also showed ASR signs. The gel contained only sodium, with very small traces of potassium which may suggest the influence of the NaCl on the alkali-silica potential. The exact role of NaCl and sodium and potassium from aggregates still needs to be investigated in more detail.Physical effects were typically showed by cracking.The larger cracks, which were parralel to the surface, were present in the cover zone and were empty. The presence of these cracks is the result of wetting and drying cycles.

It can be concluded from the visual observations of concrete specimens that the most harmful physical and chemical exposure conditions were for specimens made with granite aggregate and aggregate from glacial deposits as against those made with basalt aggregate.
The concretes with (G) and (Gr) aggregate subjected to sodium chloride revealed deterioration, including surface scaling and cracking.

The formations of Friedel's salt and cracks in the matrix were observed in all analyzed concrete specimens, thus the reaction mechanisms appear to be similar in all concretes.

The investigation was funded by National Centre of Research and Development in Poland and General Directorate for National Roads and Motorways, as a part of R&D Project: Development of Road Innovations (RID) concerning "The alkali reactivity of domestic aggregates", 2016-2019

References

1. D.M Ramakrishna, T. Viraraghavan, Water, Air, and Soil Pollution **166**(1-4), 49-63 (2005)
2. M.A. Glinicki, R. Jaskulski, M. Dąbrowski, Roads and Bridges - Drogi i Mosty **15**(1), 21-43 (2016)
3. J. Jain, A. Janusz, J. Olek, D. Jóźwiak-Niedźwiedzka, Transportation Research Record: J. Transportation Res. Board **2290**, 69-75 (2012)
4. X. Shi, L Fay, C. Gallaway, K. Volkening, M.M. Peterson, T. Pan, A. Creighton, C. Lawlor, S. Mumma, Y. Liu, T.A. Nguyen, Report No. CDOT-2009-01 (Colorado Department of Transportation, Denver, 2009)
5. L. Sutter, T. van Dam, K.R. Peterson, D.P. Johnston, Transportation Res. Record **1979**, 60-68 (2006)
6. K. Wang, D.E. Nelson, W.A. Nixon, Cem. Concr. Comp. **28**(2), 173-188 (2006)
7. M.C. Santagataa, M. Collepardi, Cem. Concr. Res. **30**, 1389- 1394 (2000)
8. X. Shi, L. Fay, M.M. Peterson, Z. Yang, Mater. Struct. **43**, 933–946 (2010)
9. P. Desai, Alkali silica reaction under the influence of chloride based deicers, 106 (M.Sc. Thesis, Clemson University, 2010)
10. F. Rajabipour, E. Giannini, C. Dunant, J.H. Ideker, M.D.A. Thomas, Cem. Concr. Res. **76**, 130–146 (2015)
11. C. Giebson, K. Voland, H.-M. Ludwig, B. Meng, Struct. Concr. **18,** 528-538 (2017)
12. C. Müller, I. Borchers, E. Eickschen, Beton **62**(10), 397-404 (2012)

13. D. Jóźwiak-Niedźwiedzka, K. Gibas, M.A. Glinicki, Roads and Bridges - Drogi i Mosty **16**(3), 223-239 (2017)
14. Z. Naziemiec, E. Pabiś-Mazgaj, Roads and Bridges - Drogi i Mosty **16**(3), 203-222 (2017)
15. Z. Owsiak, J. Zapała, P. Czapik, Cem. Wapno Beton **3**, 49-154 (2012)
16. M.A. Bérubé, J.F. Dorion J.F. Duchesne, J.B. Fournier, D. Vézina, Cem. Concr. Res. **33**, 77–84 (2003)
17. A. Heisig, L. Urbonas, R. E. Beddoe, D. Heinz, Mater. Struct. **49**, 4291–4303 (2016)
18. R.G. Sibbick, C.L. Page, Magazine Concr. Res. **50**(2), 147-159 (1998)
19. M. Balonis, B. Lothenbach, G. Le Saout, F.P. Glasser, Cem. .Concr. Res. **40**, 1009-1022 (2010)
20. L. Sutter, K. Peterson, G. Julio-Betancourt, D. Hooton, T. van Dam, K. Smith, Report No. SD2002-01-X, 56 (2008)

Impact of the air-entrained concrete with the blast-furnace slag cement on the intensity of reinforcement corrosion process

Wioletta Raczkiewicz[1] , *Wioletta* Grzmil[1,*], and *Justyna* Zapała – Sławeta[1]

[1] Kielce University of Technology, al. Tysiąclecia Państwa Polskiego 7, 25-314 Kielce, Poland

Abstract. The concrete cover has significant impact on the intensity of reinforcement corrosion process in reinforced concrete elements. Depending on the type of corrosion risk different types of cement are recommended for concrete. In conditions of chloride corrosion, due to, for example, the use of de-icing agents in winter, it is recommended to use concrete with cement containing granulated blast-furnace slag. The risk of chloride corrosion due to the use of de-icing agents is often additionally associated with the frost and repeated freezing and thawing actions of structural elements. The liquid freezing in the pores of concrete increases its volume, which causes the increase of internal stresses and leads to cracks in the concrete cover increasing the diffusion of chlorides. In order to reduce this phenomenon the introduction of air-entraining admixture is beneficial. The paper presents the experimental results that allow to determinate the effect of the air-entraining admixture addition on the intensity of reinforcement corrosion in concrete with blast-furnace slag cement. The tests were carried out on two groups of reinforced concrete specimens subjected to freezing and thawing cycles in 3% NaCl solution. One group of specimens was prepared with the air-entraining admixture addition and the other without it. The electrochemical tests of the reinforcement corrosion activity, based on the measurements of the corrosion current density were curried. The phase composition by X-ray diffraction and microstructure of concrete under the scanning electron microscope (SEM) coupled with the X-ray microanalyser (EDS) were examined. The results allowed to comprehensively assess the effect of air-entrained concrete with blast-furnace slag cement on the degree of reinforcement corrosion risk in concrete caused by the simultaneous action of chlorides and frost.

1 Introduction

The protection of reinforcement against corrosion is of decisive importance in ensuring the durability of reinforced concrete elements [1-4]. Such protection is ensured by proper and properly made concrete cover, i.e. a cover with appropriate thickness and properties of components [1-6]. Its task is to limit or slow down the penetration of aggressive substances deep into the concrete structure that may damage the passive layer on the reinforcing bars

* Corresponding author: wgrzmil@tu.kielce.pl

and initiate the corrosion process. In conditions of chloride corrosion (due to, for example, the use of de-icing agents in winter), it is recommended that reinforced concrete elements are made of concrete with cement containing granulated blast-furnace slag [7-9]. As a result of using blast-furnace slag cement during the hydration a larger amount of the amorphous C-S-H gel is formed with a reduced C_3A content compared to commonly used Portland cement [3, 8-10], the concrete bonding and hardening processes are slower.

The structure of concrete with metallurgical cement is more compact and tight with a lower capillary pore content, which reduces the diffusion of chlorides and due to the lower hydration heat, the shrinkage strains are smaller and slower [3, 8-10]. As a result, the penetration of aggressive substances into the concrete structure is effectively limited. This is particularly important in the case of extremely aggressive and pitting chloride corrosion. The diffusion of chloride ions, which are the strongest steel depassivating agent, occurs faster than CO_2 diffusion and the reinforcement corrosion is rapid [2, 3, 5, 11-13].

The risk of chloride corrosion due to the use of de-icing agents is often additionally associated with the frost and repeated freezing and thawing actions of structural elements. The liquid freezing in the pores of concrete increases its volume, which causes the increase of internal stresses and leads to cracks in the concrete cover increasing the diffusion of chlorides [2, 3, 14-15]. In order to reduce this phenomenon the introduction of air-entraining admixture is beneficial [3, 14-17]. As a result of air-entraining in the concrete the air bubbles occur (which are not completely filled with water) which absorb the water excess from the capillary pores during ice crystallization. In addition, the air voids are the place where certain chemical reaction products are deposited, so that no stresses can be created that can damage the concrete structure [14].

Often, durability tests are limited either to material tests of the concrete itself or to assess the degree of reinforcement corrosion in the structural elements [18-20]. This article presents both the results of electrochemical tests of the reinforcement corrosion activity based on the measurements of the corrosion current density and the results of material tests, including the determination of the mineral concrete composition. The tests allow comprehensively to assess the effect of air-entrained concrete with blast-furnace slag cement on the degree of reinforcement corrosion risk in concrete caused by the simultaneous action of chlorides and frost.

2 Materials and test methods

Twelve reinforced concrete specimens with dimensions 210 × 228 × 100mm were prepared for the tests. Based on the standard recommendations placed in Eurocode 2 [21] regarding the design of the structure in accordance with the accepted construction class and exposure class, it was assumed that the elements subjected to 50 years of durability, exposed to chlorine corrosion and aggressive impact of freezing - thawing. In connection with the above, specimens were made of concrete class C30/37, consistency S-3, with a water / cement ratio w/c = 0.43. The components of a concrete mix for 1 m^3 are included in table 1.

Table 1. Components of the C30/37 concrete mix for for 1 m^3.

Ingredients	Amount
blast-furnace slag cement CEM III/A 42.5 N-LH/HSR/NA	384 [kg/m^3]
aggregate I – basalt grit f 2/8 mm	600 [kg/m^3]
aggregate II – basalt grit f 8/16 mm	650 [kg/m^3]
mine sand	680 [kg/m^3]
water – water supply	166 [l]

plasticizer – Adva Flow 440	0.6% of cement volume
air-entraining admixture – Darex AEA W (LP)	0.5% of cement volume

The tests were carried out on two groups of reinforced concrete specimens subjected to freezing and thawing cycles in 3% NaCl solution to initiate chloride corrosion, one group (6 pieces) were specimens with air-entraining admixture and the other group (6 pieces) were specimens without air-entraining admixture. The air content in the specimens was 3.5%. (~0.5% of cement volume). Two parallel ribbed bars with a diameter of mm 8 mm made of BST 500 steel were placed in each specimen. The bars were placed in a spacing of 70 mm from the specimen side edges. The specimens after 7 days of ripening were disassembled and kept in water for a period of 14 days. After being removed from the water, the specimens were stored at a temperature of 20°C ± 2°C, and a relative humidity of 50% ± 5%. Then, selected electrochemical and material tests were performed.

The electrochemical measurements were performed in two stages: stage I - recursive measurements made on all specimens S0 and SA (reference to later measurements); stage II - measurements on the same specimens made after 120 cycles of freezing and thawing in 3% NaCl solution in order to initiate corrosion processes on the reinforcement. Specimens S0 and SA were subjected to freeze-thaw cycles in a frost resistance test chamber with an automatically controlled test program. During the freezing cycles, the specimens were completely immersed in the solution. The temperature range was + 18°C ÷ -18°C. During the day ~ 3 cycles took place. During the freeze-thaw cycles the reinforcement bars protruding from the specimens were insulated and had no contact with the solution. The protruding rods were tightly wrapped with several layers of insulation (with parameters guaranteeing its effectiveness at a temperature of -40°C to + 105°C and in an aggressive environment) and sealed with grease. The photo of the specimen with the connected electrochemical measurement set is shown in Figure 1.

Fig. 1. Photograph of the reinforced concrete specimen with the galvanostatic pulse apparatus.

Electrochemical tests were carried out using the galvanostatic pulse method. This is one of the non-destructive testing methods described e.g. in [4, 13, 18-20, 22]. This method uses the fact that the corrosion process of reinforcement in concrete is an electrochemical process, and the measurements of some electrical parameters referenced to the base results [18, 22] allow to indirect assess the progress of steel corrosion in concrete. As part of the galvanostatic pulse tests, the measurements of the corrosion current density, i.e. the most important and the most reliable parameter were made. In both stages measurements of the corrosion current density on the specimens were carried out in accordance with the guidelines given in [22].

On each specimen measurements were made at four measurement points located regularly over reinforcing bars (two points above each bar), and the obtained results were compared to the base results [22] according to which reinforcement corrosion activity is determined as follows:

- $i_{cor} < 0.5$ [$\mu A/cm^2$] – not forecasted corrosion activity,
- $i_{cor} = 0.5 \div 2.0$ [$\mu A/cm^2$] – irrelevant corrosion activity,
- $i_{cor} = 2{,}0 \div 5.0$ [$\mu A/cm^2$] – low corrosion activity,
- $i_{cor} = 5.0 \div 15.0$ [$\mu A/cm^2$] – moderate corrosion activity,
- $i_{cor} > 15.0$ [$\mu A/cm^2$] – high corrosion activity.

Material tests included determining the phase composition of the concrete by X-ray diffraction analysis of specimens S0 and SA after freeze-thaw cycles dissociated from the layer at the reinforcing bar. The phase composition of concretes was performed by X-ray powder diffraction using PANanalytical Empyrean diffractometer. In addition, the microstructure of both specimens (S0 and SA) was observed under scanning electron microscopy (SEM). The microstructure examination was conducted under scanning electron microscope FEI Company Quanta FEG 205 coupled with the X-ray microanalyser (EDS).

3 Research results and discussion

On the basis of electrochemical tests, after the results of measurements of the corrosion current density to the base of the standard results, it was possible to determine the corrosion activity of the specimens' reinforcement. The results from all measuring points on all specimens were pre-analysed [23]. Then, one measurement point was separated from each specimen, in which the values of corrosion current density were the highest after the second phase of measurements. This point in each specimen determined the degradation degree of the whole specimen. In the further part of the research, the results obtained at these points were analyzed and compared as representative for the whole specimen.

In stage I of the tests, the values of corrosion current density measured on all specimens S0 and SA did not exceed $i_{cor} = 2$ $\mu A/cm^2$ (Fig. 2), which in reference to the given criterion, indicated "irrelevant corrosion activity of the reinforcement". Based on the results of the corrosion current density made in stage II (i.e. after 120 cycles of freezing and thawing the specimens in 3% NaCl), it can be seen that in all specimens S0 (without air-entraining admixture) the density of the corrosion current density increased noticeably, and, $i_{cor} = 6.42 \div 9.73$ $\mu A/cm^2$, which was indicative of "moderate reinforcement corrosive activity" (Fig. 2). Whereas in specimens SA (with air-entraining admixture), although subject to the same freezing and thawing cycles and chlorides, the density of corrosion current density increased only slightly: in four specimens it did not exceed $i_{cor} = 2$ $\mu A/cm^2$, and in two specimens it reached $i_{cor} = 3.86$ $\mu A/cm^2$ (Fig. 2) and nowhere exceeded the limit of" low reinforcement corrosion activity".

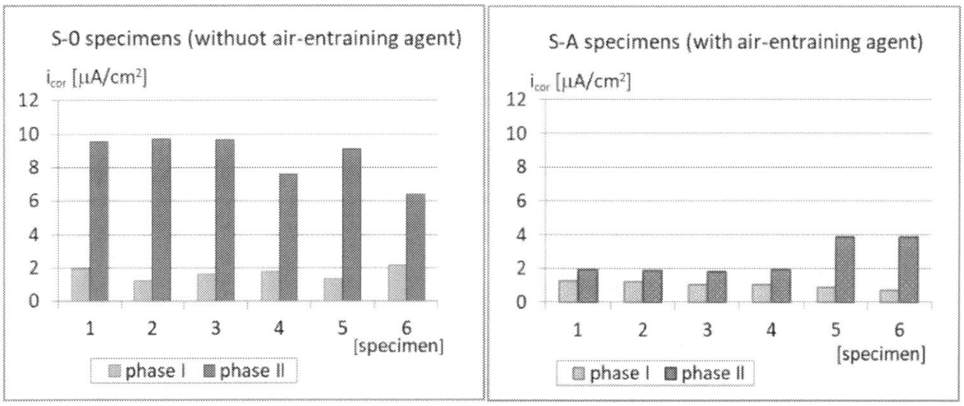

Fig. 2. Results of corrosion current density measurements made on specimens S0 (without air-entraining admixture) and SA (with air-entraining admixture) in two stages of research: stage I - recurrent measurements, stage II - measurements carried out after 120 freezing and thawing cycles in 3% NaCl solution.

Analysis of the phase composition of concrete taken from the specimen indicated a low Friedel salt content. Intensity of peaks at an angle of 11,2°2Θ in specimens taken from air-entrained and non-air-entrained concrete were comparable. It can therefore be assumed that chloride ions were associated to a similar degree. The low Friedel salt content in the analysed concretes may have resulted from the limitation of the C_3A phase content in the cement – high sulphate resistant (HSR). Friedel salt in this case is formed as a result of the reaction of aluminates being components of blast furnace slag and chloride ions derived from NaCl dissociation [3]. In addition, it should be noted that a small number of chloroaluminate phases are the result of penetration of chlorides into hardened concrete, i.e. the C_3A phase has already been bound in the form of sulfoaluminate. In this case, chloride ions penetrating the concrete structure can be absorbed by the C-S-H phase, which is formed in increased amounts when blast-furnace slag cement is used. An analysis of the phase composition of air-entrained and non-air-entrained concretes is shown in Figure 3.

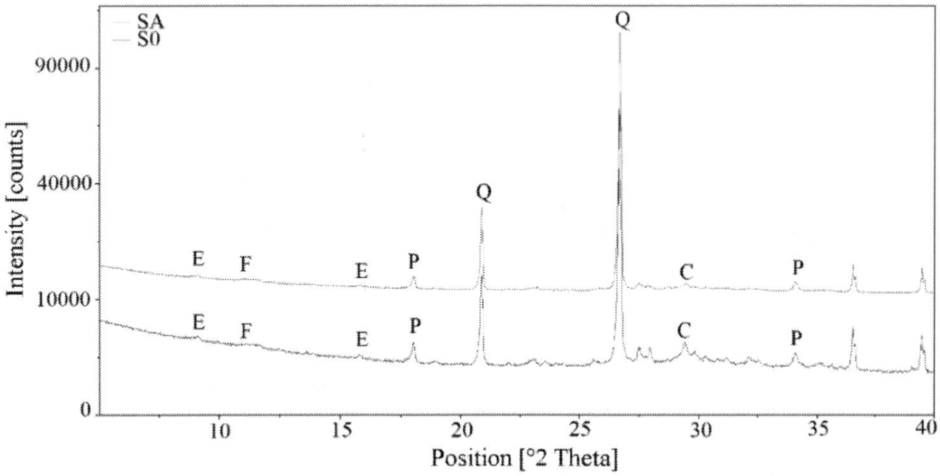

Fig. 3. Diffractogram of the concrete specimens with aeration admixture SA and without aeration admixture S0, after freeze-thaw cycles in the reinforcing bar, E-ettringite, F- Friedel salt, P-portlandyt, C- calcite, Q- quartz.

The presence of Friedel salt in the analysed concretes was also confirmed by the SEM-EDS analysis. Figure 4 shows a photo from a scanning microscope with visible Friedel salt crystals in concrete together with EDS analysis.

Fig. 4. Friedel salt crystals in concrete together with EDS analysis.

Increased susceptibility of non-aerated concrete, subjected to freezing and thawing cycles in chloride solution, can be explained by changes in the microstructure resulting from the susceptibility of concrete to frost corrosion. Observations of the microstructure of non-air-entrained concrete specimens indicate a higher number of micro cracks compared to air-entrained concretes (Fig. 5).

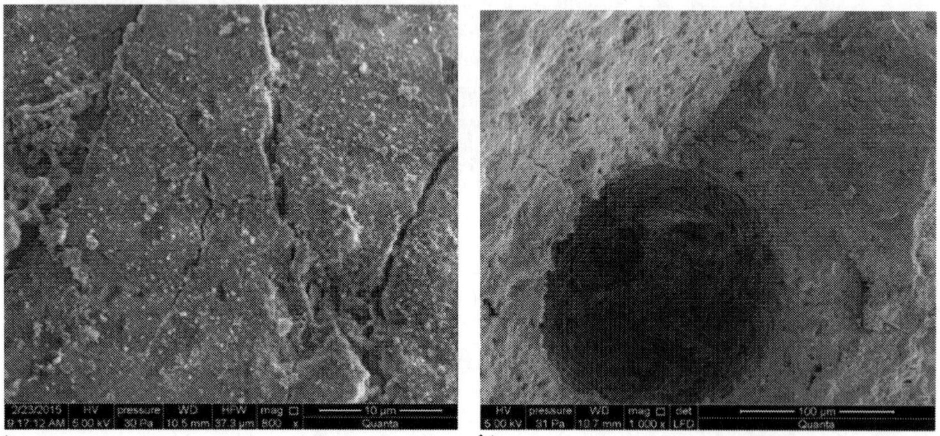

a) b)

Fig. 5. Microstructure of concrete after freezing and thawing cycles: a) non-air-entraining concrete, b) air-entraining concrete.

Based on electrochemical analysis it was found that in non-air-entraining concrete after freeze-thaw cycles the density of the corrosive current increased more than in the aerated concrete, despite the small differences in phase composition of air-entrained and non-air-entrained concrete. In aerated concrete, Cl^- ions could be adsorbed on the pore walls, some chloride ions are immobilized, so it does not participate in the diffusion process [24]. In addition, the absorbed chlorides have noncrystalline nature so they would be not show in X-ray diffraction [25]. Analysis of microstructure shows that the damage caused by freezing and thawing in aerated concrete was smaller. This is due to the fact that the dispersed air bubbles absorb the internal pressure increase caused by freezing, which reduces the damage. A smaller number of micro cracks in aerated concrete means less penetration of chlorides into the concrete interior, due to the reduction of the number of potential penetration routes and the same reduction of steel corrosion, which was observed by other researcher [26]. As Zhang and co-workers indicate, the absorption coefficient of chlorides in aerated concrete is smaller compared to non-aerated concrete. In non-aerated concrete, this coefficient increases in proportion to the number of freezing and thawing cycles. In aerated concrete, this ratio is reduced due to the material's characteristic structure [27].

4 Conclusions

On the basis of tests carried out on reinforced specimens with aerated and non-aerated CEM III/A cement, which were subjected to 120 freezing and thawing cycles in 3% NaCl, it was concluded that:

1. Based on the measurements of the corrosion current density, it was found that in S0 specimens (without air-entraining admixture) the corrosion activity of the reinforcement after 120 cycles of freezing and thawing specimens in 3% NaCl increased by 7,07 $\mu A/cm^2$, on average, while in SA specimens (with air-entraining admixture) only by 1,51 $\mu A/cm^2$.

2. Analysis of phase composition of hardened concrete did not show significant differences in aerated and non-aerated concrete.

3. Increased corrosion activity of reinforcement in non-air-entrained concrete specimens, subjected to cycles of freezing and thawing in a chloride solution, results from changes in the microstructure of concrete caused by frost corrosion.

4. In order to increase the resistance of reinforced concrete to the simultaneous action of chloride and frost corrosion, it is necessary to use both blast-furnace slag cement as well as an air entraining admixture.

References

1. L. Czarnecki, P.H. Emmons, *Naprawa i ochrona konstrukcji betonowych* (SPC, Kraków, 2002)
2. Z. Ściślewski, *Ochrona konstrukcji żelbetowych* (Warszawa, Arkady 1999)
3. W. Kurdowski, *Chemia cementu i betonu* (SPC, Kraków, 2010)
4. A. Zybura, M. Jaśniok, T. Jaśniok, *Diagnostyka konstrukcji żelbetowych. Badania korozji zbrojenia i właściwości ochronnych betonu* (PWN, Warszawa 2011)
5. X.M. Shi, N. Xie, K.J. Fortunate, J. Gong, Constr. Build. Mater. **30**, 125 (2012)
6. S. Ahmad, Cem. Concr. Comp. **25**, 459 (2003)
7. J. Małolepszy, *Trwałość betonów z cementów żużlowych*, 225–243, (Con., Wisła, 2002)
8. J. Deja, Cement Lime Concrete **74**, 280-283 (2007)
9. Z. Giergiczny, Eng. Bull. **67**, 275-279 (2010)
10. W. Grzmil, Z. Owsiak, Cem. Lime Concr. **80**, 137-144 (2013)
11. W. Kurdowski, Cem. Lime Concr. **69**, 56-60, (2002)
12. T. Jaśniok, B. Słomka-Słupik, A. Zybura, Cem. Lime Concr. **81**, 158-165 (2014)

13. W. Grzmil, W. Raczkiewicz, Cem. Lime Concr. **84**, 311-319 (2017)
14. Z. Rusin, *Technologia betonów mrozoodpornych* (SPC, Kraków, 2002)
15. L. Czarnecki, J. Deja, K. Flaga, W. Kurdowski, J. Małolepszy, W. Radomski, J. Śliwiński, Budownictwo, Technologie, Architektura **69**, 66-69 (2015)
16. J. Deja, Cem. Lime Concr. **74**, 280-283 (2007)
17. J. Deja, Cem. Concr. Comp. **25**, 357-361 (2003)
18. W. Raczkiewicz, D. Michałowska - Maziejuk, Structure and Environment **4**, 21-27, (2012)
19. B. Elsner, et al, *International Conference Repair of Concrete Structures* (Svolvaer, Norway, 1997)
20. H.E. Sørense, T. Frølund, *Int. Conference on Concrete in Marine Environment* (Hanoi, 2002)
21. EN 1992-1-1 (2004)
22. http://www.germann.org/TestSystems/GalvaPulse/GalvaPulse.pdf (January, 2018)
23. W. Raczkiewicz, Advances in Materials Science (to be published)
24. R. Cerny, Z. Pavlık, P. Rovnanıkova, Cem. Concr. Compos. **26**, 705 (2004)
25. L. Bertolini, B. Elsener, P. Pedeferii, R. Polder, *Corrosion of Steel in Concrete: Prevention, Diagnosis, Repair* (WILEY-VCH Verlag GmbH&Co.KGaA, 2005)
26. Z. Fangzhi, M. Zhiming, Z. Tiejun, Adv. Mat. Sc. Eng. **16** (2016)
27. P. Zhang, Y. Cong, M. Vogel, Z. Liu, H.S. Muller, Y. Zhu, T. Zhao, Con. Build. Mat. **148** (2017)

Application of acrylic-based coatings for concrete protection

Aleksander Kozak[1,*]

[1]Cracow University of Technology, Warszawska Street 24, Cracow, 31-155, Poland

Abstract. The paper presents a description of acrylic-based polymers used in civil engineering for concrete protection. Some information on chemistry and properties of the acrylic coatings is included together with current trends in coating technology. Moreover, the paper shows test results, that is barrier properties and SEM observations of a commercial elastic acrylic-based polymer subject to artificial weathering for 1000h. The studies showed that the weathering did not change carbon dioxide permeability of the tested coating and its influence on water vapour permeability of the material was also very small. The test results show that an assessment of coating performance based only on barrier properties is not enough.

1 Introduction

In the middle of XX century the first acrylic water-borne emulsions were developed. Later on, a wide range of modifications have been made in order to produce durable coatings. Co-monomers like styrene are commonly used to ensure appropriate hardness, gloss, and high water absorption resistance. Further modifications include addition of monomers containing functional groups like carbonyl or hydroxyl. The obtained polymers can be used to produce two-component coatings containing water-borne acrylic and a crosslinking material used as a hardener. Moreover, single-component and self-crosslinking systems are also used. They set by air oxidation [1]. Water-borne acrylic coatings are known for some sought-after properties, such as very good weathering resistance and toughness, thus these types of coatings find a wide range of applications in civil engineering. In addition, they are a very good alternative for solvent-born systems keeping in mind safety of workers and environment protection. Current tendency is to significantly reduce volatile organic content (VOC) in coating, and the acrylic coating follows the trend very well. The VOCs are present in the acrylics but in very small amounts. They play a very important role in formation of film. During the process, fusion of the polymer particles occurs in the liquid coating, and, initially, the channels among the latex particles are formed which later on disappear when the particles merge together. In order to accelerate the process, coalescing agents are added. These volatile solvents optimize the coalescence of the particles. For any polymers exists a limiting temperature above which full coalescence is possible. This temperature is called minimum film-formation temperature (MFFT). If a dispersion paint

* Corresponding author: akozak@pk.edu.pl

film dries below the MFFT of its polymer, the film will be quite brittle, and the formation of a coherent film will be impossible. By adding a coalescing agent the temperature can be reduced enabling full coalescence. The most effective coalescing agents are partially water-soluble coalescing agents [1,2].

One of the more important additions to coatings is titanium dioxide, which ensures gloss and chalk resistance. Moreover, nano-TiO_2 has catalytic properties which are very useful in removing impurities [3]. Acrylic coatings exhibit good adherence to concrete substrate and, also, show good alkali resistance, albeit to mild ones. It is recommended to apply an acrylic coating on fresh concrete when it is at least one month old to achieve very good long-term performance properties. Moreover, this type of coating ensures good balance between water vapour permeability and carbon dioxide resistance. The barrier properties are often expressed in a diffusion equivalent air layer thickness S_d [m]. In the case of S_d for CO_2, the value should be more than 50 meters, whereas S_d for H_2O ought to be less than 2 meters [4,5]. Another very beneficial feature of acrylic coatings is their high weathering resistance, thus they are often chosen as protective materials on concrete structures. All above features are very desired while choosing appropriate coatings for long-term concrete protection [4,6-9].

The paper presents test results of barrier properties of a highly elastic acrylic coating subject to artificial weathering for 1000h. Moreover, some SEM observations of the material have also been made.

2 Materials and methods

The elastic acrylic-based coating was tested for 1000h, i.e. 126 cycles in a weathering chamber according to ASTM G 154a [10]. The tested coating was either applied onto concrete substrate or studied in the form of foils. One weathering cycle consisted of irradiating with the use of UV lamps at 60 ± 3^0C for 6h and, then, sprinkling with distilled water at 40 ± 3^0C for 2 h. The weathering cycles started with the UV irradiation, as proposed by Aragon and Frizzi [11]. The radiation ranged from 295nm to 400nm, which corresponded to UVA+B. The radiation intensity was 40 ± 2 W/m^2 (measured with a Solarmeter) and was of the same value as the one recorded on a very sunny summer day in Cracow. The studies focused on barrier properties of the material and some SEM observations, whereas mechanical properties (adherence to concrete substrate and elongation tests) together with a structural analysis (FTIR) of the material were described in [12]. The permeability to carbon dioxide before and after weathering was determined using the RILEM-Cembureau method [13], which was described in detail in [14]. Cylindrical concrete samples 150 mm in diameter and 50mm thick were used in all tests. Before the tests, they were earlier cut from 150mmx300mm cylinders (Fig.1). Prior to the tests, the coatings were applied onto the concrete substrates and seasoned for a period given in the data sheet of the material.

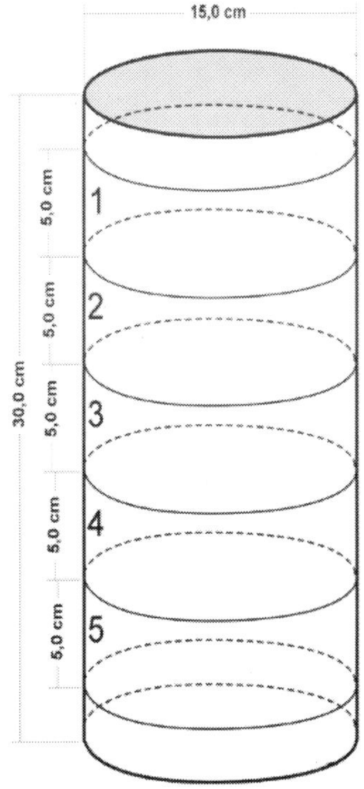

Fig. 1. Concrete cylinder for preparing samples for CO_2 permeability tests.

Water vapour permeability was determined using coatings in the form of foils. The tests were done according to ASTM E-96 [15].

3 Results and discussion

3.1 Carbon dioxide permeability

The test results of both initial and weathered samples showed that the coating was completely impermeable to carbon dioxide under the test conditions and fulfilled the requirements concerning S_d for CO_2 [4,5].

3.2 Water vapour permeability

The permeability to water vapour was determined, and the equivalent air layer thickness S_d for H_2O was calculated. Table 1 shows water vapour permeability of the coating foils.

Table 1. Water vapour permeability.

Aging time	Sample no.	Δm under steady state	H_2O flux	Diffusion coefficient	Diffusion resistance	S_d
		[g]	[kg/m^2·s]	[m^2/s]	[s/m]	[m]

1	2	3	4	5	6	7
0 cycles	1	0.180	6.94E-07	9.98E-09	23042	0.65
	2	0.174	6.71E-07	9.65E-09	23837	0.67
	3	0.186	7.18E-07	1.03E-08	22299	0.63
					mean	0.65
126 cycles	1	0.058	2.24E-07	3.22E-09	70299	2.02
	2	0.059	2.28E-07	3.27E-09	61905	1.98
	3	0.067	2.58E-07	3.72E-09	76808	1.75
					mean	1.92

The study showed that the acrylic coating before weathering had the mean S_d for H_2O value 0.65, thus the material fulfilled the requirements [4,5]. A slight increase in S_d value was noted after 126 cycles of weathering, but the samples still fulfilled appropriate requirements. Such changes could be explained by further curing of the polymer while aging. It appears that assessing coating behaviour while weathering based only on barrier properties does not show the actual condition of the material. As given in [12], after 126 weathering cycles the coating showed significant changes in deformation and strength.

3.3 SEM observations

The use of a Zeiss EVO MA 10 microscope allowed observing changes in the coating surface after weathering. Fig.2 and Fig.3 show coatings before and after weathering for 1000h, respectively. SEM pictures showed that after aging some destruction processes have taken place in the coating.

Fig. 2. SEM picture before weathering.

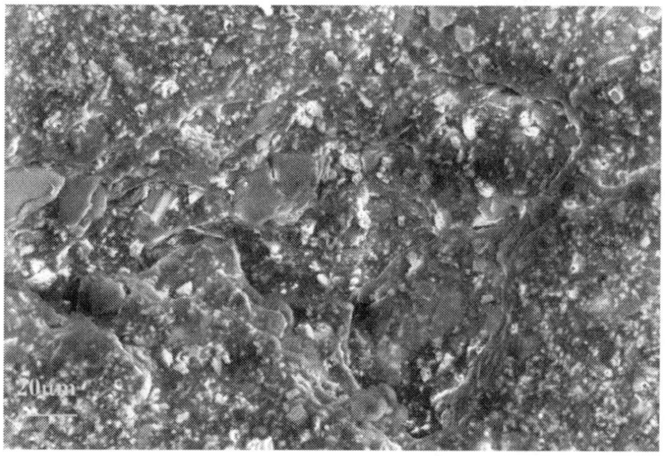

Fig. 3. SEM picture after weathering.

4 Conclusions

Based on the obtained test results, the following conclusions can be drawn:
1. The studies showed that the artificial weathering carried out according to the proposed research programme did not change resistance to carbon dioxide permeability. The samples were completely impermeable to carbon dioxide before and after accelerated weathering.
2. Water vapour permeability tests showed a slight increase in S_d for H_2O after artificial weathering, but the value still fulfilled the requirements. It seems it could be due to additional curing of the material while weathering.
3. Scanning electron microscopy showed some degradation in the coating after 126 cycles.
4. Assessing coating performance based only on the examination of its barrier properties is not enough to determine the actual performance of the material.

References

1. L.M. Smith, *Generic coating types. An introduction to industrial maintenance coating materials*. (TPC Books, 1996)
2. https://coatings.specialchem.com/selection-guide/coalescing-agents-in-coatings (January, 2018)
3. E. Stanaszek-Tomal, A. Kozak., *Energy efficient, sustainable building materials and products* (Ed. by I. Hager, Cracow, 2017)
4. R. Bassi, S.K. Roy, *Handbook of coatings for concrete* (Whittles Publishing, 2002)
5. T. Zafeiropoulou, E. Rakanta, G. Batis, Prog. in Org. Coat. **72** (2011)
6. N.L. Thomas, Prog. in Org. Coat. **19** (2) (1991)
7. P.C. Harwood, *Surface coatings – specification criteria. Protection of concrete: Proceedings of the international conference,* 176-184 (1990)
8. R.N. Swamy, S. Takinawa, *Surface coatings to preserve concrete durability. Protection of Concrete: Proceedings of the international conference,* 129-144 (1990)
9. O. O. Knudsen, A. Forsgren, *Corrosion control through organic coatings. 2nd edition* (CRC Press, 2017)
10. ASTM G154a (2000)
11. M.A. Aragon, E. Frizzi, Prot. Coat. Europe, 8-14 (2002)
12. A. Kozak, Przem. Chem. **94**(8), 1348-1350 (2015)

13. RILEM Technical Recommendation: Tests for gas permeability of concrete TC (1999)
14. A. Kozak, Proced. Eng. **108**, 664–672 (2015)
15. ASTM E-96 (2000)

Evaluating the Influence of Polymer Modified Asphalt Binders on Low Temperature Properties

Amal Abdelaziz[1,*], *Chun-Hsing* Ho[1], and *Matthew* Snyder[1]

[1]Northern Arizona University, 2112 S Huffer Ln, Flagstaff AZ 86011, USA

Abstract. Low temperature cracking is one of the most common distress types in asphalt concrete pavements, particularly in cold regions. Many factors influence the behaviour of asphalt concrete pavements at low temperatures, such as the applied traffic load, environmental conditions and material characteristics. Asphalt binders are one of the primary factors that influence material properties. The purpose of this study is to compare the performance of two types of asphalt binders: styrene-butadiene-styrene (SBS) modified asphalt binder and unmodified asphalt binder in resisting low temperature cracking. The study was conducted in Flagstaff, located at the area of Northern Arizona, in the United States. Asphalt samples were collected from the paving sections and were compacted and trimmed into small beams. Bending Beam Rheometer tests were performed, using the trimmed specimens at temperatures of -6°C, -12°C and -18°C. Based on the results of the study, it was concluded that, SBS modified asphalt binder performs better in resisting low temperature cracking, compared to the unmodified binder. Based on the study outcomes, it is recommended to use SBS polymer modified polymers in areas subjected to severe cold weather events to maximize the life span of asphalt concrete pavements.

1 Introduction

Low temperature cracking or thermally induced cracking of asphalt pavements is one of the most predominant distress types in cold regions, particularly in Canada and northern United States [1, 2]. This problem leads to the early deterioration of asphalt concrete pavements, due to the inability of pavement to dissipate the stress from excessive thermal contraction [3]. Previous research showed that the addition of polymers to asphalt mixtures enhances its performance, by increasing the ability of pavement to resist thermal cracking, fatigue distress and rutting. Moreover, the addition of polymers has shown to improve the performance of asphalt binders by increasing the characteristics of elastic recovery, viscosity, cohesive strength and ductility [4]. The research on the behaviour of polymer modified bitumen can be found in extensive studies since the 90s. The concept of modifying virgin bitumen was based on increasing the viscoelastic range of virgin asphalt to improve its flexibilty at low temperatures [5].

Styrene-butadiene-styrene (SBS) polymer is one of the most common modifiers used in asphalt mixtures around the world. Most of the current research is focused on the impact of

* Corresponding author: aa2399@nau.edu

SBS polymers on high temperature properties of asphalt mixtures. However, the factors affecting low temperature properties have not been well studied yet. Some of the previous research showed that SBS polymers could improve the low temperature cracking properties of asphalt mixtures. For example, a research done at Tongji University showed that the addition of SBS polymer reduces the stiffness of asphalt mixtures at low temperatures, indicating that SBS modifiers have a positive influence on the performance of asphalt pavements at low temperatures [6]. On the other hand, other research projects showed that the addition of SBS polymer does not have a significant impact on the stiffness of asphalt mixtures. For example, a research performed at Cooper Union University on evaluating the influence of the addition of SBS polymer, showed that modified SBS polymers are more prone to low temperature cracking, compared to unmodified asphalt binders [7]. Thus, due to the inconsistent research results, it is necessary to conduct further investigations to better understand the influence of SBS modified asphalt binders on low temperature cracking properties.

Currently, there are multiple testing methods that can be used to determine the low temperature properties of asphalt mixtures which include, the Indirect Tensile Test (IDT), the Thermal Stress Restraint Specimen Test (TSRST), and the Bending Beam Rheometer (BBR) test. The IDT method is a common test used in evaluating thermal cracking. The method is a combination of three tests which are, the static creep, resilient and modulus and strength tests [8]. While, the TRST method characterizes low temperature cracking taking into account the thermal and mechanical effects. The test is performed by maintaining a constant axial strain and as the temperature decreases, to allow the specimen to contract, while the servo-hydraulic press in the device prevents it. This process increases the thermal stress in the specimen until it fractures [9]. The Bending Beam Rheometer (BBR) test, also known as the standard test for binder at low temperatures, is a simple and cost-effective method used to determine the creep behaviour of asphalt mixtures. The results of BBR tests showed to be comparable to those of the IDT test [10]. The test measures the deflection of thin asphalt beams, at a constant load and constant temperature. The advantages of using BBR test over the other testing methods include, the equipment availability at reasonable costs and the reduced specimen size, as the beams tested in the BBR machine are only 12.5 x 6.75 x 127 mm [11].

The purpose of this study is to provide a laboratory investigation method using BBR test, to study the impact of adding SBS polymer on the low temperature properties of asphalt mixtures. A paving project was implemented along the area of Northern Arizona, in the United States. Two types of asphalt mixtures were used in the project which are: unmodified asphalt binder and SBS modified binder. BBR tests were performed to determine the low temperature cracking properties of both asphalt binders. In addition, viscoelastic analysis was conducted to evaluate and compare the relaxation modulus of both asphalt binders. Based on the laboratory results and viscoelastic analysis, conclusions and recommendations for future research needs are provided.

2 Methodology

2.1 Sample Preparation

The study was completed using two types of asphalt mixtures: SBS polymer modified asphalt mixture and unmodified asphalt mixture. A PG 70-22 asphalt binder was used for both asphalt mixtures. A ¾ inches nominal material aggregate size was used for the unmodified and modified asphalt binders. The characteristics of the mix design used for asphalt mixtures are summarized in Table 1.

Table 1. Mix design characteristics.

Mix Design	**Unmodified Asphalt Binder (Non- SBS)**	**Modified Binder (SBS)**
Nominal material aggregate size	¾ inches	¾ inches
Asphalt binder type	PG 70-22	PG 70-22 SBS
Percent of asphalt binder (%)	5.7	5.7
Bulk specific gravity (G_{mb})	2.385	2.379
Maximum theoretical gravity (G_{mm})	2.509	2.499
Percent of air void, %	5	4.8
Voids in mineral aggregate filled (VMA), %	15.4	15.5
Percent of void filled (VFA), %	67.8	69

To study the low temperature cracking properties, asphalt samples were collected from the paving project and got reheated and compacted at the Materials Laboratory of Arizona Department of Transportation (ADOT), using a Superpave gyratory compactor (SGC). Then, asphalt samples were cut into six faced blocks. After that, the asphalt blocks were trimmed into small specimens to get them prepared for BBR testing. The testing procedures used to perform BBR test are based on The American Association of State Highway and Transportation Officials (AASHTO) standard TP125: Determining the Flexural Creep Stiffness of Asphalt Mixtures Using the Bending Beam Rheometer (BBR) [12]. Figures 1 through 9 illustrate the process of obtaining the asphalt samples from the paving site until they were tested in the BBR machine.

Fig. 1. Getting samples from site

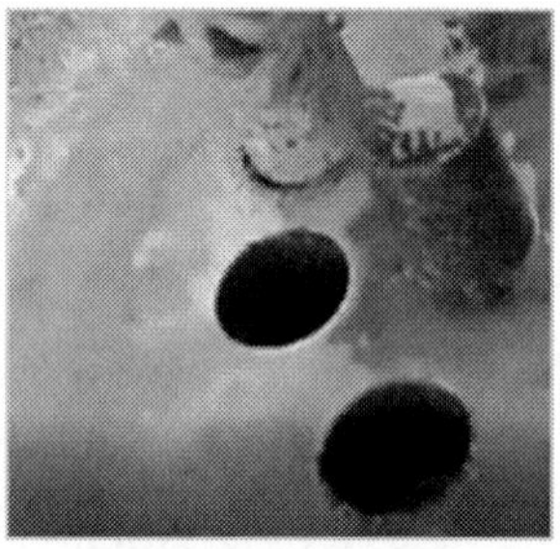

Fig. 2. Asphalt samples

Fig. 3. Transporting samples to lab

Fig. 4. Six faced asphalt block

Fig. 5. Asphalt block trimming

Fig. 6. Trimmed flat beams

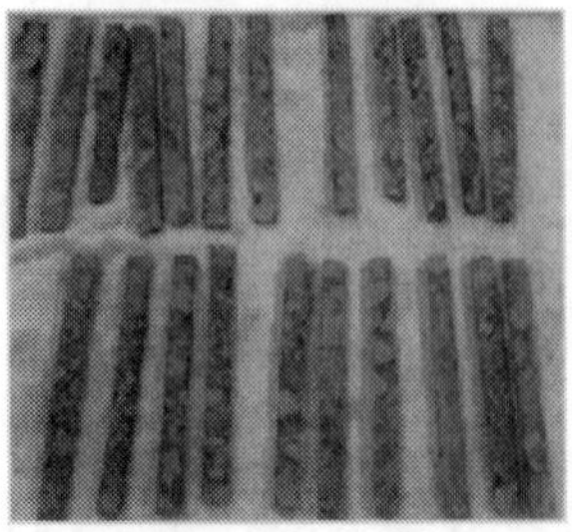

Fig. 7. Thin beams for BBR tests

Fig. 8. Beams were conditioned in the bath of the BBR machine for 60 minutes

Fig. 9. BBR test

2.2 Determination of Air Void Ratio

Measuring the air void ratio of the asphalt blocks is necessary, to ensure that the air void ratio did not significantly change when the SGC samples were cut into six faced asphalt blocks. Moreover, inconsistency in the air void ratio could lead to misleading BBR test results. Air void ratios were calculated based on Equation (1):

$$\% \text{ air voids} = (1 - D/E) * 100 \quad (1)$$

Where:
D = the bulk specific gravity; and
E= the maximum rice specific gravity.

The calculated air voids percentages for the asphalt blocks were approximately 2% to 3% less than the percent of air voids for the SGC asphalt samples. The consistent reduction in air voids among the different asphalt blocks, indicates that the effect of the change in air voids on BBR tests is negligible.

2.3 Bending Beam Rheometer Tests

Thin asphalt beams were tested in the BBR machine at three different temperatures: -6°C, -12 °C and -18 °C, based on the used binder and in accordance with AASHTO T313 Standard [13]. A minimum of 5 beams from each asphalt mixture were used to run BBR tests at the desired temperatures [14, 15]. The BBR test measure the deflection and the stiffness of the beams at time increments of 0.5 seconds. The linear viscoelastic stiffness modulus obtained from the BBR tests can be calculated using Equation (2):

$$S(t) = PL^3/4bh^3\delta(t) \quad (2)$$

Where:
S(t) = the creep stiffness as a function of time in MPa;
P = constant applied load in N;
L= span length in mm;
b = beam width in mm;
h = beam thickness in mm; and
$\delta(t)$ = beam deflection in mm.

The measured stiffness values using BBR tests for both asphalt mixtures at a time of 60 seconds are summarized in Table 2. As illustrated Figure 10, the average elastic modulus values of the unmodified asphalt binder are higher than the SBS modified binder, indicating that unmodified binders have higher stiffness and strength compared to the modified asphalt binder. Determining the stiffness or the elastic modulus of asphalt mixtures is critical, as a reduction in the stiffness, increases the critical strains exerted on pavement, as a result making the pavement more prone to surface deformation [16]. In addition, it can be noted that all of the calculated coefficients of variances are below 20%, indicating low variability among the different beams [17].

Table 2. Statistical information on the stiffness of asphalt mixtures

Temperature	**-6˚C**			**-12˚C**			**-18˚C**		
Binder Type	AVG (MPa)	STD (MPa)	CV	AVG (MPa)	STD (MPa)	CV	AVG (MPa)	STD (MPa)	CV
Unmodified binder	6158	409	0.066	9624	1305	0.136	12396	2109	0.17
SBS modified binder	4243	407	0.096	6138	190	0.03	9857	1237	0.13

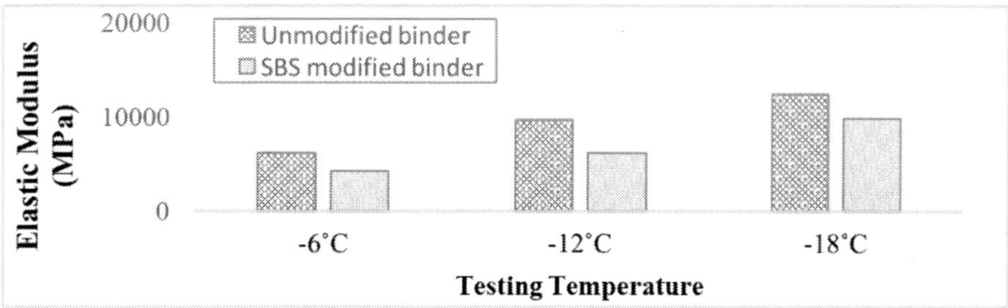

Fig. 10. Elastic Modulus at 60 seconds for non-SBS and SBS mixtures

3 Results and Discussion

Viscoelastic analysis was performed to determine creep compliance values of the unmodified asphalt binder and SBS modified binder at the three different temperatures: -6˚C, -12 ˚C and -18 ˚C. The time-temperature superposition principle (TTSP) was used to shift the creep compliance curves at the three temperatures for both asphalt mixtures, as shown in Figures 11 through 14. The shift factor used in the adopted TTSP model of the was calculated using an exponential function of time. The viscoelastic analysis showed that SBS modified asphalt binder has higher creep compliance values compared to the unmodified binder.

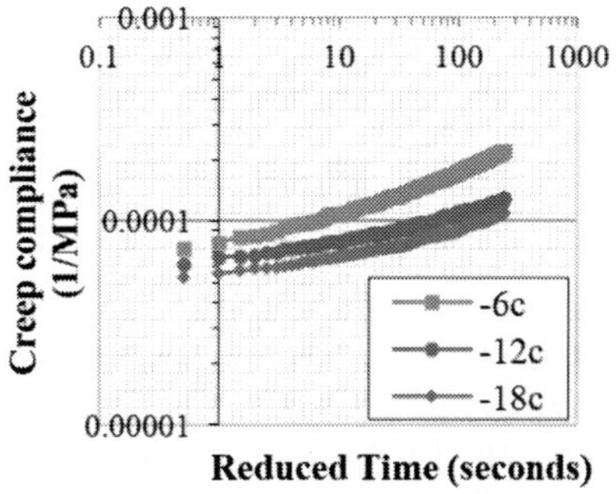

Fig. 11. Creep Compliance Curves (Non-SBS)

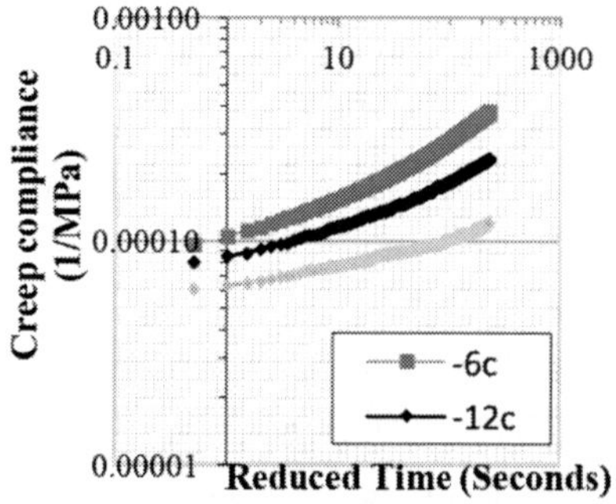

Fig. 12. Creep Compliance Curves (SBS)

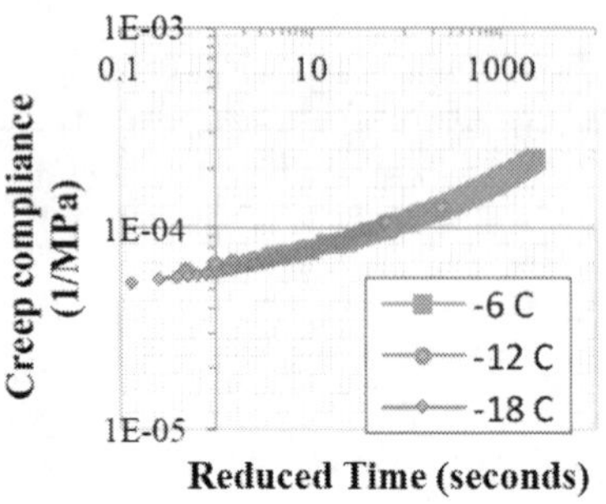

Fig. 13. Master Curve (Non-SBS)

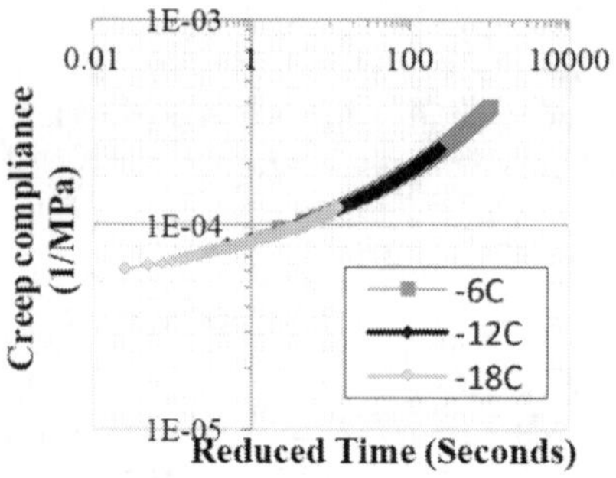

Fig. 14. Master Curve (SBS)

The power law function, provided in Equation (3), was used to generate a fitting curve for the master curves of both asphalt mixtures, as shown in Figures 15 and 16. A reference temperature of -12°C was used. The parameters of the power function: were determined using the fitting curves, and are summarized in Table 3.

$$D(t) = D_0 + D_1 t^n \tag{3}$$

Where:
D(t) = creep compliance values in 1/MPa at reduced time;
t= reduced time in seconds; and
D_0 , D_1 and n are power law parameters.

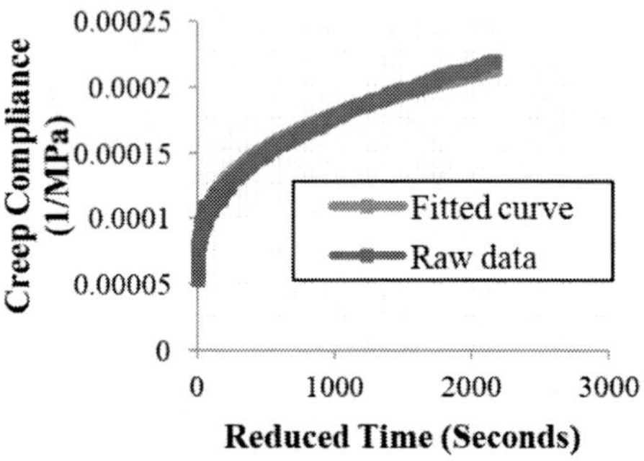

Fig. 15. Power Law funtion (non-SBS)

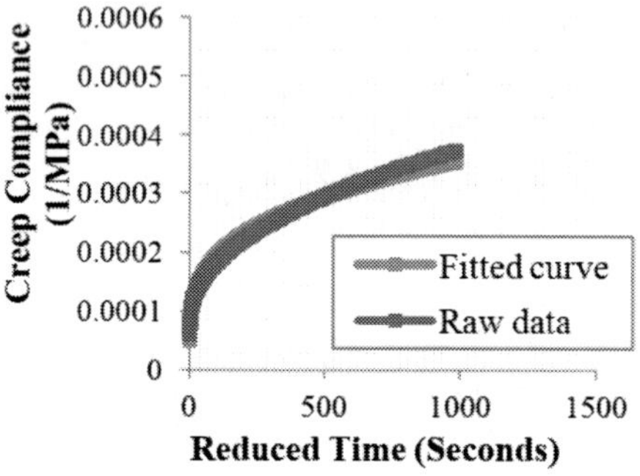

Fig. 16. Power Law funtion (SBS)

Table 3. Power law function parameters

Asphalt Binder	D_0 (1/MPa)	D_1 (1/MPa)	n
Unmodfied asphalt binder	4.3461E-05	1.69881E-05	0.299999998
SBS modified binder	3.8132E-05	3.7428E-05	0.310000515

The relaxation modulus of the unmodified binder and the SBS modified binder are provided in Figure 17. The SBS modified asphalt binder showed to have higher relaxation capabilities compared to the unmodified binder. Thus, indicating that the SBS asphalt mixture have better abilities in relaxing thermal induced stress, compared to the unmodified binder, as it showed to have better relaxation properties.

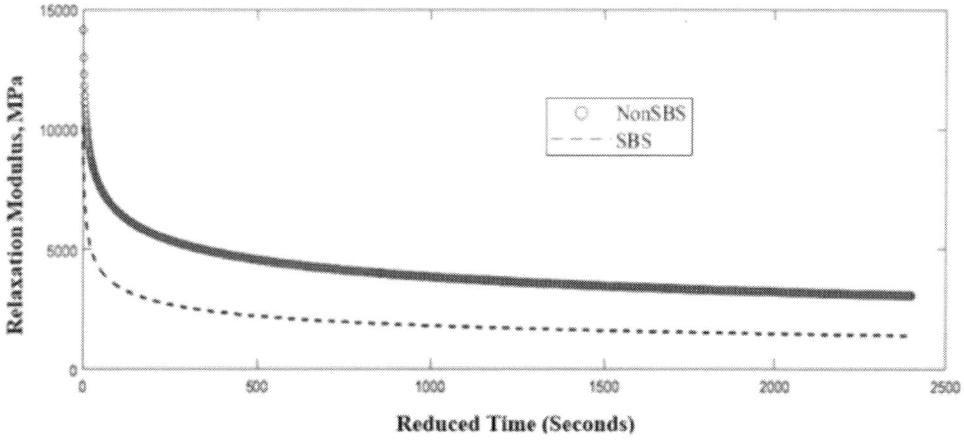

Fig. 17. Relaxation modulus curves.

4 Conclusions and Recommendations

In this study the influence of adding SBS polymer on the low temperature properties of asphalt mixtures was investigated. The study was completed by comparing the thermal cracking properties of unmodified asphalt mixtures and SBS modified mixture, through laboratory tests and viscoelastic analysis. Based on the outcomes of the study the following can be concluded:

- The Bending Beam Rheometer (BBR) test is capable of evaluating low temperature properties of asphalt mixtures efficiently, with a minimum variability among different asphalt beams.
- The unmodified asphalt binder showed to have higher stiffness values compared to the SBS modified binder at the three testing temperatures: -6 °C, -12 °C and -18°C.
- The relaxation capabilities of the SBS modified asphalt binder were higher than the unmodified binder, indicating that the SBS modified binder has better abilities in resisting induced thermal stress.
- The addition of SBS modifier to asphalt binder lead to an improvement in low temperature cracking properties of asphalt mixtures.
- Field visits need to be done to further verify the results obtained from laboratory methods and viscoelastic analysis.

References

1. M.O. Marasteanu, X. Li, T.R. Clyne, V. Voller, D.H. Timm, D. Newcomb, *Low Temperature Cracking of Asphalt Concrete Pavement* (2004)
2. X-J. Li, M. O. Marasteanu, Experimental Mechanics **50**(7), 867-876 (2010)
3. J. Liu, S. Zhao, L. Li, P. Li, S. Saboundjian, Cold Regions Science and Technology **141**, 78-85 (2017)
4. Y. Yildirim, Construction and Building Materials **21**(1), 66-72 (2007)
5. K.R. Wardlaw, S. Shuler, *Polymer modified asphalt binders* (1992)
6. P. Lin, W. Huang, Y. Li, N. Tang, F. Xiao, Constr. Build. Mater. **154,** 609-622 (2017)
7. K.W. Ng, Masters Abstracts Int. **46**, 6 (2008)

8. P.K. Das, Y. Tasdemir, B. Birgisson, Constr. Build. Mater. **30**, 643-649 (2012)
9. N. Tapsoba, H. Baaj, C. Sauzéat, H.D. Benedetto, M. Ech, Constr. Build. Mater. **120**, 393-402 (2016)
10. X. Gong, P. Romero, Z. Dong, Y. Li. Constr. Build. Mater. **156**, 56-62 (2017)
11. P. Romero. *Using the bending beam rheometer for low temperature testing of asphalt mixtures* (2016)
12. AASHTO TP125 (2016)
13. AASHTO T313 (2008)
14. P. Romero, C.H. Ho, K. VanFrank, *Development of Methods to Control Cold Temperature and Fatigue Cracking for Asphalt Mixtures* (2011)
15. CH. Ho, P. Romero, Road materials and pavement design **12**(2), 293-314 (2011)
16. X. Hu, A.N. Faruk, J. Zhang, M.I. Souliman, L.F. Walubita, Int. J. Pavement Res. Techn. (2017)
17. N. Nielsen, B. Hauser, T. Preber, P. Sebaaly, D. Johnston, D. Huft, S. Bang, L. Roberts. *Quality Base Material Produced Using Full Depth Reclamation on Existing Asphalt Pavement Structure* (2007)

Resistance to external sulfate attack - Comparison of two alkali-activated binders

Miroslav Komljenović[1,*], *Nataša* Džunuzović[1], and *Violeta* Nikolić[1]

[1]Institute for Multidisciplinary Research, University of Belgrade, Kneza Višeslava 1, 11000 Belgrade, Serbia

Abstract. Durability of binders, mortars and concretes in aggressive environments is of crucial importance for their commercial application. In this paper the resistance to external sulfate attack of two different alkali-activated binders (AABs), based either on blast furnace slag (BFS) or fly ash/blast furnace slag (FA/BFS) blend, was compared with two different commercially available Portland cement (CEM II) blended either with BFS or with FA and BFS. Comparison of sulfate resistance was based on compressive strength testing (the loss of strength) of mortar samples exposed to sodium sulfate attack up to 180 days and samples cured under controlled conditions for the same period of time. Furthermore, the evolution of microstructure of alkali-activated binders and pH of sodium silicate solution during testing were also analyzed. Despite different gel chemistry being involved, both alkali-activated binders based either on BFS or FA/BFS blend showed excellent resistance to external sulfate attack and even better than selected Portland cements tested under the same experimental conditions.

1 Introduction

Durability of concrete structures during long-term exploitation is of key importance for their safe and efficient functioning and it is determined by its ability to resist chemical attack, weathering action, abrasion, or any other process of deterioration. Chemical reactions that can alter the cement paste properties are one of the fundamental causes of concrete degradation.

The external sulfate attack, as an important process of concrete degradation, might occur due the presence of sulfate ions in soils, underground waters, sea water or industrial waste waters [1-2]. Generally, sulfates are harmful to Portland cement-based structures, but their detrimental impact depends on the type of cement used, quality of concrete, concrete exploiting conditions, nature and concentration of aggressive sulfate solution, as well as presence of different salts and/or cations in sulfate solution. Expansion, cracking, spalling, strength loss, and ultimately disintegration are usually recognized as consequences of external sulfate attack.

In the last few decades alkali-activated binders (AABs) have drawn attention of scientific and professional community as promising alternative to Portland cement. AABs

*Corresponding author: miroslav.komljenovic@imsi.rs

are formed as products of chemical reaction between calcium silicate- or alumino-silicate-based solid precursors and alkali activator. The entire range of natural and synthetic materials can be used as solid precursors for this purpose [3]. The most promising solid precursors for large-scale applications are industrial wastes/by-products such as fly ash (FA) from thermal power plants and blast furnace slag (BFS) from pig iron production. In order to underpin the predicted future role, AABs are expected to meet the requirements specified for Portland cements in terms of quality and durability [4-5].

Despite the fact that the external sulfate attack on Portland cement was intensively studied, the experimental conditions applied were usually incomparable due to the type of samples used (paste, mortar and/or concrete), w/c ratio, sample shape (prisms, cubes or cylinders) and dimension, curing conditions prior to sulfate attack (water, saturated limewater, temperature, time, humidity), aggressive medium type and concentration, as well as conditions of sulfate attack (full or partial immersion of samples, pH and SO_4^{2-} concentration control, time, solution agitation, recycling, etc.) [6].

Different methods for monitoring the process and kinetics of sulfate attack were also used: expansion, mass loss, porosity, as well as flexural and/or compressive strength. The ASTM C1012 method was frequently used as internationally recognized and standardized test procedure. However, this method was often criticized due to its limitation i.e., orientation to expansion effect. On the other hand, there is no appropriate European (EN) standard available. Therefore, the experimental conditions for testing the resistance of Portland cement systems to external sulfate attack or the pass/fail criteria that should be fulfilled are still elusive.

The main goal of this paper is to compare the resistance to external sulfate attack (5% Na_2SO_4) of two chemically different alkali-activated binders with commercially available Portland cements. Alkali-activated binders are based either on blast furnace slag (BFS) or fly ash/blast furnace slag (FA/BFS) blend. The method used within this investigation is focused on the strength loss, as recommended by the State of the art report CEN/TR 15697 [7] related to the performance testing for sulfate resistance of cement. Furthermore, the evolution of microstructure of alkali-activated binders and pH of sodium silicate solution during testing are also analyzed.

For comparison of mechanical characteristics after exposure to a sulfate solution, Portland cement (CEM I - Ordinary Portland Cement, OPC) was usually used as a reference material. However, CEM I is well-known as highly susceptible to the influence of Na_2SO_4 solution. Therefore, commercially available Portland cement (CEM II), blended either with BFS or with FA and BFS, were used as benchmark materials in this work. Finally, all binders were tested under the same experimental conditions [6, 8] making this comparison more reliable.

2 Materials and methods

2.1 Materials

In this study the following materials were used:

1. Blast furnace slag (BFS) *from pig iron production* - "Železara Smederevo" d.o.o., Smederevo, Serbia
2. Fly ash (FA) *from thermal power plant* - "Morava", Svilajnac, Serbia
3. Portland-slag cement (CEM II/A-S 42.5N) - *from cement producer* Kosjerić, Serbia.
4. Portland-composite cement (CEM II/B-M (S-L) 42.5N) - *from cement producer* "CRH Srbija", d.o.o., Novi Popovac, Serbia

5. Sodium silicate – water glass ($Na_2O{\cdot}nSiO_2$) - "Galenika-Magmasil" d.o.o., Zemun, Serbia.
6. Sodium hydroxide (99% NaOH) - VWR, Germany.
7. Sodium sulfate (99% Na_2SO_4) – Superlab, Beograd, Serbia.

Table 1. Chemical composition of initial materials.

	SiO_2	Al_2O_3	Fe_2O_3	CaO	MgO	SO_3	S	Na_2O	K_2O	LOI at 1000 °C
BFS	37.50	7.27	0.73	38.48	10.86	0.39	1.51	0.54	0.26	2.13
FA	55.23	21.43	7.42	7.94	2.61	0.81	-	-	-	1.66
CEM II/A-S	21.70	4.62	2.98	62.53	1.56	2.44	0.00	0.45	0.60	2.80
CEM II/B-M	21.33	5.49	2.62	58.67	3.01	2.27	-	-	-	4.41

2.2 Preparation of mortar and paste samples

Alkali-activated binders (AABs), based either on blast furnace slag (BFS) or fly ash/blast furnace slag (FA/BFS = 1/1) blend, were prepared with sodium silicate solution. Modulus (SiO_2/Na_2O mass ratio) of sodium silicate used was 1.5 and it was adjusted by adding sodium hydroxide. The concentration of the activator with respect to the dry mass of solid precursors (SP: BFS or FA/BFS) was 4% of Na_2O in the case of BFS and 10% of Na_2O in the case of FA/BFS blend.

Mortars of alkali-activated binders were prepared by mixing sodium silicate solution with SP and standard sand [SRPS EN 196-1 (2008)]. SP/sand ratio was 1/3. For Portland cement mortars water/cement (w/c) ratio was 0.5. In order to have the same consistency as benchmark material (CEM II) the optimal amount of water required for alkali-activated binder mortars was determined by flow table test. Therefore, water/binder (w/b) ratio was 0.43 in the case of BFS and 0.39 in the case of FA/BFS blend. Water in w/b ratio included water from sodium silicate solution and water added for appropriate consistency, while binder in w/b included SP and solid part of the activator. Three mortar prisms (40 mm × 40 mm × 160 mm) were prepared for each mixture according to the Serbian standard SRPS EN 196-1 (2008), which is in compliance with the European EN 196-1 standard.

Pastes of alkali-activated binders were prepared by mixing sodium silicate solution with SP (in the same proportion as in the case of mortars) and water. Water/binder (w/b) ratio was 0.28 in the case of BFS and 0.34 in the case of FA/BFS blend. Dimensions of paste samples were 25 mm × 25 mm × 30 mm. For SEM/EDS analysis pastes were prepared by immersing the samples in isopropyl alcohol for 24 h and and then dried at 50 °C for 60 minutes.

For the first 24 h mortar samples were cured in molds in a humid chamber (20±2°C and R.H. ≥ 90%). and after demolding the initial curing was continued up to 28 days also in a humid chamber. After the period of initial curing mortar samples were either exposed to external sulfate attack (5% Na_2SO_4 solution) or further cured in a humid chamber (reference samples). Paste samples were cured/exposed under the same experimental conditions.

5% Na_2SO_4 solution was prepared by dissolving Na_2SO_4 in distilled water and plased in 10 l plastic containers. Mortar samples were placed on plastic carriers and completely immersed in the solution thus allowing all surfaces to be in contact with the aggressive sulfate solution. Sulfate solution/mortar samples mass ratio was ~4. Complete replacement of Na_2SO_4 solution was performed every 30 days over anticipated period of external sulfate attack testing.

2.3 Methods of characterization

Chemical composition of the initial FA and BFS samples was determined by classic chemical analysis – alkali melting, and CEM II according to the Serbian standard SRPS EN 196–2 (2008).

Compressive and flexural strengths of alkali-activated binders and CEM II were tested according to the SRPS EN 196-1 (2008) standard which is in compliance with the European EN 196-1 standard, using Controls model 65-L1300/P (Italy) device.

Microstructural characterization was carried out on alkali-activated binders pastes by scanning electron microscopy (SEM) method using VEGA TS 5130 MM Tescan electron microscope. Energy dispersive X-ray spectroscopy (EDS) was done by INCAPentaFET-x3 (Oxford Instruments). Prior to the SEM-EDS analysis, samples were polished and Au-coated. In order to obtain more accurate binder matrix composition during the EDS analysis the unreacted FA or BFS particles visible on the sample surface were deliberately excluded. The results of EDS analysis represent the average values of 20–30 individual EDS analysis for each examined sample.

The pH values of Na_2SO_4 solution during the testing period were measured by pH-meter (pHTestr30, Eutech Instruments, Holland).

3 Results and discussion

3.1 Compressive strength

The results of mortar compressive strength testing of alkali-activated binders and corresponding benchmark CEM II are given in Fig. 1. Alkali-activated BFS did not show any sign of deterioration after exposure to external sulfate attack over a period of 90 days (Fig. 1a). On the contrary, the samples of alkali-activated BFS exposed to sodium sulfate solution showed higher strength increase when compared to the reference samples cured in a humid chamber for the same period of time. This means that the sodium sulfate solution actually acted as an alkali activator rather than the harmful substance. At the same time, the corresponding benchmark material Portland-slag cement (CEM II/A-S 42.5N) showed significant strength decrease after 60 days of exposure to external sulfate attack i.e. the beginning of significant deterioration (Fig. 1c).

Alkali-activated FA/BFS blend also did not show significant sign of deterioration even after exposure to external sulfate attack over a period of 180 days (Fig. 1b). The samples of alkali-activated FA/BFS blend exposed to sodium sulfate solution showed strength increase but at somewhat slower rate when compared to the reference samples cured in a humid chamber for the same period of time. This means that the sodium sulfate solution only inhibited the reaction of alkali activation to a certain extent, but definitely did not act as the harmful substance. At the same time, the corresponding benchmark material Portland-composite cement (CEM II/B-M (S-L) 42.5N) showed significant strength decrease after 90 days of exposure to external sulfate attack i.e. the beginning of significant deterioration (Fig. 1d).

Better behavior of alkali-activated binders in aggressive sulfate environment is also visible when relative mortar compressive strength (samples exposed to sulfate attack in respect to the reference samples of same age) of alkali-activated binders and benchmark

CEM II are compared (Fig. 2). Therefore, it can be concluded that both alkali-activated binders based on BFS or FA/BFS blend showed remarkable resistance to external sulfate attack and better than corresponding Portland cement counterparts tested under the same experimental conditions.

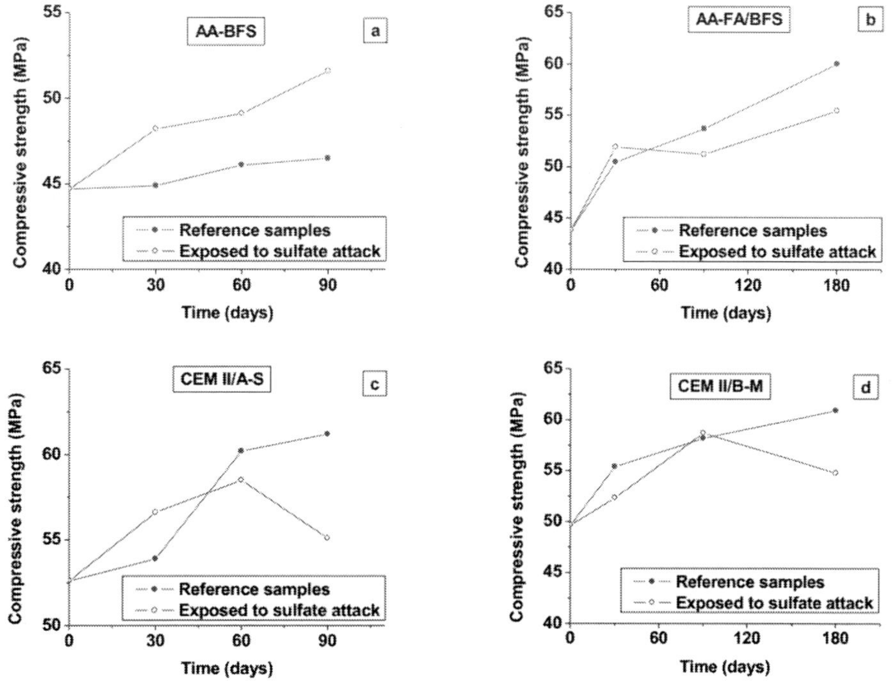

Fig. 1. Mortar compressive strength of alkali-activated binders and benchmark CEM II (reference samples and samples exposed to sulfate attack; a) Alkali-activated BFS; b) Alkali-activated FA/BFS blend; c) Portland-slag cement (CEM II/A-S 42.5N); d) Portland-composite cement (CEM II/B-M (S-L) 42.5N); (adapted from [6, 8]).

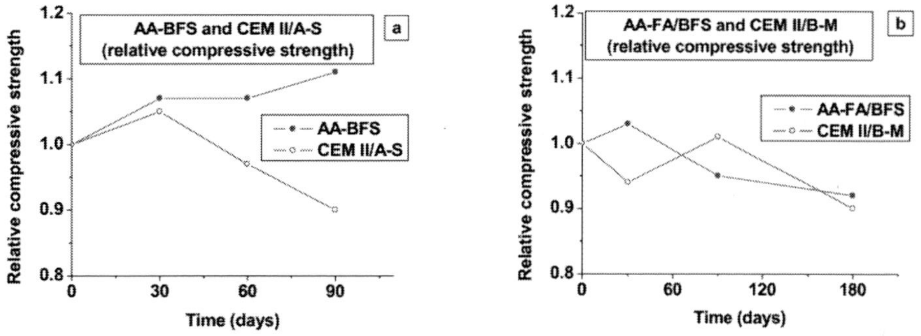

Fig. 2. Relative mortar compressive strength of alkali-activated binders and benchmark CEM II (samples exposed to sulfate attack in respect to the reference samples of same age); (adapted from [6, 8])

3.2 EDS analysis

The results of microstructural (EDS) analysis of alkali-activated binders are given in Fig. 3.

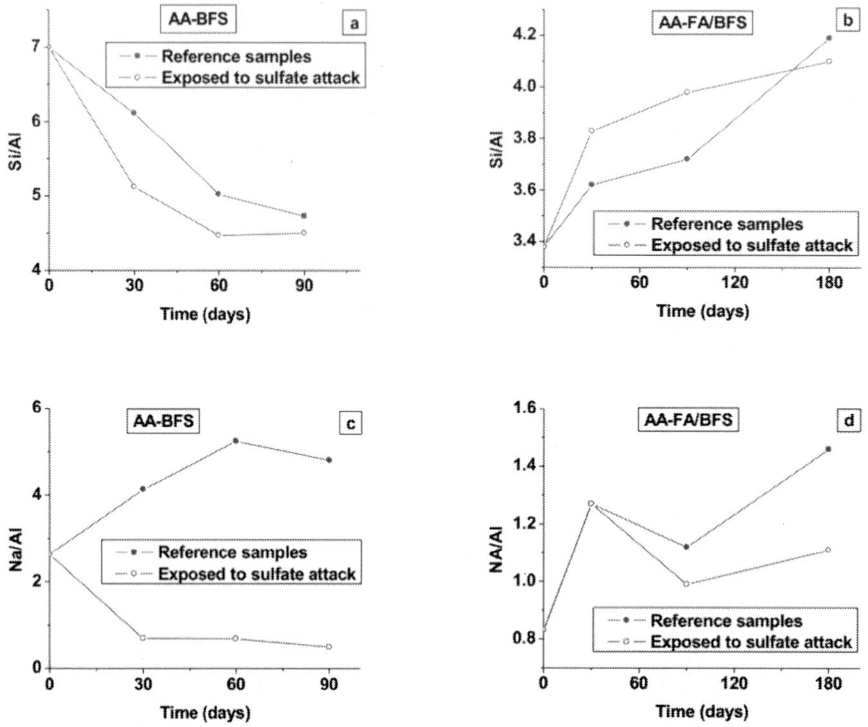

Fig. 3. EDS analyses of reference samples and samples exposed to sulfate attack (the average ratios of the main elements of alkali-activated binder matrix): a) AA-BFS; b) AA-FA/BFS blend; (adapted from [6, 8])

The Si/Al atomic ratio of alkali-activated BFS decreased throughout the whole period of 90 days of testing, whereby this ratio decreased at faster rate in samples exposed to external sulfate attack (Fig. 3a). In both cases this ratio was tending to the Si/Al value of 4.5-4.7. Obviously, the decrease of the Si/Al ratio was connected with the strength increase (Fig. 1a). On the other hand, the Si/Al atomic ratio of alkali-activated FA/BFS blend increased throughout the whole period of 180 days of testing, whereby in the initial period of 90 days this ratio increased at faster rate in samples exposed to external sulfate attack (Fig. 3b). In both cases this ratio was tending to the Si/Al value of 4.1-4.2. In the case of FA/BFS blend the strength increase (Fig. 1b) was connected with the increase of the Si/Al ratio, confirming completely different gel chemistry being involved when compared with the alkali-activated BFS. It is well known that in the case of alkali-activated BFS the predominant reaction products are C-S-H or C-(A)-S-H gels [6], while in the case of FA/BFS blend the predominant reaction products are N-A-S-H or N-(C)-A-S-H gels [8].

The Na/Al atomic ratio of alkali-activated BFS samples exposed to external sulfate attack decreased throughout the whole period of 90 days of testing (Fig. 3c), most probably due to the sodium leaching. Obviously, sodium leaching did not cause any strength deterioration (Fig. 1a) since sodium can be replaced with H+ or H_3O^+ ions in gel structure without recordable structural damage. At the same time, the Na/Al atomic ratio in the reference samples increased, probably due to the higher sodium uptake in the binder matrix in the same period. On the other hand, the Na/Al atomic ratio of alkali-activated FA/BFS blend increased after a period of 180 days of testing, both in reference samples and samples exposed to sulfate attack (Fig. 3d), confirming sodium being more tightly bound in the N-A-S-H or N-(C)-A-S-H gels than in the C-S-H or C-(A)-S-H gels.

3.3 pH of Na_2SO_4 solution

The results of pH of the initial sodium sulfate solution and throughout the whole period of testing are given in Table 2.

Obviously the slightly acidic pH (6-6.7) of the initial sodium sulfate solution shifted to highly alkaline pH (> 11) after first 30 days of testing and remained almost constant after further testing. One should have in mind that sodium sulfate solution was completely replaced each 30 days, whereby continuous sodium leaching from alkali-activated binder matrix was confirmed by highly alkaline pH.

Table 2. pH of Na_2SO_4 solution during the investigation.

Sample	Initial pH	After 30 days	After 60 days	After 90 days	After 180 days
AA-BFS	6.63	11.44	11.54	11.60	-
AA-FA/BFS	6.04	13.11	-	11.30	11.25

4 Conclusions

Based on the results presented the following conclusion can be drawn:

Despite different gel chemistry being involved, both alkali-activated binders based on BFS or FA/BFS blend showed exceptional resistance to external sulfate attack and even better than corresponding Portland cement CEM II counterparts tested under the same experimental conditions.

Alkali-activated binders showed better resistance to external sulfate attack than Portland cements due to unavailability of aluminum to react with sulfates since aluminum was tightly bound in the C-(A)-S-H, N-A-S-H, N-(C)-A-S-H, and/or hydrotalcite gel.

This research was financially supported by the Ministry of Education, Science and Technological Development of the Republic of Serbia through project TR 34026.

References

1. F.P. Glasser, J. Marchand, E. Samson, Cem. Concr. Res. **38**, 226–46 (2008)
2. A. Neville, Cem. Concr. Res. **34**, 1275–96 (2004)
3. M. Komljenović, *Handbook of Alkali-Activated Cements, Mortars, and Concretes*, 171-217 (Woodhead Publishing , 2014).
4. M. Whittaker, L. Black, Adv. Cem. Res. **27**(9), 532-545 (2015)
5. W.G.V. Saavedra, D.E. Angulo, R.M. de Gutiérrez, J. Mater. Civ. Eng. **28**(12), 04016148 (2016)

6. M. Komljenović, Z. Baščarević, N. Marjanović, V. Nikolić, Constr. Build. Mat. **49,** 31-39 (2013)
7. CEN/TR 15697 (2008)
8. N. Džunuzović, M. Komljenović, V. Nikolić, T. Ivanović, Constr. Build. Mat. **157** 737–747 (2017)

Study of an alkali-activated binder based on tungsten mining mud and brick powder waste

Naim Sedira[1], and *João* Castro-Gomes[1,*]

[1]Centre of Materials and Building Technologies (C–MADE/UBI), Department of Civil Engineering and Architecture, University of Beira Interior (UBI), 6201–001 Covilhã, Portugal

Abstract. Blends of Tungsten mining waste mud (TMWM) and brick waste powder (BP) with different dosages were used as precursors for the study of a new binder obtained by alkali-activation. The synthesis was obtained at 60°C curing during the first 24 hours and at 20°C during the remaining period. A combination of sodium hydroxide (SH) and sodium silicate (SS) solutions was used with SS/SH weight ratio equal 1.5. The solid precursors/liquid activators weight ratio equal to 4. And the modules SiO_2/Na_2O increase with the increasing of BP dosages 5.21 and 5.59 for dosages 10% and 50%, respectively. Mineralogical characterisation of raw materials was carried out by X–ray diffraction (XRD). The effect of the dosage of BP on the compressive strength and pore size distribution of the new binder was investigated from 24 hours up to 28 days. The pore size distribution was obtained mercury intrusion porosimetry (MIP). The increase in the dosages of BP, between 10 to 50%, was followed by an increase in compressive strength, from 25 to 59 MPa, for all the tested ages. The binder matrix become more dense and compact with the gradually increase of BP dosages, as found out by MIP.

1 Introduction

The cement manufacturing process is responsible for about 6% of the global CO_2 emissions. The increase in carbon dioxide (CO_2), greenhouse gases and the disposal problems of industrial waste caused serious impact on the environment and climate changes. The alkali–activated binders technology offers other feasible cementing materials that could reduce the cement industries CO_2 emissions by about 80% to 90% [1], [2]. Mining and/or construction and demolition activities accounted for 3.0 tonnes out of a total of 3.2 tonnes per inhabitant of mineral waste, equivalent to 93.5% of the total mineral waste generated across EU. The construction waste generated 45% of total waste and its production is growing at a rapid pace [3]. The increase in the amount of industrial wastes such as waste glass, plastics and fired clays cause serious environmental problems [4], [5].

In Europe, the activities of mining and quarrying generate approximately 55% of total industrial wastes according to Eurostat [3]. Thus, the waste generated from mining and quarry industry accumulated in large deposits present a potential risk. The storage of these

* Corresponding author: castro.gomes@ubi.pt

wastes directly on land may lead to various environmental issues [6]. Different mine tailings have been used in various previous studies as precursors for alkali–activation [7]. Tungsten tailing is the waste generated from a tungsten mine. It can lead to environmental risk, like contamination of soil, pollution of water and air in the surrounding areas [8]. According to Castro–Gomes et al. [6], the mineralogy of tailings from Panasqueira tungsten mining waste was found to be mainly quartz and muscovite. However tungsten mining waste can be re-used in several innovative applications [9]. Recently, Sedira and Castro-Gomes [10] obtained an alkali–activated binder where 50% of the tungsten mining waste was replaced by electronic arc furnace slag (EAFS) to increase the overall amorphous SiO_2 and calcium content. The compressive strength of alkali–activated binder was above 30 MPa after 90 days curing.

Brick waste powder (BP) is one of the different types of ceramic waste used to develop alkali–activated binders [11]. Several studies and research were made to develop either alkali–activated materials or hybrid cement using BP as a source of alumina silicate materials and other minerals. Early studies focused on producing alkali–activated materials based on itself or on the combination of BP with and other minerals; namely, different ceramic waste types [12], [13], BP with concrete waste [14] and BP with GGBFS [15], brick waste powder with fly ash [16], red clay brick waste only [17], [18] and ordinary Portland cement with BP [18].

In the present study, we evaluated the synthesis of an alkali-activated binder (AAB) obtained by blending Tungsten mine waste mud (TMW) and brick powder (BP), using as activators a mix solution of sodium silicate (SS) and sodium hydroxide (SH). This research work is part of REMINE research program (H2020 RISE-Marie Curie Action) that aims to develop new ideas for the reuse of mining wastes into innovative alkali-activated-based materials for structures and buildings.

2 Experimental

2.1 Material

The materials used in this work are Tungsten mining waste mud (TMWM) and brick waste powder (BP). The mud waste was collected from Panasqueira mine located in Covilhã, Portugal. The mud was dried in oven at 60°C temperature for 24 hours. Afterwards the dried mud was mechanically disaggregated using a crushing machine and sieved under 500 μm particle size.

Regarding BP the following processing procedures were adopted: First, brick coarse waste was submitted to washing to remove contaminations and to separate the other construction impurities; Afterwards it was placed in the oven at a temperature of 60°C, for 24 hours for complete drying; Finally, the coarse brick waste was crushed using a crushing device and was sieved to obtain particles with size under 250 μm.

The chemical compositions of these materials, were determined by scanning electron microscopy (SEM-EDX, HITACHI S-3400N).

The bulk powder densities of TMWM and BP were determined using a gas displacement pycnometer (model AccuPyc 1340, Micromeritics, Norcross, Georgia) and were determined as 3.03 and 2.73 g/cm^3, respectively. The Blaine fineness of the different powders was determined according to EN 196-6, by using a Blaine air permeability apparatus (model ACMEL BSA1) and were determined as 3339 and 2954 cm^2/g, respectively. Loss on ignition (LOI) was obtained by TGA analysis by using TA Instrument SDT Q 50, TMWM loss about 11.6% of weight and BP loss about 3.6% of weight at the temperature 1000 °C. The physical characteristics and chemical compositions of the TMWM and BP materials are presented in Table 1.

Table 1. Physical characteristics and chemical composition (% mass) of waste materials.

Composition/Properties	**Materials**	
	BP	**TMWM**
O	47.32	38.26
Si	27.39	18.2
Al	13.09	7.51
Ti	0.61	0.3
S	0.15	2.64
K	2.40	3.39
Ca	0.57	0.41
Fe	5.52	9.02
Mg	0.87	2.43
Na	0.37	0.80
P	0.06	–
Zn	–	1.09
Other	1.66	15.89
LOI*	3.8	11.6
Density g/cm^3	2.7339	3.0319
Blaine specific area cm^2/g	2954	3339
*LOI, Loss on ignition at 1000 °C		

For the alkaline activators, the sodium hydroxide solution was prepared by dissolving sodium hydroxide pellets (98% purity, obtained from Fisher Scientific, Schwerte, Germany) in deionized water and allowed to cool before use. Sodium silicate (obtained from Solvay SA, Póvoa de Santa Iria, Portugal) presented the following chemical composition: SiO_3/Na_2O = 3.23 (8.60% by weight Na_2O, 27.79% by weight SiO_2, 63.19% by weight H_2O, and 0.4% by weight Al_2O_3).

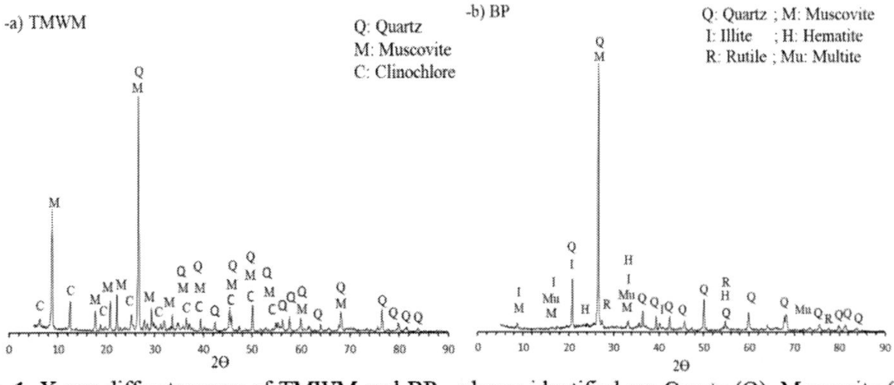

Fig. 1. X–ray diffractograms of TMWM and BP - phases identified are Quartz (Q), Muscovite (M), Clinochlore (C), Hematite (H), Illite (I), Rutile (R), Mulite (Mu).

TMWM and BP X-ray diffractograms are presented in figure 1. The X-ray diffractogram of the TMWM indicates it has a crystalline nature and is consists mainly of muscovite and quartz, which were identified by their characteristic, as follows: muscovite (Ref. PDF#46–1409), quartz (Ref. PDF#46–1045) and clinochlore (Ref. PDF#29–0701) as shown in figure 1-a) [19]. X–ray diffractogram spectra of BP, as presented in figure 1-b) indicates the semi–crystalline degree of the waste material and the presence of quartz SiO_2 (Ref. PDF#46–1045) as the main phase.

2.2 Synthesis and analysis

The binder mixes were produced by mixing of TMWM and BP as solid precursors, with sodium silicate and sodium hydroxide as alkaline activator solutions. A solid precursors/activator solutions weight ratio of 4 was used, the volume ratio between the two waste materials and the modules SiO_2/Na_2O, SiO_2/Al_2O_3 weight ratio are shown in the Table 2.

Table 2. Mixes compositions.

Specimen labels	Volume proportions for the blended waste materials	Variables	Ratios
H1	10% BP and 90 % TMWM	SiO_2/Na_2O	5.21
		SiO_2/Al_2O_3	7.1
		Precursor/Activator ratio (P/A)	4
H2	20% BP and 80 % TMWM	SiO_2/Na_2O	5.29
		SiO_2/Al_2O_3	6.9
		Precursor/Activator ratio (P/A)	4
H3	30% BP and 70 % TMWM	SiO_2/Na_2O	5.38
		SiO_2/Al_2O_3	6.75
		Precursor/Activator ratio (P/A)	4
H4	40% BP and 60 % TMWM	SiO_2/Na_2O	5.49
		SiO_2/Al_2O_3	6.58
		Precursor/Activator ratio (P/A)	4
H5	50% BP and 50 % TMWM	SiO_2/Na_2O	5.59
		SiO_2/Al_2O_3	6.4
		Precursor/Activator ratio (P/A)	4

The blended mixtures were activated with a combination of the following solutions: 67% of Sodium Silicate (SS) - Na_2SiO_3 and 33% of 10M of Sodium Hydroxide (SH) - NaOH. Before beginning to prepare the mixtures, both SS and SH activators, with a mass ratio of 3:2 were mixed together. Afterwards, the liquid activators solution was mixed the precursors and poured in the curing moulds. Then, the filled moulds were placed in an oven at a temperature of 60°C for 24 hours, for synthesis. To avoid water evaporation the TMWM–BP alkali-activated binders, the filled moulds were wrapped with plastic film during the synthesis to avoid water evaporation. After the initial 24 hours curing period, the specimens were demoulded and were left to cure in laboratory conditions (about 20 °C) for 3, 7, 14 and 28 days.

The compressive strength tests were performed using a 3000kN electro–hydraulic mechanical testing machine (ADR Touch 3000 BS EN Compression Machine with Digital Readout and Self Centring Platen), in accordance with EN 196-1. Compressive strength data was obtained using 25 mm cubic size specimens.

Mercury intrusion porosimetry (MIP) was adopted to study the pore structure of BP-TMWM alkali–activated binders. Using Microporometrics AutoPore IV 9500 V1.07. Mercury porosimetry data is obtained by recording the volume of mercury that penetrates the porous specimen as a function of pressure. MIP measurement was applied to determine the pore structure difference i.e., difference in porosity and pore size distribution.

3 Results and discussion

3.1 Compressive strength

The compressive strength results of TMWM–BP alkali–activated binders (average of testing 5 specimens per each mixture) were obtained at the ages of 1,3,7, 14 and 28 days. Figure 2 shows the development of compressive strength in terms of the dosage of BP used in the TMWM–BP alkali-activated binders. Mixture H1 reached 10 MPa compressive strength after the first day curing. It was the lowest value obtained regarding all mixes tested. At 28 days, the compressive strength obtained for the same mix (H1) was 25 MPa. Thus, after one day curing the compressive strength represented about 40% of 28 days compressive strength. Mixture H5 obtained a compressive strength of 22.4 MPa, after one day curing, being the highest result corresponding to the highest BP (50%) content. This initial result represents 32,6 % of the 28 days curing H5 compressive strength, that was about 59 MPa, and the highest result of all mixtures. The increase of BP dosage the increase in compressive strength from 25 MPa after one day curing to 59 MPa after 28 days curing, corresponding to H1 and H5 samples, respectively. The increase of BP content from 10 to 50% resulted of a 136% compressive strength increase. Significant development in compressive strength was observed from the first day up to 14 days of curing. In comparison to the 28 days maximum compressive strength, the compressive strength relative increase from the first up to 14 days ranges between 95 % and 99% for all mixtures. Generally, the higher the dosage of BP in the mixes, the higher the compressive strength obtained in all the tested ages [20].

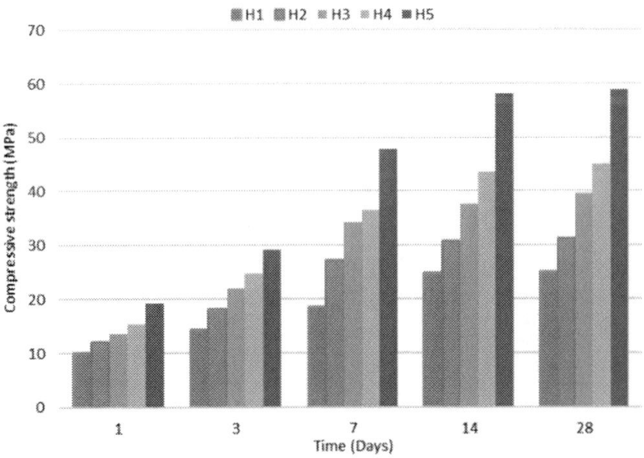

Fig. 2. Development in the compressive strength of alkali-activated mixtures (H1 to H5) containing 10, 20, 30, 40 and 50 % of BP, as a function of time.

3.2 Pore size distribution

MIP analysis revealed that the higher dosage of BP the higher percentages of pores with smaller diameter (< 20 nm and diameter between 20 and 50 nm), namely in the samples H3, H4 and H5. The samples H1, H2 presented the lesser percentages of smaller pores (< 20 nm and diameter between 20 and 50 nm) and higher percentages of larger pores size (50 nm to up to 800 nm), as show in figure 3. Thus, it was verified that with increasing BP content in the mixes the hardened paste becomes denser, since with higher BP dosages more quantity of alkali-activated gel is formed that can fill in the micro pores in TMWM–BP hardened binder, leading to further increase in the density of final matrix phase [21]. The changes in pore size distribution for different H mixes can explain the increase in the compressive strength for higher dosages of BP, and along time of reaction.

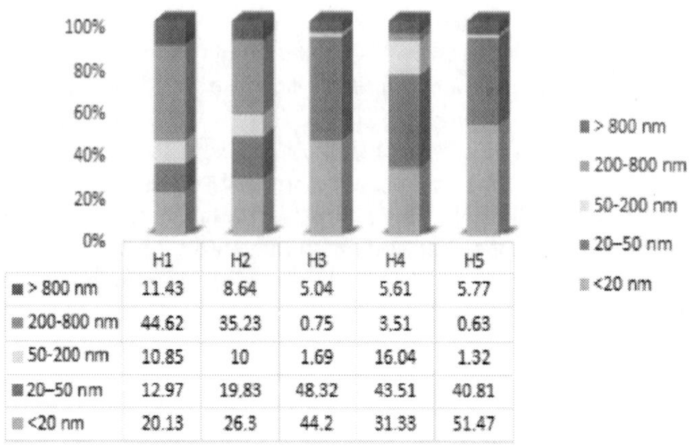

	H1	H2	H3	H4	H5
> 800 nm	11.43	8.64	5.04	5.61	5.77
200-800 nm	44.62	35.23	0.75	3.51	0.63
50-200 nm	10.85	10	1.69	16.04	1.32
20–50 nm	12.97	19.83	48.32	43.51	40.81
<20 nm	20.13	26.3	44.2	31.33	51.47

Fig. 3. Pore size distribution (%) of different alkali-activated binder samples.

The pore size distribution and the critical pore size was determined for different BP dosages. As well, known the pore width corresponding to the highest rate of mercury intrusion per change in pressure is known as the "threshold," "critical," or "percolation" pore width. After achieving this highest rate of intrusion, mercury can penetrate the interior of the sample [22]. The relationships between critical pore size and the dosages of BP contained in the five samples are presented in Figure 4. The differential distribution curves for the samples H1, H2 and H3 have several peak shifts. While, for the samples H4 and H5 only one rounded peak dominates the differential distribution curves. The initial peak corresponding to lower pore size may correspond to the intrusion of mercury through a connected capillary network, while the rounded peak may correspond to the crushing of interposed reaction products, as found by other authors [23].

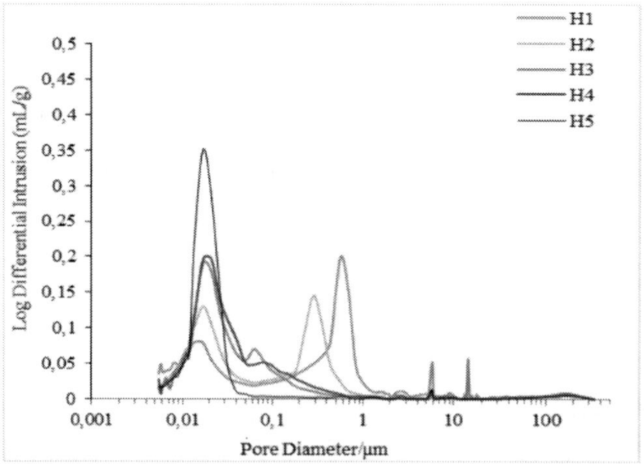

Fig. 4. Overlay of the differential distribution curve for the specimens cured for 28 days.

4 Conclusions

In this study the synthesis of an alkali-activated binder (AAB) obtained by blending Tungsten mine waste mud (TMW) and brick powder (BP), using as activators a mix solution of sodium silicate (SS) and sodium hydroxide (SH), was demonstrated. It was demonstrated that brick waste powder has a positive effect on the mechanical properties of alkali-activated binders when blended with other mineral wastes, such as Tungsten mining waste mud.

The blend of 50% of brick waste powder and 50% of TMWM resulted in an alkali-activated material with high compressive strength and dense structure.

This research was partially supported by European Commission under Horizon 2020, Marie Skłodowska–Curie Actions, Research and Innovation Staff Exchange (RISE), by REMINE - "Reuse of Mining Waste into Innovative Geopolymeric–based Structural Panels, Precast, Ready Mixes and Insitu Applications". Project no 645696. (https://reminemsca.wordpress.com). This work was also partially financed by Portuguese national funds through FCT – Foundation for Science and Technology, IP, within the research unit C–MADE, Centre of Materials and Building Technologies (CIVE–Central covilhã–4082), University of Beira Interior, Portugal.

References

1. S. Usha, D.G. Nair, S. Vishnudas, Int. J. Civ. Eng. Technol. **5**, 219–225 (2014)
2. J. Davidovits, World Resour. Rev. **6**, 263–278 (1994)
3. *Waste statistics Main statistical findings -Total waste generation* (Eurostat ,2016).
4. V. Loryuenyong, T. Panyachai, K. Kaewsimork, C. Siritai, Waste Manag. **29**, 2717–2721 (2009)
5. P. P. Peña et al., Adv. Mater. Sci. Eng. **2016**, 2842969 (2016)
6. J. P. Castro-Gomes, A. P. Silva, R. P. Cano, J. Durán Suarez, A. Albuquerque, J. Clean. Prod. **25**, 34–41 (2012)
7. N. Sedira, J. Castro-Gomes, G. Kastiukas, X. Zhou, A. Vargas, Min. Sci. **24**, 29–58 (2017)
8. B. Wilson, F.B. Pyatt, Sci. Total Environ. **370**, 401–408 (2006)
9. J. Castro-Gomes et al., *International Congress on Engineering UBI*, (Portugal, 2017)
10. N. Sedira, J. Castro-gomes, *REMINE - International Conference & Brokerage Event (RICON17) – UBI* (Portugal, 2017)
11. L. Reig, M.M. Tashima, L. Soriano, M.V. Borrachero, J. Monzó, J. Payá, Waste and Biomass Valorization **4**, 729–736 (2013)
12. F. Puertas, A. Barba, M.F. Gazulla, M.P. Gómez, M. Palacios, Mater. Construcción **56**, 73–84 (2006)
13. Z. Sun et al., Constr. Build. Mater. **49**, 281–287 (2013)
14. A. Allahverdi, E.N. Kani, Int. J. Civ. Eng. **7**, 154–160 (2009)
15. N.R. Rakhimova, R.Z. Rakhimov, Mater. Des. **85**, 324–331 (2015)
16. P. Rovnanik, B. Reznik, P. Rovnanikovà, Procedia Eng. **151**, 108–113 (2016).
17. K. Komnitsas, D. Zaharaki, A. Vlachou, G. Bartzas, M. Galetakis, Adv. Powder Technol. **26**, 368–376 (2015)
18. R. A. Robayo, A. Mulford, J. Munera, R. Mejia de Gutiérrez, Constr. Build. Mater., **128**, 163–169 (2016)

19. F. Pacheco-Torgal, J. Castro-Gomes, S. Jalali, Constr. Build. Mater. **23**, 200–209, (2009)
20. N. Sedira, J. Castro-Gomes, M. Magrinho, Constr. Build. Mater. (to be published) (2018)
21. X. Guo, H. Shi, X. Wei, Cem. Concr. Compos. **79,** 53–61 (2017)
22. D.N. Winslow, S. Diamond, ASTM J. Mater. **5**, 564–585 (1969)
23. R.A. Cook, K.C. Hover, Cem. Concr. Res. **29**, 933–943 (1999)

Influence of water to solid ratio on mechanical properties of GBFS-based geopolymer foam concrete

Tomasz Piotrowski[1,*], and *Piotr* Prochoń[1]

[1]Warsaw University of Technology, Faculty of Civil Engineering, Al. Armii Ludowej 16, 00-637 Warsaw, Poland

Abstract. The development of sustainable building materials with reduced environmental footprint in both, manufacturing and operational phases of the material lifecycle, is attracting increased interest in the construction industry worldwide. A recent innovation, the geopolymer foam concrete, combines the performance benefits and operational energy savings achievable through the use of lightweight foam concrete, with the cradle-to-gate emissions reductions obtained through the use of a geopolymer binder derived from granulated blast-furnace slag (GBFS). In this study mechanical properties of GBFS-based foam concrete were investigated for samples of different water to solid ratio (0.252, 0.287 and 0.321). According to ASTM C 796-97 both mass of the foaming solution and water in sodium silicate solution was considered as part of the total amount of mixing water. As a solid part, GBFS and solid part of activators (NaOH and sodium silicate) was accounted. A group of specimens (40x40x160 beams and 100x100x100 cubes) have been prepared and volume density, bending, compressive strength tests have been performed. In a result an optimized lightweight GBFS-based geopolymer foam concrete was obtained, characterized by 1.8 kg/dm3 volume density, 2.6 MPa bending strength and 51.8 MPa compressive strength measured on beams and 44.1 MPa compressive strength on cubes.

1 Introduction

The name "geopolymers" was introduced by J. Davidovits to describe "mineral polymers" obtained in the way of geosynthesis or geochemistry 0. Those amorphous polymers are formed by mixing an aluminosilicate raw material with aqueous alkali solution 2. To obtain geopolymer composites of certain properties, one needs to select particular raw material, mix it with alkaline solution of appropriate molar concentration and Si/Al ratio and then cure it either in ambient temperature or increased one. As a binder researchers use different pozzolanic materials: metakaolin [3,4], fly ashes [5,6,7], slags 8 or mixes of abovementioned [2,9]. The reaction process needs an alkaline activator, among which the most popular are sodium or potassium silicates and sodium or potassium hydroxides [5,6,7]. The type of used

[*] Corresponding author: t.piotrowski@il.pw.edu.pl

activator has a high influence on the raw materials dissolution of Al and Si ions. Jaarsveld and Deventer pointed out that using NaOH solutions in geopolymers binding process in comparison to KOH solutions results in higher degree of condensation, higher compressive strength, lower degree of crystallinity and acid resistance 10. Using NaOH together with water glass, which increases the Si content in geopolymer mix, provides improvement in mechanical properties of the composite 11. The most common NaOH molar concentration used in research is not higher than 20M with suggestion that sodium silicate to sodium hydroxide ratio should not exceed 1.0 12.

Despite the fact, that foam concrete is not a subject of European standards, it is used in civil engineering. For example it is applied in sub-structures of industrial concrete floors [13] or as contact layer between concrete and subsoil [14]. The geopolymer foam composites present high potential as a low toxicity thermal insulation [15,16] of low thermal conductivity (around 0.1 W/(m·K)) depending on the used alkaline agent, foaming agent type and foaming agent mass ratio [17,18]. By foaming the geopolymer binder it is possible to obtain a composite with density lower than 1309 kg/m^3 [19]. Just like as in regular geopolymer binders, the curing temperature has an influence on the geopolymerization rate, enabling to produce more dense composite with higher compressive strength, lower water absorption and lower porosity [20].

In this study, the concept of geopolymer foam concrete is presented and analyzed. The granulated blast-furnace slag (GBFS) was alkali-activated and examined using compressive and flexural strength testing. The research seems to be important in terms of understanding possible production of lightweight foam concretes with geopolymer binders.

2 Materials and Methods

2.1 Granulated Blast Furnace Slag (GBFS)

The GBFS with high amorphous content (97.42%), the chemical composition presented in tab. 1 (in conformity with EN 15671-1) was used as a binding material. This raw material with density of 3.001 g/cm^3 consists of coarse and fine particles with sharp edges. The median value of the particle size distribution adopted value of 21.44 μm (see Fig. 1).

Table 1. GBFS description.

Chemical composition (%) and physical characteristic	GBFS
SiO_2	37.82
Al_2O_3	7.40
Fe_2O_3	0.85
MnO	0.17
MgO	6.34
CaO	44.05
Na_2O	0.20
K_2O	0.28
TiO_2	0.26
P_2O_5	0.05
LOI	0.00
Total of XRF	97.42
$\sum (SiO_2 + Al_2O_3 + Fe_2O_3)$	46.02
Real density (g/cm^3)	3.001
Median particle size (μm)	21.44
Amor.	92.79

Fig. 1. SEM image of the GBFS.

2.2 Alkaline activators

Two types of alkaline activators were used in geopolymer foam concrete production: the sodium silicate and the sodium hydroxide. The aqueous hydroxide solution was prepared with sodium hydroxide pellets and water. Additionally sodium silicate of molar module: 2,5 +/- 0,1 and minimum oxides content 39% was used.

3 Mixture Properties and Experimental Methodology

All the samples were made with variable amount of water with other parameters ceteris paribus (mixture proportions presented in Table 2.). Initial assumptions for mix composition based on J. Provis and H. Wang article [22], where the amount of raw material and activators was constant in the geopolymer production, changing only the water content. Following the standard ASTM C796-97 the foaming solution in the foam was considered as part of the total mixing water. The calculated total mixing water to slag ratio (w/s) was from 0.29 to 0.37. The pre-foaming method, a common one for foamed concrete production, was used. The foam was prepared separately with Meex AG foaming agent and Idrobase foam nebuliser (PS 8 bar). The NaOH water solution, sodium silicate and GBFS were mixed in a hand held mixing stirrer to create geopolymer paste. Then the foam was gently added and mixed to avoid occurrence of bubbles with large diameter. 10 cm cubes and 4x4x16 cm beams were casted and left for 48h before demolding. The samples were cured in the ambient conditions till testing after 90 days. The compressive and flexural strength were measured in accordance with EN 196-1 and EN 12390-3.

Table 2. Mix composition.

Symbol	GBFS [g]	NaOH pellets [g]	Water [g]	Sodium silicate[g]	Foam [g]	w/s ratio
GFC_029	8000	432	801.33	2116.8	240	0.29
GFC_033	8000	432	801.33+320.59	2116.8	240	0.33
GFC_037	8000	432	801.33+640.59	2116.8	240	0.37

4 Experimental Results and Discussions

Figure 2 presents influence of w/s ratio on the volume density. It is observed that increase of total water amount decompose foam during mixing and forming. The specimens with high w/s have density close to standard geopolymer paste [21] approx. 1.8 g/cm^3 and do not present porous structure as it is clearly visible on the cross section of the GFC_029 (see Fig.3).

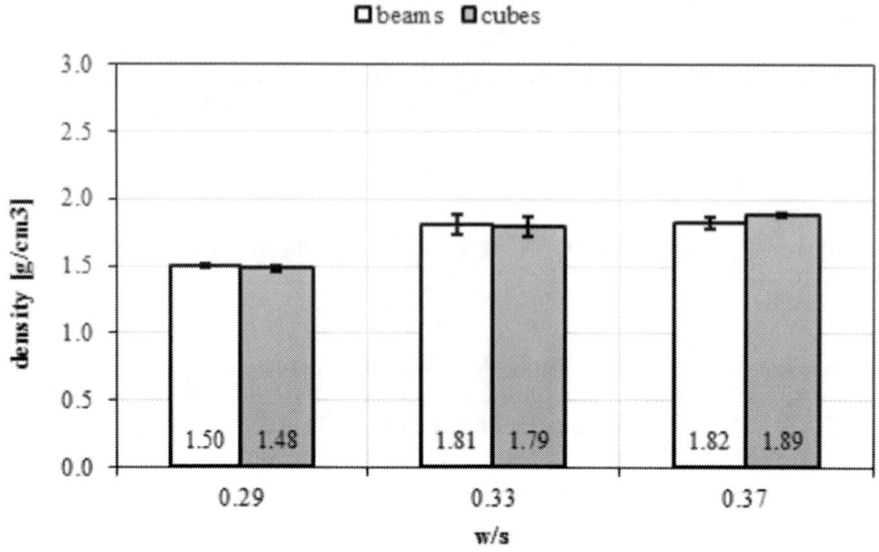

Fig. 2. Density of the cubes and beams with different w/s ratio.

Fig. 3. Cross section of specimen w/s=0.29.

Figures 4 and 5 present changes in flexural and compressive strength depending on the w/s ratio. The level of flexural strength value for all the specimens were between 2 and 3 MPa that is comparable to regular cement pastes. The GFC_029 specimen of the most porous structure obtained the lowest compressive strength value (22.2 MPa for beams and 18.1 MPa for cubes). The maximum compressive strength was gained for GFC_033 (51.8 MPa for beams and 44.1 MPa for cubes). It should be stressed that their densities were only 1.8 g/cm^3. When comparing compressive strength results for the GFC_033 and GFC_037 samples with less porous structure, the common relation for cement based composites is observed – a water to binder ratio increase makes a compressive strength decrease. It is probably caused by higher shrinkage, causing cracks and microstructure defects. Nevertheless their compressive strengths are higher than for regular cement pastes.

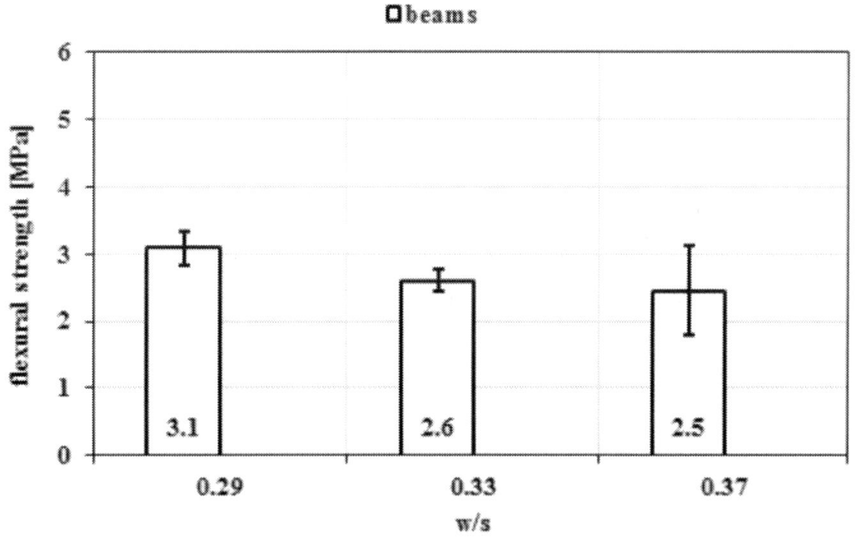

Fig. 4. Flexural strength of the beams with different w/s ratio.

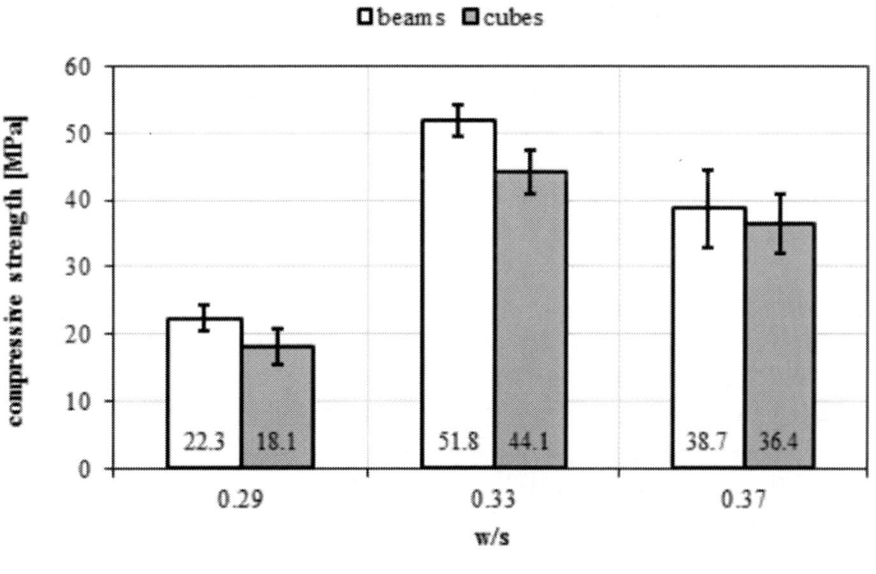

Fig. 5. Compressive strength of the beams and cubes with different w/s ratio.

5 Conclusions

The foam concrete is a material sensitive to water overdosing that can decompose foam during mixing and forming. Specimens with w/s=0.29 ratio have porous structure, the lowest density 1.500 g/cm^3 and relatively high compressive strength (around 20 MPa). The highest values of compressive strength were obtained in case of specimens of w/s=0.33 ratio but at the same time the porous structure was not observed. Further increase of w/s ratio does not influence the density but creates a decrease in compressive strength value. This preliminary investigation showed a big potential of geopolymer foam concrete furthermore, the research on this subject will be continued.

The results were obtained in the Karol Nizio bachelor thesis. Research was supported by Warsaw University of Technology.

References

1. J. Davidovits, *Geopolymers. Chemistry & Application* (Institut Geopolymere, France 2011)
2. S. Puligilla, P. Mondal, Cem. Concr. Res. **43**, 70 (2013)
3. M. Zhang, Cem. Concr. Comp. **78**, 21 (2017)
4. P. He, M. Wang, S. Fu, D. Jia, S. Yan, J. Yuan, J. Xu, P. Wang, Y. Zhou, Ceram. Int. **42**, 14416 (2016)
5. Y. Hou, D. Wang, W. Zhou, H. Lu, L. Wang, J. Wuhan Univ. of Technology- Mater. Sci. Ed. **24**, 771 (2009)
6. M.G. Canfield, J. Eicher, K. Griffith, D.J. Hearn, J. Mater. Sci. **49**, 5922 (2014)
7. G.S. Ryu, Y.B. Lee, K.T. Koh, Y.S. Chung, Constr. Build. Mater. **47**, 409 (2013)
8. M.A. Qidwai, J. Eng. Res. Studies **VI**, 6 (2015)
9. P. Nath, P.K. Sarker, Constr. Build. Mater. **66**, 163 (2014)
10. J. Van Jaarsveld, J. Van Deventer, Eng. Chem. Res. **38**, 3932 (1999)
11. F. Pacheco-Torgal, J. Castro-Gomes, S. Jalali, Constr. Build. Mater. **22**, 1315 (2008)
12. M.S. Morsy, S.H. Alsayed, Y. Al-Salloum, T. Almusallam, Arab. J. Sci. Eng. **39**, 4333 (2014)
13. M. Kadela, M. Kozłowski Procedia Eng. **161**, 468 (2016)
14. M. Kadela, B Pol Acad. Sci-Tech. **64**, 751 (2016)
15. Z. Zhang, J.L. Provis, A. Reid, H. Wang, Constr. Build. Mater. **56**, 113 (2014)
16. D. Shi, L. Shi, J. Zhang, J. Cheng, Adv. Mater. Res. **450-451**, 1504 (2012)
17. Z. Abdollahnejad, F. Pacheco-Torgal, T. Félix, W. Tahri, J.B. Aguiar, Constr. Build. Mater. **80**, 18 (2015)
18. V. Vaou, D. Panias, Miner. Eng. **23**, 1146 (2010)
19. J.G. Sanjayan, A. Nazari, L. Chen, G.H. Nguyen, Constr. Build. Mater. **79**, 236 (2015)
20. M.M. Al Bakri Abdullah, K. Hussin, M. Bnhussain, K.N. Ismail, Z. Yahya, R.A. Razak, Int. J. Mol. Sci. **13**, 7186 (2012)
21. D.L.Y. Kong, J.G. Sanjayan, Cem. Concr. Res. **40**, 334 (2010)
22. J. Provis, H. Wang, Cem. Concr. Comp. **62**, 97 (2015)

Effect of High Temperature on Mechanical Properties of Geopolymer Mortar

Mateusz Sitarz[1], *Izabela* Hager[1,*], and *Joanna* Kochanek[1]

[1]Cracow University of Technology, ul. Warszawska 24, 31-155 Kraków, Poland

Abstract The advent of modern aluminosilicate binders has resulted in practical uses for fly ash, which belongs to the group of secondary mineral raw materials; this provides an interesting alternative to cementitious binders, which have a large carbon footprint. Geopolymers mineral binders with a growing range of applications in construction. Their inorganic, amorphous structure endows them with better resistance to high temperature compared to cementitious binders. In this study, the alkaline activation of aluminosilicate Połaniec Fly Ash made it possible to obtain non-cementitious material with compressive strength of 18.5 MPa. The article presents the results of a comparison of characteristics of geopolymer mortar and Portland cement mortar after thermal exposure.

1 Introduction

Currently, the effects of human interference with the natural environment, the depletion of natural resources, the excessive exploitation of forest resources as well as our impact on the environment through the production of greenhouse gases are becoming increasingly felt. As a result, the idea of sustainable development has gained currency. In the construction industry, this extremely important trend consists in introducing new materials that exhibit a reduced environmental impact and lower energy consumption during their production while maintaining all desirable properties. Researchers are looking for binders consisting mainly of post-process waste that would provide alternatives to cement, and are making efforts to efficiently utilise this type of waste, using it to produce new, safe building materials, which include geopolymers based on active alkali-activated aluminosilicates.

The synthesis of geopolymers enables a significant reduction in greenhouse gas emissions and is a less energy-intensive process compared to the production of traditional cement [1]. Recently, we have seen growing interest in geopolymer binders owing to their favourable mechanical properties and durability, low cost of manufacturing and high resistance to chemical agents as well as to high temperature [1–3]. Thanks to these properties, geopolymer materials may provide an interesting alternative to traditional cementitious binders.

* Corresponding author: ihager@pk.edu.pl

2 Effect of temperature on cementitious and geopolymer binders

It is known from literature that changes in the properties of heated mineral composites with cementitious binders depend on the composition and type of the aggregate used, the water-cement ratio as well as the content of pozzolanic additions. The rate at which the element is heated is an important factor as well. As the temperature rises, the water gradually evaporates, the C-S-H phase becomes dehydrated and portlandite decomposes [4].

During the heating of concrete, the volume of aggregate increases and the cementitious binder shrinks, which results in a significant loss in material strength [4]. A similar mechanism was observed in the case of geopolymers. Differences in coefficients of thermal expansion between aggregate and cement paste cause damage to composites under elevated temperature conditions. The extent of damage depends on the coefficient of thermal expansion of the aggregate used and on the properties of the cement paste such as the amount of shrinkage and its ability to absorb the resulting strains [5, 6].

Tests of temperature effects carried out by various teams have confirmed the geopolymers' robust resistance to high temperature. Davidovits [7] demonstrated that properly designed geopolymer binders can exhibit fire resistance up to 1200°C. Barbosa and Mackenzie [8] analysed the behaviour of a metakaolin geopolymer with an activator that contained sodium silicate and sodium hydroxide. At 100–200°C, the material shrank only slightly due to water loss. In the 250–800°C range, the specimens maintained dimensional stability. At 800°C, shrinkage was observed due to increasing density and changes in volume caused by the crystallisation of new phases. At temperature ranging from 880°C to 900°C, this process was halted and the specimens maintained dimensional stability (depending on differences in their composition) up to 1000–1300°C, and after that they eventually melted. When Bakharev [9] investigated fly ash-based geopolymer with an activator containing Na, she observed shrinkage cracks and a sudden decrease in strength at 800°C.

According to many researchers [6, 10–14] dimensional stability and the geopolymer's resistance to high temperature depend on binder composition, including without limitation the Si/Al ratio, alkali content and the liquid/solid relationship. Alkali content has a dual effect. On the one hand, high alkaline substance content accelerates the activation reaction and makes it possible to obtain a material with high initial strength. On the other hand, alkali content must be limited in order to ensure the thermal stability of the material. A high alkali content in a composite which is exposed to high temperature accelerates the melting of the material. Another important factor is the SiO_2/Al_2O_3 ratio. The most desirable ratio ranges from 2 to 4 – within this range, the initial strength of the material increases and thermostable zeolites are formed. Higher ratios hinder the formation of these structures and promote sintering. At lower ratios, excessive formation of zeolites may occur at high temperature, increasing internal strain in the material. In their work concerning the effect on elevated temperature on cement paste, mortar and light geopolymer concrete, Abdulkareem et al. [15] noted that materials with highest amounts of activator solution exhibited the largest decrease in strength when heated to 400, 600 and 800°C.

Fan et al. [16] analysed the impact of the water/ash ratio on the strength of a geopolymer made from class F fly ash, KOH and Na_2SiO_3. On the basis of the tests conducted, it was found that the optimal water-to-ash ratio was 0.3. At this ratio, the highest residual strength after heating to 500°C was observed. In addition, no spalling was observed during the tests. The authors indicated that differences in coefficients of thermal expansion between the matrix and the aggregate were the main reason for the loss in geopolymer concrete strength under high temperature conditions [17,18].

3 Experimental details

Twelve specimens in the form of 40x40x160mm beams were made from each tested mortar. Cement mortar specimens were marked H1 to H12 and geopolymer mortar specimens were marked G1 to G12. After demoulding, cement mortar specimens were first cured in water for 7 days and subsequently under air-dry conditions until the tests were conducted. After forming, geopolymer mortar specimens were kept at 70°C for 24 hours and then cured for 27 days under air-dry conditions.

In this investigation siliceous fly ash from Połaniec power station was used as source material to prepare the geopolymer mortar. Total content of silicon dioxide, alumina and iron (III) oxide exceeds 70%, loss of ignition at around 5%. The fractions 0-15 μm and 15-30 μm represent almost half of the composition of the fly ash. The activating alkali liquids consisted of Na_2SiO_3 and NaOH beads. Na_2SiO_3 solution had a modulus ratio (Ms) equal to 2.5 where $Ms=SiO_2/Na_2O$, $Na_2O=11.1\%$, $SiO_2=27.9\%$ and water-61.0%. The NaOH beads and Na_2SiO_3 solution were mixed together with Na_2SiO_3/ NaOH mass ratio of 3. Composition of tested materials is given in Table1

Table 1. Composition of cement and geopolymer mortar.

Component	Cement mortar		Geopolymer mortar	
	Mass [g]	Vol. / 1 m^3	Mass [g]	Vol./1 m^3
Ożarów CEM I 32.5R cement	450.0	0.167	-	-
Sand 0-2 mm	1350.0	0.578	1350.0	0.367
Water	225.0	0.225	150.0	0.108
Sodium silicate	-	-	337.5	0.165
Sodium hydroxide	-	-	112.5	0.038
Połaniec fly ash	-	-	800.0	0.303
Plasticiser	-	-	30.0	0.019

The purpose of the conducted research was to compare the strength characteristics (flexural tensile strength and compressive strength) of standard Portland cement mortar and of mortar based on geopolymer binder after heating specimens in a furnace at a rate of 1°C/min to 200, 400, 600 and 800°C and then cooling them down to room temperature. All tests were conducted after 28 days of curing. Compressive strength tests were carried out on the specimens remaining after flexural strength tests. Three specimens were tested for each heating temperature.

4 Results and discussion

After the specimens reached 28-day, their bulk densities and strength characteristics were tested at 20°C. The following results were obtained for geopolymer G and cement mortars (H): average density $\rho_{oG}=1.77$ g/cm³, $\rho_{oH}=2.23$g/cm³, average flexural strength $f_{tG}=5,00$MPa, $f_{tH}=8.07$MPa and compressive strength $f_{cG}=20.14$MPa, $f_{cH}=52.05$MPa. Analogous analysis was repeated for specimens subjected to heating to 200°C, 400°C, 600°C and 800°C and subsequently cooled to room temperature.

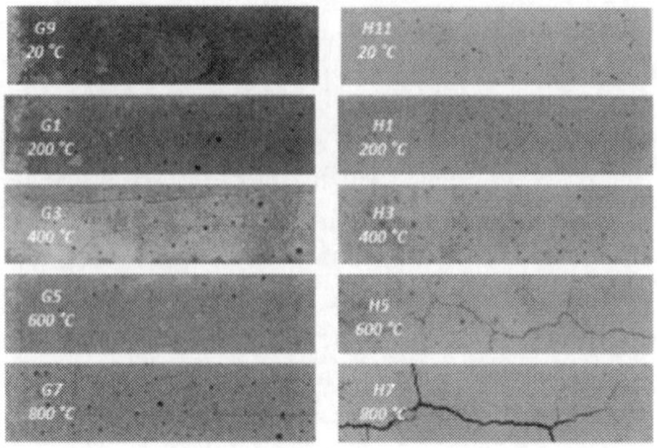

Fig. 1. Mortar specimens aspect after heating: geopolymer (left) and cement (right)

Fig. 1 presents the appearance of specimens after heating. Clear differences can be noticed in the thermal resistance of the materials compared. On cement mortar specimens, there are visible cracks which emerged at 600°C.

Based on the visual inspection of specimens, the geopolymer mortar exhibited significantly better thermal resistance. On the photographs below, changes in the colour of mortars based on geopolymer and cementitious binders as a result of heating are shown. The photographs show specimen cross-sections after flexural strength tests.

a)

b)

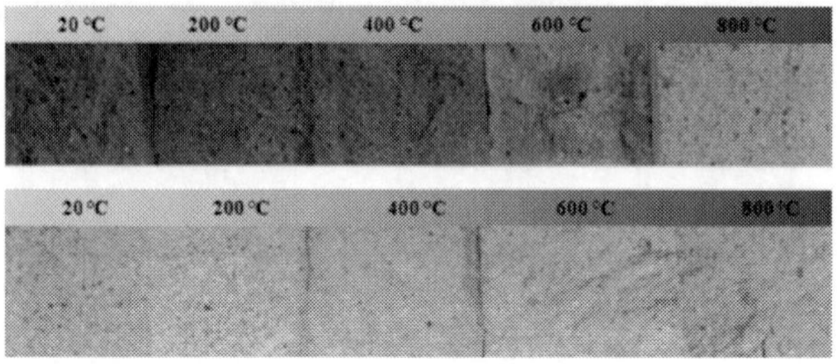

Fig. 2. Colour change of heated mortars: a) geopolymer; b) cement

The chart below presents changes in the bulk density of the materials tested; relative values are compared, i.e. the ratio of the density of the heated specimen to the density of a specimen which was not subjected to high temperature.

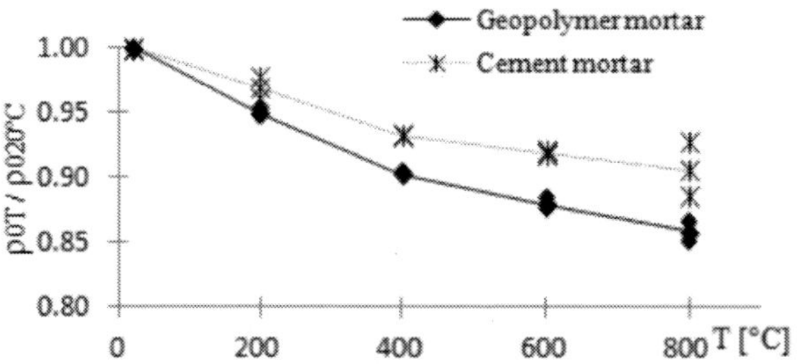

Fig. 3. Relative density of heated geopolymer and cement mortar specimens.

The chart shows that both materials exhibit a decrease in density with the increase of temperature, which results mainly from loss of water. After analysing the weight loss of the materials analysed, it can be concluded that geopolymer mortar specimens showed greater weight loss.

In order to determine changes in mechanical properties under high temperature conditions and to compare the thermal resistance of geopolymer mortar and cement mortar, flexural strength and compressive strength tests were conducted both on unheated specimens and on specimens exposed to temperature of 200°C, 400°C, 600°C and 800°C. Strength tests on specimens were conducted after 28 days, in each case for two unheated specimens and two specimens exposed to high temperature. The charts in Fig. 4 present a comparison of the relative flexural and compressive strength of the materials tested, i.e. the ratio of the strength of the heated specimen to the strength of the specimen not exposed to high temperature.

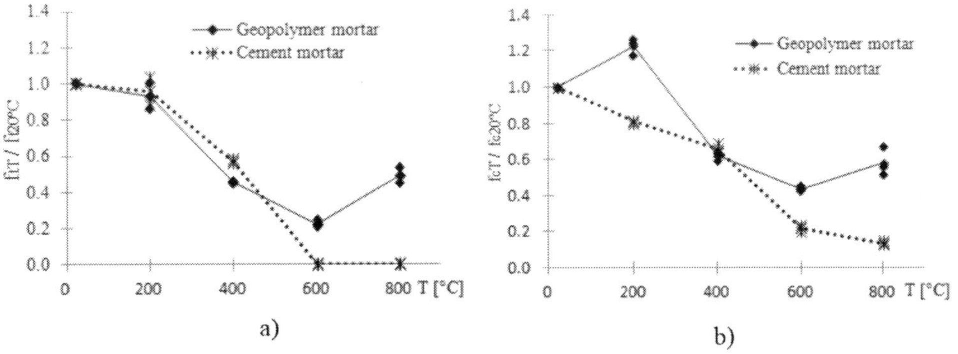

Fig. 4. Changes in relative values of a) flexural and b) compressive strength as a result of heating

As a result of exposure to high temperature, the mechanical properties of both materials deteriorate gradually. Flexural tensile strength decreases by approximately 7% for both materials; after heating to 400°C, the decrease already amounts to 43% and 55% for geopolymer and cement mortar, respectively. After heating to 600°C, cement mortar strength drops to zero, which is the result of the visible cracking of the specimen, the dehydration of cement paste and the decomposition of portlandite. The tensile strength of geopolymer mortar heated to the same temperature is still at 20% of the initial value. Interestingly, heating to 800°C results in a slight increase in strength.

The compressive strength of cement mortar as a result of heating exhibited a quasi-linear decrease within the entire tested temperature range, which was the result of the C-S-H phase dehydration process, portlandite decomposition and the cracking, which was clearly visible on the surface of the specimens heated to 600°C and 800°C.

In the case of the geopolymer mortar, an increase in compressive strength was observed at 200°C. This increase is most probably caused by the progressive geopolymerisation process, which results in the strength increase of the material. Further geopolymer heating to 400°C and 600°C caused a progressive decrease in strength by 38% and 58%, respectively, compared to the initial value recorded at 20°C. On the other hand, after the material had been heated to 800°C, an increase in strength was observed again, probably resulting from the sintering of mineral ingredients.

5 Conclusion

The resistance of mineral composites to high temperature depends on many factors. In geopolymer mortars, just as in cement mortars, the paste becomes dehydrated and shrinks while aggregate volume increases. However, the number and size of the cracks visible to the naked eye in cement mortar after heating to 600°C and 800°C are significantly greater than in geopolymer mortar.

The compressive strength of geopolymer mortars increases after heating to 200°C, reaching 123% of the initial value, which is most probably the result of the continued polymerisation process. Heating to 400°C and 600°C results in compressive strength decreasing by 38% and 58%, respectively, relative to the initial values recorded at 20°C. At

800°C, the material in fact becomes stronger as a result of the sintering of the sodium-aluminosilicate phase, which yields an increase of strength.

As concerns the appearance of specimens after heating, geopolymer materials exhibit greater resistance to high temperature. Despite their lower initial strength values, the geopolymer mortars tested showed greater absolute strength after being heated to 600°C and 800°C compared to cement mortars.

The research conducted has demonstrated that geopolymers exhibit better thermal resistance than cement mortar, which loses its mechanical properties when heated to a temperature higher than that at which portlandite decomposes. An interesting phenomenon which requires further research is the behaviour of geopolymer binders at temperature higher than 800°C.

References

1. L. Provis, J.S.J. Van Deventer, *Geopolymers: structure, processing, properties and industrial applications* (Woodhead Publishing, Cambridge, 2009)
2. P. Duxon, A. Fernandez-Jiminez, JL. Provis, GC. Luckey, A. Palomo, J.S.J. Van Deventure, J. Mater. Sci. **42**, 2917–2933 (2007)
3. C. Li, H. Sun, L. Li, Cem. Concr. Res. **40**, 1341–9 (2010)
4. I. Hager, Bull. of the Pol. Academy of Sci. Tech. Sci. **61**, 145-154 (2013)
5. S. Subaer, A. van Riessen, J. Mater. Sci. **42**, 3117–3123 (2006)
6. D.L.Y. Kong, J.G. Sanjayan,, Cem. Concr. Compos. **30**, 986–991 (2008)
7. J. Davidovits, *Geopolymer, chemistry and applications* (Institut geopolimere, Saint-Quentin, 2011)
8. V.F.F. Barbosa, K.J.D. Mackenzie, Mater. Res. Bull. **38**, 319–331 (2003)
9. T. Bakharev, Cem. Concr. Res. **36**, 1134–1147, (2006)
10. A. Fernández-Jiménez, J.Y. Pastor, A. Martín, A. Palomo, J. Am. Ceram. Soc. **93,** 3417–3417 (2010)
11. J. Ye, W. Zhang, D. Shi, Constr. Build. Mater. **69**, 41–48 (2014)
12. P.V. Krivenko, G.Y. Kovalchuk, J. Mater. Sci. **42**, 2944–2952 (2007)
13. D.L.Y. Kong, J.G. Sanjayan, K. Sagoe-Crentsil, Cem. Concr. Res. **37**, 1583–1589 (2007)
14. J. Temuujin, W. Rickard, A. van Riessen, Adv. Powder. Technol. **24**, 495–498 (2013)
15. O. A. Abdulkareem et al., Constr. Build. Mater. **50**, 377–387 (2014)
16. F. Fan et al., Constr. Build. Mater. **160**, 66–81 (2018)
17. D.L.Y. Kong, J.G. Sanjayan, Cem. Conc. Res. **40**, 334–339 (2010)
18. W.D.A. Rickard, G.J.G. Gluth, K. Pistol, Cem. Concr. Res. **80**, 33–43 (2016)

MATEC Web of Conferences **163**, 06005 (2018) https://doi.org/10.1051/matecconf/201816306005
MATBUD'2018

The Effect of Additives on the Properties of Metakaolin and Fly Ash Based Geopolymers

Michał Łach[1], *Kinga* Korniejenko[1,*], *Maria* Hebdowska-Krupa[1], *Janusz* Mikuła[1]

[1] Institute of Materials Engineering, Faculty of Mechanical Engineering, Cracow University of Technology, Jana Pawła II 37, 31-864 Cracow, Poland

Abstract. The main motivation of research work is connected with environmental issues. The production of the most important building material of the 20th century - Portland cement technology is associated with significant environmental pollution. The process requires very high temperature and it is energy consuming. During the manufacturing also takes place emission of significant amounts of carbon dioxide and highly toxic nitrogen oxides into the atmosphere These factors show that new solution in this area is required. The most promising alternative is inorganic polymer (geopolymer) technology.

The main objective of the presented research work was to design a new composite for practical applications, especially in construction industry. The paper presents the results of research of geopolymer composites based on geopolymer binders made of metakaolin and fly ash with the addition of titanium oxide and aluminum-calcium cements (including mainly calcium monoglinate) in amount of 4 and 6% by weight. Research methods applied: tests for mechanical properties (compressive strength tests), scanning microscopy investigations (SEM) and X-Ray Diffraction (XRD).

The results show that the addition of aluminum-calcium cements (including calcium monoglinate) significantly increases the compressive strength of geopolymers. Geopolymers based on fly ash with the addition of 6% calcium-aluminum cement with a calcium monoglinate content above 69% are characterized by compressive strength above 50 MPa, while geopolymers from metakaolin with the same additive were characterized by compressive strength above 80 MPa.

1 Introduction

Geopolymers are a group of materials defining as inorganic aluminosilicate polymers with specific composition and properties. The name 'geopolymer' was first used by the French scientist Professor Joseph Davidovits in 1970. These materials have very high mechanical properties and excellent resistance to chemically aggressive environments, especially resistance to a variety of acids and salts [1,2]. They were initially developed as a fire resistant alternative to organic polymers. They have good fire resistance (up to 1000 °C) and no

* Corresponding author: kinga.korniejenko@mech.pk.edu.pl

emission of toxic fumes when heated [3,4]. Nowadays, the geopolymers are mainly applied in construction industry [5,6].

The one of the limitations for wide use geopolymers is a common problem with efflorescence caused by excess sodium oxide remaining unreacted in materials [7]. It is usually indirect effect of the properties of the raw materials and manufacturing process. Raw materials for the production of geopolymers may be natural materials such as volcanic tuff or metakaolin as well as fly ash from a combined heat and power plant and blast furnace slag [8,9]. Geopolymers are created by reaction between alkaline solutions and solid aluminosilicates. The activation process requires high pH. Research into the possibility of creating and developing geopolymer compositions for the best mechanical properties as well as without efflorescence effects are one of the most important area [10]. The limitation of efflorescence will allow to spread this material in industrial applications [11].

The appearance of efflorescence is caused by the elimination of sodium or potassium ions from the geopolymer structure. If products have contact with water or are in high moisture environment, the water leaches alkali ions through capillary action and moves them to the surface. After evaporation of water, these ions react with CO_2 from the atmosphere and hydrated sodium or potassium carbonates are created ($Na_2CO_3.nH_2O$, $NaHCO_3.nH_2O$, K_2CO_3). This process will take a place in the materials when there are unreacted sodium or potassium oxides (cations) and the materials are characterized by high porosity [11,12]. The white efflorescence on the surface – it is sodium carbonate heptahydrate $Na_2CO_3 \cdot 7H_2O$ [11]. The identified carbonate crystallizes from inside the pores and 'grows' on the outside of the sample [11,13]. It can be a large problem, because it decreases not only aesthetic value for the final products, but also mechanical properties of the material [10,11].

Exemplary research on additives for controlling efflorescence were made by Kani et al. [10]. They introduce aluminium rich mineral admixtures, such as: metakaolin (Zigma International, India), blast furnace slag (Isfahan Steel Complex, Iran), and three different calcium and aluminum cements Secar 71 (Kerneos, France), Secar 80 (Kerneos, France) and Koracast (Iran Refractory Products, Iran) into composition. These additives were introduced in the amounts of 2, 4, 6, 8 % by mass. In the case of the addition of slag and metakaolin, a very slight decrease in the tendency to efflorescence was observed, but in the case of calcium-aluminum cements, much better positive effects in the reduction of efflorescence were observed.

According to the authors the reason for such effects may be the form of aluminium. In cements with a high aluminum content, aluminum is in the form of calcium aluminates, whereas in the slag and metakaolin it is in the form of aluminosilicates [10]. This important difference not only has an influence on results - different behaviours in the geopolymerization reactions, but also different compositions create the various reaction products [10,14]. The availability of calcium, not only its presence, is the most important point, because aluminium is more easily accessible from the $CaAl_2O_4$ phase dominating in Secar 71 than from $CaAl_4O_7$ and Al_2O_3 phases that are present in Secar 80. The most important effect is that geopolymerization reactions in the presence of admixtures cement with a high content of alumina lead to a more cross-linked aluminosilicate gel [10,14].

However, the slag addition has a very slight influence on decreasing the tendency to efflorescence, but using it as an admixture has a positive influence on the mechanical properties of the composition. The use of slag as a reinforcing admixture and parameter modifier is mainly based on the introduction of calcium into the geopolymerization process. Calcium-containing gels help to create a more compact microstructure than those that consist only of fly ash. This is one of the main reasons for high strength achieved in slag containing geopolymers [10,14]. Other researches show that it is also necessary to investigate this problem long-term, because the pH and electrical conductivity values show different effects in two different curing conditions. The addition of slag increases the pH and electrical conductivity of the solution eluting for the hydrothermal heated samples, while for samples kept at room temperature, the addition of slag affects the lowering of pH and electrical conductivity. This means that the addition of calcium plays a different role depending on the temperature of annealing. In addition, it was found that the addition of slag, reduces the rate of alkali leaching to the solution, and thus reduces the rate of efflorescence, but only in the early phase, which suggests no impact on the potential for efflorescence in long-term [8,9].

2 Experimental

2.1 Materials

The geopolymers were made from fly ash from the CHP plant in Skawina (Poland) and sand (ratio 1:1) with the various additions and from metakaolin form the Czech Republic and sand (ratio 1:1) with various additions.

The fly ashes were thoroughly investigated as a possible raw material for the production of the geopolymers. The chemical composition of the samples was determined by X-ray fluorescence (WD-XRF). They oxide composition is shown in Table 1.

Table1. Chemical composition of fly ash and metakaolin.

	Oxide composition (wt.%)							
	SiO_2	TiO_2	Fe_2O_3	Al_2O_3	CaO	MgO	K_2O	Na_2O
Fly ash	55.9	1.09	5.92	23.49	2.72	2.61	3.55	0.59
Metakaolin	53.01	0.74	1.34	41.54	0.27	0.38	0.71	0.82

Additives were introduced in the form of aluminum-calcium cements. They were the source of aluminum and calcium ions. The contents of the main ingredients in cements are shown in Table 2. The composition of additives was given by the manufacturer.

Table 2. The contents of the main ingredients in the additives introduced into the geopolymer compositions.

	Al_2O_3	CaO	SiO_2	Fe_2O_3	Main phase	Other phases
CCA 1	> 41%	> 35.5 %	< 4 %	< 15 %	CA	C_4AF,$C_{12}A_7$, C_2AS
CCA 2	> 69 %	> 28%	< 0.5 %	< 0.3 %	CA,CA_2,	$C_{12}A_7$,αA

The different variants of samples have been prepared for the tests (Table 3.). Geopolymer samples based on fly ash or metakaolin with various additions such as: 6% titanium dioxide and aluminum-calcium cements - 4% and 6% by mass.

Table 3. Compositions of samples prepared for testing.

	Base		Addition		
	Fly ash	Metakaolin	CCA1	CCA2	Titanium dioxide
FA.0	+	-	-	-	-
FA.1	+	-	4%	-	-
FA.2	+	-	6%	-	-
FA.3	+	-	-	4%	-
FA.4	+	-	-	6%	-
MET.0	-	+	-	-	-

MET.1	-	+	4%	-	-
MET.2	-	+	6%	-	-
MET.3	-	+	-	4%	-
MET.4	-	+	-	6%	-
TD	+	-	-	-	6%

Geopolymers were prepared using a 12M aqueous NaOH solution together with aqueous sodium silicate (ratio 1:2).

2.2 Geopolymer synthesis

To produce geopolymers flakes of technical sodium hydroxide were used and an aqueous solution of sodium silicate (R-145) whose molar module was 2.5 and density was about 1.45 g/cm3. The tap water was used instead of the distilled one. The alkaline solution was prepared by means of pouring the aqueous solution of sodium silicate over the solid sodium hydroxide. The water/binder ratio was 0.52 in case of metakaoline and 0.5 in case of fly ash. The solution was thoroughly mixed for 15 minutes in a low-speed mixer and allowed to equilibrate until a constant concentration and temperature. Next, the obtained paste was poured into cubic molds 150x150x150 mm. The solidification was conducted on the vibratory table. Then the molds were heated for 24 hours at 75°C in the molds closed by foil, cooled to the ambient temperature, taken out of the molds and stored for 28 days and 90 days. The samples were stored in controlled conditions: in the temperature 20°C and humidity 60%.

2.3 Examination methods

The morphology of the samples was examined by means of the scanning microscope JEOL JSM 820. The ashes had been properly prepared before i.e. dried to the constant mass, put on the coal base to carry the electric charge of the sample away, and covered with thin layer of gold with the JEOL JEE-4X vacuum evaporator.

Compressive strength tests according to EN 12390-3, using Matest 3000 kN, were conducted on cubic samples 150x150x150 mm conditioned at room temperature for 28 or 90 days. Five repetition were carried out for each type of sample.

The phase composition was determined using powder X-ray diffraction Debye-Scherrer-Hull method. Diffractograms of samples W1A and W2A were recorded with the use of a Rigaku SmartLab X-ray diffractometer using the following parameters: CuKα radiation, a graphite reflection monochromator, tube voltage 45 kV, tube current 200, step scan mode: step size = 0.05°2θ, count time per 1 step = 1 s. The values of interplanar distances obtained from diffractograms were used for the identification of phases present in the samples based on data contained in the ICDD (International Centre for Diffraction Data 2014) Catalogue and XRAYAN software.

3 Results and discussion

The results from the compressive strength test are shown in Table 4. The increasing of the compressive strength for the samples with additives is significant. The reference samples based on fly ash, without additives has about 37.8 MPa. The best results were achieved for geopolymer with a 6% addition of CCA2 - 53.5 MPa. The addition of CCA1 have also positive influence on mechanical properties. However, this addition requires the optimization in reference to amount, the better results have been obtained for 6% addition by mass - 50.0 MPa.

In the case of metakaolin-based geopolymers, a significant increase in compressive strength is observed up to the value of 89.2 MPa for the composition containing 6% CCA1.

Table 4. The results of compressive strength tests.

	Compressive strength [MPa]
FA.0	37.8
FA.1	50.0
FA.2	44.0
FA.3	50.0
FA.4	53.5
MET.0	66.0
MET.1	73.8
MET.2	89.2
MET.3	75.9
MET.4	77.6
TD	45.0

The XRD studies for samples have been made. Chosen results are presented in Fig. 1 and 2. XRD analysis allowed the identification of the following minerals in the studied materials:

- in sample FA.4: feldspars, illite, leucite, mullite, plagioclase, quarz;
- in sample MET.4: feldspars, illite, leucite, plagioclase and quartz.

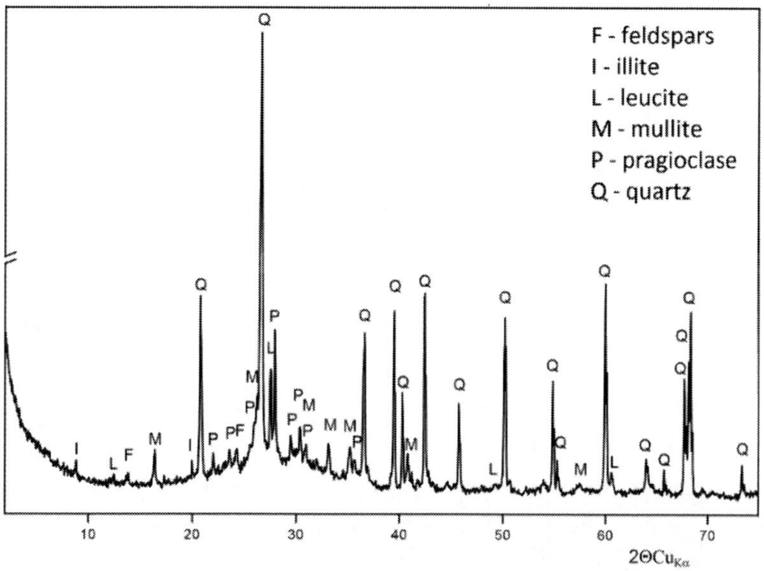

Fig. 1. XRD pattern of the fly ash based geopolymer (sample FA.4).

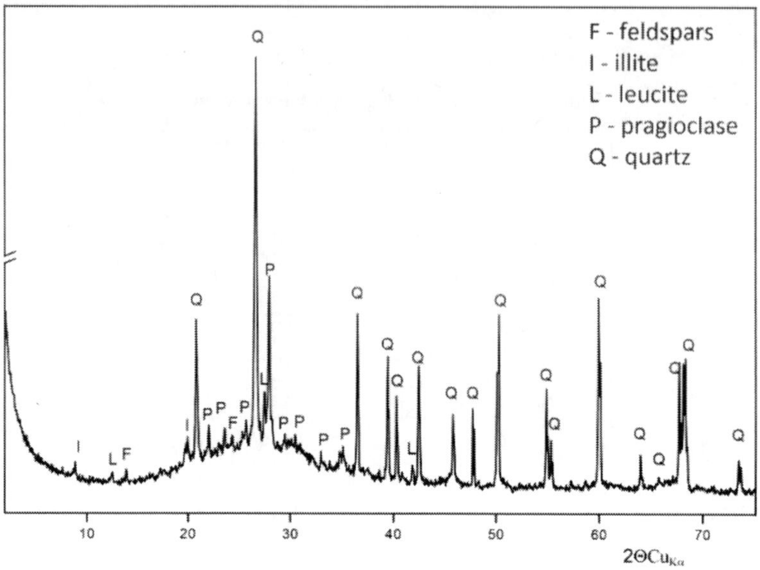

Fig. 2. XRD pattern of the metakaolin based geopolymer (sample MET.4).

The SEM observations were made for each composition. The results are presented in Fig. 3, 4, 5 and 6. A different character of the morphology for various samples is visible. Pores and cracks are noticeable.

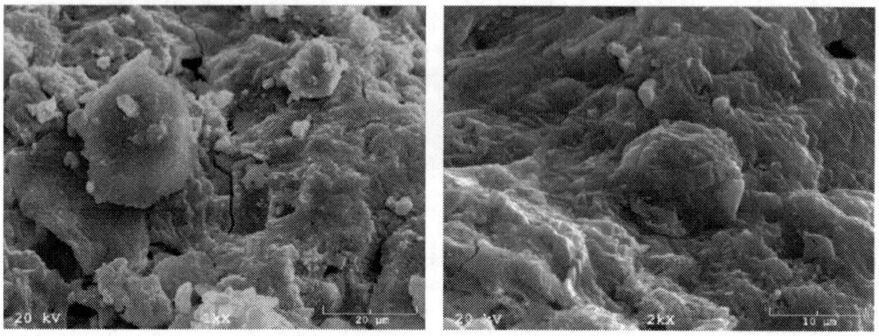

Fig. 3. SEM images of fly ash based geopolymer with 6% CCA 1.

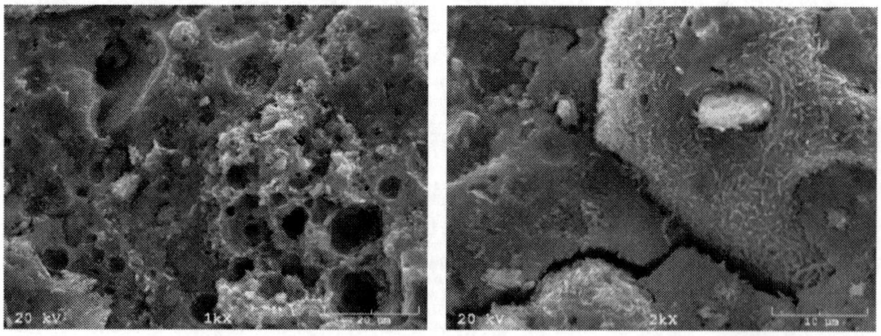

Fig. 4. SEM images of fly ash based geopolymer with 6% CCA 2.

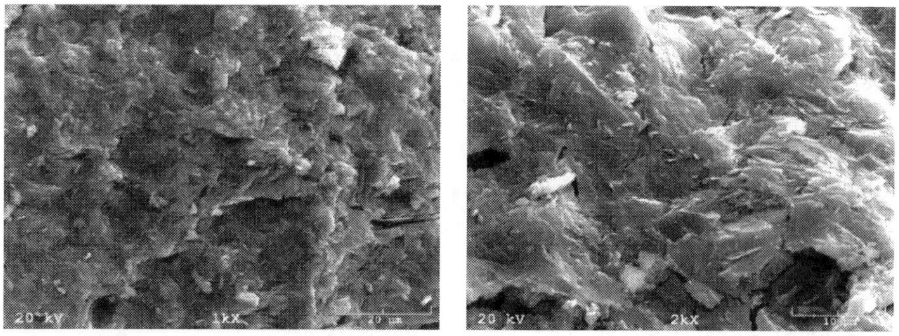

Fig. 5. SEM images of metakaolin based geopolymer with 6% CCA 1.

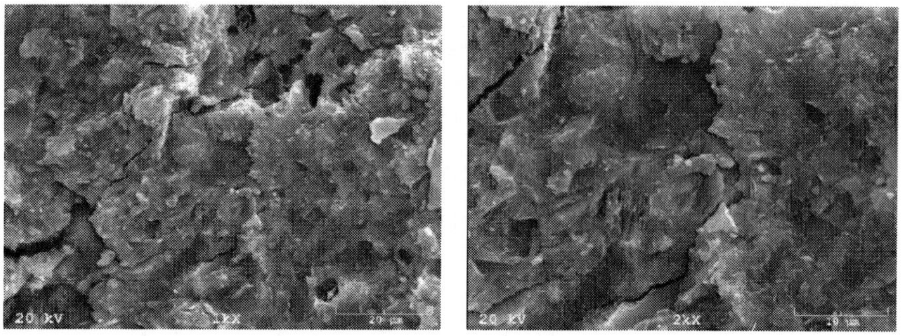

Fig. 6. SEM images of metakaolin based geopolymer with 6% CCA 2.

The conducted research are confirmed by literature reports about the increase of compressive strength by additives such as aluminum-calcium cements and titanium oxide on geopolymers. The admixtures of aluminum-calcium cements increase compressive strength values for both fly ash based and metakaolin based geopolymers. The addition of 6% calcium-aluminum cement with an calcium monoglinate content above 69% results in a 40% increase in compressive strength in geopolymers based on fly ash. In the case of metakaolin-based geopolymers, the increase in compressive strength was also observed – about 35% (in case of 6% admixture containing: above 41% of Al_2O_3 and above 35% CaO).

4 Conclusions

The main objective of the presented research work was to design a new composite for practical applications. The results shows that the addition of calcium monoglinate significantly increases the compressive strength of geopolymers. Geopolymers based on fly ash with the addition of 6% calcium-aluminum cement with a calcium monoglinate content above 69% are characterized by compressive strength above 50 MPa, while geopolymers from metakaolin with the same additive were characterized by compressive strength above 80 MPa.

The available literature also mention the favourable effect of aluminum-calcium cements on the process of limiting efflorescence on geopolymers. However, this phenomena is not fully described and not completely explained. It requires further detailed research. The authors of this article are in the process of conducting such research and determining the influence of calcium-aluminum additives on the process of limiting efflorescence.

This work was supported by the ERANet-LAC 2nd Joint Call (http://www.eranet-lac.eu) and funded by National Centre for Research and Development, Poland, under grant: "Development of eco-friendly composite materials based on geopolymer matrix and reinforced with waste fibers".

References

1. J. Davidovits, *Geopolymer chemistry and application* (Geopolymer Institute, Saint-Quentin, France, 2015)
2. A. Palomo, M.T. Blanco-Varela, M.L. Granizo, F. Puertas, T. Vazquez, M.W. Grutzeck, Cem. Concr. Res. **29**, 7 (1999)
3. J. Davidovits J., J. Therm. Anal. **37**, 8 (1991)
4. R.E. Lyon, P.N. Balaguru, A. Foden, U. Sorathia, J. Davidovits, M. Davidovits, Fire Mater. **21**, 2 (1997)
5. L. Provis, J.S.J. van Deventer, *Geopolymers: structure, processing, properties and industrial applications* (Elsevier, New York, 2009)
6. M. Łach, K. Korniejenko, J. Mikuła, Procedia Eng. **151** (2016)
7. G.F. Huseien, J. Mirza, M. Ismail, S.K. Ghoshal, A.A. Hussein, Renewable and Sustainable Energy Reviews **80** (2017)
8. J. Davidovits, *Market Trends and Potential Breakthroughs; Geopolymer 2002 Conference* (2012)
9. Xu H., J. van Deventer, Int. J. Miner. Process. **59**, 3 (2000)
10. E.K. Kani, A. Allahverdi, J.L. Provis, Cem. Concr. Compos. **34** (2012)
11. Z. Zhang, H. Wang, J.L. Provis, A. Reid, *Efflorescence: A Critical Challenge for Geopolymer Applications?*, (University of Southern Queensland, Toowoomba, Australia, 2013)
12. J. Mikuła, M. Łach, *Geopolimery – nowa przyjazna środowisku alternatywa dla betonów na bazie cementu portlandzkiego. Przykłady zdrożenia i wyniki badań* [in:] Rozwiązania proekologiczne w zakresie produkcji, J Mikuła rd. (Politechnika Krakowska, 2014)
13. A. Palomo, M.W. Grutzeck, Cem. Concr. Res. **29**, 8 (1999)
14. N. Sedira, J. Castro-Gomes, G. Kastiukas, X. Zhau, A. Vargas, Mining Sci. **24** (2017)

Influence of Soil-Cement Composition on its Selected Properties

Krystian Brasse[1,*], *Tomasz* Tracz[1], *Tomasz* Zdeb[1], *Piotr* Rychlewski[2]

[1]Cracow University of Technology, Faculty of Civil Engineering, ul. Warszawska 24, 31-155 Kraków, Poland
[2]Road and Bridge Research Institute, ul. Instytutowa 1, 03-302 Warszawa, Poland

Abstract. The paper discusses the results of mechanical and technological tests of soil-cement composites made with cohesive soil. The compositions of analysed soil-cement mixtures differed in terms of their cement paste volume fractions and water-cement ratios. Limiting values of these technological parameters that enable the application of the soil-cement mixtures obtained in real life conditions for the purposes of the Deep Soil Mixing (DSM) method were determined. Based on the test results obtained, it was found that mechanical properties of the materials analysed were very sensitive to changes in their compositions. Variations in the volume fraction of cement paste within the range analysed caused mechanical properties to change even by an order of magnitude.

1 Introduction

Unfavourable soil conditions for erecting structures are an increasingly frequent difficulty associated with new investment projects. One solution is to reinforce the underlying soil so that it can become an integral part of new civil engineering structures. New soil improvement techniques and the development of technologies related to geotechnical engineering materials have resulted in many methods that can be used to reinforce weak soil substrate. However, an important consideration is that the modification of geotechnical properties of soils should allow for their efficient and rational use as a foundation for new construction projects. Thus desired characteristics of the solutions to be adopted are simplicity, lower implementation costs and environmental friendliness [1-3]. These requirements can be met by in-situ mixing of soil, since it enables the direct modification of its properties, above all strength, plasticity and tightness, without the need for bringing the soil to the surface. As a result of the soil being mixed with binding material, it is strengthened through the chemical reactions that occur. In this way, a new material with the desired mechanical properties is obtained after adding the stabiliser. Most commonly used stabilisers are mineral binders such as cement and lime [1, 2], but the range of materials which can be used is wide – from fly ash through slags, bitumens and resins to bentonite. Specially developed mixtures that consist of several of the aforementioned materials can also be used for reinforcing weak soils [2].

* Corresponding author: krystian.brasse@pk.edu.pl

These methods have undoubtedly seen more use and have been steadily developed in recent decades thanks to the numerous advantages of the in-situ deep soil mixing technology. The choice of the reinforcement method as well as the choice of the type of stabiliser used depends on the type of the soil to be reinforced. It should be stressed that these stabilisers can be used both with non-cohesive and cohesive soils [3-6]. Soil reinforcement itself may be conducted on the surface or in an in-depth manner. Surface stabilisation is mainly used in road construction, but also in general and industrial construction. It consists in mixing the upper soil layer with a stabiliser (usually cement) and subsequently compacting it [2, 7]. More often, however, deep soil reinforcement technology is used; in many countries, it is synonymous with the very popular Deep Soil Mixing (DSM) method developed in the U.S. in the 1980s [1]. Additionally, in the in-situ soil mixing technology there are two ways of adding the binding material – dry or wet mixing [1, 2]. "Dry" deep soil mixing is used to reinforce highly hydrated or organic soils [5, 6]. The reinforcement is in the form of columns formed by mixing the soil with the stabiliser in powdered form; they are referred to as DSM-dry columns [7]. It should be stressed, however, that the DSM-dry technology is not widely used in Poland, mostly due to the difference in the quality of the final product compared to the one obtained using the DSM-wet deep soil mixing method [5]. This technology differs primarily in the method of adding binder to the soil – it is fed together with water in the form of paste. This method is described in detail, inter alia, in [4, 8, 9, 10].

Nevertheless, in spite of its many advantages, the DSM technology does have some disadvantages, e.g. the degree of reinforcement obtained often varies considerably [11]. Among the factors that affect the reinforcement effect are the physical and chemical properties of the reinforced soil, the properties of the binder, additives and water as well as the mixing technique and conditions and the setting time and conditions [12]. Therefore, new projects require not just experience and practice, but also proper engineering design. For this purpose, one must know local soil conditions and the assumed strength criteria for the proposed reinforcement as well as be aware of technical capabilities of the equipment used and economic aspects of the project [13]. In the case of the DSM technology, based on practice and experience, attempts are made to document various principles and aspects of designing DSM columns [11, 14].

In practice, there are no standardized requirements for the strength of soil-cement composites, but their mechanical parameters result directly from the loads occurring in the structure. It should be emphasized that, for example, DSM columns cooperate with the surrounding soil and their strength must be only greater than the soil. The most common compressive strength of discussed composites ranges from 2 to 5 MPa. Higher values can be achieved in case of non-cohesive soils and smaller ones in cohesive. In engineering practice, the minimum strength is about 2-2.5 MPa, however, it cannot be ruled out that the required strength will be less as a result of structural calculations [10].

Observations from engineering practice indicate that technological considerations often determine the composition of soil-cement. This means that the amount of cement paste pumped in during soil mixing and its liquidity very often result from the need to obtain a homogeneous mixture, and its consistency must be adapted to the efficiency of mixing devices. The very rapidly growing segment of deep soil mixing machinery and the increasingly common trench-mixing method offer greater possibilities of controlling this process. In summary, modern mixing methods make it possible to obtain homogeneous mixtures with a broader range of acceptable cement paste amounts or liquidities.

2 Study objective

This paper presents the results of research aimed at determining the influence of the composition of soil-cement designed and produced under laboratory conditions on its selected properties. The information obtained should help determine the relationships that can be used when designing soil-cement compositions in order to achieve the desired properties of hardened composites. Designing the composition of soil-cement and studying its effect on (mainly mechanical) soil properties is important insofar that it allows to obtain appropriate levels of virgin soil reinforcement.

3 Materials

The research programme involved producing a series of soil-cement composites whose compositions varied in terms of the amount of cement paste V_z used and water-cement (w/c) ratios. Cohesive soil identified as clay was selected for producing the composites. The soil-cement compositions tested were created as a result of mixing the virgin soil described in Table 1 with CEM II/B-S 32.5 R cement paste with the properties stated in Table 2.

Table 1. Properties of the clay selected.

Properties	Value
Particle size distribution [% by mass] - gravel fractions 2/40 mm - sand fractions 0.05/2 mm - dust fractions 0.002/0.05 mm - clay fractions <0.002 mm	 - 40 42 18
Bulk density ϱ [g/cm3]	2.10
Bulk density of the soil skeleton ϱd [g/cm^3]	1.81
Maximum bulk density of the soil skeleton ϱdmax [g/cm^3]	1.91
Natural moisture content $w_{n\ mean}$ [%]	15.81
Optimal moisture content w_{opt} [%]	11.1
Yield point w_P [%]	15.32
Liquid limit w_L [%]	25.8
Plasticity index I_L [%]	0.05
Approximate content of $CaCO_3$ [%]	1-3

Table 2. Properties of CEM II/B-S 32.5 R cement.

Properties	Value
Compressive strength after 2 days [MPa]	18.0
Compressive strength after 28 days [MPa]	49.0
Initial setting time [min]	190
Water at normalised consistency [%]	28
Specific surface (Blaine) [cm^2/g]	3.570

4 Methods and test results

To date, engineering activities involving the production of soil-cement structures have not been preceded by laboratory tests. The amount of paste and the w/c ratio of the soil-cement mix were conditioned by the need to obtain the appropriate consistency. When designing the composition of soil-cement, the workability of the mixture during mixing and its consistency after mixing should be adopted as boundary conditions. Mixtures were required to be self-compacting with no segregation of ingredients. Hence different amounts of paste are used for different types of soil so as to obtain the consistency that allows the material produced to be used at the target location. In this manner, when designing soil-cement compositions, workability maps were developed for cement mixtures (Table. 3), showing the ranges of cement paste amounts V_z and water-cement (w/c) ratios within which interesting compositions in terms of the properties expected can be sought. In the case of cohesive soils with the characteristics stated in Table 1, the range of cement paste volumes used was from 550 to 650 dm^3/m^3. w/c ratio values ranged from 1.2 to 3.8. Changes in the amount of paste added and in the w/c ratios adopted involved differences in cement content for each mixture (in kg/m^3). In the case of the soil-cement mixtures tested, cement content ranged from 133 to 427 kg/m^3. In engineering practice, the most common cement mass fraction ranges from 300 to 400 kg/m^3. This amount results from a compromise between the achieved mechanical properties of soil-cement composite and the costs of its production. Summing up, 32 variants of soil-cement compositions were analysed.

Table 3. Map of workability of soil-cement composites made with cohesive soil.

	w/c	4.0	3.8	3.6	3.4	3.2	3.0	2.8	2.6	2.4	2.2	2.0	1.8	1.6	1.4	1.2	1.0
V_z [dm^3/m^3]	650	-	-	-	-	-	-	-	-	-	+	+	+	+	+	+	X
	600	X	+	+	+	+	+	+	+	+	+	+	+	+	+	+	X
	550	X	+	+	+	+	+	-	-	-	-	-	-	-	-	-	-

-	mixture consistency – too liquid, significant sedimentation, significant bleeding
-	mixture consistency – too thick, not self-compacting
+	mixture consistency – proper, no sedimentation, no bleeding
X	strength too low, w/c > 3.8
X	cement consumption not cost-effective, w/c ratio < 1.2, cement content > 427 kg/m^3

The scope of soil-cement tests covered the following properties: consistency determined on the basis of the mixture flow diameter measured by means of a flow table, flexural tensile strength determined for 100x100x500 mm prisms and compressive strength determined for 100 mm cubes cut from prism halves after bending. Tests of mechanical properties were conducted after 28 days of curing under laboratory conditions that prevented the evaporation of mixing water.

Fresh mix consistency tests were carried out in accordance with the standard EN 1015-3. This method consists in measuring the flow diameter of a portion of the mixture tested after shaking it several times on a table. Analysing the measurements conducted, it can be stated that in the case of the soil tested, mixtures with the appropriate consistency exhibited flow diameter values ranging from 170 to 230 mm. The photographs below (Fig. 1, Fig. 2, Fig. 3) show three levels of mixture consistency obtained.

Tests of mechanical properties of hardened composites were conducted for 25 soil-cement compositions selected on the basis of the consistency assessment conducted, after the mixtures that did not meet the required criteria had been rejected.

The flexural strength test was conducted in accordance with the standard EN 12390-5, which concerns testing concrete specimens placed on supporting pins with the constant spacing l equal to 3 d, i.e. 300 mm. The only deviation from the guidelines included in the standard was the reduction in the stress increase rate during the test to 0.01 MPa/s due to the differences in the expected strength of the material tested compared to the flexural strength of normal concrete. Flexural strength (f_{cf}) results are summarised in Table 4. Owing to the number of tests performed, partial results are not presented, and the results presented are limited to mean values from three measurements.

Fig. 1. Mixture consistency too thick:
- flow table = 155 mm;
- V_z = 550 dm^3/m^3; w/c = 2.2

Fig. 2. Mixture consistency proper:
- flow table = 200 mm;
- V_z = 600 dm^3/m^3; w/c = 2.4

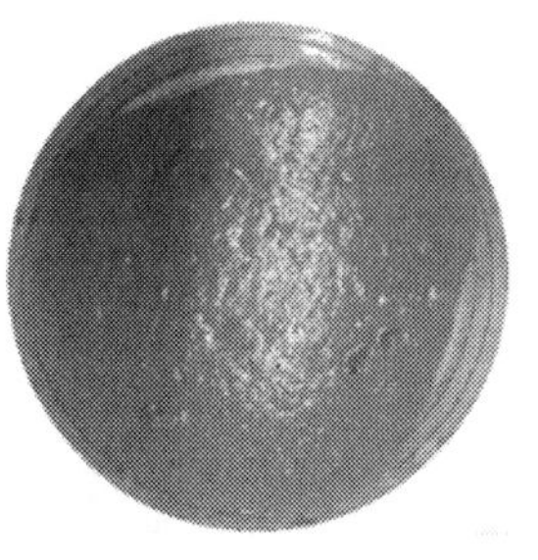

Fig. 3. Mixture consistency too liquid:
- flow table = 270 mm;
- V_z = 650 dm^3/m^3; w/c = 2.6

Table 4. Results of flexural strength (f_{cf}) tests for specimens after 28 days of curing in MPa.

	w/c	4.0	3.8	3.6	3.4	3.2	3.0	2.8	2.6	2.4	2.2	2.0	1.8	1.6	1.4	1.2	1.0
V_z [dm^3/m^3]	**650**	-	-	-	-	-	-	-	-	-	0.8	0.8	1.0	1.2	1.5	2.0	X
	600	X	0.4	0.4	0.4	0.4	0.4	0.7	0.8	0.8	1.0	1.1	1.2	1.5	1.8	2.0	X
	550	X	0.5	0.5	0.6	0.6	0.7	-	-	-	-	-	-	-	-	-	-

The chart below (Fig. 4) shows the effect of cement paste amount and of the w/c ratio on the flexural strength of soil-cements based on cohesive virgin soil after 28 days of curing.

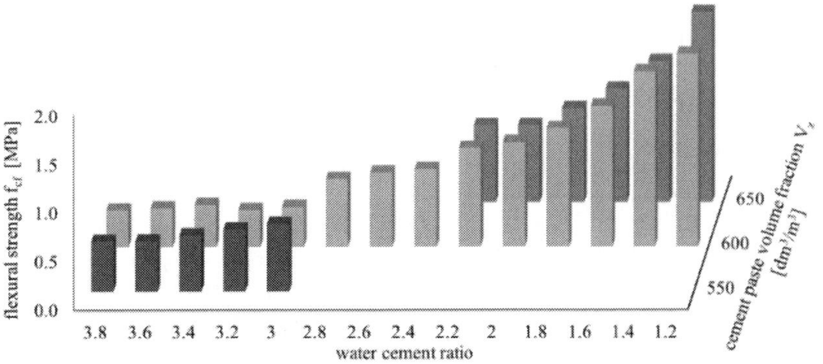

Fig. 4. Flexural strength chart depending on the water-cement ratio and cement paste volume fraction after 28 days of curing.

The flexural strength values obtained after 28 days of curing ranged from 0.4 to 2.0 MPa. The analysis of results points to an increase in flexural strength achieved by soil-cement as the w/c ratio decreases. Additionally, it was observed that as the amount of cement paste added V_z decreased for a given w/c ratio, there was a slight increase in the strength of the soil-cement mixtures tested.

The compressive strength test was conducted in accordance with the standard EN 12390-3, which also concerns testing concrete specimens. Again, the only deviation from the guidelines included in the standard was the reduction in the stress increase rate during the test to 0.05 MPa/s. Compressive strength (f_c) results are summarised in Table 4. Similarly as before, the results presented are mean values from three measurements.

The chart below (Fig. 5) shows the effect of variable cement paste amounts and various values of the w/c ratio on the compressive strength of soil-cements based on cohesive virgin soil after 28 days of curing.

Table 5. Results of compressive strength (f_c) tests for specimens after 28 days of curing in MPa.

	w/c	4.0	3.8	3.6	3.4	3.2	3.0	2.8	2.6	2.4	2.2	2.0	1.8	1.6	1.4	1.2	1.0
V_z [dm³/m³]	650	-	-	-	-	-	-	-	-	-	2.9	3.7	4.3	4.9	6.5	10.2	X
	600	X	1.2	1.4	1.5	1.5	1.7	2.8	2.6	3.2	3.6	4.6	5.0	6.5	6.1	10.8	X
	550	X	1.6	1.8	1.7	2.2	2.5	-	-	-	-	-	-	-	-	-	-

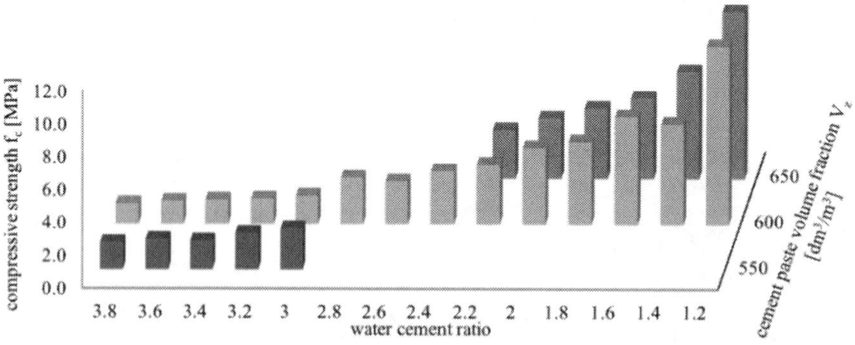

Fig. 5. Compressive strength chart depending on the water-cement ratio and cement paste volume fraction after 28 days of curing.

For compressive strength after 28 days, the range of results obtained ranged from 1.2 to 10.8 MPa. Analysing the effect of changes in cement paste content V_z and w/c ratio on the results obtained, it can be stated again that reducing both the cement paste content V_z and the w/c ratio results in an increase in the compressive strength of the composites tested.

In order to make the test results obtained useful for the practical design of soil-cement from cohesive soil exhibiting the properties presented in Table 1, an attempt was made to determine the correlation between soil-cement composition and the properties of the composite obtained. The chart below (Fig. 6) shows the effect of w/c ratio on compressive strength after 28 days of curing. It has been observed that the power function matches the variation presented very well. Owing to the clear impact of the amount of cement paste on the strength of soil-cement, it was decided that three such functions should be determined:

- $f_c = 18.48\ w/c^{-1.81}$ is applicable to w/c 3.8÷3.0 with $V_z = 550$ dm³/m³;
- $f_c = 16.12\ w/c^{-1.92}$ is applicable to w/c 3.8÷1.2 with $V_z = 600$ dm³/m³;
- $f_c = 13.23\ w/c^{-1.93}$ is applicable to w/c 2.2÷1.2 with $V_z = 650$ dm³/m³.

The power functions determined reflect the results obtained very accurately as evidenced by very high values of determination coefficients, which range from 0.976 to 0.995.

Seeking to simplify the description by making the compressive strength of soil-cement dependent on a single composition parameter, the chart in Fig. 7 only shows the effect of the amount of cement on strength. It turns out that this relationship is indeed present and can be successfully described by using a linear function, since the amount of cement results from the assumed w/c ratio and the amount of cement paste and thus is a universal parameter of soil-cement composition.

Moreover, flexural strength was compared to compressive strength (Fig. 8) and it was observed that the proportion between these two values is around 0.2 and is independent of the strength figures compared.

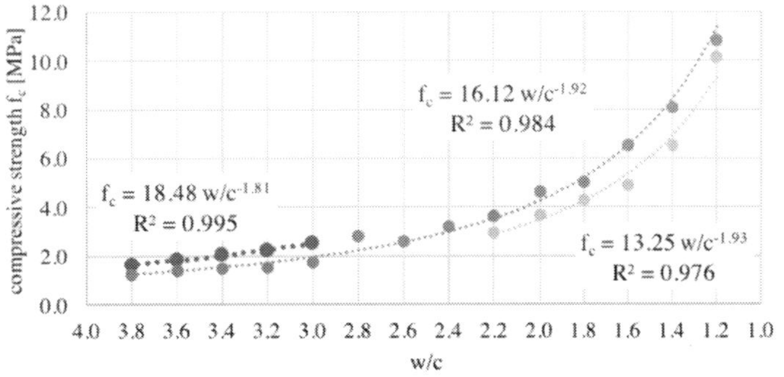

Fig. 6. Compressive strength chart depending on the water-cement ratio and cement paste volume fraction after 28 days of curing.

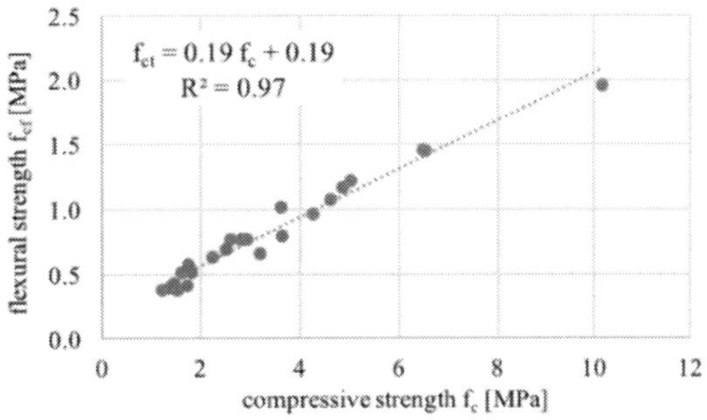

Fig. 7. Comparison of flexural and compressive strength after 28 days of curing.

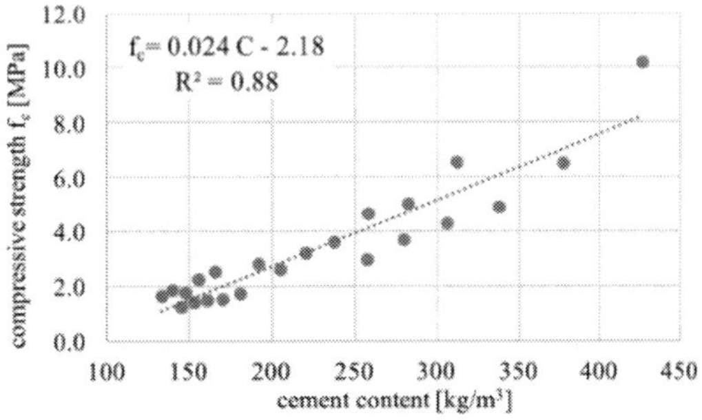

Fig. 8. Impact of cement content on compressive strength after 28 days of curing.

5 Conclusions

Soil-cement is a composite made by mixing cement paste with virgin soil. The amount of cement paste and its liquidity, and hence the w/c ratio, have a very significant effect on the liquidity of the homogenised soil-cement mix. In the case of cohesive soil (clay) with the properties described in Table 1, the demand for cement paste fluctuated from 550 to 650 dm^3/m^3 depending on the w/c ratio, and the w/c ratio varied from 3.8 to 1.2.

The variability range of the quantitative composition of the soil-cements analysed, which is described above, resulted in differences in the flexural strength achieved, which ranged from 0.4 to 2.0 MPa, and also in compressive strength, which ranged from 1.2 to 10.8 MPa after 28 days of curing.

Flexural strength is equal to 19% of compressive strength and this relationship is essentially independent of the recorded strength levels.

Compressive strength after 28 days of curing depends primarily on the w/c ratio, with the obvious upward trend in strength as the w/c ratio decreases. The strength of the hardened composite is also dependent on the cement paste amount – the less cement paste, the higher the strength, of course within the range considered. The power functions presented above reflect the relationship between the w/c ratio and compressive strength very accurately. They are a very useful tool for designing the composition of such composites.

A very universal soil-cement composition parameter is the amount of cement used, which linearly affects the compressive strength of the composite after 28 days of curing. With a cement amount as low as approximately 150 kg/m^3, compressive strengths of slightly less than 2 MPa were achieved, while 420 kg/m^3 of cement resulted in a composite with a compressive strength above 10 MPa.

The research results presented above, which of course have a limited application, since they concern cohesive soil (clay) and CEM II/B-S 32.5R cement, enable the precise design of soil-cement compositions, and the composites obtained exhibit properties that make them highly useful as construction materials.

References

1. M. Topolnicki, Inżynieria Morska i Geotechnika **6**, 385-398 (2003)
2. J. Madej, *Ulepszanie podłoża* (Politechnika Koszalińska, Koszalin, 2015)
3. S. Topoliński, *Badanie wytrzymałości kompozytów grunt organiczny-spoiwo cementowe formowanych metodą mieszania,* (Uniwersytet Technnologiczno-Przyrodniczy im. Jana i Jędrzeja Śniadeckich, Bydgoszcz, 2014)
4. A. Leśniewska, *Wytrzymałościowe i technologiczne aspekty wzmacniania gruntu metodą wgłębnego mieszania na mokro* (Politechnika Gdańska, Gdańsk, 2007)
5. P. Kanty, J. Rybak, D. Stefaniuk, Nowoczesne Budownictwo Inżynieryjne **2**, 70-75 (2017)
6. Menard Polska Sp. z o.o., Kolumny DSM, www.menard.pl, (January, 2018)
7. P. Łęcki, P. Dojcz, (Instytut Techniki Budowlanej, 2008)
8. M. Topolnicki, *XXIV Ogólnopolskie Warsztaty Pracy Projektanta Konstrukcji*, 265–280 (2009)
9. P. Rychlewski, Inżynier budownictwa **5**, 98-102 (2015)
10. P. Rychlewski, Inżynier budownictwa **2**, 92-95 (2013)
11. P. Kanty, Nowoczesne Budownictwo Inżynieryjne **1**, 68-73 (2017)
12. M. Topolnicki, Ground Improvement **9**, 331-428 (2004)
13. J. Chu, S. Varaksin, U. Klotz, P. Mengé, Construction processes **4**, (2009)
14. M. Topolnicki, *XXXII Ogólnopolskie Warsztaty Pracy Projektanta Konstrukcji*, 1–29 (2017)

Toward a better understanding of the effects of cement treatment on microstructural and hydraulic properties of compacted soils

Harifidy Ranaivomanana[1,*], and *Andry* Razakamanantsoa[2]

[1]Institut de Recherche en Génie Civil et Mécanique (GeM), UMR-CNRS 6183, Université de Nantes (IUT de Saint-Nazaire), 58, Rue Michel Ange, 44606, Saint-Nazaire, France
[2] IFSTTAR, Institut Français des Sciences et des Technologies des Transports, de l'Aménagement et des Réseaux, CS 4 Route de Bouaye, 44344, Bouguenais, Cedex France

Abstract. This study deals with the problem of the experimental characterization of cement-treated compacted soils in terms of microstructural and hydraulic properties. Some tests are conducted on two different types of soil: silty sand and clay as fine soils and gravelous sand and alterite as granular soil. Some samples are mixed with 5% of cement and compacted at different levels (i.e., 85%, 95%, 100% and 105% of the maximum dry density, respectively, as achieved using the standard compaction method). The results of the mercury intrusion porosimetry (MIP) tests performed on these cement-treated soils reveal significant changes as regards macropores due to the combined effects of treatment and compaction. Consequently, a decrease in the permeability is clearly observed for all the tested soils when the degree of compaction increases. This decrease is significantly greater in fine soils, which are more sensitive to compaction effects than granular soils.

1 Introduction

Soil stabilization is a widespread technique used to improve workability and hydromechanical properties of soils. It consists in pulverizing natural soil, mixing it with a chemical additive like a hydraulic binder (hydrated lime, for instance) or chloride salts, and, finally, compacting the mixture obtained thoroughly. Lime is mainly used for the treatment of clayey soils and more specifically of expansive clays. Lime treatments being well-known and well-documented, the present study is focused on the stabilization with cement. Unlike lime, stabilization using cement can be applied to different types of soils and not only clayey soils. A better understanding of the interactions between the microstructure of cement-treated soils and their mechanical and hydraulic properties is discussed. The literature on cement-based soils is poor compared with lime-treated soils [1]. Hydraulic binders are known to govern change in treated soil microstructure. This change is amplified by the compaction process that takes place after treatment. The influence of the compaction rate on microstructural and hydraulic properties of compacted soils has been addressed in a

* Corresponding author: harifidy.ranaivomanana@univ-nantes.fr

recent paper by Ranaivomanana et al. [2]. The authors highlight the decrease in macropore volume and the formation of micropores within clayey soils when the degree of compaction increases. The water permeability, therefore, also decreases by two to three orders of magnitude when the compaction rate increases. Consequently, our understanding of the combined effects of soil and compaction treatments on microstructural and hydraulic properties of treated soils still needs to be improved.

2 Materials and methods

2.1 Soils tested and cement used

Four types of soil are tested: Silty sand, gravelous sand, clay and alterite denoted by SS, GS, C and A, respectively. These different soils come from the northwest of France and have been sampled from the current extension project area for the high-speed railway line "LGV Bretagne-Pays de la Loire", which requires a geotechnical investigation of the soils of the site. Physical properties of soils are presented in reference [2]. The cement used for the treatment is a Portland limestone cement (CEM II/B-LL 32,5R CE CP2 NF) in compliance with the European Standards [3].

2.2 Sample preparation and characterization of soils after compaction

The soil water content is adjusted up to the desired compaction water content. The moistened soil is then blended in an airtight container with an ageing cycle of 48 hours for moisture homogenization. The cement content is expressed as dry weight of the soil. The homogenized soil is mixed with 5% of cement in a mechanical mixer during 5 min. The compaction phase is performed according to the standard Proctor compaction procedure described in ASTM D698 [4]. The optimal density and water content are calculated from the compaction curves (Table 1) and used as reference values for the following preparation step.

Table 1. Physical properties of soils studied at OPN.

Soil reference	**SS**	**GS**	**C**	**A**
Water content (%)	13.2	12.9	19.0	13.1
Dry density (kN/m^3)	19.3	19.6	17.5	19.5

The compaction densities chosen for this study are 85, 95, 100, and 105% of the SPO for the permeability measurements. It should be noted that for all the tests conducted here, the samples are prepared at the optimum water content. The tests are carried out using cylindrical samples 50 mm in diameter and 100 mm in height. Once compacted, the samples are prepared and immediately wrapped in a cellophane film. They are finally left at the laboratory ambient temperature (20 ± 1 °C) for 28 days to ensure a sufficient hydration of the cement. Two characterization tests are performed on compacted soils: Mercury Intrusion Porosimetry (MIP) and water permeability measurements. MIP tests consist in injecting mercury into a dried and degassed material, and are carried out using a Micromeritics Autopore IV porosimeter, which can characterize pores between 0.06 and 350 μm in diameter. In order to maintain the microstructure, the samples are dried by sublimation after a 24-hour freezing phase. The volume of mercury injected corresponds to the cumulative pore volume accessible to mercury. Assuming that the pores are cylindrical, the relationship between the accessible diameter d_c of the pores and the mercury injection pressure P is given by Washburn's law as:

$$P = -\frac{4\sigma_{Hg}\cos\theta}{d_c} \quad (1)$$

σ_{Hg} is the mercury surface tension (σ_{Hg} = 0.485 N/m at 25 ° C) and θ the contact angle of the meniscus solid to mercury (θ = 130 °> 90 ° because mercury is a non-wetting fluid). Four permeability test series are carried out on compacted soils. Those tests are duplicated to ensure more representative results. Thus, two samples are tested for each chosen degree of compaction for a total of 32 permeability tests with an average duration of two months. The samples tested are prepared by static compaction using a hydraulic press with compaction force up to 50 kN. The permeability tests are carried out using flexible-wall permeameters.

3 Results obtained on compacted soils

3.1 MIP tests results

During the soil compaction phase, an association of particles and aggregations forms. Granular soils retain a lower tendency to form aggregates in contrast to fine soils, where they cluster into larger aggregates. When the aggregates form, the pore size distribution can be divided into micropores, *i.e.*, intra-aggregate pores and macropores *i.e.*, inter-aggregates pores. It has been previously observed by Ranaivomanana et al. [2] that the compaction process mainly affects the macropores. Moreover, compaction also affects pore interconnections by blocking the percolation paths and modifying the morphological parameters. The pore size distribution curves of the tested soils (natural and treated) are presented in Figure 1 and Figure 2.

Because of the formation of larger aggregates in addition to the hydrophilic performances of cement, the addition of cement to fine soils is characterized by the distribution of cement particles around the agglomerates without penetrating them [5]. Randomly, in certain zones, cement serves as a link between two adjoining agglomerates. Cement also contributes to fill the macropores whose number increases with the formation of hydration products. This suggests that during the compaction process of cement-treated soils, both size and volume changes in macropores can be assumed as the result of the addition of cement and of the compaction process. This assumption is confirmed by the MIP results displayed in Figure 1a to 1d and Figure 2a to 2d. At 85% of the SPO values, which is the lower degree of compaction used in this study, both the volume and the size of the macropores are virtually the same (pore diameter around 60 μm) as those observed in natural materials (Figure 1a and Figure 1c). The influence of the treatment on the macropores, therefore, remains limited as long as sample compaction is low. An increase in the micropore volume in soils treated at 85%, 95%, 100% and 105% of the SPO values, however, is observed. These new micropores are associated to the voids between the cement grains and/or between the hydration products because their size and volume do not change with the compaction rate. Beyond 85% of the SPO value, a substantial decrease in macropore volume as well as size is observed on the treated soils when the degree of compaction increases (Figure 2a and Figure 2c). Although some similar trends are obtained on natural materials, the size and the volume of the macropores are smaller in cement-treated samples than in natural ones. At 105% of the SPO values, the macropores are unobservable in treated silty sand while they are visible in natural samples. Those observations corroborate the fact that the influence of the treatment is enhanced by the quality of the compaction and point to the possibility of optimizing treatments by using suitable degree of compaction. As mentioned previously, granular soils retain a low tendency to form aggregates.

Contrary to fine soils, the cement-based treatment of granular soils generates new aggregates resulting from the development of many strong bonds between the initial aggregates by cementation effect. Cementation obviously affects the macropores as shown by the MIP results obtained in gravelous sand (Figure 1b and Figure 2b). For all the degrees of compaction examined in this research (85%, 95%, 100% and 105% of the SPO values), a decrease in both size and volume of the macropores in the cement-treated gravelous sand is observed. However, the MIP results reveal another interesting aspect of the modification of the macropores. The changes appear to be independent of the degree of compaction. This interesting finding is a consequence of the cementation effect that stabilizes the microstructure by creating strong and durable bonds between the initial aggregates. Gravelous sand becomes, therefore, less sensitive to the density of compaction after treatment using cement (Figure 1b and Figure 2b). As in fine soils, an increase in the micropores volume associated with the voids between the cement grains and/or between the hydration products is also observed here. The MIP results obtained with cement-treated alterite (Figure 2d) reveal two apparently conflicting aspects: the formation of new macropores at 85% and 95% of the SPO values, respectively and the disappearance of these macropres at 100% and 105% of the SPO values, respectively, as in gravelous sand. This can be explained by the non-uniform homogenization of the tested samples during the preparation phase due to the destructuration of the alterite. A range of low porosity modes occurring between 0.05 and 1μm related to macropores is also observed in Figure 2d. These macropores appear to result from the destructuration of the alterite during the compaction phase despite the development of strong bonds between the initial aggregates, as with gravelous sand. This is confirmed by the fact that the newly formed macropores are independent of the compaction quality. Their size and volume, indeed, remain the same whatever the degree of compaction. This result satisfactorily agrees with that obtained using some natural materials, in which no variation of the macropore characteristics is noted when the degree of compaction increases (Figure 1d and Figure 2d).

3.2 Water permeability results

The permeability results are plotted on a semi-log scale as a function of the compaction rate (Figure 3). For both natural and treated fine soils, the permeability values (log. scale) linearly decrease almost identically (same slope) as the degree of compaction increases (Figure 3a and Figure 3c). These results are consistent with the MIP results and highlight the fact that the decrease in permeability values observed here is mainly the consequence of the macropore change caused by the compaction effects for natural materials and by both the compaction and treatment effects for treated materials. According to the MIP results, the macropore changes observed in the treated materials are greater than in natural ones as stated in the previous section. As a result, the fewer the macropores, the lower the permeability as shown in Figure 3a and Figure 3c. Notice that the difference between the permeability values of natural and treated materials is in the order of one to two orders of magnitude. As regards gravelous sand, permeability values also decrease linearly as the degree of compaction increases, but with a less marked slope for treated samples than for natural ones (Figure 3b). As mentioned above, the cementation effect stabilizes the microstructure of treated materials because of the development of strong and durable bonds between the initial aggregates. Therefore, the evolution of the permeability values obtained in treated gravelous sand is less sensitive to the density achieved in compaction than in natural sand, is not a surprise. However, according to the MIP results, the treated samples show greater variations in both the size and volume of the macropores than in the natural ones. This explains the difference between the permeability values of natural and treated materials of 2-6 orders of magnitude. For both the natural and the treated alterite, the permeability results also decrease linearly (Figure 3d). However, above 95% of the SPO, the permeability values obtained in the treated materials are 2-2.5 orders of magnitude greater than those measured in natural materials. This can be explained by the formation of new macropores resulting from the destructuration of the alterite during the compaction phase and despite the development of strong bonds between the initial aggregates. According to the MIP results, newly formed macropores are independent of the compaction rate, which may account for the fact that the permeability values obtained in the treated materials are less sensitive to the degree of compaction, as shown in Figure 3d.

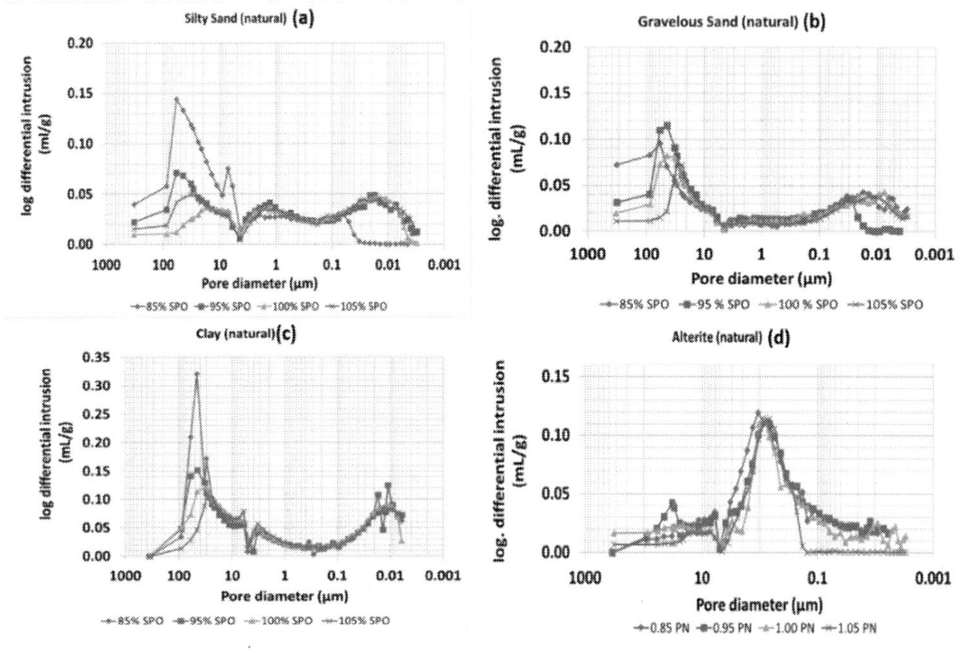

Fig. 1. Pore size distributions of natural soils for compaction rates in the present study.

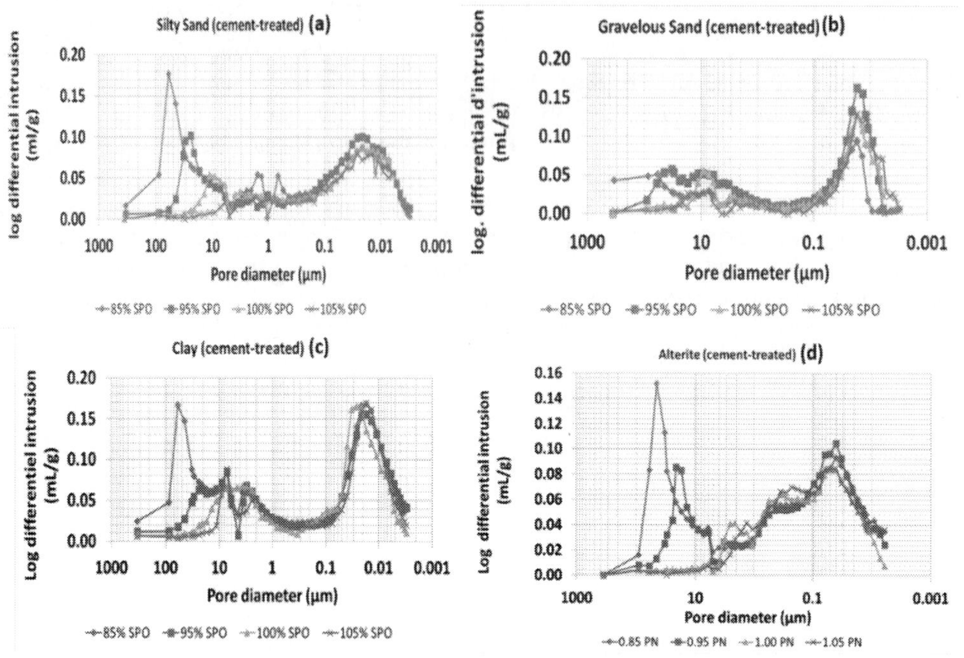

Fig. 2. Pore size distributions of treated soils for compaction rates in the present study.

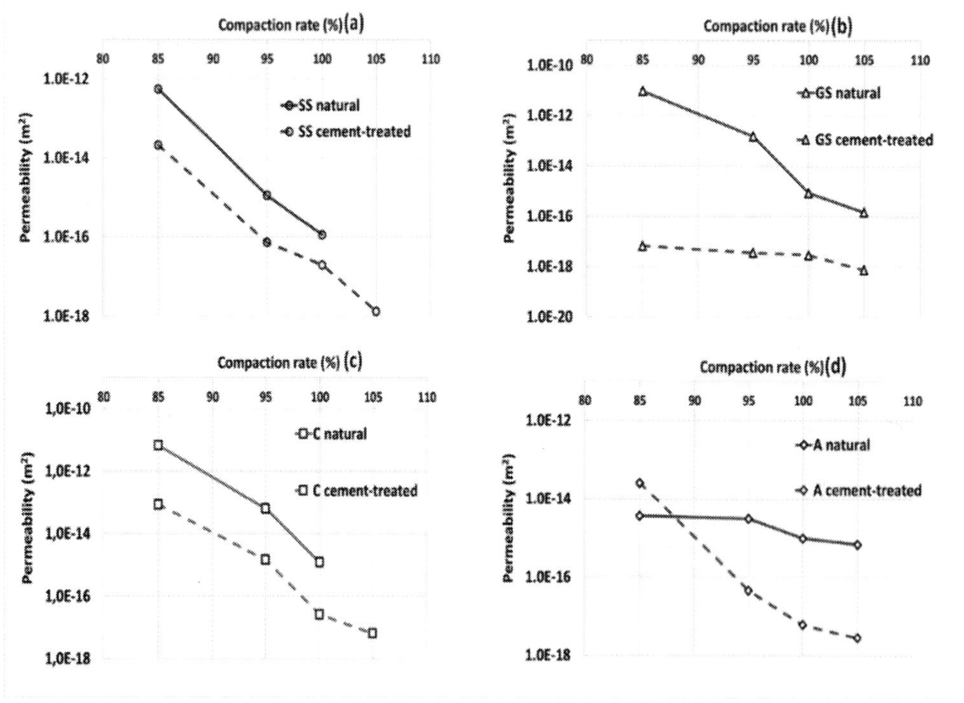

Fig. 3. Water permeability vs degree of compaction for natural and cement-treated soils.

4 Conclusion

An experimental characterization of cement-treated compacted soils as regards the microstructural, hydraulic and mechanical properties has been conducted in this paper. Some tests have been conducted on two different fine soils (silty sand and clay) and on two granular soils (gravelous sand and alterite). The soils tested have been mixed with 5% of cement and compacted at 85%, 95%, 100% and 105% of the SPO values, respectively. The results of the mercury intrusion porosimetry (MIP) tests performed on cement-treated soils after compaction highlight the substantial decrease in both the size and volume of the macropores. This decrease is due to the combined effects of treatment and of compaction for the fine soils, and mainly to treatment as regards the granular soils. Such a process, indeed, initiates a cementation effect, which stabilizes the microstructure through the development of strong and durable bonds between the initial aggregates. It should be noted, however, that the formation of new macropores within cement-treated alterite is probably a result of its destructuring during compaction phase. The water permeability results are in accordance with the MIP results because liquid flows mainly takes place within the macropores. By plotting the water permeability evolution *versus* the degree of compaction with a semi-log scale, a linear decrease of this parameter is observed for all the tested soils as the degree of compaction increases. This decrease is particularly significant with fine soils, which are more sensitive to the effects of compaction than granular ones.

References

1. S. Bin S., L. Zhibin, C.Yi, Z. Xiaoping Z, J. Mater. Civ. Eng. **19** (2007)
2. H. Ranaivomanana, A. Razakamanantsoa, O. Amiri, Int. J. Geomech. **17**(4), (2016)
3. EN 197-1 (2011)
4. ASTM D698 - 12e2 (2012)
5. K. Lemaire, D Deneele, S. Bonnet, M. Legret. Eng. Geol. **166** (2013)

The influence of an applied standard test method on a measurement of concrete stabilized secant modulus of elasticity

Lucyna Domagała[1,*], and *Justyna* Dobrowolska[1]

[1]Cracow University of Technology, ul. Warszawska 24, Cracow 31-155, Poland

Abstract. The paper focuses on the influence of the standard test method applied to determine the concrete stabilized secant modulus on a specified value. The new European Standard EN 12390-13 for testing hardened concrete accepts two methods (A and B) for the determination of the secant modulus of elasticity in compression. The aim of the research was to establish how different testing procedures affect a measured value of modulus of elasticity. Four structural concrete series: two lightweight aggregate concretes and two normal-weight ones were subject to tests of moduli of elasticity determined by both standard methods, as well as compressive strength and density. The carried out tests revealed that the procedure of testing modulus of elasticity influenced a measured value. Method A led to higher values of modulus in relation to Method B, irrespective of concrete density and strength. Nevertheless, a certain relationship between the concrete structure homogeneity and the difference in results of moduli determined by both methods may be observed.

1 Introduction

Modulus of elasticity is one of the most important mechanical properties of concrete which has a significant impact on the structural design process. There are many types of concrete moduli of elasticity (dynamic, static initial tangent, tangent, stabilized secant, initial secant), however it is the stabilized secant one which is fundamentally taken into consideration at structural member designing. Until recently, in many European countries various guidelines for this modulus determination were in force. These documents differed in testing procedures, especially in terms of the moisture condition of specimens to be tested, the lower and the upper stress values and a number of loading and unloading cycles. As a result, the values of modulus determined by different testing methods may have varied more than those specified by one method for concretes of different strength classes [1-9].

Fortunately, in 2013, a new European Standard, EN 12390-13 [10], dedicated for the determination of the secant modulus of elasticity in compression, was published at last. This long awaited standard on one hand unifies to some extent the procedures of modulus testing, on the other hand it leaves a certain freedom in selection of the specimen type and number, as well as the testing method.

* Corresponding author: ldomagala@pk.edu.pl

2 Standard test methods for determination of the secant modulus of elasticity

The new standard EN 12390-13 [9] accepts two methods of determination of concrete modulus of elasticity: Method A and Method B. Method A is more complex and requires longer time for testing, but it allows the determination of two secant moduli of elasticity: the initial secant modulus ($E_{C,0}$) and the stabilized secant modulus ($E_{C,S}$). It should be noticed that it is the stabilized modulus which value is used in the structure designing process. However, the degree of variation of these two moduli may be helpful for assessment of the material susceptibility to stress induced micro-cracking or micro-cracking caused by drilling in the case of cores taken from a structure. The second method (Method B), intended only for the determination of the stabilized secant modulus, is much simpler, requires shorter time and as a result it is much more often applied by researchers.

Nevertheless, there is no information neither in the standard nor in the scientific literature on the difference in the modulus values determined by these two standard methods.

2.1 Method A

The testing procedure according to Method A consists of three initial preloading cycles in the stress range from σ_p = 0.5 MPa ÷ σ_b up to σ_b = 0.10 ÷ 0.15·f_{cm} (where f_{cm} – mean compressive strength) and three main loading cycles in the stress range from the lower stress σ_b up to the upper stress $\sigma_a = f_{cm}/3$ (Fig. 1). The initial secant modulus is determined at the first main loading cycle, while the stabilized secant modulus is measured at the last, the third cycle. Finaly the stabilized secant modulus is calculated according to formula (1), where σ_a^m and σ_b^m are the stress values measured at the third cycle while $\varepsilon_{a,3}$ and $\varepsilon_{b,2}$ are the corresponding strains.

$$E_{c,s} = \frac{\sigma_a^m - \sigma_b^m}{\varepsilon_{a,3} - \varepsilon_{b,2}} \quad (1)$$

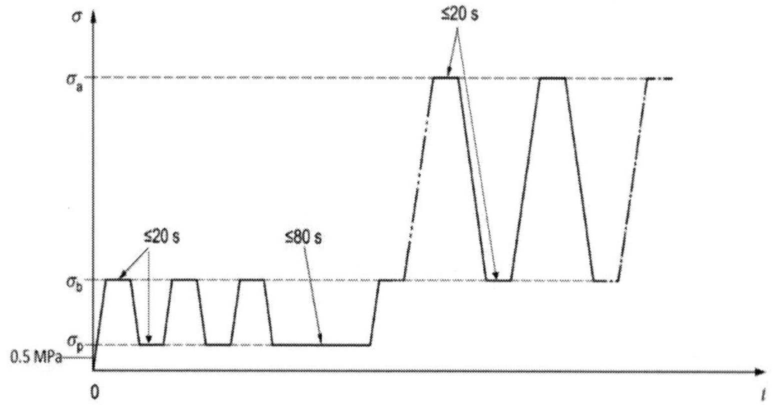

Fig.1. Cycle for the determination of initial and stabilized modulus of elasticity according to Method A.

2.2 Method B

The testing procedure according to Method B consists only of three main loading cycles in the stress range from the lower stress σ_p up to the upper stress σ_a. (Fig. 2). The values of σ_p and σ_a are assumed as for Method A. As a result, the stabilized secant modulus is calculated according to formula (2), where $\sigma_a{}^m$ and $\sigma_b{}^m$ are the stress values measured at the third cycle while $\varepsilon_{a,3}$ and $\varepsilon_{p,2}$ are the corresponding strains.

$$E_{c,s} = \frac{\sigma_a^m - \sigma_p^m}{\varepsilon_{a,3} - \varepsilon_{p,2}} \tag{2}$$

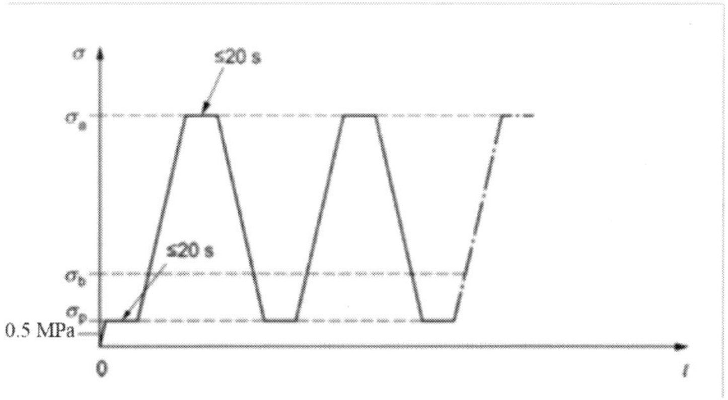

Fig.2. Cycle for the determination of stabilized secant modulus of elasticity according to Method B.

3. Experimental details

The research aim was to check whether both standard methods for determination of modulus of elasticity (Method A and Method B) lead to the same test result. In the case of different measurements, the scope was to analyse whether the difference of moduli specified by these two methods is dependent on the concrete strength and type.

3.1 Materials

Four series of structural concretes of different strength and density classes were subject to tests. All series were made of Portland cement CEM I 42.5 R, natural sand as fine aggregate and tap water. Two types of coarse aggregate were used: natural sandstone gravel 4/16 mm as normal-weight aggregate and sintered fly ash 4/8 mm as lightweight one. Sintered fly ash was initially wetted to the moisture content of 17 % by mass to eliminate the risk of workability loss. Two water - cement ratios were taken into consideration: 0.37 and 0.55. The constant volume proportion of cement mortar to coarse aggregate (55 % : 45 %) was assumed for all concrete series. In the case of w/c = 0.37 additionally superplasticizer was applied to ensure workability similar to concrete mixtures of w/c = 0.55. The composition of all prepared mixtures are presented in Table 1.

6 standard cylinders (150/300 mm) and 6 standard cubes (150 mm) were moulded from each series of concrete mixtures.

Table 1. Composition of normal-weight concretes (NC) and lightweight concretes (LC).

Series	w/c	Constitiuent Materials [kg/m^3]					
		Coarse aggregate	Water for initial wetting of coarse agg.	Natural sand	Cement (c)	Water (w)	Super-plasticizor
NC1	0.55	1140	0	570	377	208	0
NC2	0.37	1140	0	570	470	174	1.88
LC1	0.55	576	98	570	377	208	0
LC2	0.37	576	98	570	470	174	1.88

3.2 Testing methods

All tests were carried out on specimens cured for 28 days in water in standard conditions according to EN 12390-2 [10]. At the beginning for each concrete series 3 cylindrical specimens were tested according to EN 12390-3 [11] to determine the mean value of compressive strength ($f_{cm,cyl}$). The rest 3 cylindrical specimens of each concrete series were applied for determination of modulus of elasticity according to Method A and B of EN 12390-13 [8]. In both methods the stress of 0.5 MPa was assumed as σ_p and the value of 1/3 $f_{cm,cyl}$ was taken as the upper stress σ_a. However, in the case of Method A the value of 0.1 $f_{cm,cyl}$ was taken as the lower stress σ_b, while for Method B the lower stress was assumed as σ_p. The testing equipment applied for determination of modulus of elasticity consisted of a programmable compression testing machine and a strain meassuring set of 3 extensometers (Fig. 3).

Fig. 3. A concrete specimen with the measuring equipment for modulus determination.

In order to eliminate the influence of material heterogeneity of a specimen, the arrengment of a specimen in relation to machine platens and the location of strain sensors in relation to a specimen on a measurement of modulus of elasticity, tests according to procedures A and B were carried out in sequence on each specimen. According to well-known rules, repeated or cyclic loading of concrete up to the stress as low as 1/3 f_{cm}, which is considerable lower than the fatique strength, should not affect the measurement of modulus of elasticity [1,2,12]. Nevertheless, just in case, the measurement repeatability was checked prior to main tests of modulus of elasticity. Therefore selected specimens, ultimately predicted for initial compressive strength determination, were subject to earlier tests of modulus in sequince according to following methods order: A-B-A-B-A or B-A-B-A-B. Before each next test a specimen was completely unloaded, however its position in relation to the compressive testing machine platens was reteined.

Moreover, saturated and oven-dry densities were tested on cube specimens, according to EN 12390-7 [13]. The results of density tests made it possible to assess the moisture content of specimens during modulus testing.

4 Results and discussion

Application of various type of coarse aggregate and cement matrix resulted in different properties of all four hardened concretes. Their compressive strength ranged from 33.5 MPa up to 59.3 MPa, while corresponding oven-dry density varied from 1690 kg/m^3 up to 2290 kg/m^3. The test results of compressive strength, oven-dry and saturated density, as well as moisture content are given in Table 2.

Table 2. Basic properties of normal-weight concretes (NC) and lightweight concretes (LC).

Series	Mean compressive strength ($f_{cm,cyl}$) [MPa]	Standard deviation of strength [MPa]	Mean saturated density [kg/m^3]	Mean oven-dry density (ρ) [kg/m^3]	Mean moisture content [%]
NC1	42.9	1.5	2310	2220	4.1
NC2	59.5	1.0	2340	2290	2.2
LC1	33.5	0.6	1840	1690	8.9
LC2	45.3	1.6	1890	1780	6.2

Repeatability tests of modulus of elasticity, carried out for several specimens randomly selected from normal-weight and lightweight concretes, proved that there is no visible difference in results determined on the same specimen according to one method (A or B). Results of moduli tested in sequence several times by the same method revealed usually no changes or varied at the most by 0.3 GPa. This value is much lower than the difference of results usually achieved for various specimens of the same concrete. All values of moduli measured according to methods A and B on the same specimen, regardless of the type of concrete, varied visibly. Nevertheless, it should be stated that the order of applied methods (method A after or before B) did not reflect the measurement of the stabilized secant modulus ($E_{C,S}$) obtained in the same testing procedure. However, in any sequent test a higher value of the initial secant modulus ($E_{C,0}$) was reached in comparison to the first test. It was caused by a certain durable compression of concrete structure during the previous loading. Consequently a specimen behaved as stiffer at the first loading cycle of next tests. Therefore, in order not to disturb the measurement of $E_{C,0}$, in the main tests all specimens were firstly subject to Method A and then to Method B. The relationship between initial and stabilized moduli, with respect of the types of concrete considered hereby, are discussed in [14].

The obtained test results of the stabilized secant moduli of elasticity determined according to both standard methods as well as their interrelationships are presented in Table 3.

Generally, application of less stiff lightweight aggregate instead of normal-weight one caused the reduction of modulus of elasticity by 30 % on average. This decrease was slightly more pronounced for the concrete made of weaker cement matrix strength (w/c = 0.55). However, in the case of the same coarse aggregate type the application of the weaker mortar resulted in moduli lower by ca 18 % in relation to concretes prepared with stronger matrix (w/c=0.37). The decrease resulting from the weaker matrix application was insensibly higher for lightweight concrete.

Table 3. Stabilized secant moduli of elasticity determined according to Methods A and B for normal-weight concretes (NC) and lightweight concretes (LC) and their relationships.

Series	Method A			Method B			$E_{C,SA}/E_{C,SB}$ [%]	$E_{C,SAm}/E_{C,SBm}$ [%]	$E_{C,SA}-E_{C,SB}$ [GPa]	$E_{C,SAm}-E_{C,SBm}$ [GPa]
	$E_{C,SA}$ [GPa]	$E_{C,SAm}$ [GPa]	σ_E [GPa]	$E_{C,SB}$ [GPa]	$E_{C,SBm}$ [GPa]	σ_E [GPa]				
NC1	26.8	27.1	0.3	24.6	25.0	0.3	1.09	1.08	2.2	2.1
	27.5			25.1			1.10		2.4	
	26.9			25.2			1.07		1.7	
NC2	32.2	32.2	0.7	30.3	30.1	0.7	1.06	1.07	1.9	2.1
	31.4			29.2			1.08		2.2	
	33.0			30.8			1.07		2.2	
LC1	17.9	18.1	0.4	17.3	17.5	0.2	1.03	1.04	0.6	0.6
	17.8			17.4			1.02		0.4	
	18.7			17.8			1.05		0.9	
LC2	22.8	22.6	0.3	21.8	21.6	0.2	1.05	1.05	1.0	1.0
	22.8			21.6			1.06		1.2	
	22.2			21.4			1.04		0.8	

Analyzing the influence of the applied standard test method on a determination result of concrete modulus of elasticity it should be clearly stated that in the case of all tested specimens, independently from the type of used aggregate and cement matrix, Method A led to higher results than Method B. There are two probable reasons of such a phenomenon. Firstly, it may be the effect of strengthening and compaction of concrete structure during preloading cycles applied in Method A. Secondly, the higher value of the lower stress used in main loading cycles of Method A may be favourable for better concrete compression.

Nevertheless, the difference between values measured by Method A and B was more pronounced for normal-weight concretes than for lightweight ones. For both normal-weight series (NC1 and NC2) Method A resulted in modulus higher by 2.1 GPa, (by ca 8 %) in relation to Method B, while for lightweight concretes LC1 and LC2 it gave values higher by 0.6 GPa and 1.0 GPa respectively (for both by ca 4 %). The smaller difference between moduli determined by both standard procedures for lightweight concretes probably results from more homogenous structure of these materials in comparison to normal-weight concretes. Generally, better structural homogeneity of lightweight concrete is caused by more regular shape and size of artificial particles, stronger bond between cement paste and porous aggregate and better compatibility of their stiffness, that means smaller difference of their moduli of elasticity. The direct prove of better homogeneity of lightweight concrete is its behaviour under compression up to failure, which leads to more linear stress-strain relation with a considerably reduced post-peak descending part in comparison to normal-weight concrete [15-17]. When concrete is loaded only up to one third of the ultimate stress, as it happens at modulus testing, it is harder to notice deviation from linearity of stress-strain relation. Nevertheless, charts of σ-ε at cycled loading during modulus testing differ significantly for lightweight and normal-weight concretes. The examples of typical stress-strain behaviour of both types of concrete subject to modulus testing according to Methods A and B are presented in Figure 4. At the first main loading in Method A a certain curvilinearity of stress-strain relation is observed in the case of normal-weight concretes, while the relation for lightweight concretes is almost linear from the beginning. In Method B, which followed Method A, these differences are not so visible because of earlier compression of concrete structure, mentioned above. Moreover, hysteresis loops, formed in sequent loading and unloading cycles, overlap more in the case of lightweight concretes than for normal-weight ones. As a result the strain stability of lightweight concretes was achieved already at the first loading in Method B and at the second loading in Method A, while for normal-weight concretes it was obtained at the second and the third loading respectively.

The shown differences between moduli determined in both standard methods should be considered as important, especially for normal-weight concretes. The values measured by Methods A and B for the same concrete differ more than moduli tested by the same method but for concretes differing by ca 2 strength classes. In the case of concrete made of stiffer aggregate, in comparison to sandstone applied in this research, the difference between moduli determined in both standard methods would probably be even higher. It cannot be ruled out that the application of a higher value of the lower stress in Method A, e.g. $\sigma_b = 0.15 \cdot f_{cm}$ instead of $\sigma_b = 0.1 \cdot f_{cm}$ used in this research, may also increase this difference.

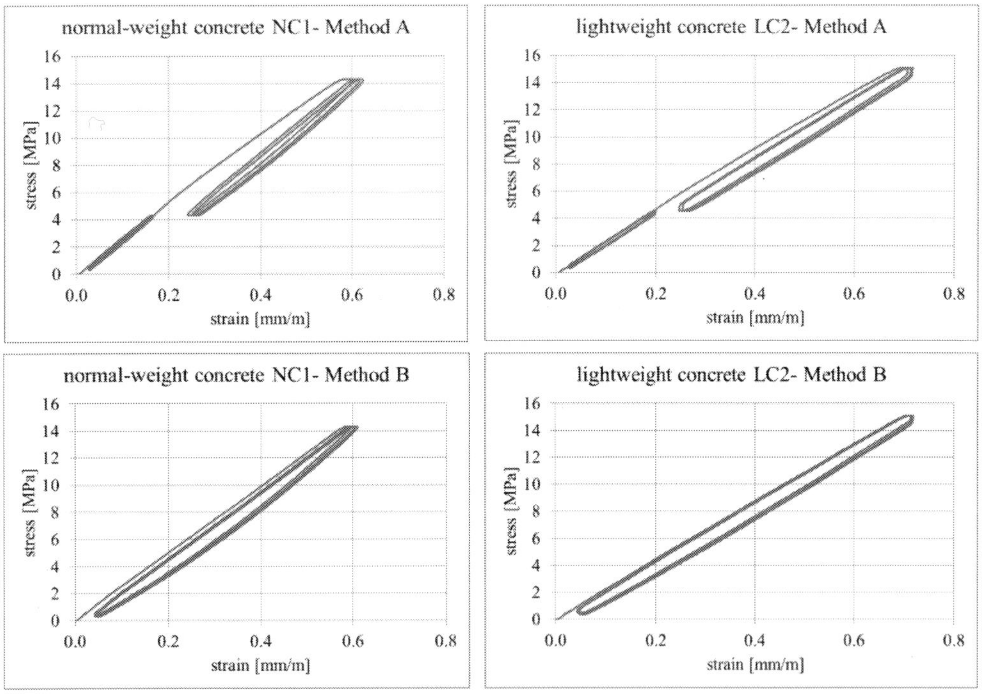

Fig. 4. Typical stress-strain relations at loading cycles of procedures A and B for normal-weight concrete (NC) and lightweight concrete (LC).

The measured values of modulus of elasticity differ significantly from those calculated according to designing standard Eurocode 2 (EN 1992-1-1[5]). The standard estimation for normal-weight concretes NC1 and NC2 gives the values of 23.8 GPa and 26.3 GPa respectively, while for lightweight concretes LC1 and LC2 corresponding moduli are 18.7 GPa and 22.7 GPa. It means that the values estimated for normal-weight concretes are lower even by 5.9 GPa and 3.4 GPa in relation to those determined by Method A and Method B respectively. However, in the case of lightweight concretes there is only a small difference between the values estimated and the specified according to Method A, while Method B leads to the measurements lower (by up to 1.2 GPa) than the estimated moduli.

5 Conclusions

The carried out tests and the analysis of the obtained results show that independently from type of used aggregate and cement matrix, two standard methods for determination of the secant modulus of elasticity lead to different measurements. The effect of strengthening and compaction of concrete structure during preloading cycles and a higher value of the lower stress used in the main loading cycles are probable reasons of the higher results achieved in tests performed according to Method A.

The difference between the values measured by Method A and B was more pronounced for normal-weight concretes than for lightweight ones. In the case of less homogenous concrete, e.g. made of stiffer aggregate than sandstone applied in this research, the difference would probably be even more significant. It cannot be ruled out that the application of a higher value of the lower stress in Method A, e.g. $\sigma_b = 0.15 \cdot f_{cm}$ instead of $\sigma_b = 0.1 \cdot f_{cm}$ used in this research, may also increase this difference. These two cases require to be proved in additional tests.

Although the new European Standard EN 12390-13 for determination of the secant modulus of elasticity sets a lot of problems related to testing procedures in order, the possibility of the application of two different standard methods, which give different test results, does not help researchers and structure designers. Perhaps one of these two standard methods should be indicated as the recommended one.

References

1. K. Mehta, P. Monteiro, *Concrete microstructure, properties and materials* (MCGraw Hill Education, 2013)
2. A. Nevile, *Properties of concrete* (Pearson, 2011)
3. L. Domagała, *5th Int. Conf. Concrete and concrete structures,* 85 (2009)
4. L. Domagała, Materiały Budowlane (in Polish) **8**, 55 (2017)
5. EN 1992-1-1 (2008)
6. DIN 1048-5 (1991)
7. Instrukcja ITB nr 194 (in Polish) (1998)
8. EN 12390-13 (2013)
9. B.D. Liu, W.J. Lv, P.F. Li, Constr. Build. Mater. **69**, 133 (2014)
10. EN 12390-2 (2009)
11. EN 12390-3 (2009)
12. J. Newman, B.S. Choo, *Advanced Concrete Technology* (Elsevire, 2007)
13. EN 12390-7 (2009)
14. L. Domagała, Solid State Phenomena **258**, 566 (2017)
15. M-H Zhang, O. Gjørv, ACI Mater. J. **88**, 240 (1991)
16. H. Cui H, T. Lo, S. Memon, F. Xing, X. Shi, Constr. Build. Mater. **36**, 845 (2012)
17. L. Domagała, J. Civ. Eng. Management **17**, 36 (2011)

Usefulness of porphyry and amphibolites as a component of concrete for airfield pavements

Małgorzata Linek[1,*], *Piotr* Nita[2], *Paweł* Wolka[1], and *Wojciech* Żebrowski[1]

[1]Kielce University of Technology, Faculty of Civil Engineering and Architecture, Department of Transportation Engineering, Tysiąclecia Państwa Polskiego Street 7,25-314 Kielce, Poland
[2]Air Force Institute of Technology, Księcia Bolesława Street 6, 01-494 Warsaw, Poland

Abstract. Coarse aggregate used as part of cement concrete is of primary significance for the obtained parameters of hardened concrete. In case of concretes intended for airfield pavements, the application of granite grit is recommended. Alternative to this type of aggregate in the form of porphyry and amphibolites aggregate was suggested. In order to assess the suitability of different aggregate types for concrete mixes, their bulk density, absorbability, polishing resistance, abrasion and crushing resistance were determined. Also, the internal structure of the suggested aggregates and its influence on changes of hardened concrete composite structure were subject to the assessment. The influence of aggregate type on the structure of cement matrix and contact areas between the matrix and aggregate grains were specified. The observed changes, in case of the internal structure of concretes based on porphyry and amphibolites aggregates , with reference to granite aggregate, resulted in changes of mechanical and physical parameters. Analyses included the determination of bulk density, absorbability, compression, bending and splitting resistance. According to the obtained laboratory test results, the significant influence of the aggregate type applied to the mix on parameters of hardened concrete, with regard to the application thereof to the airfield pavements was proved.

1 Introduction

The article is aimed at the assessment of the suitability of the suggested aggregates in the composition of cement concrete intended for airfield pavements. The basic properties of granite, porphyry and amphibolites aggregate were determined. It was assessed whether the applied aggregate influences the change of parameters of internal structure of hardened cement concrete, the change of physical and mechanical properties of concrete intended for air field pavements.

* Corresponding author: linekm@tu.kielce.pl

2 Materials

Analyses focused on three series of various types of aggregate. Series labelled A was amphibolites aggregate fraction of 2/8mm and 8/16mm, series P was porphyry aggregate of grain-size distribution of 2/8mm and 8/16mm, while series G was the reference aggregate – granite aggregate of grain-size distribution of 2/8mm and 8/16mm. In case of the selected types of aggregate the following were determined: volume density (ϱ_a) according to [1], absorbability (W) according to [2], abrasion resistance (M_{DE}) according to [3] and crushing resistance (LA) according to [4].

The influence of the type of the applied aggregate on the change of concrete mix parameters was determined. The experiment assumed to perform three series of concrete mix. The first series was reference concrete mix (MG). Two consecutive series included concrete mix with granite aggregate substitute in the form of porphyry aggregate (series MP) and amphibious (series MA). The tests included: mix density using cylinder of 8 dm^3 capacity according to [5], consistency using concrete slump method according to [6], air contents using manometer method according to [7] The prepared concrete mix series MG, MP and MA after being arranged into (cubic moulds 150x150x150 mm and 150x300 mm according to [8]) were subject to standard curing [9] for the assumed research periods (7 and 28 days).

The experiment assumed to perform three series of concrete. The first series was reference concrete (CG), in accordance with the requirements of [10]. Two consecutive series included concrete with aggregate substitute in the form of porphyry aggregate (series CP) and amphibious aggregate (series CA). Concretes for airport pavements were analyzed. The determined concrete parameters included the following: volume density (ϱ_o) according to [11], absorbability (N) according to [10], compressive strength (f_{cm}) according to [12], splitting tensile strength (f_{ct}) according to [13] and concrete bending strength (f_{cf}) after diversified curing periods.

Selected samples of Cg, CP and CA concrete were observed in scanning electron microscope. Preparation surface subject to observations by means of scanning electron microscope was not less than 1.0cm^2, and the scope of magnifying power was assumed between 200x to 100000x. [14]

3 Characteristics of aggregates

The fraction is distinguished by irregular, angular and sharp shape. It was concluded that fractions of porphyry grit do not contain quartz crystals - fig. 2a. Based on the chemical microanalysis of the analysed porphyry grit, the occurrence of the following elements, i.e. oxygen, sulfur, aluminum, potassium, carbon, sodium, calcium, magnesium was proved (figure 2).

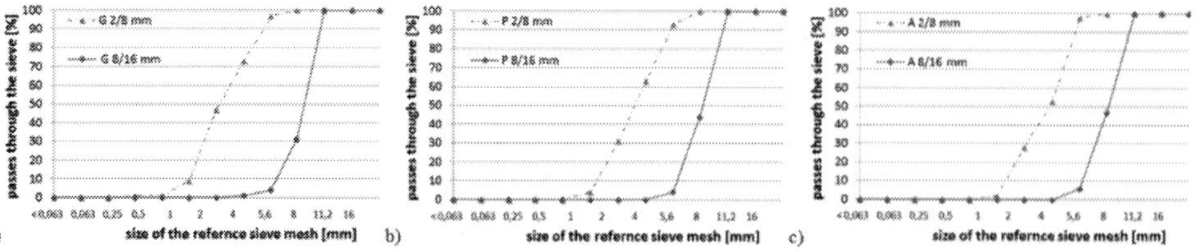

Fig. 1. Designed aggregate mixture curve, together with limit curves (lower and upper) - a) MG, b) MP, c) MA.

a) SEM image b) chemical microanalysis

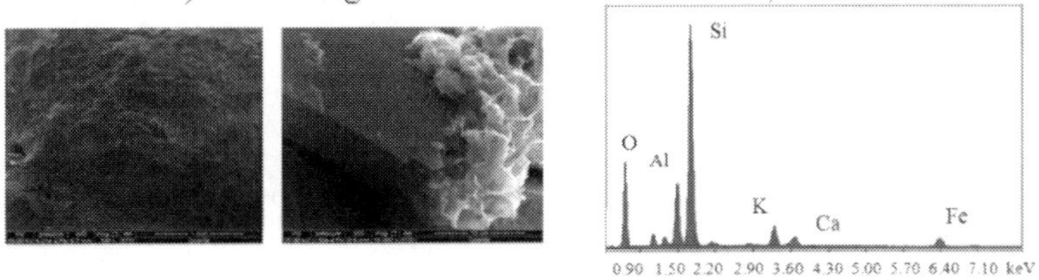

Element	Wt [%]	At [%]
C	3.46	6.19
O	37.87	49.74
Na	3.26	3.04
Mg	1.61	1.42
Al	10.37	8.25
Si	35.53	27.16
K	3.67	2.02
Ca	1.73	0.93
Ti	0.00	0.00
Fe	3.28	1.26
Σ	100.00	100.00

Fig. 2. Structure of the examined porphyry aggregate: a) SEM image of cement grain from SEM and b) image of chemical microanalysis.

According to SEM observations it was proved that grit fraction is distinguished by angular shape and medium-blastic structure - fig. 3a. Based on the chemical microanalysis of the analysed amphibolite grit, the occurrence of the following elements, i.e. oxygen, silicon, iron, aluminum, calcium, titanium, carbon, magnesium and sodium was proved (figure 3b).

a) SEM image b) chemical microanalysis

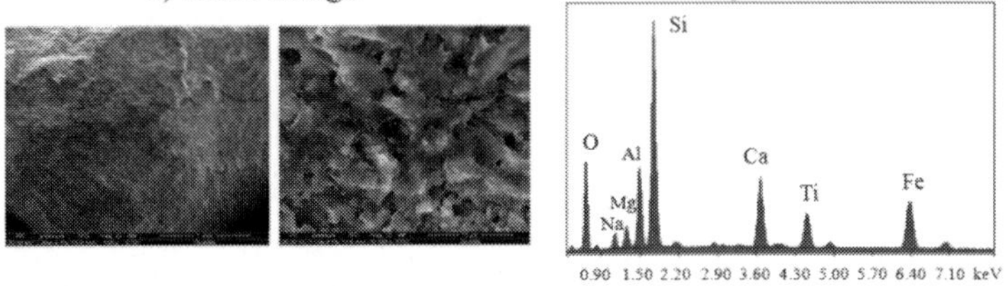

Element	Wt [%]	At [%]
C	4.57	8.95
O	29.47	43.31
Na	3.16	3.23
Mg	3.27	3.17
Al	9.13	7.95
Si	25.23	21.12
K	0.00	0.00
Ca	7.93	4.65
Ti	5.22	2.56
Fe	12.03	5.06
Σ	100.00	100.00

Fig. 3. Structure of the examined amphibolite aggregate: a) SEM image of cement grain from SEM and b) image of chemical microanalysis.

4 Methodology and test results

Composition of concretes included: cement CEM I 42.5 MSR/NA (fig. 4), sand, water according to [15], air entraining agent and plasticizing admixture. Aggregate grading composition of mix series MG, MP and MA was selected according to the guidelines of [10] taking into consideration limit curves of good grain size distribution (Figure 5).

a) SEM image b) chemical microanalysis

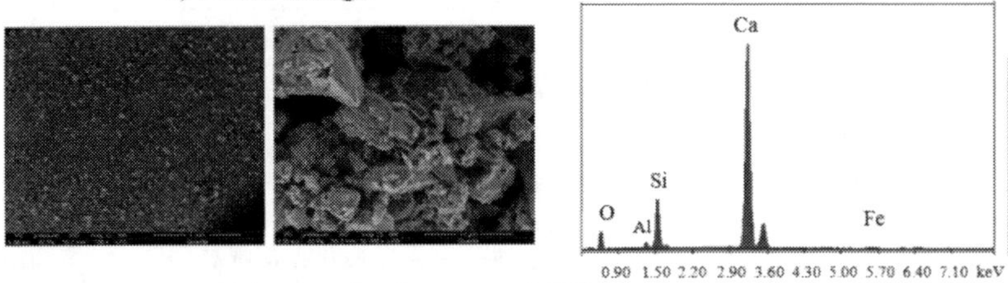

Element	Wt [%]	At [%]
C	3.22	6.76
O	30.23	47.61
Na	0.00	0.00
Mg	0.98	1.02
Al	1.49	1.39
Si	11.41	10.23
S	0.59	0.46
Ca	50.85	31.97
Ti	0.00	0.00
Fe	1.23	0.55
Σ	100.00	100.00

Fig. 4. Structure of the examined cement: a) SEM image of cement grain from SEM and b) image of chemical microanalysis.

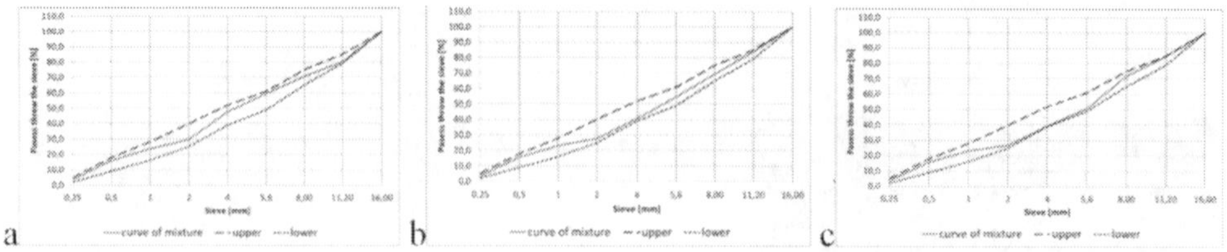

Fig. 5. Designed aggregate mixture curve, together with limit curves (lower and upper): a) MG, b) MP, c) MA.

Quantity selection of the remaining components (table 1) was based on experimental methods maintaining the consistency of reference mix at S1 level and taking into consideration the exposure class XF4, with the air amount between 4,5-5,5%.

Table 1. Compositions of the designed mixes of the following series MG, MK and MS.

Components	**MG**	**MP**	**MA**
	[kg/m^3]	[kg/m^3]	[kg/m^3]
Cement CEM I 42,5N MSR/NA	375	375	375
Fine aggregate 0/2mm	518	512	555
Granit 2/8mm	845	-	-
Granit 8/16mm	557	-	-
Porphyry	-	834	-
porphyry	-	550	-
amphibolite	-	-	905
amphibolite	-	-	596
Water	150	150	150
plasticizer	1.69	1.69	1.69
Aerating agent	0.94	1.13	1.13

4.1 Assessment of parameters of the analyzed aggregates

According to the conducted tests (table 2) in case of the analyzed aggregate types it was proved that amphibolites aggregate is distinguished by the greatest volume density and porphyry aggregate by the lowest volume density. In case of absorbability, comparable parameters were obtained in case of granite and amphibolites aggregates, while in case of porphyry aggregate this value is higher and exceeds 2%.

Table 2. Parameters of aggregates.

parameters	**j.m.**	**G**		**P**		**A**	
ϱ_a	Mg/m^3	2.65		2.60		2.95	
W	%	0.60		2.70		0.60	
M_{DE}	-	8.46	M_{D10}	8.96	M_{D10}	5.16	M_{D10}
LA	-	34.8	L_{A35}	9.49	L_{A15}	15.08	L_{A20}

According to the conducted tests (table 2) in case of the analyzed aggregate types it was proved that amphibolites aggregate is distinguished by the greatest volume density and porphyry aggregate by the lowest volume density. In case of absorbability, comparable parameters were obtained in case of granite and amphibolites aggregates, while in case of porphyry aggregate this value is higher and exceeds 2%.

4.2 Assessment of the influence of aggregate type on the properties of concrete mixes

According to the obtained laboratory test results (table 3) it was proved that the air content in all mixes complies with standard requirements [10]. In case of MG mix, the air content was the lowest and it was 4.5%. Significant diversification was proved in case of volume density. Mix of MA series was distinguished by the highest parameter, while the mix of MP series was distinguished by the lowest parameter. The difference was 7%. Based on the obtained results, no significant influence of the aggregate type on the concrete mix consistency was proved. In case of all analyzed mixes, the consistency defined in experimental test corresponded to S1 class.

Table 3. Parameters of mixtures.

mixture	**Density [kg/m^3]**	**Consistency [mm]**	**Air content [%]**
MG	2392	11	4.5
MP	2280	10	5.5
MA	2460	10	5.5

4.3 Assessment of influence of the suggested aggregates on the selected physical and mechanical parameters of hardened concrete

During scientific research, the amount of the required samples was determined using student's T-distribution assuming the significance level of 0.05. The minimum essential number of samples ranged between 4 and 5, depending on the type of the conducted test. In case of such assumptions, 6 samples were selected, which, each time, were intended for the laboratory tests. According to the obtained experimental test results (table 4) it was proved that the type of coarse aggregate significantly influences volume density of hardened concrete. Concretes of CG series were distinguished by the highest volume density, while concretes of CP series of the lowest volume density. The examined feature is also strictly dependent on the length of curing period. The diversification of the examined feature in case of CG and CP concrete type was approx. 113 kg/m^3 and 130 kg/m^3 after 7 and 28 days accordingly. Type of aggregate also influences the amount of water absorbed by concrete. In case of the examined aggregate type, it was proved that concrete series CA and CP are distinguished by the lowest absorbability, the amount of which does not exceed 3.5%. Concrete of CG series is distinguished by the absorbability of 4%. It should be emphasised that all of the analyzed concrete types complied with the requirements of [10] standard with regard to absorbability.

Table 4. Average hardened concrete density and absorbability of series CG, CK and CS during diversified curing periods (7 and 28 days).

concrete	Density, [kg/m^3]				Absorbability, [%]	
	7		28		28	
	$\overline{X}$	σ	$\overline{X}$	σ	$\overline{X}$	σ
CG	2459	11.048	2460.38	9.033	4.0	0.090
CP	2346	11.860	2330.12	11.214	3.4	0.168
CA	2450	2.534	2446.93	3.572	3.2	0.096

The type of the coarse aggregate applied to the mix significantly influences the obtained mechanical parameters of hardened concrete. The growth of resistance in case of concretes analyzed after 7 and 28 days independent of the aggregate type retained on the comparable level of 7MPa. Concrete of CP series was distinguished by the highest compressive strength, while concrete of CA series the lowest one. In case of the analysed concrete type of CP and CA series the diversification was approx. 20 MPa, both after 7 and 28 days of curing. Comparable tendency retains in case of the analyzed concretes with regard to splitting tensile strength (table 5). Concrete of CP series was distinguished by the highest resistance, while concrete of CA series of the lowest resistance.

Table 5. Average hardened concrete tensile splitting strength and compressive strength of series CG, CK and CS during diversified curing periods (7 and 28 days).

concrete	Compressive strength, [MPa]				Tensile splitting strength, [MPa]			
	7		28		7		28	
	$\overline{X}$	σ	$\overline{X}$	σ	$\overline{X}$	σ	$\overline{X}$	σ
CG	50.1	1.522	57.0	1.300	3.9	0.136	4.3	0.350
CP	59.2	1.281	67.9	1.917	4.2	0.372	4.8	0.091
CA	40.6	1.464	47.3	2.182	3.4	0.218	3.8	0.224

According to the analysis of the obtained test results it was proved that concrete of CP series is distinguished by the most favourable parameters. Compared to the recommended concrete of CG series, according to [10] standard, concrete of CP series has higher resistance parameters and is distinguished by comparable absorbability. Therefore, it can be the alternative for granite aggregate. The results obtained after the frost resistance test (table) are also the confirmation thereof. In case of the applied amphibolites aggregate the parameters obtained by concrete are significantly lower. The intent of using this type of aggregate should be preceded by detailed tests.

4.4 Assessment of influence of aggregate type on concrete microsructure

According to the conducted SEM observations it was proved that the type of aggregate contributes to the change of internal structure of cement matrix (fig. 6-8). The internal structure of concrete of CA series is distinguished by non-continuous contact areas between aggregate grains and cement matrix (fig. 8a). Hydrated calcium silicates occurs in the fine-grained form. Contact areas between aggregate grains and cement matrix undergo cracking of the width up to 8μm. Cement matrix undergo cracking of the width up to 6-7μm (fig. 8c). Internal walls of air voids undergo cracking of the width up to 2-3μm (fig. 8b). Cement matrix, in case of concrete of CP series, is consistent and crystallization of hydrated calcium silicate occurs in fine-grained form (fig. 7c). Crystallization of ettringite with the length of single crystals of which is up to 3μm, occurs in cement matrix. Contact areas between aggregate grains and cement matrix undergo cracking of the width up to 3μm (fig. 7a). Crystallization of ettringite (length up to 1μm) and cracked walls of air void interior were proved (fig. 7b). Crystallization of ettringite (length up to 10μm) and cracked walls of air void interior were proved. Crystallization of ettringite (length up to 3μm) and cracked walls of air void interior were proved (fig. 6b). Cement matrix, in case of concrete of CG series, is consistent and crystallization of hydrated calcium silicate occurs in fine-fibrous form (fig. 6c). Contact areas between aggregate grains and cement matrix undergo cracking of the width up to 3μm (fig. 6a).

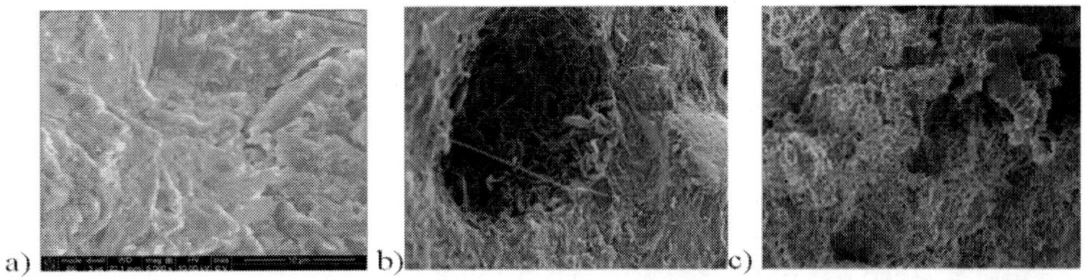

Fig. 6. Internal structure of cement concrete series CG: a) contact areas between cement matrix and aggregate grains, b) air pores, c) cement.

Fig. 7. Internal structure of cement concrete series CP: a) contact areas between cement matrix and aggregate grains, b) air pores, c) cement.

a) b) c)

Fig. 8. Internal structure of cement concrete series CA: a) contact areas between cement matrix and aggregate grains, b) air pores, c) cement.

5 Conclusions

According to the conducted laboratory tests, the following conclusions have been reached:
- porphyry aggregate composition contains oxygen, sulfur, aluminum, potassium, carbon, sodium, calcium, magnesium O, S, Al, K, C, Na, Ca, Mg and amphibolite contains O, Fe, Si, Al, Ca, Ti, C, Mg and Na;
- applied aggregates favourably affect the changes of internal structure of concrete composite (contact areas between aggregate grains and cement matrix, crystallization of cement matrix, air void);
- significant influence of the used aggregates on the change of parameters of cement concrete intended for the airfield pavements was proved;
- aggregates porphyry used in concrete mix contribute to the increase of concrete ultimate compressive strength after standard curing period by 20%, concrete tensile splitting strength after standard curing period by 9% and decrease of concrete absorbability by 0.6%;
- aggregates amphibolite used in concrete mix contribute to decrease of concrete ultimate compressive strength after standard curing period by 17%, tensile splitting strength after standard curing period by 8% and decrease of concrete absorbability by 0.8%.

References

1. PN-EN 1097-7 (2008)
2. PN-EN 1097-6 (2013-11)
3. PN-EN 197-1 (2012)
4. PN-EN 1097-2 (2010)
5. PN-EN 12350-6 (2011)
6. PN-EN 12350-2 (2011)
7. PN-EN 12350-7 (2011)
8. PN-EN 12390-1 (2013)
9. PN-EN 12390-2 (2011)
10. NO 17-A204 (2015)
11. PN-EN 12390-7 (2011)
12. PN-EN 12390-3 (2011)
13. PN-EN 12390-6 (2011)
14. M. Linek, IOP Conf. Series: Mat. Sc. Eng. **245**, 032032 (2017)
15. PN-EN 1008 (2004)

Probability calibration of deformation factor for timber roofs in the Polish mountain zones

Tomasz Domański[1,*], and *Kamil* Kmiecik[1]

[1] Cracow University of Technology, Warszawska St. 24, 31-155, Kraków, Poland

Abstract. The resistance parameters of timber material structures decrease with time, depending on the type of load and timber classes. Strength and Modulus of Elasticity reduction effects, referred to as creep-rupture effects, due to long term loading at high stress ratio levels are known for many materials. Timber materials are highly affected by this reduction in strength and deflection with duration of load. Characteristic values of load duration and deformation factors are calibrated by means of using probabilistic methods. The reliability is estimated by means of using representative short- and long-term limit states. Time variant reliability aspects are taken into account using a simple representative limit state with time variant strength and simulation of whole life time load processes. The parameters in these models are fitted by the Maximum Likelihood Methods using the data relevant for Polish structural timber . Based on Polish snow data over 45 years from mountain zones in: Zakopane – Tatra , Świeradów – Karkonosze, Lesko – Bieszczady, the snow load process parameters have been estimated. The reliability is evaluated using representative short – and long –term limit states. The deformation factor k_{def} is obtained using the probabilistic model.

1 Introduction

Timber is a general name which comprises some of the elements of timber materials presented in Fig.1.

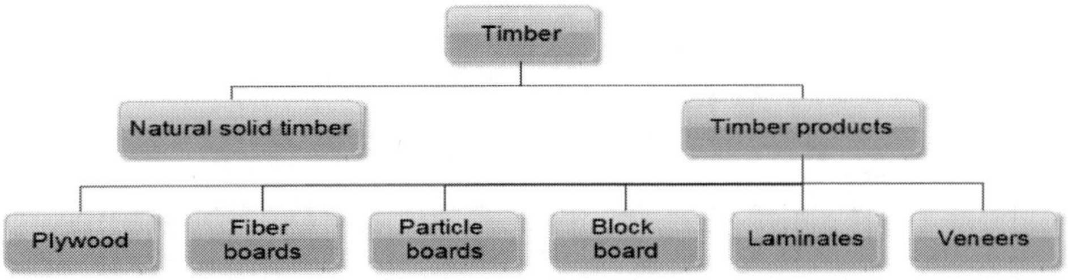

Fig. 1. Timber materials.

* Corresponding author: doman@pk.edu.pl

The scope of work is the infuence of variable loads (snow) to timber resistance parameters. Characteristic values of timber material parameters have to be factored by the deformation/ creep factor k_{def} / k_c. The final mean values of modulus of elasticity, $E_{mean,fin}$, shear modulus $G_{mean,fin}$, and slip modulus, $K_{ser,fin}$ can be obtained as [1]:
for serviceability limit state:

$$E_{mean,fin} / G_{mean,fin} / K_{ser,fin} = \frac{E_{mean} / G_{mean} / K_{ser}}{1+k_{def}}, \tag{1}$$

and for ultimate limit state:

$$E_{mean,fin} / G_{mean,fin} / K_{ser,fin} = \frac{E_{mean} / G_{mean} / K_{ser}}{1+\psi_2 k_{def}}. \tag{2}$$

2 Deformation factor / Creep factor (k_{def}/k_c)

The k_{def} was derived from creep testing in accordance with EN 1156 [2,3]. The creep factor k_c is defined as the ratio of increase in deflection with time under load to the initial elastic deflections. The creep factor k_c in time t is given as follows:

$$k_c = \frac{(a_T - a_0) - (a_1 - a_0)}{a_1 - a_0}. \tag{3}$$

Where: a_T is the total deflection -in millimeters at time t minutes, a_1 is the deflection in millimeters at 1 minute, a_0 is the deflection in millimeters of unloaded test. The illustration of the creep factor for two load scenarios is presented in Fig. 2.

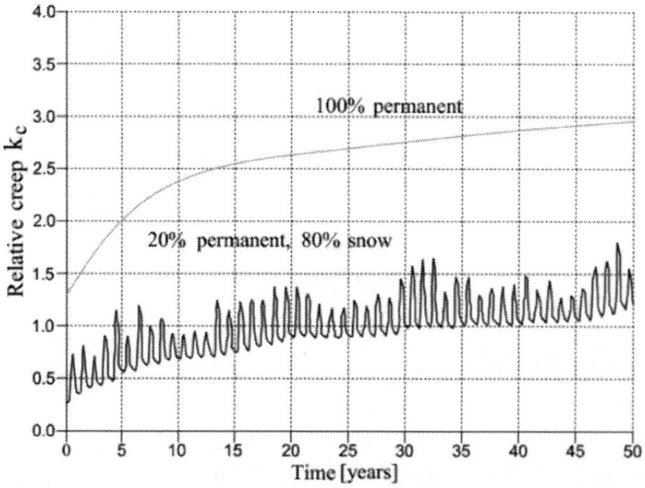

Fig. 2. Relative deflection factor for gluam beam for permanent load and for combination with a random snow load [4, 5].

The strain of wood in time $\varepsilon(t)$ is assumed to consist of the following additive parts: plastic ε_E, normal creep ε_c, mechano-sorptive creep ε_{ms} and free shrinkage strains ε_u [4, 5]:

$$\varepsilon(t) = \varepsilon_E + \varepsilon_c + \varepsilon_{ms} + \varepsilon_u . \tag{4}$$

The deformations are calculated parallel to the grain, a linear strain distribution and Bernoulli hypothesis are assumed.

The **elastic strain** depends on the elastic compliance, which is the function of the current moisture content.

The **normal creep strain**, which is due to the deformation induced by the duration of load is given in a Kelvin series form.

The **mechano-sorptive creep strain**, which is due to additional creep deformation caused by the coupled effect of variable moisture and load duration. In constant stress case this takes the following form:

$$\varepsilon_{ms}(t) = J^{\infty}\sigma\left[1 - \exp\left(-c\int_0^t |du(t')|\right)\right]. \tag{5}$$

Where: J^{∞} is the limit mechano-sorptive compliance, u is the moisture content and c is the material parameter.

The **shrinkage strain**, the reversible moisture induced strain, is assumed to be affected by the total strain. This strain component does not cause creep, but in cyclic humidity conditions it accounts for the oscillation of the creep curve.

$$\varepsilon_u(t) = \{\alpha_s - b\varepsilon(t)\}[u(t) - u(0)]. \tag{6}$$

Where: α_s is the shrinkage coefficient, b is a material parameter and $\varepsilon(t)$ is the current total strain excluding the shrinkage strain.

The output from the analysis consists of the following distributions at time t:

Moister content distribution history $u(x,y,t)$, strain history $\varepsilon(x,y,t)$, stress history $\sigma(x,y,t)$

Service class 2 will be considered in this paper. Strain history will be dependant on the stress history.

3 Stress history

The effect of sustained stress on timber bending strength and timber deflections has been recognized since at least 1850. Wood developed a time strength curve that was incorporated into the wood design procedure. This curve, often referred to as the "Madison curve", is still in use.[6]. During the last decades structural reliability methods have been further developed, improved and adopted and are now at a level, where they are being applied in practical engineering problems.

This knowledge has now reached a level where it enables designers to take into account uncertainties in material properties and actions in assessing the load carrying capacity, serviceability and service life of timber structures and connections. Nowadays, most building codes are based on a probabilistic safety approach. The code formats are deterministic with connections to reliability design achieved by failure probability, partial safety factors and characteristic values. Partial safety factors are calibrated for standard cases against probabilistic analysis for similar cases. The condition for calibration that is the probabilistic analysis and deterministic code should be fulfilled with the same safety requirements. Therefore, the required safety is usually not accomplished by using probabilistic theories in everyday designing process. Timber is a rather complex building material. The timber material characteristics depend on the specific wood species, the geographical location where the wood has been grown and furthermore on the local growing conditions. Timber is an orthotropic material and it consists of high strength grains which are predominantly orientated along the longitudinal axis of a tree. Material characteristics – the ultimate bending stress and the bending stiffness depend on the orientation of the moment axis to the grain direction. Irregularities in regard to grain direction, knots and fissures become highly decisive for load bearing capacity of a timber structural elementIt.It is possible to avoid inregulaties with small test specimen only, with so-called clear wood specimens.

The Gerhards [6] damage accumulation model is taken into consideration. The mechanism leading to the reduction of strength of a timber member under sustained load is a creep rupture. This could arise from propagation of voids in the microstructure of the timber at a stress level, lower than the short-term strength. A number of models of creep rupture, involving a damage state variable (similar to that used in the analysis of metal fatigue), have been proposed to assess damage accumulation in wood structural members subject to loading histories, typically modeled as renewal pulse process. The basic model [7-11] has the following form Eq. (7)

$$\frac{d\alpha}{dt} = F[s(t)], \tag{7}$$

where t – time, α - the damage state variable which ranges from 0 (no damage) to 1 (failure),the function $F(.)$ has two constants that must be determined from test data, and $s(t)$ – the ratio of the applied stress to the failure stress under short-term ramp loading.

Damage models are used for mathematical description of the long term strength reduction as a function of stress level and duration of loading.. The characteristics of damage models are that α is defined as the degree of damage, i.e. $\alpha = 0$ stands for no damage and $\alpha = 1$ stands for total damage or failure.

The damage accumulation model presented by Gerhard, [6] is

$$\frac{d\alpha}{dt} = \exp(-A + B\frac{\sigma}{fo}) . \tag{8}$$

Where A and B are constant, σ is the stress and fo is short term strength of member. Solution of equation (8) is :

$$\frac{\sigma}{fo} = \frac{A}{B} - \ln\frac{10}{B}\log t = a - b\log t, \tag{9}$$

where;
$$a = \frac{A}{B} + \delta \qquad b = \frac{\ln 10}{B}, \tag{10}$$

end δ models the uncertainty related to the formula (7). δ is assumed to be Normal distributed with expected value equal 0 and standard deviation 1.

Assuming constant load and considering f as the residual strength the solution to (9) is as follows:

$$\frac{f}{fo} = \frac{1}{B} ln(1 + (1 - \alpha)(\exp B - 1)). \tag{11}$$

This expression is used in simulating the damage due to load duration.

4 Snow loads

A stochastic model of snow load in the mountain zone in Zakopane – Tatra , Świeradów – Karkonosze, Lesko – Bieszczady is established on the basis of meteorological data [12, 13]. Examples of season snow loads in Zakopane , Świeradów, Lesko are shown in Figs 3, 4, 5.

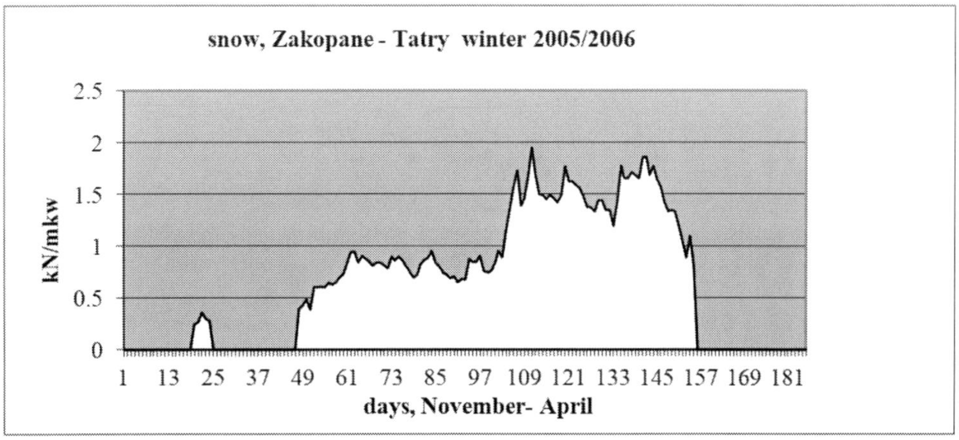

Fig. 3. Snow loads in Zakopane, Poland.

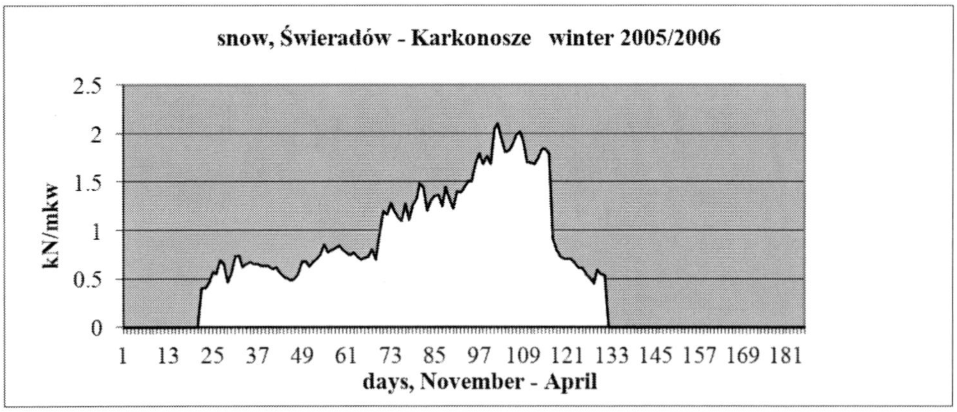

Fig. 4. Snow loads in Świeradów, Poland.

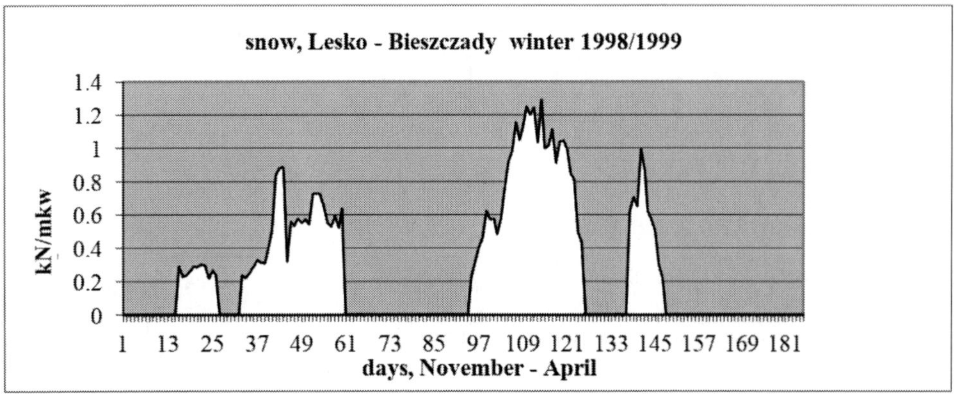

Fig. 5. Snow loads in Lesko, Poland.

A model calibrated against direct measurements of snow load is issued to transform the meteorological data into snow loads model. The load model is illustrated in Fig 6. rectangular and triangular [14].

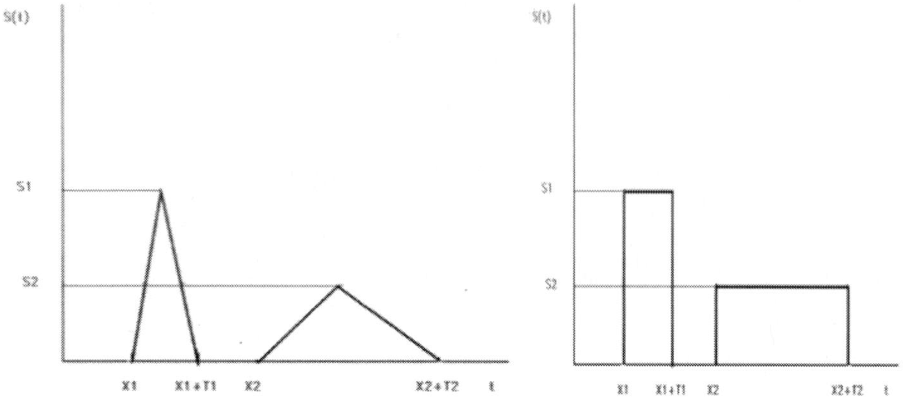

Fig 6. Snow load models – rectangular and triangular.

The snow load on terrain *S(t)*, and duration *T* of snow packages (snow pulses) are modelled as follows:
- The occurrence of snow packages at times *X1, X2,...* is modeled by a Poisson process.
- The duration between snow packages is exponential distributed with expected value $1/\lambda$, where λ is expected number of snow packages per year.
- The magnitude of the maximum snow load P_m in one snow package (snow pulse) is assumed to be Gumbel [15] distributed with expected value μ_p and standard deviation σ_p.

The Table 1. shows probabilistic parameters of snow packets in Zakopane (Z), Świeradów (S) and Lesko (L).

Table 1. Probabilistic parameters of snow packets

	Zakopane	**Świeradów**	**Lesko**
$\mu_P[kN/m^2]$	1.47	1.12	0.80
$\sigma_P[kN/m^2]$	0.49	0.65	0.37
μ_{XT}[days]	65.58	40.97	53.62
λ	1.43	1.85	1.64

5 Deformation factor estimation

In the code format *EC5* load duration effect and deformation factor are represented by the modification factor k_{mod} and k_{def} The following design equation for long-term situation with one variable load can be found in the code [16,17].

$$\frac{zf_k k_{\text{mod}}}{\gamma_m} - [(1-\kappa)\gamma_g G_k + \kappa\gamma_Q Q_k] = 0, \tag{12}$$

$$u_{tar} - u_{fin,G_k(1-\kappa)} + u_{fin,Q_k\kappa} = 0,$$

$$u_{fin,G_k(1-\kappa)} = u_{inst,G_k}(1+k_{def}), \tag{13}$$

$$u_{fin,Q_k\kappa} = u_{inst,Q_k\kappa}(1+\psi_2 k_{def}).$$

Where: u_{tar} is the limiting design value of the relevant serviceability criterion (EC1990), z is the design variable, f_k is the characteristic value for short-term strength (5% quantile), Q_k is the characteristic values of variable – snow load (2% quantile), G_k is the characteristic value of permanent load mean value) γ_G is the partial safety factor for permanent load (=1.35), γ_Q is the partial safety factor for variable load (=1.5), γ_G is the partial safety factor for material parameter for glued materials (=1.2) and the coefficient κ which represents proportion between variable and total loads. The corresponding long-term limit state equations is:

$$g = z(1-\alpha)f_o[(1-\kappa)G + \kappa Q)]. \tag{14}$$

Where α is the damage state variable, f_o is the short term strength, G, Q are the permanent and variable – snow loads.

For short-term situation conditions design and limit state equation the formulas can be described as follows:

$$\frac{zf_k}{\gamma_m} - [(1-\kappa)\gamma_G G_k + \kappa\gamma_Q Q_k) = 0, \tag{15}$$

$$g = zf_o - [(1-\kappa)G + \kappa Q]. \tag{16}$$

Where z is the design parameter, f_o – short term strength, G – permanent load, Q variable load .

The reliability indexes are calculated by simulation according to FORM method on the basis of (14), (16) and stochastic model.

By utilizing Monte Carlo simulation for generating random variables the modification factor $k_{def}$is determined according to the following procedure [16,17].

1. Calculate the short term reliability index βs for a 50 year reference period using the limit state function (14) and the design equation (1). βs is calculated as function of κ by simulation(γ_G and γ_Q are fixed).
2. Calculate the long term reliability index βL for a 50 year reference period using the limit state function (12) and the design equation (11) and k_{def}= 1. βL is calculated as function of κ by simulation (γ_G and γ_Q are fixed).

Then k_{def} factor can be estimated as follows:

$$k_{def} = \frac{\gamma_m^S(\beta s)}{\gamma_m^L(\beta L)}, \tag{17}$$

where $\gamma_m^S(\beta s)$ is the short term partial safety factor as function of β and $\gamma_m^L(\beta L)$ is the short term partial safety factor as function of β.

The results of calculations of deformation factor for timber materials such as: Solid Timber, Medium Density Fibreboard (MDF), Oriented Strand Board (OSB) and Plywood are presented in Fig. 7.

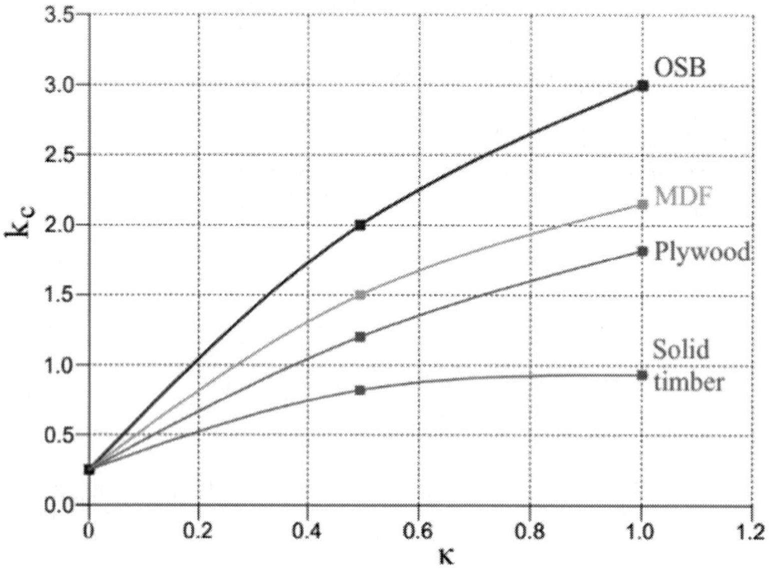

Fig. 7. Deformation/creep factor as function of parameter κ

6 Conclusions

The probabilistic models for the graded timber material properties have been formulated so that they my be readily applied in structural reliability analysis. It has been noted that a significant effect of the time variation of snow impulses-packages on the accumulated damage model has been found. Therefore the observed snow packages are quite different and the triangular and rectangular time variations are included in the present probabilistic calibration of deformation factor. The deformation factor k_{def} value depends signifcantly on the ratio between permanent and variable (snow) loads. More research is needed on the probabilistic resistance parameters of wood planks and another types of timber boards to describe timber deformaction factors.

References

1. EN 1995 (2005)
2. EN 1156 (1997)
3. M. Fan, E. Vahik, The Open Constr. Build. Tech. J., 224-232 (2008)
4. T. Toratti, Engineering Structures **33**, 2958-2961 (2011)

5. T. Toratti, *Service Limit States: Effects of Duration of Load and Moisture on Deformations* (Rapport Cost E24, Florence, 2004)
6. C. Gerhard, Wood Sci. **11**(3), 139-44 (1979)
7. J.D. Barret, R.O. Foschi, Can J. Civil Eng. **5**(40), 505-14 (1978)
8. R.O. Foschi, B.R. Folz, F. Yao, *Reliability-based design of wood structures*. (Rep. no 34. Dep. Of Civil Eng., Univ. of British Columbia, Vancouver, Canada, 1989)
9. B.R. Ellingwood, Prog. Eng. Mater. (2001)
10. J. Kohler, J. Sorensen, M. Faber, Structural Safety **29**, 255-267 (2007)
11. J. Sorensen, S. Svensson, B. Stang, Structural Safety **27**, 153-169 (2005)
12. IMiGW Kraków (2014)
13. IMiGW Wrocław (2014)
14. M. Faber, J. Kohler, J. Sorensen, Structural Safety **26**, 295-309 (2004)
15. EN 1990 (2004)
16. S. Woliński, T. Pytlowany, Archives of Civ.Eng. **LVIII**, 2 (2012)
17. T. Domański, *Wybrane zagadnienia niezawodności konstrukcji drewnianych* (Wydawnictwo Politechniki Krakowskiej, Kraków, 2016).

Material properties and the energy balance in standardized fire testing

Wojciech Węgrzyński[1,*], and *Piotr* Turkowski[1]

[1]Building Research Institute, Filtrowa St. 1, Warszawa 00-611, Poland

Abstract. The origins of standardised fire testing can be traced back to 1870's, and the origin of the standard temperature-time curve to 1917. This approach, based on a 19th-century intuition is still in use up to this day, to design the 21st-century structures. Standardized fire-testing ultimately disregards the conservation of energy in the fire, as in every test the resulting temperature of the test must be the same (precisely as the temp.-time curve). To maintain this, different amount of heat is required in every test, which means that every time a different fire is modelled within the furnace. The differences between furnace fire sizes are ignored in the certification process, but can be interesting for fire researchers to understand how different materials behave in fire conditions. In this paper, Authors explore this topic by investigating the energy balance within the furnace, and comparing different fire tests together.

1 Introduction

The fire testing was a strong force shaping the civil engineering development [1], accelerating the transition from wood and stone based structures into the modern world of concrete, steel and glass. The dissemination of fire-proof solutions with code-based prescriptive requirements resulted in observable change in the damage caused by fires to the buildings. As this approach was successful in limiting the costs of fires, it was not significantly changed since the first fire tests of XIX-th Century, and the standardised fire resistance tests are still the ultimate rating for fire performance of structures.

The main problem with the idea of temperature-time index evaluation of construction products is the lack of inclusion of a feedback loop between the fire and the structure. In the standard fire resistance test, the temperature will always be the same, and will depend only on the time. In reality however, the thermal properties of the building materials will be a determining factor for the resulting temperature field in the fire event [2]. Thermal insulators will require less heat than a concrete block to reach the same temperature, which means, that these two materials will have a significantly different behaviour in a real fire. This will affect not only temperature of the construction element, but also the way how fire develops in the building.

The aim of this paper is to underline this feedback loop, which also can be observed and measured within the standardized fire testing. To present this, the authors will discuss the

* Corresponding author: w.wegrzynski@itb.pl

factors influencing the energy balance within a test furnace, and compare a few different furnace fire resistance test results. A focus is put on the heat (fuel) consumption in the tests. This exercise shows, that standardized fire testing does essentially lead to completely different fire scenarios for the "same" standardized test, dependant on the material being tested.

1.1 Early history of fire testing

The first 100 years of the history of fire testing was thoroughly reviewed by Babrauskas and Williamson, the period between 1880 and 1918 in [3], and 1918 – 1978 in [4]. Some useful information on the historical context of fire testing are also gathered in [5,6].

The goal of the first fire tests was to assess the differences in fire behaviour of popular materials of late 19th Century – wood, masonry, light bricks, terra-cotta, iron beams and concrete. Heavy masonry provided generally good protection against fire, however, it was costly for multi-storey buildings that emerged in the 1870's. Their replacement – terra-cotta arches, lighter bricks and iron beams had variable fire resistance. Observations of the real fires (mainly related to the steel columns) shown, that these solutions were not always reliable, and required additional fireproofing. In some cases, the additional protection was also unreliable, as the terra-cotta protection often felt off shortly once the fire bursted. Dangerous situations were also observed when the firefighters approached the fire with water, as the rapid cooling often led to structure collapse.

Lack of sound basis for fire protection caused an emergence of new testing procedures. Babrauskas [3] reports that the first fire tests date back to 1790's in London, however, these tests preceded the availability of equipment to measure fire temperatures. First "real" fire resistance tests were performed in the 1870's for floors and 1880's for columns. The test procedures of that time were straightforward – they involved building a mock-up furnace compartment, filling it with combustible material (wood or coal), and performing a free burn for a prolonged time (in some cases multiple hours). As high temperature measuring thermocouples and pyrometers were still not known at that time (first high-temperature pyrometer is attributed to Le Chatelier in 1890's [7]), the temperature measurements in the tests were carried through observation of melting of various alloys with a known melting point, or with Seger cones. This only gave qualitative information on what temperature was reached, not the progression of temperature in function of time. The limitations of the temperature measurement in the early tests had a surprisingly profound effect on further, 150-year long history of fire testing.

In the early 1900's, the measurement of temperature through platinum wire devices was possible, Researchers reported [3] temperatures from ~800°C - 1100°C depending on the test, measurement method and type and size of the sample. This temperature range is similar to the temperatures observed today in fully developed fires. In the 1900's and 1910's furnaces were built to replace the ad-hoc testing setups used previously, and more reports on the results of fire testing were described in the literature. Besides floors and columns, more building elements underwent testing – doors, walls, glazing and theatre curtains.

1.2 Standardisation

Babrauskas reports [4] that first standardized methods emerged in Great Britain in 1903, and described three classes of endurance: (a) full, (b) partial and (c) temporary. These classes were related to (a) 120-240 min endurance to 982°C, (b) 90-120 min endurance to 982°C and (c) 45-60 min to 816°C. What is interesting is that this standard also had load-bearing requirements, in the form of a substitute load for floors and geometrical (thickness or area) requirements for partitions and doors. After some changes to the classes, temperature ratings and the standard recommendations this standard was about to be adopted by NFPA in USA (1914). However, this didn't happen and in 1916-17 strong effort was taken to prepare a U.S. fire standard. The effect of this work was the new standard ASTM C19 (later renumbered E119[8]), issued on the Feb 24th, 1917. This was the first standard that proposed the fire load in the furnace with a concept of a temperature-time curve, shown in the Fig. 1. In fact, this curve is still in use today, widely known as the "standard temperature-time curve", or "ISO 834 temperature-time curve" [9]. The origins of this curve can be tracked down to an approximation of a dozen temperature-time curves, obtained in various fire tests. Also, some of the basic ideas of fire testing, such as substitute load with weights are also in use to this day [10].

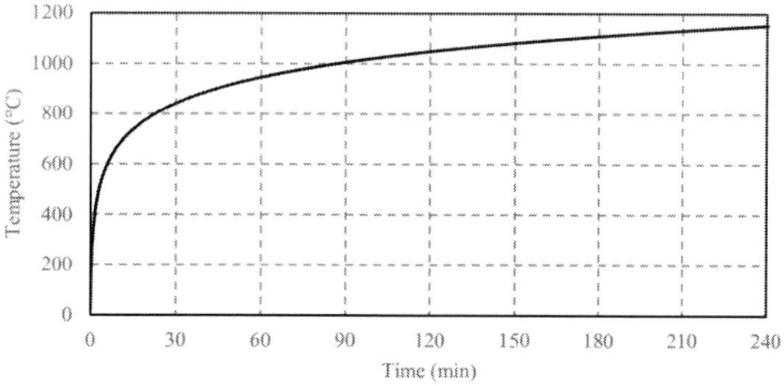

Fig. 1. Standard temperature-time curve.

Analysing the history of the fire testing, and the origins of the standard temperature-time curve, three significant problems can be brought up by modern fire science:

a) Different materials require a different amount of heat, to heat up the surrounding gas to the temperature of the standard curve, which means that different power of the fire is required in each of the tests. This means that the test standardizes the effect of the fire, and not the fire itself (and any positive effect that material may have on the fire in a compartment is generally not included in the test). Early furnaces, for which the standard temperature-time curve was created were a rather good representation of real compartments, as they used natural fuel and were not insulated, while later furnaces (insulated, gas-fuelled) were artificial constructs meant to efficiently perform the tests, and not mimic the natural fires;

b) The growth of the fire in a compartment will depend on its size, height, materials and fuels. It is not possible to average all of the fires into same growth function, as it was done for the standardized fire. Furthermore, the temperatures will be different in various spatial locations in the compartment, while the standard temperature-time curve is used universally for multiple kinds of partitions and structural elements.

c) Temperature estimations in the tests were done in majority through melting-point of alloys, or new at that time – platinum-rhodium thermocouples, which may yield a significant error in the measurements. These values were also not corrected for radiation. The 926°C temperature point was chosen arbitrarily based on a limited amount of experiments (furnace tests, not real compartment fires).

2 Energy balance in a furnace

The modern knowledge of the fire phenomena is profoundly different from that of XIX-th century. We understand that the heat and mass transfer in the compartment is what dictates the combustion process, its length and severity, while the temperature is only its result. For a very small compartment, it is easy to write a generalized form of energy balance equation (1), Fig. 2:

$$\dot{q}_c = \dot{q}_L + \dot{q}_W + \dot{q}_R + \dot{q}_B \tag{1}$$

where q_c is the heat flux released by the fire, q_L is the flux lost through the openings (convection), q_W is the heat flux lost to the walls, q_R is the heat flux lost through radiation out of the compartment and q_B is the flux lost to heat the fuel.

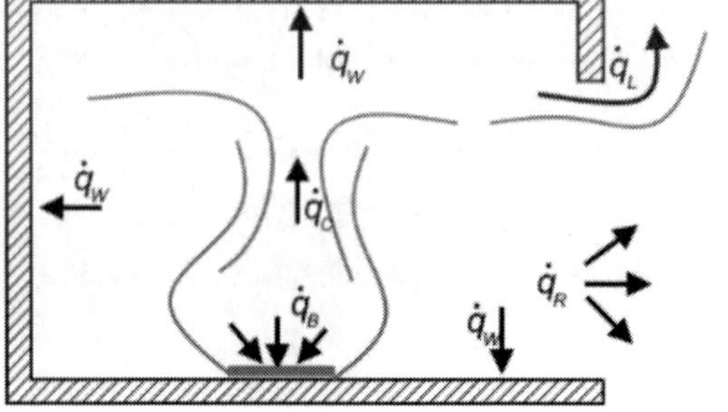

Fig. 2. Heat loses in a fire in a small compartment.

In the fire testing, the losses through the openings are only due to the ventilation of the furnace, which is used to maintain a constant pressure in the furnace. In an idealized case, the mass removed from the furnace will be equal to the mass of the products of the stoichiometric combustion in the furnace (2).

$$CH_4 + 2O_2 + 7.52N_2 \rightarrow CO_2 + 2H_2O + 7.52N_2 \tag{2}$$

Assuming perfect reaction kinetics, one mole (approx. 16 g) of methane gas burnt in the furnace will produce 0.889 MJ of energy, and this reaction will yield 290,56 g of products. For ease of further considerations, it may be assumed that for every 1 MW of energy emitted in the fire, 327 g of products have to be removed from the furnace. One additional simplification taken here is the assumption that the mixture introduced to the furnace is in a stoichiometric ratio, while in reality, it will be fuel lean (the exact ratio is adjusted on the fly by the furnace automation system). The approximated specific heat of the mixture of the products was calculated in function of the temperature, and based on the conservation of energy, the energy required to heat the products of reaction releasing 1 MJ of energy, from their initial temperature of 15°C is shown in Fig. 3. To heat up the products of reaction (2) to 1000°C approximately 440 kJ will be used. As the temperatures in the furnace are close to 850°C – 1100°C for most of the standardized test, this means that out of every 1 MW introduced to the furnace, 35 to 45% is inherently lost through ventilation to maintain the pressure balance in the furnace.

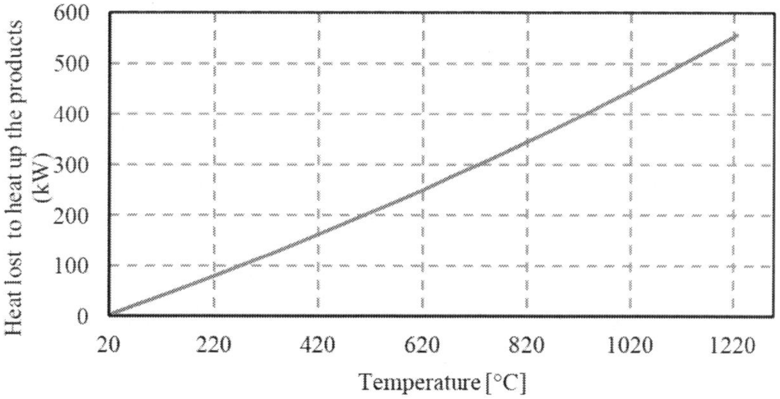

Fig. 3. Heat required to heat up the products of reaction (2), which released 1,00 MJ of energy to a given temperature (heat lost with the ventilation of the furnace at given temperature calculated with average specific heat of products in function of temperature).

The remaining 55 - 65% of the heat generated in the furnace is transferred through other ways – primarily to heat up the furnace walls, test sample and the mounting structure. For practical considerations, the losses to the furnace walls can be omitted, due to their insulation. The radiation and heat loss to the source of fire can be omitted as well, as the furnace is a closed system. The dominant factor for the heat transfer in the furnace is the heat transfer through the sample, which will be determined by the thermal properties of the material being tested. This means, that the fire tests of different materials must employ a different heat release rate, to make up for the different losses in each test. As only the temperature-time relation is conserved between the fire tests, each of them goes through a completely different fire. Yet, the rating obtained by the elements, is the same.

An additional factor to consider in solving the energy balance of the furnace, is the combustibility of the building material used in the fire test [11,12]. If the material can burn inside of the furnace, this will change the heat balance, as the additional source of heat will be introduced, Fig. 4 - 5. In such an event, the furnace automation will adapt to the new condition inside, by limiting fuel and increasing the ventilation rate, to maintain the prescribed temperature-time curve. This means, that if combustible material is tested inside the furnace, the system limits the fire, to maintain the prescribed temperature-time curve.

a) b)

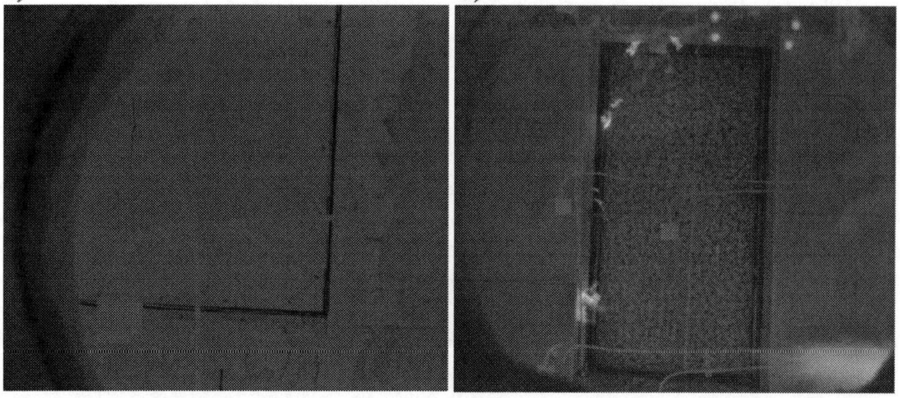

Fig. 4. View of test samples (incombustible) inside of a furnace – (a) gypsum plasterboard, (b) steel doors with interior insulation.

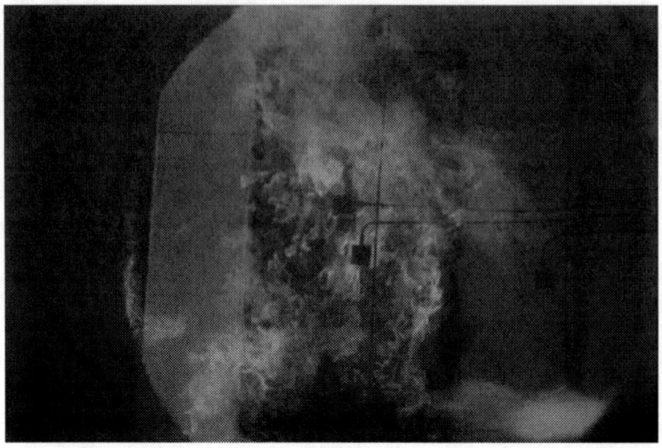

Fig. 5. View of test sample inside of a furnace – wooden doors (combustible).

3 Comparison of chosen furnace tests

Given the following assumptions:

a) the temperature in the furnace is defined as a function of time, T = f(t);

b) the furnace automation chooses the amount of fuel introduced to the furnace to follow the temperature-time curve as closely as possible;

c) the due to furnace ventilation are a function of the temperature and the pressure inside the furnace;

the amount of heat introduced to the furnace during every test will depend mainly on the material properties of the tested sample. This means that the "fire" of the test is different for every material. Comparison of the temperature-time curves for chosen furnace tests for the standard temperature-time curve, and corresponding heat release in the furnace is shown in Fig. 6 – 7. The reader must be advised, that the reading of the heat release rate in the furnace is used only as an informative feature for the furnace operator, and is recorded with a low resolution, with ~0.1 MW uncertainty.

The materials presented in this comparison are (a) gypsum plasterboard wall, (b) wooden doors (1.8 x 2 m) in AAC wall, (c) glazed aluminium doors in AAC wall and (d) array of approx. 30 different penetration seals of various sizes in AAC wall. All of these materials were tested for the same temp.-time curve (ISO 834), and all of them were tested on the exact same furnace.

As shown on Fig. 6, the temperature-time was maintained by the furnace close to the ISO 834, however, some differences are visible for sample (b) – wooden doors. This is due to the combustion of the sample within the furnace (Fig. 5). What is more interesting, is that the sample (b) required the least amount of heat from all of the presented. When the peak temperature was reached in this test, the furnace did not supply any fuel, and was actually cooling down the sample. In case of penetration seals, a large initial peak is observed, much higher than for the other materials. The largest amount of heat was consumed by the glazed door test (c), and for this experiment the heat release rate within the chamber was close to uniform during the whole duration of the experiment. Similar effect, but requiring lower amount of heat, was observed for the plasterboard wall (a).

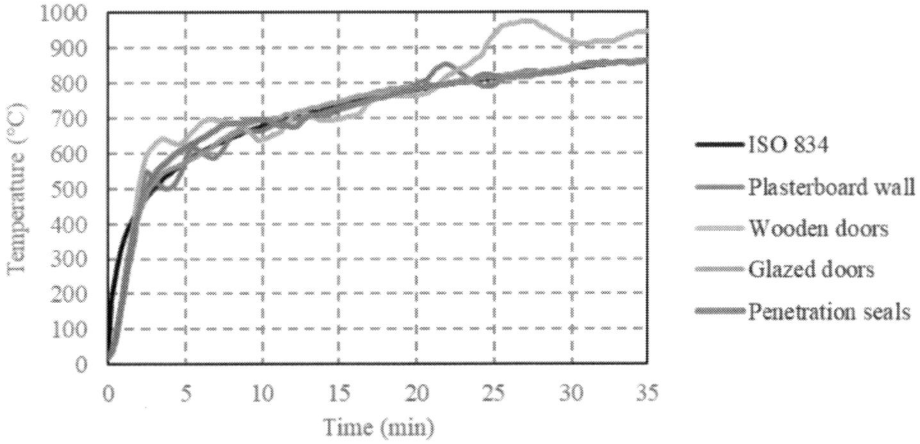

Fig. 6. Mean temperature in the furnace measured in chosen fire resistance tests of walls and doors.

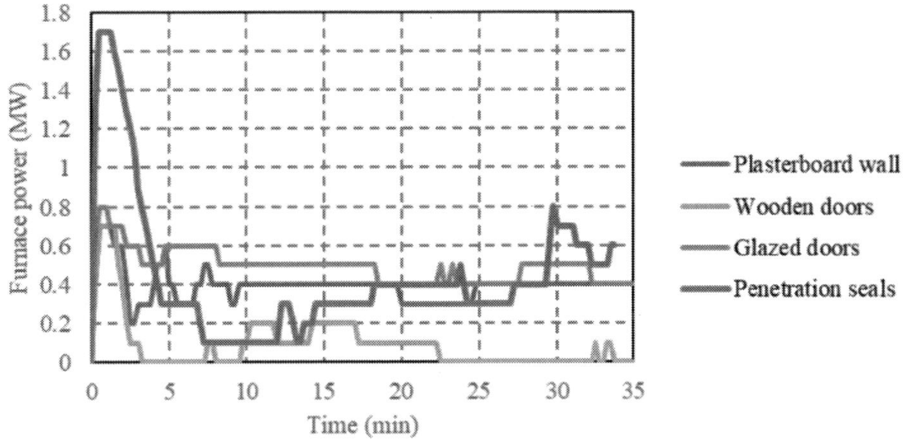

Fig. 7. Heat release rate in the furnace, for the tests presented in Fig. 6.

4 Conclusions

The goal of standardized fire testing was to provide an uniform benchmark for various building materials, that would allow to measure their resistance to the high temperatures observed in real fires. However, due to limited testing capabilities of late XIX-th century laboratories, the only way to quantify the fire was through the temperature measurements. This led to emergence of a test, that standardized the effects of a fire, and not the fire itself. While omitted by the early fire-researchers, we know that the feedback loop between building materials and fires in real buildings is profound. This feedback loop was demonstrated in this paper through a discussion of the energy balance within a furnace, and presentation of real fire tests results for various construction materials.

The materials used in the building will change the results of the fire, and may even change its behaviour. Now they are chosen solely based on the rating index they receive, which promotes materials that either contains structural water (such as gypsum, concrete) or incombustible materials that have low thermal conductivity (thermal insulations like stone mineral wool). While being perfect against the standardized test, these materials may not necessarily be the best solutions for a real case scenario, as shown in literature [2]. More to that, behaviour of combustible materials, such as structural wood, is not well represented within the standard furnace tests, as their combustion during the test does significantly influence the energy balance within the furnace.

This short paper presents another proof of how the standard fire cannot be taken as a representative scenario for real fires. Yet, it is still the driving force in the research and development of new fire-protection materials and techniques. The fire science community is aware of this problem, and have voiced these issues multiple times, however, for the building industry the paradigm shift from 19th century intuition into 21st century science has not yet occurred.

References

1. W. Węgrzyński, P. Sulik, Bull. Polish Acad. Sci. Tech. Sci. **64**, 719–730 (2016)
2. D.D. Drysdale, *An Introduction to Fire Dynamics, 3rd Edition* (Wiley, 2011)
3. V. Babrauskas, R.B. Williamson, Fire Technol. **14**, 304–316 (1978)
4. V. Babrauskas, R.B. Williamson, Fire Technol. **14**, 304–316 (1978)
5. H.E. Nelson, *Conference on Fire Safety Design in the 21st Century* (1991)
6. I.S.F. Marshall, *History of Fire and Fire Codes* (Iowa Dep. Public Work, 116)
7. B.L.B. Hunt, *The Early History of the Thermocouple,* 23–28 (1964)
8. ASTM E119 (1917)
9. ISO 834-1 (1999)
10. G. Kimbar, P. Roszkowski, P. Sulik, AIP Conf. Proc. **1922**, 80009 (2018)
11. J. Fangrat, Fire Mater. **41**, 99–110 (2017)
12. J. Fangrat, Bull. Polish Acad. Sci. Tech. Sci. **64**, (2016)

MATEC Web of Conferences **163**, 07005 (2018) https://doi.org/10.1051/matecconf/201816307005
MATBUD'2018

Rational application of hot finished rectangular hollow sections in steel structures

Krzysztof Kuchta[1,*], and *Izabela* Tylek[1]

[1]Cracow University of Technology, Faculty of Civil Engineering, ul. Warszawska 24, 31-155 Kraków, Poland

Abstract. Steel hollow sections are manufactured as both welded and seamless elements. The influence of manufacturing technology causes tubes made from the same carbon-manganese steel to differ significantly in terms of their mechanical properties. The cold-formed welded tubes are produced from a steel strip that is roll formed and welded; both technological processes take place at room temperature. This profile type is characterised by less favourable mechanical properties and by much higher inhomogeneity within the cross-section than the seamless tube. However, due to the lower cost of manufacturing, welded tubes are much more often used in building structures than seamless ones. After thermomechanical treatment, the welded tubes, which are referred to as "hot finished", have almost the same mechanical properties as seamless tubes. However, hot finished sections are more expensive than welded ones; therefore if they are to be applied in the structure in a reasonable way, higher unit cost should translate into an appropriate increase of the load bearing capacity. The aim of the paper is to indicate the application areas of hot finished rectangular hollow sections in which their use is economically justified.

1 Introduction

Steel hollow sections are characterised by very favourable weight distribution in the cross section, resulting in a large ratio of radius of inertia and torsional stiffness to their self-weight. This type of profiles is used in contemporary steel structures not only due to low self-weight but also to a number of other favourable technical and operational features, such as low drag coefficient or smaller painting area.

Despite the fact that designers started using riveted steel tubes for truss bridges as early as in the second half of the 19th century, the application of steel hollow sections in other types of steel bearing structures was initially quite limited because of their high manufacturing costs, relatively poor quality, high labour consumption required for the execution of connections and lack of knowledge concerning the design of direct joints, i.e. tube to tube connections without a gusset plate. Interestingly, progress in this area was not initiated by the building industry but by the machinery and mining industries. The popularisation of steam engines and the beginning of the extraction and transport of oil on an industrial scale caused a strong increase in demand for steel line pipes. This, in turn, led to the need to find a

* Corresponding author: kkuchta@pk.edu.pl

technology that would enable relatively cheap mass production of pipes with low self-weight and high strength, allowing for the use of larger pipe diameter and higher operational pressure. The first method of industrial manufacturing of steel pipes was patented in 1825 by James Whitehouse. In this method, welded pipes were produced from steel stripe sections heated in a furnace, forged over a mandrel, then heated again and drawn through the drawing die where the stripe edges were pressure welded. However, the pressure weld had low strength and the method itself was too costly for mass production.

A significant improvement in pipe strength was achieved due to the development of technology for production of hot-rolled seamless steel pipes, which was firstly patented in 1886 by the Mannesmann brothers. This technology was modified and improved by several engineers and inventors: Stiefel, Assel, Diescher, Ehrhardt, and others. As a result of the continuous development of steel manufacturing and joining techniques, operational pressure in oil pipelines increased 60 times over 90 years: from 2 bars in 1910 to 120 bars in 2000 [1] The total length of pipelines transporting oil and gas is currently estimated as 2 000 000 km [2].

Despite the initial failures, the production of welded pipes was continued. The manufacturing technology was developed because the process was, and still is, much less energy-consuming and cheaper than the production of seamless pipes. Improvement of the production methods had two main goals: to increase the strength of the seam and to enhance production efficiency. Welding seam technology was developed in two ways: using electric resistance welding and using electric arc welding.

In the early 1950s, the production of square and rectangular hollow sections began [3]. These sections, as opposed to circular hollow sections, were used mainly as structural elements. Due to the flat cutting line and flat wall surface, they facilitate the execution of welded and bolted joints. In 1962 the CIDECT (Comité International pour le Développementet l'Etude de la Construction Tubulaire), an international association, was founded to promote hollow steel section usage in construction industry by initiating and participating in appropriate research projects. Contemporary European Standard provisions for steel hollow sections are based on the results of CIDECT works.

The manufacturing technology of seamless rectangular hollow sections was also developed. These profiles are characterised by much smaller values of residual stress and higher homogeneity of mechanical properties in the cross section.

The development of manufacturing technology results in the elaboration of the technique for the second hot-rolling of the welded hollow sections. As a result of this additional treatment, welded sections have practically the same parameters as seamless ones. The welded hollow sections after thermomechanical treatment and the seamless hollow sections are both called "hot finished sections". The production of both types of profiles is regulated by the provisions of the PN-EN 10210 Standard [4]. These profiles have better technological properties than profiles that are cold formed at ambient temperature pursuant to PN-EN 10219 [5]. However, they are more expensive, so their application in the structure should be reasonably justified by the designer.

2 Material properties

Hot finished and cold formed hollow sections are made of carbon-manganese non-alloy quality steels, with minimum yield strength R_{eH} in the range of 235 MPa to 355 MPa. Hot finished profiles can also be produced from fine grain steels, normalized or normalized rolled to increase impact energy at low temperatures. The reduction in grain size also increases steel yield strength. The introduction of the following alloy additions: Ni, Cr, Cu, V, Mo, Nb allows it to achieve two additional strength grades (420 MPa and 460 MPa) of fine grain alloy special steels, which are included in PN-EN 1993-1-1 [6]. Hot finished tubes are formed at 830-900°C, which corresponds to the temperatures of the full annealing process, during which steel microstructure changes causing the reduction of its hardness and increase in ductility and toughness.

As far as cold formed profiles are concerned, two types of manufacturing are allowed: normalizing or normalizing rolling and thermomechanical rolling. Both processes provide the same minimum impact energy at low temperatures, required by appropriate Standards [4-5], however thermomechanically rolled steels are characterised by 10-15% lower carbon equivalent values. It should also be noted that stress relieving annealing, sometimes carried out after the cold forming, which is normally conducted at a temperature of 400-450°C does not increases steel toughness [7].

3 Geometrical and technological properties

Geometrical dimensions of square hot finished hollow sections manufactured in compliance with PN-EN 10210-2 [8] vary from 40 to 400 mm for side dimensions and from 2.6 to 20 mm for wall thickness. As far as rectangular hollow sections are concerned, the side dimensions are slightly different: from 50x30 to 500x300 but the range of wall thickness is the same. The side dimensions of the cold formed square hollow sections listed in PN-EN 10219-2 [9] range from 20 to 400 mm with the wall thickness of 2.0 to 16.0 mm, while rectangular hollow sections have side dimensions from 40 x 20 to 400 x 300 with the thickness from 2.0 to 16.0 mm. However, some European manufacturers offer hot finished sections with dimensions that exceed the values given in [8-9]: up to 700x700 mm for the side dimension and the wall thickness up to 40 mm [10].

Despite the same side size and wall thickness for hot finished and cold formed hollow sections, their sectional parameters differ due to the slightly different corner shaping (Fig. 1).

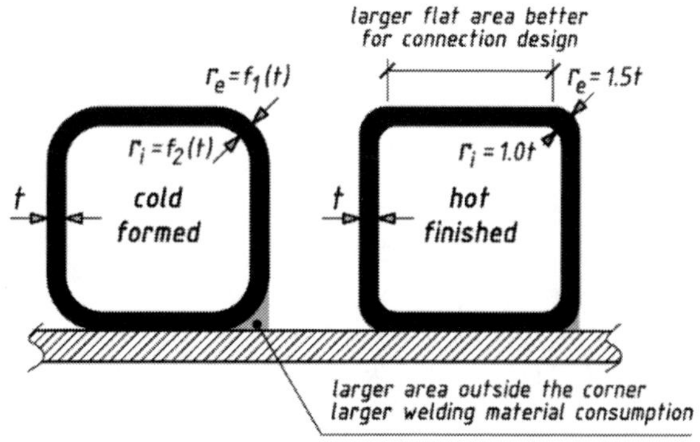

Fig. 1. Comparison of the cross section geometry for cold formed and hot finished hollow sections.

Hot finished profiles have smaller corner radii and the thickness of the corner is not uniform. Their external corner radius r_e is $1.5t$, where t is wall thickness, and the internal corner radius r_i is $1.0t$. Additionally, the surfaces of the corner and the flat part of the wall do not have the same tangent line at the place of their connection. For cold formed hollow sections, the corner radius depends on the wall thickness t and equals: $r_e = 2.0t$ and $r_i = 1.0t$ for $t < 6$ mm, $r_e = 2.5t$ and $r_i = 1.5t$ for 6 mm $< t \leq 10$ mm, $r_e = 3.0t$ and $r_i = 2.0t$ for $t > 10$ mm. Comparing hot finished and cold formed hollow sections with the same side dimension and wall thickness, one can notice that hot finished profiles have slightly larger (by about a few percent) self-weight and more favourable properties of the cross section: its area, moment of inertia, radius of gyration, elastic and plastic section moduli.

As for cold formed sections that were not normalized after cold forming, the value of corner radius significantly affects the technological conditions for weld execution. When the internal corner radius to wall thickness ratio r_i/t is lower than the minimum value provided in Table 4.2 of PN-EN 1993-1-8 [11] then, to avoid brittle cracking, welding is permitted only in zones $5t$ away from the beginning of the flat part of the external surface. Table 1 below compares values of the r_i/t ratio for hot finished and cold formed hollow sections manufactured according to [8-9] with the provisions of [11]. The wall thicknesses for non-normalized cold formed profiles that are not allowed to be welded in the corner zone were indicated. This restriction is especially bothersome for profiles with the smallest wall thickness in the series, due to significant reduction of the permitted welding zone. In extreme cases, welding may even be impossible.

There are no similar welding restrictions that apply to hot finished sections, even if they do not fulfil the conditions given in Table 1, because hot-rolled sections do not have a cold-formed zone and the hot finished products made from cold formed sections are subject to metallurgical conditions equivalent to those for hot-rolled profiles.

Table 1. Nominal and minimum values of r_i/t for rectangular hollow sections.

Wall thickness [mm]	r_i/t **[-] PN-EN 10210-2**	r_i/t **[-] PN-EN 10219-2**	**min.** r_i/t **[-] PN-EN 1993-1-8**
$t \leq 4$	1.0	1.0	1.0
$4 < t \leq 6$		1.0	1.5
$6 < t \leq 8$		1.5	1.5
$8 < t \leq 10$		1.5	2.0
$10 < t \leq 12$		2.0	2.0
$12 < t \leq 24$		2.0	3.0

According to the amendment to PN-EN 1993-1-8 [11], published in September 2009, welding in the corner zones of cold formed rectangular hollow sections may be allowed for profiles that do not fulfil the requirements given in Table 1, provided that: wall thickness $t \leq 12.5$ mm, steel is fully-killed, sub-grades are limited to: J2H, K2H, MH, MLH, NH, NLH, with the content of carbon, phosphorus and sulphur following restrictions: C ≤ 18%, P ≤ 0.020% and S ≤ 0.012%. In most cases, these limits are lower than the qualification requirements for steel grades specified in PN-EN 10219-1 [12] and none of the steel grades fails to fulfil them all. Consequently, welding in the profile corner zone cannot be allowed only on the basis of thickness and steel grade verification – additional analyses of the very restricted requirements for steel chemical composition are necessary.

Larger thickness range of hot finished profiles can be an advantage in case of architecturally exposed steel structures, where the aesthetic aspects require a constant size of the elements and at the same time the economic reasons require the reduction of element self-weight in order to adapt to the variation of the internal forces (Fig. 2). Changes of profile wall thickness are hidden inside the element, which allows to keep its external surface aesthetic and smooth.

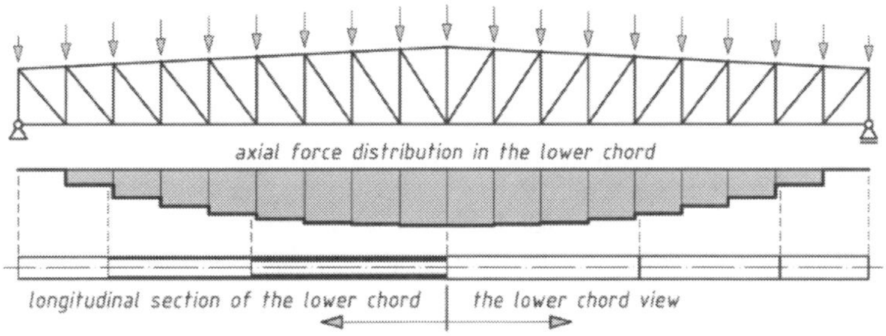

Fig. 2. Realisation of the idea of the cross-section adaptation to axial force distribution.

Larger wall thickness of hot finished hollow sections may also be relevant in the case of steel supporting structures of façades and glass roofs, where increasing of the element optical slenderness has aesthetic and practical (more daylight) advantages. Due to smaller internal corner radii the longitudinal fillet welds require less weld deposit and therefore are cheaper (Fig. 1). At the same time, the lower amount of heat transferred to the base metal reduces the residual welding stresses.

Seamless hot finished hollow sections without longitudinal weld – which is a structural notch – have also higher fatigue strength than cold formed profiles. Pursuant to PN-EN 1993-1-9 [13] hot-rolled sections have the highest detail category $\Delta\sigma_c$ = 160 N/mm^2. Fatigue strength for cold formed hollow sections should be determined using lower detail categories [13]: $\Delta\sigma_c$ = 140 N/mm^2 for the profile wall thickness $t \leq 12.5$ mm and $\Delta\sigma_c$ = 120 N/mm^2 for $t > 12.5$ mm.

The unique properties of hot finished hollow sections allow to design innovative steel structures such as the heavy dynamically loaded crane runway beam shown in Fig. 3 [14].

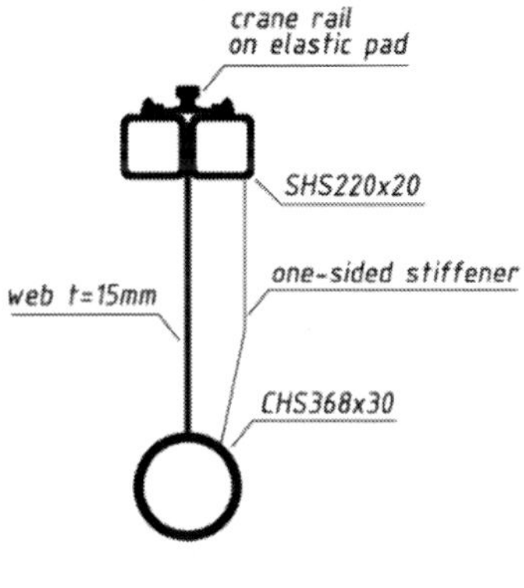

Fig. 3. Innovative crane runway beam with thick-walled hollow section flanges without bracing of lower flange.

4 Residual stresses and their influence on the resistance of axially loaded elements

The manufacturing technologies for hot finished and cold formed hollow sections are completely different, so, consequently, distributions and values of residual stresses also differ significantly. Figure 4 shows the distribution of longitudinal residual stresses on the external surface of the cold formed sections: without (Fig. 4a) and after (Fig. 4b) the hot finishing process [15]. One may notice that heat treatment results in almost two times smaller residual stresses. Apart from that, the differences between the global maximum and minimum residual stresses are also reduced. The maximum values of the stresses, obtained at weld seam axis, are: 414.9 MPa for cold formed and 196.7 MPa for hot finished section. These values correspond to 79.8% and 41.7% of the actual steel yield strength, respectively.

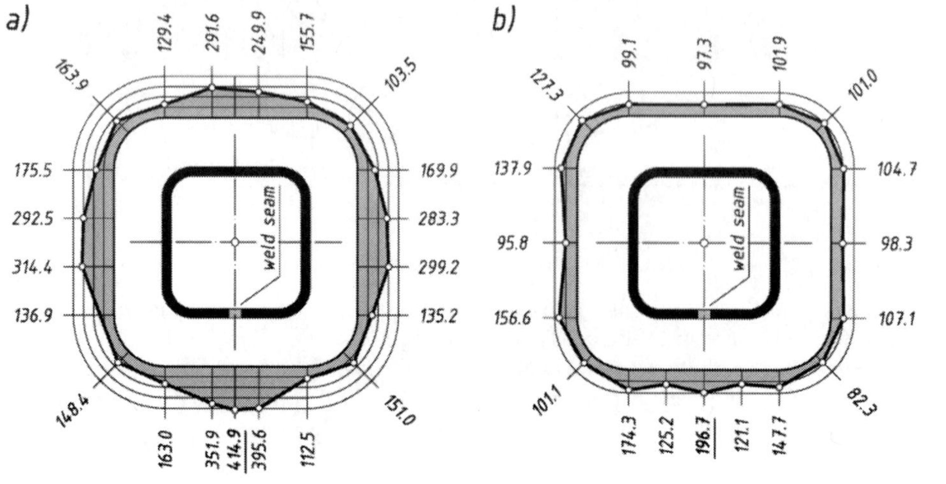

Fig. 4. Residual stresses distribution in [MPa] for: a) cold formed and b) hot finished hollow sections.

Such more favourable distribution of residual stresses results in higher resistance of axially loaded elements made of hot finished hollow sections. The differences between residual stresses distribution in these two types of hollow sections were taken into account in the provisions of PN-EN 1993-1-1 [6]. In the design procedure of axially loaded elements, the cold formed sections are associated with the buckling curve *c* and the imperfection factor $\alpha = 0.49$, while for hot finished profiles there are two buckling curves: curve a_0 with the imperfection factor $\alpha = 0.13$ for steel grade S460 and curve *a* with the imperfection factor $\alpha = 0.21$ for other steel grades. Relative differences between the values of reduction factor (buckling coefficient) for hot finished χ_{HF} and cold formed χ_{CF} hollow sections

$$\delta\chi = (\chi_{HF} - \chi_{CF}) / \chi_{CF} \tag{1}$$

are presented in Fig. 5.

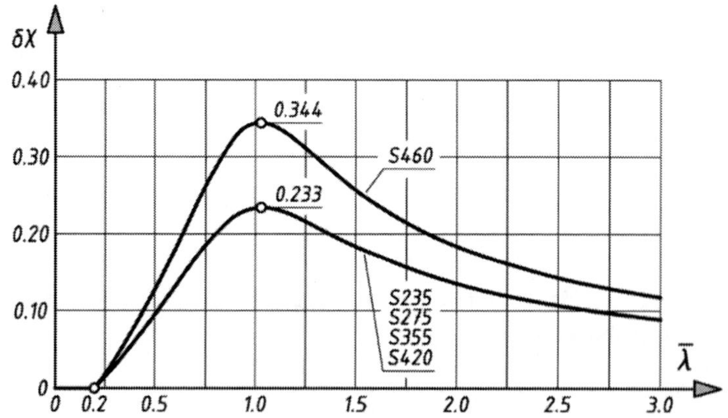

Fig. 5. Relative differences between reduction factor χ for hot finished and cold formed hollow sections.

The lower the value of buckling coefficient χ, the larger the reduction of the design buckling resistance of the compression member.

The largest relative differences in buckling coefficient χ are achieved in the range of medium non-dimensional slenderness. The functions presented in Fig. 5 reach their extremes for non-dimensional slenderness of approximately 1.0. Maximum increment of the buckling coefficient value steel grades S235 to S420 is 23.3% and for steel grade S460 – 34.4%.

5 Summary

Contemporary methods of manufacturing steel rectangular hollow sections enable their production as both hot rolled and cold formed elements. Additional thermomechanical treatment of cold formed profiles causes changes in the microstructure of steel, which becomes more homogeneous. This improves steel toughness and significantly decreases residual stresses. After this process, the metallurgical properties and residual stresses of cold formed sections are practically equivalent to those for hot finished sections. Therefore, they are covered by the same EN 10210 Standard [4]. The increase in the carrying capacity of axially loaded elements, resulting from lower residual stresses and improved mass distribution in cross section for the hot finished sections, can reach up to about 40% and depends on the steel grade. These values cannot, however, fully compensate for the increased unit price of hot finished sections, which depends on market fluctuations, profile size or the tonnage order and may vary in the range from 5% to 54% [16]. However, the decision-making criterion may be the favourable technological properties of hot finished sections, mainly the lack of restrictions for directly (tube to tube) welded connections that exist in the case of cold formed sections with wall thickness t in the ranges: $\langle 4.0$ mm, 6.0 mm$\rangle$, $\langle 8.0$ mm, 10.0 mm$\rangle$ and $\langle 12.0$ mm, 24.0 mm$\rangle$. An additional advantage of hot finished profiles made from cold formed welded tubes is the high quality of external surface finishing, which may be important e.g. in the case of architecturally exposed steel structures of such public buildings as: airport terminals, railway stations, shopping centres, etc.

References

1. P. Hopkins, *Proceedings of the 5th Asian Pacific IIW International Congress*, 1-27 (2007)
2. Y.F. Cheng, *Stress corrosion cracking of pipelines* (John Wiley & Sons, 2013)
3. J. Wardenier, *Hollow Sections in Structural Applications* (CIDECT, 2001)
4. PN-EN 10210
5. PN-EN 10219
6. PN-EN 1993-1-1
7. N. Kosteski, J.A. Packer, R.S. Puthli, J. Struct. Eng. **131**, 279-286 (2005)
8. PN-EN 10210-2
9. PN-EN 10219-2
10. D. Dutta, J. Wardenier, N. Yeomans, K. Sakae, Ö. Bucak, J. A. Packer, *Design Guide for fabrication, assembly and erection of hollow section structures* (CIDECT, TÜV-Verlag GmbH, 1998)
11. PN-EN 1993-1-8
12. PN-EN 10219-1
13. PN-EN 1993-1-9
14. C. Dittmann, S. Herion, O. Josat, P. Sunder, Stahlbau **77**, 775-780 (2008)
15. X.-Z. Zhang, S. Liu, S.-P. Chiew, Case Studies in Struct. Eng. **6**, 115-129 (2016)
16. R. Puthli, J. A. Steel Constr. **6**, 150-157 (2013)

MATEC Web of Conferences **163**, 08001 (2018) https://doi.org/10.1051/matecconf/201816308001
MATBUD'2018

Energy and hygrothermal performance of built-in mineral wool thermal insulations

Balázs Nagy[1,*], and *Tamás K.* Simon[1]

[1]Budapest University of Technology and Economics, Faculty of Civil Engineering, Department of Construction Materials and Technologies, Műegyetem rkp. 3. K.I.85., 1111 Budapest, Hungary

Abstract. The paper analyses the monitoring measurement results on energy performance and hygrothermal behaviour of built-in rock and glass mineral wool thermal insulations in an energy efficiently refurbished typical single-family dwelling house in Hungary. The experimental monitoring measurements included an external weather station; internal air condition measurements; temperature, relative humidity and heat flux sensors in the layers of the facade and attic. An energy meter was installed into the heating systems. Apart from the insulated one, the energy consumption of an uninsulated but identical family house close by to the refurbished one was monitored as control. In the paper, apart from comparing the energy performance of the monitored buildings, the conjugated heat and moisture behaviour of the built-in mineral wool insulations were examined and compared to their design values and conditions. The temperature and moisture correction factors were evaluated. The paper presents laboratory measurements on the thermal conductivity of built-in and etalon mineral wool insulation samples also. The circumstances were set according to the experienced built-in conditions in the monitored buildings and compares the energy and hygrothermal performance of new and aged insulations.

1 Introduction

In 2013, a Hungarian mineral wool manufacturer and distributor company initiated a new experiment and publicity campaign [1]. The primary aim was to demonstrate the energy consumption reduction which resulted by intensive thermal insulation in a clear and easy to understand manner. For this purpose, two families and their two detached family homes were selected by a public competition. The two buildings were almost identical in size, layout and construction, and were located within a few kilometres from each other in the Northern Alföld (Lowland) region of Hungary. The first building, henceforth building 'A', in the town of Hajdúnánás was insulated at the company's expense, while building 'B' in the nearby town of Hajdúdorog was left uninsulated as a control building (Fig. 1). A monitoring system was planned and set up to measure and record the relevant quantities. An additional research was proposed to comparatively measure the built-in hygrothermal

* Corresponding author: nagy.balazs@epito.bme.hu

performance of the stone mineral wool external thermal insulation system (ETICS) on the facade wall and the glass mineral wool attic insulation, which were used in building 'A'.

2 Methodology

2.1 Measurement of the boundary conditions

A weather station mounted on the unused chimney of building 'A', at every 15 minutes measured the external boundary conditions like: dry-bulb temperature, relative humidity (RH), atmospheric pressure, wind speed and direction, as well as the precipitation above the roof. In both buildings, internal boundary conditions like: dry-bulb temperature and relative humidity were recorded in all of the rooms, and the families were instructed to follow similar heating and ventilation patterns for the duration of the project.

2.2 Measurement of the attic and walls

The attic and facade wall insulation variations and the process of sensor installations are presented in Fig. 1 and 2. The measurements lasted for a whole heating season, from the 27th of September in 2013 until 31st of March in 2014. At the facade wall a generally used construction method was applied (only patches of adhesive mortar was used instead of continuous mortar strips). This study shows the adverse effects of the lack of a proper adhesive mortar layer between the wall and the insulation, i.e. if a thin layer of air is formed behind the insulation. The measured facade wall was oriented to south-east direction, and measured across its cross-section with sensors in between each layer and one sensor in the middle of the 20 cm thick insulation as well. As in case of many other existing detached family houses, the insulation of the slab between the interior and the attic space was made without refurbishment of the roof and so the lack of addition of a reflecting foil underneath the roofing tiles, the humidity (as well as the wind) was able to penetrate into the attics. Therefore, vapour permeable underlay was laid on the top of the glass mineral wool insulating layer, to protect it. Sensors were installed into each accessible layer of the insulated attic slab and in the middle of the 25 cm thick insulation as well. The heat flow densities were measured with an Omega HFS-3 thin plate heat flux meter and a 16-bit ADC with an integrated surface temperature sensing K-type thermocouple at the interior surface of the wall. The temperature and relative humidity sensors were Sensirion SHT 75 type, having sensitivity in temperature measurement of ±0.5 K and in humidity measurement of ±2 %. The external surface temperature was measured with a digital temperature sensor, having a similar accuracy.

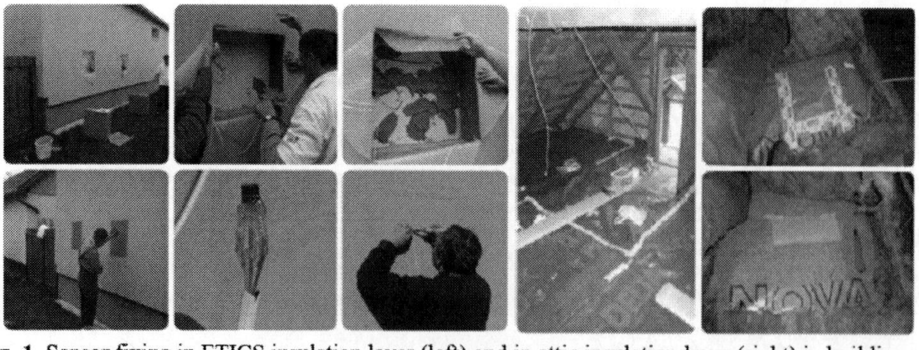

Fig. 1. Sensor fixing in ETICS insulation layer (left) and in attic insulation layer (right) in building 'A'.

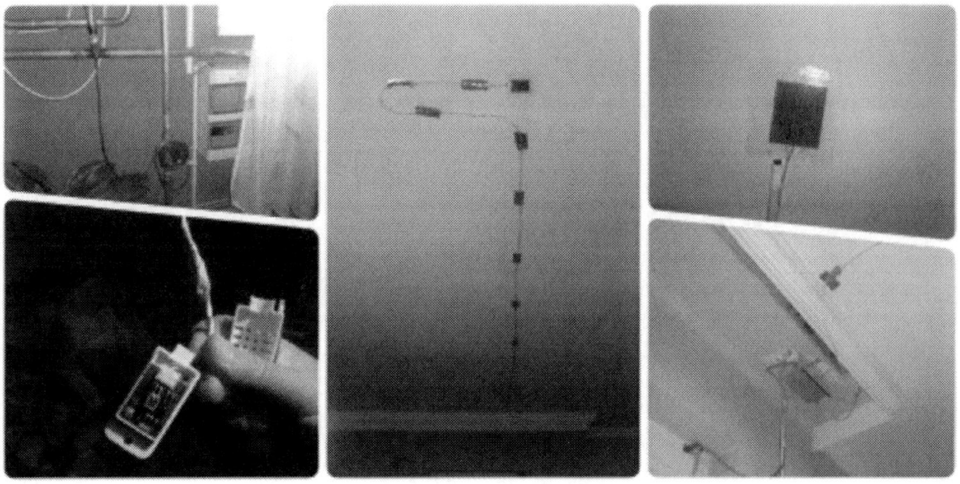

Fig. 2. Energy meter, temperature and RH sensor placement inside building 'A'.

2.3 Measurement of the energy used for heating

The energy used for heating was also logged in every 15 minutes with a thermal energy meter (Fig. 2.), which was installed into the heating system (both houses had a gas boiler and radiators using water as the heating medium), while gas consumption was calculated by using the nominal values of the boilers efficiency to provide an easier comparison. The energy monitoring took place for the same time period as the monitoring of the materials in 2.2 above. A summary of the results can be found on a Hungarian webpage [1].

2.4 Temperature and relative humidity data evaluation

To illustrate our measured dataset from the 12th of October till 20th of May, we used data contouring. We represented the values in a 3D map to show the temperature and relative humidity distributions of the building structures. To determine the values between the measured points Kriging interpolation method was applied. Kriging is a geostatistical interpolation technique that considers both the distance and the degree of variation between known data points when estimating values in unknown areas. It can be seen as a point interpolation which requires a point map as an input, results in a raster map with estimations as the output, and optionally an error map may be obtaind.

2.5 Calculation of the momentary thermal conductivities

The constantly changing value of thermal conductivity can be determined from temperature and relative humidity data, which was recorded during the measurements. The value of the thermal conductivity of each layer can be determined according to ISO 10456:2007 [2] based on the temperature and relative humidity values which are measured in the layers and at the layer boundaries:

$$\lambda_2 = \lambda_1 \cdot F_T \cdot F_m \cdot F_a = \lambda_1 \cdot e^{f_T(T_2-T_1)} \cdot e^{f_u(U_2-U_1)} \cdot 1 \quad (1)$$

where λ_2 is the adjusted (design) thermal conductivity; λ_1 is the declared (catalogue) value of thermal conductivity; F_T, F_m and F_a are the temperature, moisture and age conversion factors of which the last one was set to 1, since no further ageing was considered. f_T is temperature conversion coefficient; (T_2-T_1) represent the temperature difference between the real life application and laboratory mean temperature of the material. The factor f_u is moisture conversion coefficient; (U_2-U_1) is the moisture content difference between the real life application and laboratory measurement of declared values. For moisture conversion, the connection between the measured RH in a material and the moisture content is the moisture storage function.

2.6 Calculation of the thermal transmittances

Under steady-state conditions the thermal transmittance (U-value) can be obtained by measuring the heat flow density through an element, and recording the temperature on both sides of the element. Since steady-state conditions are never achieved in practice, there are many different methods to eliminate a difficult calculation. In this case we assumed that, the mean values of heat flow densities and temperatures over a long-time period give a good estimate for the steady-state condition. There are several stages of U-value measuring process, including the selection and calibration of heat flow meters and temperature sensors, then the measurements and analysis of the data. For calculating the U-value from measured temperature data and heat flow densities, averaging method according to ISO 9869-1:2014 was used [3]. There are three conditions, and the test can be completed when these conditions are fulfilled.

The first condition is that the duration of the test must exceed 72 hours, while the temperature is stable around the heat flow meter. The second is that the R-value obtained at the end of the test cannot deviate more than ± 5 % from the value obtained 24 hours before. The last condition is that the R-value obtained by analysing the data from the first-time period during INT(2·DT/3) days does not deviate more than ± 5 % from the values obtained from the data of the last time period of the same duration, where DT is the duration of the test in days and INT is representing the integer part. This method assumes that the U-value can be obtained by dividing the mean heat flow density by the mean temperature difference. In such a way an estimated U-value is obtained by the following formula:

$$U = \frac{\Sigma q}{\Sigma(T_i - T_e)} \tag{2}$$

where q is the heat flux, T_i is the internal temperature and T_e is the external temperature.

The U-value can be determined by using the following formulas from the surface temperatures, or from the thermal resistance values:

$$U = \frac{q}{T_i - T_e} = h_{si} \cdot \frac{T_i - T_s}{T_i - T_e} = \left(h_{ci} + \varepsilon \cdot \sigma \cdot \frac{T_i^4 - T_s^4}{T_i - T_s}\right) \cdot \frac{T_i - T_s}{T_i - T_e} = \frac{1}{\frac{1}{h_{ci} + h_{ri}} + \Sigma \frac{d_j}{\lambda_{2,j}} + \frac{1}{h_{ce} + h_{re}}} \tag{3}$$

where h_{ci} is the internal convective part and h_{ri} is the internal radiative part of the h_{si} internal surface heat transfer coefficient, ε is the emissivity factor of the surface, σ is the Stefan-Boltzmann constant, T_i and T_e stands for internal and external air temperatures, T_s is the temperature of the measured internal surface. Here λ_2 is the temperature and moisture content adjusted thermal conductivity of a d_j thick material layer. While h_{ce} is the external convective part and h_{re} is the external radiative part of h_{se} external surface heat transfer coefficient according to ISO 6946:2017 [4].

2.7 Laboratory measurement of the thermal conductivity of insulations

Under steady-state conditions the thermal conductivity can be obtained by laboratory measurements by using a guarded hot plate apparatus according to EN 12667:2001 [5] as can be seen on Fig. 3. The measurements were carried out on 3-3 pre-conditioned etalons and already built-in mineral wool samples respectively by using the following, measured parameters:

$$\lambda = \frac{V \cdot I \cdot d}{A \cdot \Delta T} \quad (4)$$

where V is the voltage of the heating plate, I is the current of the heating plate, d is the sample thickness, A is the measured area and ΔT is the temperature difference between the cooled and the heated surfaces. The samples were packed into thin PE foil to maintain their relative humidity during the measurements, and also framed by aluminium tape and glass mineral wool insulation to avoid temperature change and moisture migration from the sides. The declared thermal conductivity of the samples were also calculated by using a confident statistical tolerance interval of 90% according to ISO 10456:2007 [2].

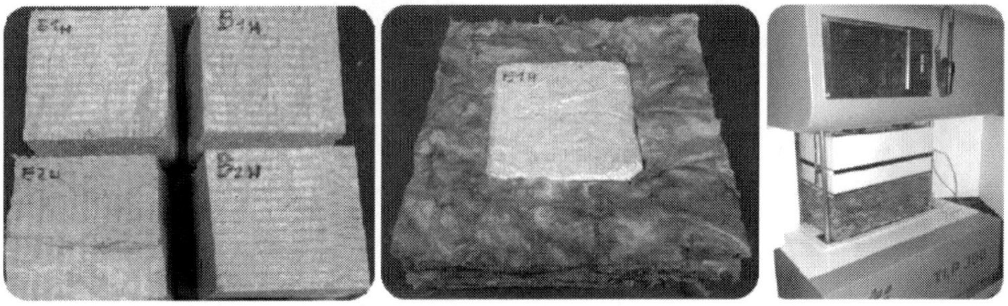

Fig. 3. Etalon and built-in samples measured in guarded hot plate apparatus.

3 Results

3.1 Energy used for heating

The heating energy was logged betwen 27th September 2013 – 31th March 2014 with a thermal energy meter built into the heating system (Fig. 4.). The energy consumption reached its peak in both cases at the end of January. During this period the maximum energy consumption was 0.58 GJ in the 'B' uninsulated house during a day, and at the same time it was less than a half of that, 0.26 GJ in the 'A' insulated house. If the overall energy consumption is summed up, the insulated house in Hajdúnánás used 23.73 GJ while the uninsulated house in Hajdúdorog used 52.6 GJ; therefore we can see that the insulated house used almost 55% less energy than the uninsulated one during the investigated period.

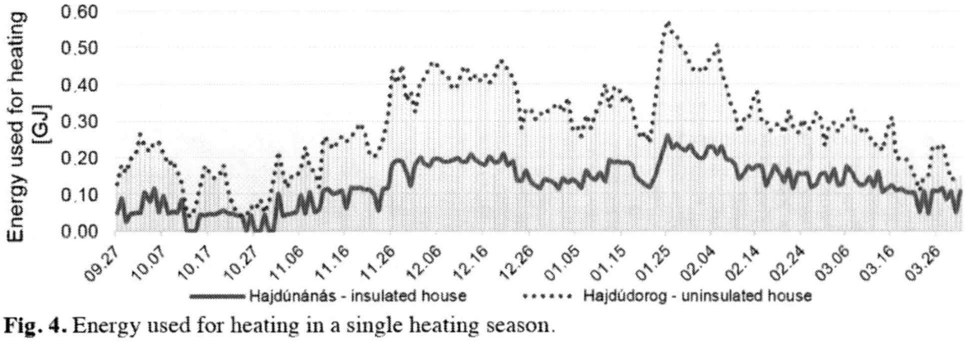

Fig. 4. Energy used for heating in a single heating season.

3.2 Hygrothermal behaviour of the insulated facade wall

Fig. 5. show the measured and Kriging interpolated temperature and relative humidity distributions in the cross sections over time. The average temperature in the ETICS was 12,51 °C, while minimum temperatures in ETICS was 2.51 °C and the maximum temperatures was 27.45 °C. The average RH was 61.34% which equals to 2.95 kg/m^3 moisture content calculated using laboratory measured sorption isotherm of the material, while the minimum was 31.8% (1,27 kg/m^3 moisture), and the maximum was 77.25% (4,27 kg/m^3 moisture).

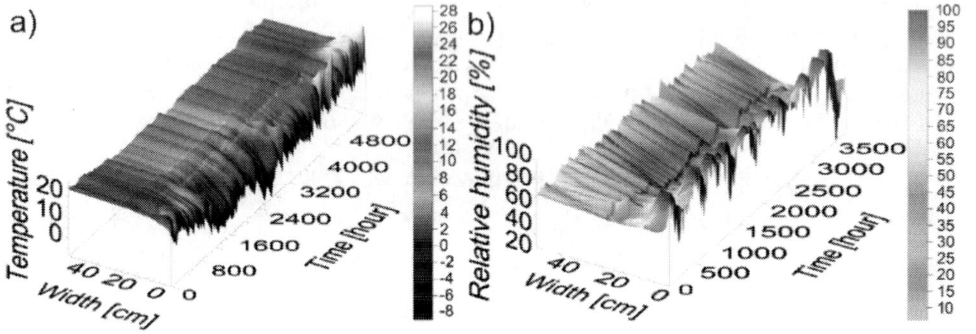

Fig. 5 a) Temperature distribution and b) RH distribution in the insulated wall.

Based on the measured values, the obtained average thermal conversion factor was 1.016, and the average moisture conversion factor was 1.011. It is observable that both temperature conversion and moisture conversion factors are in the range of 1-2% of increment, together resulting a 2,7% increase in the ETICS thermal conductivity. Therefore the average thermal conductivity of the ETICS during the whole measurement must be corrected by the conversion factors, the design value was 0.0399 W/mK (in which the thermal conductivity of the rock mineral wool was 0.035 W/mK, determined at 10 °C and using dry samples, but ETICS contains the mortar and thin plaster layer as well). In life the on-site performance of the ETICS gave 0.041 W/mK as a result for the U-value.

3.3 Hygrothermal behaviour of the insulated attic slab

The attics form a transition between the interior and outdoor spaces, the roof protects the thermal insulation on the ceiling from the impacts of the weather, but this protection is not 100% efficient, especially when reflecting roof foil is not used. The temperature and relative humidity across the sections are presented in Fig 6. The average temperature of the 25 cm thick glass mineral wool insulating layer was 14.97 °C, the minimum was 4.55 °C and the maximum was 29.54 °C. The average relative humidity in the insulating layer was 59.24% which correspond to 4.64 kg/m^3 moisture content calculated according to it's sorption isotherm, the minimum was 31.95% (2.82 kg/m^3 moisture) and the maximum was 94.16% (10.84 kg/m^3 moisture). It is observable that the mineral wool made out of glass had more moisture in absolute value compared to rock mineral wool, due to the material's different sorption isotherms.

Based on the measured temperature and RH data, the average temperature correction factor was 1.018 and the average moisture correction factor was 1.121 showing that in attics, where no roof foil is used under the tiles, the moisture accumulation could significantly change the thermal conductivity of the insulation. The adjusted thermal conductivity of the insulation was 0.044 W/mK, which was 14% higher, than the 0.039 W/mK declared glass mineral wool thermal conductivity, what is used for design.

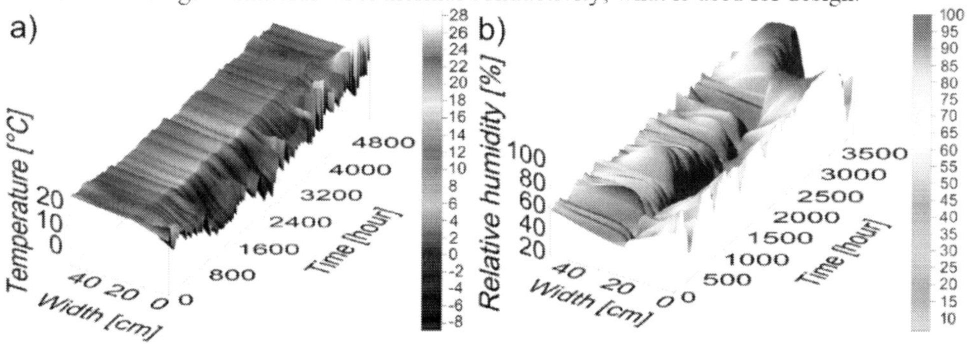

Fig. 6. a) Temperature distribution and b) RH distribution in the insulated attic.

3.4 Thermal transmittances of facade and attic

Based on the in-situ measured values, by using Eq. 2, the U-values of the wall sections and the ceiling were calculated. These U-values take the dynamic boundary conditions, layer temperatures and RH changes into account. The average U-value of the insulated facade wall was 0.16 $W \cdot m^{-2} \cdot K^{-1}$, while the steady calculated U-value was 0.15 $W \cdot m^{-2} \cdot K^{-1}$. The average U-value of the insulated attic section was 0.191 $W \cdot m^{-2} \cdot K^{-1}$, while the calculated one was 0.17 $W \cdot m^{-2} \cdot K^{-1}$.

By using the measured data, h_s surface heat transfer coefficients as well as R_s surface resistances can be obtained. The average internal horizontal surface heat transfer coefficient for the wall was $h_{si} = 8.49\ W \cdot m^{-2} \cdot K^{-1}$ ($R_{si} = 0.118\ m^2 \cdot K \cdot W^{-1}$) and the external surface heat transfer coefficient was $h_{se} = 26.82\ W \cdot m^{-2} \cdot K^{-1}$ ($R_{se} = 0.037\ m^2 \cdot K \cdot W^{-1}$) which values are close to the values given in the standard [4], however the distribution changed during the time of the measurement.

3.5 Laboratory measured thermal conductivity of new and built-in insulations

The average temperature of the rock mineral wool was 12.29 °C and the average RH was 59.06% during the on-site measurements, according to the examined monitoring results. We used new "etalon" and already built-in and cut-out 15x15x10 cm rock mineral wool thermal insulation samples to recreate the average thermal conductivity, occurred during the on-site measurements. We tried to pre-condition the samples before the guarded hot plate measurements as close as possible to the average on-site measured values. The samples were stored at a constant temperature chamber at 12 °C, where magnesium nitrate salt solution was used to create 57% relative humidity. After the samples reached mass equilibrium state, the thermal conductivity measurements were started.

We used a firm compression on the samples to ensure the continuous connection and minimal resistance between the temperature measuring plates and the samples, therefore samples were compressed around 94-95 mm in their thickness with the same 400 N compressing force, what showed that in this case built-in samples maintained their mechanical properties when compared to etalon samples. Simon [6] investigated that moisture damages or improper curing of the mineral wool could lead to compressive strength loss resulting the compression of the mineral wool layer in roof construction. In that case, however the thermal conductivity of built-in mineral wool only slightly increased, but decreasing in thickness could further decrease the insulating layer's thermal resistance.

The measured thermal conductivities are given in Table 1. The difference between etalon and built-in samples and the declared thermal conductivity was 2.86%.

Table 1. Declared thermal conductivity of etalon and built-in mineral wool insulation

Sign of sample	Size of samples [mm]	Compressed thickness [mm]	Compressed density [kg/m^3]	Thermal conductivity at 12 °C [W/mK]	Thermal conductivity $\lambda_{90/90}$ [W/mK]
E1	149.81x151.40	94.11	139.509	0.0349	0.035
E2	149.99x151.58	94.74	133.153	0.0348	
E3	150.12x150.72	94.87	127.735	0.0346	
B1	148.38x149.49	94.89	129.467	0.0360	0.036
B2	154.31x149.69	94.70	131.116	0.0361	
B3	149.14x149.29	94.54	130.265	0.0359	

4 Summary

In the study we analysed by measured datasets the energy performance of an house which is thermally insulated and which is not and the hygrothermal behaviour of the insulated building construction of a family house during an entire heating season. The measurements proved that thermally insulating a family house, significantly increase its energy efficiency and could end up saving more than 50% energy consumption. We examined the temperature and RH distributions, thermal conductivities and thermal transmittances of the facade wall and attic slab. The results showed that the rock mineral wool ETICS on the facade wall had 2.7% higher average thermal conductivity than the designed value. The attic's glass mineral wool insulation had 14% higher thermal conductivity due to increased moisture content in the insulation. However, laboratory measured built-in rock mineral wool samples had 2.86% higher declared values at the same conditions as new and etalon samples, which could mean that the differences between thermal conductivities of rock mineral wool may have occurred due to the aging of the built-in insulation material.

Supported by the ÚNKP-17-3 New National Excellence Program of the Ministry of Human Capacities. The authors are gratefully acknowledges Ms. Dóra Szagri, Dr. Dániel Bakonyi and Dr. Rita Nemes for their help and contribution to the research.

References

1. http://www.nalamszigetelnek.hu (January, 2018)
2. ISO 10456 (2007)
3. ISO 9869-1 (2014)
4. ISO 6946 (2017)
5. EN 12667 (2001)
6. T.K. Simon, L. Mlinárik, V. Vargha, J. Build. Physics **39**, 285-294 (2015)

Multifunctional decorative coatings based on cement-silicate binder

Arina Shaybadullina[1,*], *Grygory* Yakovlev[1], *Valery* Grakhov[1], *Irina* Polyanskikh[1], and *Yuliya* Ginchitskaya [1]

[1] Kalashnikov Izhevsk State Technical University, Studencheskaya Str., 7, Izhevsk 426069, Russian Federation

Abstract. A facade decorative coating has been developed that is based on sodium silicate and Portland cement and modified with a complex ultra- and nanodispersed admixture including titanium dioxide, expanded perlite sand, and multi-walled carbon nanotubes dispersion. The advantage of the produced coating is using Portland cement as a silicizer instead of the conventionally used zinc oxide. Adding ultra- and nanodispersed admixtures to the cement-silicate composition leads to the structural modification of the matrix of the binder composition along with the formation of a more durable coating (up to 4-5 times compared with the existing analogues) and the possibility of absorbing man-made electromagnetic radiation up to 38%. Expanded pearlite sand in the silicate coating provides a relief surface when applied on the base. The cement-silicate composition for coating facades is shown to have the following physical and technical characteristics: the viscosity measured with Viscometer-6 is 29 sec; the hydrogen index of the medium pH=12.33; the film resistance to static effect of water is 8 hr; the paint consumption for a two-layer coating is 200-400 g/m^2; the adhesion (the cross-cut test) is 1 point; the conventional light-fastness is 4-5 points; the frost-resistance of the silicate coating is 75 cycles. The developed facade cement-silicate coating can be applied on the surface of silicate-containing materials. The cost effectiveness of using the multifunctional protective-decorative coating has been proved in comparison with similar compositions.

1 Introduction

At present, there is a growing interest in environmentally friendly building materials that have both increased durability and aesthetic properties [1-3]. One of the best materials having these properties is ceramic brick. Facing ceramic brick has a high durability, weather resistance, and decorative value and also functions as a construction and facing material.

Using a wide color palette, texture and shape of decorative facing ceramic bricks enhances the architectural treatment of residential and civil buildings. A number of enterprises manufacturing ceramic bricks improve the decorative properties of their products by means of volumetric pigmentation and glazing [or applying polymer-based fusion bonded

* Corresponding author: arina-shaybadullina@mail.ru

paints (Fig. 1) [1]. The known methods for producing decorative brick have a number of disadvantages. In particular, the process of engobing and glazing facing brick is characterized with the complexity of technology and a relatively low durability of products and requires a lot of additional equipment [4].

Fig. 1. General view of buildings with painted facing ceramic brick (Izhevsk, Russia).

The existing technology of painting brick surface with polymer fusion bonded paints has a significant disadvantage associated with the lack of vapor and gas permeability of the obtained coating. This disadvantage causes the damage of polymer coating of brick walls due to efflorescence that appears on the brick surface and separates the polymer coating from the base [5-7]. These processes are associated with the migration of salt solutions through the brickwork, which, as a rule, are present in the composition of a masonry solution (Fig. 2).

a) b)

Fig. 2. General view of a polymer coating on the brickwork (a), a polymer coating separating from the brick surface due to efflorescence (b).

Thus, the development of a silicate coating with the improved physical and technical properties due to functional admixtures (multi-walled carbon nanotubes (MWCNTs), titanium dioxide [8], and expanded perlite [9] is a relevant task. Moreover, a facade coating, along with a high durability, acquires some additional properties, which provide special functions associated with the ability to absorb electromagnetic radiation, create the texture and color of the painted surface.

2 Materials and methods of research

This study focuses on the developed cement-silicate composition modified with ultra- and nanodispersed admixtures. The binder composition includes sodium silicate, Portland cement, an aqueous solution of sodium phosphate as a retarder, as well as ultrafine titanium dioxide, expanded perlite sand, and multi-walled carbon nanotubes suspension as modifiers of the structure. The required tint of the silicate composition was achieved by means of using a coloring paste of alkali-resistant pigments. The multi-walled carbon nanotubes suspension was first processed in a high-speed bead disperser and then treated with ultrasound in order to uniformly distribute multi-walled carbon nanotubes in the liquid phase and to avoid coagulation during storage.

The physical and technical characteristics of the developed silicate coating and the methods for their measurement meet the existing GOST and are given in Table 1. The relative viscosity of lacquer and paint is the time (in sec) of a continuous flow of a certain volume of the tested material through the calibration nozzle of viscometer. The viability of a lacquer and paint material is the maximum time during which this paint as separate components can be used after their mixing. The film resistance to the static effect of water is to determine the decorative and protective properties of coatings after their exposure to liquids during a given time. The adhesion is determined through the cross-cut test on the surface of the finished coating. Then the state of the coating is assessed visually according to the 6-point system (0 - no changes, 5 - more than 65% of the separating area). The vapor permeability and water absorption of the coating were determined similarly to the European standard ISO 7783:2011 and EN 1062-3:2008, respectively.

Table 1. Physical and technical characteristics of the facade silicate coating

Parameter	**Test method**	**Value**
The viscosity measured with VZ-6	GOST 8420-74	29 seconds
The viability (depending on the retarder concentration)	GOST 27271-2014	110 min
The hydrogen index of the medium, pH	GOST 52020-2003	12,33
The film resistance to static effect of water	GOST 9.403-80	8 h
The hiding power depending on the pigment used		200-400 g/m^2
The adhesion by the cross-cut test	GOST 15140-78	1 point
The vapor permeability of the paint layer V	GOST 33355-2015	437 $g/m^2 \cdot$ day
The thickness of the air layer with the equivalent vapor permeability, Sd	GOST 33355-2015	0.05 m
The water absorption, W	GOST 33352-2015	0.80 $kg/m^2 \cdot h^{0.5}$
The frost resistance of the silicate coating	GOST 7025-91	75 cycles

The shielding properties of carbon nanotubes are provided by their good conductivity. They have a low specific gravity and can be used for producing thin transparent films that preserve the appearance of a shielded object. This makes their usage as a protective coating more convenient [10].

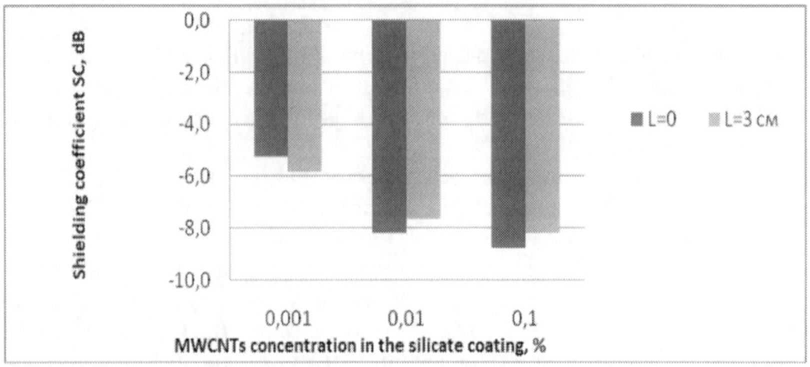

Fig. 3. The shielding coefficient of the silicate coating depending on distance L

The experiments studied the dependence of the shielding coefficient on the conditions of electromagnetic wave travel through a sample of material touching the test sample and the antennas located at a distance L (Fig. 3). The effective absorption capacity has a typical dependence and increases with the growing concentration multi-walled carbon nanotubes as a conductive filler whose content reached 7% of the mass of Portland cement. Adding carbon nanotubes into the composition of the studied silicate material leads to the formation of a binding silicate matrix and a dense, strong, and durable coating absorbing up to 38% of technogenic electromagnetic radiation affecting buildings and structures [11].

3 Testing the developed cement-silicate coating

The developed facade silicate coating, in contrast to a polymer coating, has the required porosity to ensure the vapor and gas permeability of brick walls. Moreover, unlike polymer coatings, the linear expansion coefficients of the developed silicate coating and ceramic brick have equal values, which eliminates the peeling of the silicate coating under alternating temperatures. The increased durability of the base silicate coating that the studied compositions are based on should also be noted. In particular, there are examples of buildings with the base silicate coating with a service life of more than 20 years (Fig. 4).

a) b)

Fig. 4. The building of Vneshterminalcomplex constructed in Izhevsk in 1996: a) – before painting in 1996, b) – while using the silicate coating in 2017.

The facade coating was tested in Altair, LTD, a factory producing ceramic brick (Izhevsk). Hollow ceramic bricks of M-100 were used as samples. The prepared silicate coating solution was applied with 1 layer using WESTER KP-10 air hopper gun (Fig. 5) under the air pressure created by FIAC EURO 25 compressor.

a) b)

Fig. 5. Applying the coating with a paint sprayer (a), marking the ceramic brick with the applied coating (b)

The following types of coatings were used in the experiment: 1 – a conventional composition without admixtures, 2 – a composition modified with a multi-walled carbon nanotubes dispersion. The pigments used was alkali-resistant coloring pastes produced by Novyy Dom, LTD (Izhevsk) (Fig.6).

a) b)

Fig. 6. Ceramic brick coated with the decorative silicate composition: a) - white, b) - yellow

Expanded perlite had to be added due to the demand for textured coatings applied when using colored facing brick in residential buildings. High water resistance of expanded perlite in combination with the silicate coating of increased water resistance provides a textured coating on a facade surface. In this case, the pores in the structure of expanded perlite are blocked with a cement-silicate composition and at the same time act as a damper for water that crystallizes under negative temperature. This increases the frost resistance of the silicate coating. Thus, the peeling of the textured layer under alternating temperatures is avoided.

The studied compositions of the facade coating were also tested at the construction site during the erection of a residential house in Nizhnyaya street, d.12a (Izhevsk). To apply the silicate coating, fragments of brickwork with dimensions of 1000x1000 mm and a thickness of 120 mm were built. In the process of applying, the compositions both with and without a MWCNTs dispersion were used. After being applied, the MWCNTs-modified coating had a more uniform and smoother surface in comparison with the conventional composition without nanotubes (Fig. 7).

a) b)

Fig. 7. General view of the surface without pigment (a): b) – on the right – the coating composition without nanotubes, on the left – the coating with a MWCNTs dispersion

The developed facade silicate coating was used for painting brick walls of a stab-in boiler house in a residential building in Soyuznaya Street, 1, Izhevsk. The paint was applied in 3 layers on a dry base, which was previously cleaned from dust and dirt. The facades were painted at a temperature of + 15 °C. The composition of the finishing coating (3rd layer) had a MWCNTs dispersion. The primer layer was a thin layer of facade silicate coating without pigment (8b) providing the required adhesion for the follow-up application of the finishing coating. Each next layer of the pigmented facade silicate coating was applied on the surface after thorough drying (1-2 hours).

a) b) c)

Fig.8. The design color solution of facades of a residential building (the arrow shows the place of painting) (a); applying the finishing layer of the silicate coating with a spray gun (b); general view of the walls after painting (c).

Adding a MWCNTs dispersion into the composition of the applied silicate coating led to the improved technological properties of the coating. At the same time, the viscosity of the paintwork was reduced due to the structural modification of the components of the cement-silicate composition. The curing of the silicate coating modified with carbon nanotubes stimulates the structural formation of compositions with a denser structure reinforced with nanotubes. This provides the improved physical and technical properties and a shielding effect from technogenic electromagnetic radiation.

4 Conclusion

Applying the silicate coating on the surface of freshly burned ceramic bricks under industrial conditions has shown the possibility of replacing the existing technology of applying a polymer coating with the developed silicate composition. The advantages of the developed silicate coating are the absence of thermal processes associated with paint curing, the simplicity of applying by means of spray guns, the possibility of creating a textured layer and providing equal coefficients for linear expansion of the substrate and coating.

Testing the coating at the construction site confirmed the processability of its application and the high performance of painting work. The applied silicate coating confirmed the viability of the composition under environmental conditions when exposed to atmospheric precipitation.

References

1. S. K. Ghosh, G. Waghoo, A. Kalita, D. Balgude, K. R Kumar. Progress in Organic Coatings **73**, 70-75 (2012)
2. L. X. Han, J. Han, F. M. Sun, Y. J Huo. Advanced Mater. Res.**160-162**, 880-885 (2011)
3. L. Kleerekoper, A. Dobbelsteen. van den, E. Ham. van den, T. Hordijk, C. Martin, Urban Climate **14**, 290-300 (2015)
4. H. Li, L. Dong, Z. Jiang, X. Yang, Z. Yang. J. of Cleaner Production **133**, 1017-1026 (2016)
5. G. Yakovlev, G. Pervushin, O. Kizievich, Y. Ginchickaya, P. Tajbahtina Stroitel'nye materialy **4,** 69-71 (2016)
6. R.J. Flatt, J. Crystal Growth **242**, 435-454 (2002)
7. H. Brocken, T. G. Nijland, Constr. Build. Mater. **18**, 315-323 (2004)
8. L. Graziani, E. Quagliarini, M. D'Orazio, Constr. Build. Mater. **129**, 116-124 (2016)
9. İ. B. Topçu, B. Işıkdağ, Build. Env. **42**, 3540-3546 (2007)
10. B. Wanga, Z. Guoa, Y. Hana, T. Zhang, Constr. Build. Mater. **46**, 98-103 (2013)
11. A. Shaybadullina, A. Politaeva, G. Yakovlev, G Pervushin, Adv. Mater. Res. **1122,** 121-126 (2015)

Influence of Load-bearing Structure on Size of Bonded Facade Cladding

Pavel Liška[1,*], *Barbora* Nečasová[1], and *Jiří* Šlanhof[1]

[1] Brno University of Technology, Faculty of Civil Engineering, Veveří 331/95, Brno 602 00, Czech Republic

Abstract. Architecture has been an integral part of our lives ever since people first existed. Structures are required by both investors and society to have what is considered a highly modern appearance while maintaining elements of a long service life. To meet such requirements, it is necessary to use modern technologies and materials. Bonded joints represent one of the options. Unlike with mechanical joints, it is possible to use bonded joints to anchor large format cladding panels to bearing substructures of various shapes and sizes. The design is simple, but very technically demanding to implement. One of the factors which have an impact on both the design and its implementation is the bearing substructure itself. As part of a research project, a load bearing substructure made from wood and aluminium alloy was tested. The test results prove that the mechanical properties of the materials used, especially their thermal and moisture expansion, directly influence the size of the cladding. In the case of a bearing substructure made from wood, the cladding may be larger by several percentage points than in the case of a bearing substructure made from aluminium alloy.

1 Introduction

The character of the landscape which surrounds us is shaped not only by natural processes but also the buildings we have created within it. An unprecedented degree of emphasis is placed on architecture today. Not only investors but also society demand that buildings exhibit what is considered a modern appearance without sacrificing any of the long service life that is desired of them. In order to fulfil these requirements, modern technologies as well as materials must be used. Ceramic cladding has always had an irreplaceable position in the building industry and this situation remains unchanged even today [1-3]. The range of utility properties and advantages of ceramic cladding have stayed the same but aesthetic requirements have shifted over time. At present, a significantly progressive growth in the popularity of large-format cladding elements can be observed. Large-format cladding with a minimum of joints has become unexpectedly popular not only with architects but also with the wider public. In the case of ventilated facades for buildings, the anchoring of large-format cladding to load-bearing substructure usually takes place mechanically. Stress concentration occurs at the locations of mechanical joints, lowering fatigue strength [4-5]. Apart from this,

* Corresponding author: liska.p@fce.vutbr.cz

scratches, notches and other flaws occur at these locations more frequently during the installation of screws. This can subsequently make the penetration of local pollution into the cladding material easier. When adhesives are used, these deficiencies can be prevented. Even though the anchorage of cladding to a load-bearing substructure using adhesives is not covered by existing legislation [6-7], the increasing requirements concerning the combination of the materials used has caused bonding to become an ever more widely used anchorage technique [8]. An individual approach has to be taken to the design and realization of any bonded facade system. This approach is influenced by such factors as the position (locality) of the structure or the facade, the orientation of the facade to the points of the compass, climatic conditions, the selected type of material, etc.

Research was performed concerning the influence of the material chosen to construct the load-bearing substructure for a bonded facade on the final size of the cladding material with regard to the bonding system used. This research is the source of the results and conclusions presented in this contribution.

2 Methodology

2.1 Materials

The selection of the cladding material to be used is usually carried out by the investor that commissioned the work. This step should always be taken before considering the use of bonded facade systems because material characteristics such as mechanical, physical and chemical properties should already be known before the selection of a specific type of facade adhesive and the material of the load-bearing structure. An unsuitable combination of selected materials can cause a considerable decrease in the lifespan of the whole facade system. It is therefore necessary to pay increased attention to the selection of specific glued components of the facade as early as during the design stage. Many people assume that the choice of material for the load-bearing substructure is only a formality. However, the load-bearing substructure has a significant impact on the final size of the cladding material and, in many cases, also on the selection of the bonding system. It has been determined [9] that in cases when the temperature and humidity properties of the load-bearing substructure and cladding material are significantly different, the choice of a suitable bonding system compensates for their dissimilar properties.

Four bonding systems were selected for the execution of experimental measurements. The first three systems presented in Table 1 were intended to be used directly on bonded facades. The last bonding system mentioned, Type IV, is intended only for general construction bonding. The chosen systems include examples of both one-component polyurethane adhesives and one-component MS polymers.

Table 1. Basic material properties of adhesives as stated by their producers [10].

Adhesive System	Tensile Strength [MPa]	Shear Strength [MPa]	Material Base
Type I	4.0	2.5	Polyurethane
Type II	9.0	5.5	Polyurethane
Type III	1.2	1.6	MS – Polymer
Type IV	3.5	3.0	MS – Polymer

For the purposes of the presented research, the cladding material employed was large-format ceramic cladding, which is suitable for use on ventilated building facades due to its properties, see Table 2. During the research, a load-bearing substructure made from aluminium alloy was tested along with a timber substructure (spruce, strength class C24). The basic technical parameters are shown in Table 3.

Table 2. Selected material properties of the tested ceramic facade cladding [11].

Properties	**Average Values**
Bulk density	2 800 $kg{\cdot}m^{-3}$
Thermal expansion coefficient	$6.0{\cdot}10^{-6}{\cdot}K^{-1}$
Expansion when air humidity changes from 35% to 95%	0.000 $mm{\cdot}m^{-1}$
Maximum cladding dimensions (according to the producer)	3 000 x 1 000 mm

Table 3. Selected material properties of the tested load-bearing substructures [12-13].

Properties/Load-bearing Substructure	**Average Values**	
	Aluminium Alloy	**Wood (Spruce C24)**
Tensile strength	65 – 470 MPa	14 MPa (parallel with the fibres) 0.4 MPa (vertical to the fibres)
Thermal expansion coefficient	$23{\cdot}10^{-6}{\cdot}K^{-1}$	$5{\cdot}10^{-6}{\cdot}K^{-1}$ (parallel with the fibres)
Expansion when air humidity changes from 35 % to 95 %	0.000 $mm{\cdot}m^{-1}$	0.500 $mm{\cdot}m^{-1}$

2.2 Methods

Experimental testing of the suitability of the selected adhesives in combination with ceramic cladding and the material chosen for the load-bearing substructure for use with ventilated facades was carried out based on the declaration of conformity [14].

In order to achieve the aims defined above, the adhesion of the surface finish of the building structure had to be determined first, i.e. the adhesion of the cladding to the base via the adhesive. The tests basically involve the recording of the maximum force which was able to tear off a pre-defined area of the tested surface finish from the base using a normal force, as defined by the ČSN 73 2577 standard [15]. The second test method imposed upon the test specimens consisted in the determination of the strength of a bonded assembly in shear during tensile loading according to the procedure defined in the ČSN EN 1465 standard [16]. The production, conditioning and testing of the samples from which the maximum force during the adhesion failure and shear strength of a bonded joint was determined together with displacement have already been described by the authors of this article in previous publications [17-18].

2.2.1 Determination of the tensile stress in the bonded joint

Adhesion and shear strength characterize the mechanical properties of a glued joint and their determination is essential for the performance of further calculations. The tensile strength/adhesion of the surface finish to the base and the determination of the shear strength during tensile stress are calculated using formula (1).

$$\sigma_{adh}(\tau) = \frac{F_{max}}{A_{ef}} \tag{1}$$

where

σ_{adh} (τ) adhesion / shear strength of the surface finish to the base under tensile stress, in MPa

F_{max} the maximum force required for the sample debonding, in N

A_{ef} the effective surface area of the bonded joint, in mm^2

2.2.2 Determination of the stiffness of the supports

In order to calculate the size of the cladding material, the stiffness of the supports (i.e. the bonding system) needs to be calculated both in tension and in shear according to formula (2). The bonding system is capable of eliminating the tension from the mutual shifting of facade elements within its mass.

$$k = \frac{\frac{F_{max}}{A_{ef}}}{\frac{\Delta l}{l_o}} = \frac{\sigma_{adh}(\tau)}{\varepsilon} = E \tag{2}$$

where

k the stiffness of the support in tension, in N/mm^2

E Young´s elasticity modulus, in N/mm^2

F_{max} the maximum force required for the sample debonding, in N

A_{ef} the effective surface area of the bonded joint, in mm^2

Δl the difference between the length of the bonded joint at failure and its initial length, in mm

l_0 the initial length of the bonded/cemented joint, in mm

σ_{adh} (τ) the adhesion (shear strength) of the bonded set under tensile loading, in N/mm^2

ε relative elongation, a dimensionless variable

2.2.3 Determination of the load applied to the facade system

The verification of the load-bearing capacity of the adhesive is calculated for the effects of wind according to formula (3), and the self-weight of the cladding and the thermal and moisture expansion of the cladding and load-bearing substructure, according to formula (4). The calculation and combination of individual loading states was carried out in accordance with the ČSN EN 1990 standard [19] (i.e. the same procedure as defined by Eurocode 1), a detailed description of the procedure for determining the magnitude of the load has already been presented by the authors [18].

The calculation considers the wind load to be 482.3 $N{\cdot}m^{-2}$. This corresponds to the wind strength present in a Category II wind exposure area in the Czech Republic. The calculation was carried out for a 12 m high structure. This is the maximum permitted height of a structure when the anchoring method of bonding to the load-bearing substructure is used. This requirement is related to issues connected with the fire safety of the facade and does not concern the static capabilities of the system.

$$\frac{R_z}{\sigma} \leq 1 \tag{3}$$

where

σ adhesion/tensile strength (experimental measurements), in MPa

R_z stress caused by wind effects (Nexis32 software), in MPa

The thermal and moisture expansion of the cladding is considered in the calculation along with the self weight of the cladding. Furthermore, the thermal expansion of the load-bearing substructure is also taken into account if it is made of aluminium alloy, whereas if it is made of wood, its thermal and moisture expansion is considered.

$$\sqrt{\left(\frac{R_x}{\tau}\right)^2 + \left(\frac{R_y}{\tau}\right)^2} \leq 1 \tag{4}$$

where

τ shear strength (experimental measurements), in MPa

R_x and R_y stress caused by the self weight and the thermal and moisture expansion of the cladding and the load-bearing substructure (Nexis32), in MPa

3 Results and discussion

The tensile stress in the bonded joint was determined according to formula (1). Results presented in Tables 4 and 5 show the lowest 5th percentile of the adhesion and shear stress values obtained for 18 test specimens for each adhesive and test. The best results were achieved in the case of the Type II bonding system while the worst were obtained for the Type I system.

Table 4. The lowest 5th percentile of the adhesion and shear stress values for an aluminium alloy load-bearing substructure.

Adhesive system	Adhesion σ [MPa]	Elongation [mm]	Shear Stress τ [MPa]	Elongation [mm]
Type I	1.579	6.864	1.359	10.625
Type II	2.268	3.616	5.550	11.003
Type III	1.930	1.676	1.586	5.552
Type IV	1.935	1.184	1.584	6.953

Table 5. The lowest 5th percentile of the adhesion and shear stress values for a timber load-bearing substructure.

Adhesive system	Adhesion [MPa]	Elongation [mm]	Shear Stress τ [MPa]	Elongation [mm]
Type I	1.083	4.723	1.351	10.821
Type II	1.560	2.506	5.574	11.278

Type III	1.299	1.120	1.592	5.662
Type IV	1.340	0.875	1.601	6.787

Moreover, the stiffness of the supports was determined. The results presented in Table 6 shows the highest 95th percentile of the obtained support stiffness values. These values are included in calculations carried out using Nexis32 specialized software. The lower the stiffness of the adhesive, the greater is its elongation under the same force.

Table 6. Support stiffness values for load-bearing substructures made from aluminium alloy and wood.

Adhesive System / Supporting Substructure	Aluminium Alloy		Wood	
	Stiffness of Support in Tension [N/mm^2]	Stiffness of Support in Shear [N/mm^2]	Stiffness of Support in Tension [N/mm^2]	Stiffness of Support in Shear [N/mm^2]
Type I	0.712	2.339	0.716	2.367
Type II	2.456	6.845	2.384	6.763
Type III	3.548	3.885	3.522	3.790
Type IV	3.851	5.121	3.619	5.210

The influence of individual loading states on the total stress in a bonded joint was determined and the expressions in percent of the influence of individual loading states on the total stress in a bonded joint for the two load-bearing substructures is shown in Table 7 and Table 8. The load-bearing substructure, which is firmly anchored to the base, has the greatest impact on the total stress. In the case of the aluminium alloy load-bearing substructure, the percentage is less than 82% on average for all bonding systems. In the case of the wooden load-bearing substructure, the percentage is 77 %.

Table 7. Expression of the influence of individual loading states on the total stress in a bonded joint for the aluminium alloy.

Adhesive system / Load Condition	Self-weight of the Cladding [%]	Thermal Expansion of the Cladding [%]	Moisture Expansion of the Cladding [%]	Thermal Expansion of the Substructure [%]	Moisture Expansion of the Substructure [%]
Type I	0.934	14.878	0.055	84.132	0
Type II	0.453	20.107	0.074	79.365	0
Type III	0.653	16.973	0.063	82.312	0
Type IV	0.544	18.398	0.068	80.990	0

Table 8. Expression of the influence of individual loading states on the total stress in a bonded joint for the wooden substructure.

Adhesive system / Load Condition	Self-weight of the Cladding [%]	Thermal Expansion of the Cladding [%]	Moisture Expansion of the Cladding [%]	Thermal Expansion of the Substructure [%]	Moisture Expansion of the Substructure [%]
Type I	1.212	19.530	0.072	26.395	52.790
Type II	0.551	24.184	0.090	25.059	50.117
Type III	0.837	21.260	0.079	25.941	51.883
Type IV	0.662	22.753	0.084	25.500	51.001

Table 9. Results of the calculations performed according to equations (3) and (4).

Adhesive System / Load-bearing Substructure	Aluminium Alloy		Wood	
	Equation (3)	Equation (4)	Equation (3)	Equation (4)
Type I	0.021	0.349	0.031	0.234
Type II	0.015	0.150	0.022	0.106
Type III	0.018	0.403	0.027	0.272
Type IV	0.018	0.106	0.026	0.077

Table 9 shows the results of calculations performed according to equations (3) and (4). The best results are achieved in the case of the Type IV bonding system when both aluminium alloy and wood are used. It needs to be emphasized that this system is not intended for use with bonded facades. In contrast, the worst results occur when the Type III bonding system is used. The basic static calculation took the width of one panel to be 1 000 mm. For the purposes of the research project, the term "large-format ceramic cladding element" was defined as a planar ceramic cladding element of which at least one of the dimensions exceeds 1 000 mm. The calculated maximum dimensions of the cladding for the load-bearing substructure made from aluminium alloy as well as that made from wood are 1 000 x 3 000 mm for all of the tested bonding systems.

4 Conclusions

The bonding systems differ with regard to the elongation at which tearing occurs. If the bonding system is so stiff that it does not enable elongation, failure of the joint occurs either in the cladding or within the bonded joint itself, and it subsequently falls off. For this reason, the use of an adhesive with high stiffness is completely unsuitable for less resistant materials. To a certain degree, this fact can be minimized by using a smaller board format with which such great deformation does not occur. Cement bonded particleboard is a type of cladding with low resistance yet high thermal and moisture expansion. This cladding should not be combined with adhesives featuring high stiffness and a load-bearing substructure made from aluminium alloy. When cladding with high resistance and minimal thermal and moisture expansion is used, e.g. ceramic cladding, the use of a bonding system with high stiffness is completely problem-free.

The thermal expansion coefficient, which is $18 \cdot 10^{-6} \cdot K^{-1}$ higher in the case of aluminium than in the case of wood, has an essential influence on the difference between the properties of load-bearing substructures made from such materials. On the other hand, wood is influenced by changes in the relative humidity of the air. This is a significant factor which compensates for the thermal expansion of aluminium to a large degree.

Cladding size has a significant effect on the strength of bonded joints in shear. As this system can only be used for structures with a maximum height of 12 m, the requirements for joint adhesion are significantly lower than for shear strength.

This contribution was supported by the Czech Ministry of Industry and Trade under the TRIO programme No. FV20606 and Brno University of Technology under Grant No. FAST-S-17-4255.

References

1. N. Riley, *Tile Art: A History of Decorative Ceramic Tiles* (Magna Books, 1992)
2. Ceramic Society of Japan, *Advanced Ceramic Technologie & Products* (Springer, 2012)
3. D. Watkin, *A History of Western Architecture* (Laurence King Publishing, 1986)
4. G. Krüger, R. Schneider, Otto-Graf-J. **10**, 87-98 (1999)
5. Y. Ciupack, P. Hartmut, S. Manuel, I. Erdeniz, Steel Construction **7**, 178-182 (2014)
6. IJ.J. van Straalen et. al., Int. J. Adhes. **18**, 41- 49 (1998)
7. H. Pasternak, Y. Ciupack, Int. J. Adhes. **53**, 97-106 (2014)
8. M. Švábeník, Stavebnictvi3000, http://www.stavebnictvi3000.cz/clanky/skryte-lepeni-fasadnich-obkladu-sikatack-panel-system (January, 2018)
9. B. Nečasová, P. Liška, M. Novotný, Proceedia Eng., Submitted 2017-10-05
10. Technical documentation of relevant adhesive system
11. http://www.irisceramica.com (January, 2018)
12. https://www.alumeco.cz/technické-informace/údaje-o-hliníku/vlastnosti-hliníku?s=0 (January, 2018)
13. http://people.fsv.cvut.cz/~machacek/prednaskyNNK/NNK-11.pdf (January, 2018)
14. http://cze.sika.com/dms/getdocument.get/b1f6b60b-623c-33d9-8ac7-ddfd79d1b207/Prohl.Sika%20Tack%20Panel.pdf (January, 2018)
15. ČSN 73 2577 (1981)
16. ČSN EN 1465 (2009)
17. P. Liška, J. Šlanhof, B. Nečasová, Adv. Mat. Res. **1041**, 191-194 (2014)
18. P. Liška, J. Šlanhof, B. Nečasová, B. Kovářová, Adv. Mat.s Res. **824**, 50-59 (2016)
19. ČSN EN 1990 (2004)

Impact resistance of external thermal insulation systems

Ewa Sudoł [1,*], *Dawid* Dębski[2], *Renata* Zamorowska[1], and *Barbara* Francke[1]

[1] Building Research Institute, Filtrowa St. 1, Warszawa 00-611, Poland
[2]Atlas sp. z o.o., Kilińskiego St. 2, Łódź 91-421, Poland

Abstract. In the paper the results of an experimental program intended to determine factors influencing the impact resistance of the External Thermal Insulation Composite Systems (ETICS) were presented. For the research the systems based on polystyrene have been chosen. The insulation material was faced with a rendering consisting of base coat reinforced with standard or armored glass fibre mesh and silicone or silicone-silicate binders as finishing coats. The influence of various renderings components was evaluated with respect to resistance to hard body impact and resistance to hail. The test results were discussed in the context of the possible impact level on ETICS in use.

1 Introduction

External Thermal Insulation Composite Systems (ETICS) is one of the most popular method of improving the energy efficiency and refurbishing the façades of new and existing buildings, that has been widely used in Poland and Europe for last 60 years [1]. ETICS comprises a set of products including thermal insulation material – usually these are expanded polystyrene or mineral wool boards, adhesive intended for the attachment of the thermal insulation material to the substrate, anchors (if applied), rendering system consisting of reinforcement – standard or armored glass fibre mesh, base coat and finishing coat. Alternatively, it can also include key coat, impregnating agents and decorative coat.

One of the most important properties of the façade is its durability determined by the resistance to hygrothermal factors, biological corrosion and mechanical effects which include operational factors – accidental impacts of heavy objects and acts of vandalism and factors connected with weather conditions, in particular hail [2]. It is worth noting that according to European Severe Weather Database, around 400 cases of large hail (hailstones having a minimum diameter of 2 cm) are reported in Poland annually [3].

Appropriately high impact resistance of the façade constructed with the ETICS is of significant importance not only for aesthetical reasons, but also for technical reasons. Even minor mechanical damage to the finishing coat makes it easier for the water to infiltrate the enclosure and lead to moisture accumulation within the façade. It speeds up the process of its degradation and any possible accumulation of moisture in the thermal insulation coat reduces the effectiveness of the thermal insulation.

* Corresponding author: e.sudol@itb.pl

The solutions used in the modern ETICS lead to increased impact resistance of the facing coat. They include modifications in the area of the reinforcement coat involving the use of the flexible dispersion base coat, armored glass fibre mesh, as well as the new-generation organic thin coat finishing coats – silicate, silicone or silicone-silicate, which form a flexible, durable facing coat [4, 5].

The tests on ETICS focus mainly on the properties of individual above-mentioned components of thermal insulation systems [6]. Methods established to assess the performance of ETICS systems (described i.e. in ETAG 004 [7]), include only basic evaluation of system impact resistance, not taking into account more severe conditions of use like vandalism or hail resistance. Here we present the results of the study on the influence of particular component of the ETICS system on its overall impact resistance, according to standard and modified methods.

2 Materials and methods

The most common solution in Poland – expanded polystyrene (EPS) boards that comply with EN 13163 with the perpendicular tensile strength of ≥ 100 kPa (EPS TR 100) and ≥ 80 kPa (EPS TR 80) was selected for testing from among the thermal insulation materials.

The following were selected for the reinforcement from among the products:

- mineral I base coat which contained mineral fillers, cement, redispersible polymer powder, rheology modifiers, which requires addition of 21-23% (by weight) of water,
- mineral II base coat which contained mineral fillers, cement, redispersible polymer powder, rheology modifiers, which requires addition of 20-22% (by weight) of water,
- dispersion base coat which is a ready-to-use paste based on dispersion of polymer resins, fibres, modifiers and fillers.

A standard and armored glass fibre mesh was selected for the base coat reinforcement. The mesh characteristics are given in table 1.

Table 1. Glass fibre mesh characteristic

Type	Mass per unit area, kg/m^2	Mesh size, mm	Residual resistance after alkalis ageing, N/mm	Relative residual resistance, %, (after alkalis ageing) of the strength in the as delivered state
mesh 150	150	4.5 x 5.0	≥ 20	≥ 50
mesh 300	340	12.0 x 13.0	≥ 30	≥ 50

The following finishing coats were selected:

- silicone I finishing coat which contained mineral fillers, silicone resin, hydrophobic agents, rheology modifiers, fibres and pigments,
- silicone II finishing coat which contained mineral fillers, silicone resin, hydrophobic agents, rheology modifiers, fibres and pigments,
- silicone-silicate finishing coat which contained mineral fillers, silicone resin, silicate resin, hydrophobic agents, rheology modifiers, fibres and pigments.

19 variants of complex thermal insulation systems were prepared using the above-mentioned materials. Their configuration is presented in Table 2. In the variants with the mineral base coat (U1–U5), a silicone key coat was applied before applying the finishing

coat. In the variants with the dispersion base coat (U6–U21) no key coat was applied. No decorative coats were applied.

Table 2. Configuration of thermal insulation systems

Variant	Components			
	insulation product	base coat	glass fibre mesh	finishing coat
U1	EPS TR 100	mineral I	150 – one layer	silicone I
U2	EPS TR 100	mineral I	150 – one layer	silicone II
U3	EPS TR 100	mineral I	150 – one layer	silicone-silicate
U4	EPS TR 100	mineral II	150 – one layer	silicone I
U5	ESP TR 100	mineral II	150 – one layer	silicone II
U6	EPS TR 100	dispersion	150 – one layer	silicone I
U7	EPS TR 100	dispersion	150 – one layer	silicone II
U8	EPS TR 100	dispersion	150 – one layer	silicone-silicate
U9	EPS TR 100	dispersion	150 – two layers	silicone I
U10	EPS TR 100	dispersion	150 – two layers	silicone II
U11	EPSTR 100	dispersion	150 – two layers	silicone-silicate
U12	EPS TR 100	dispersion	150 – one layer 300 – one layer	silicone I
U13	EPS TR 100	dispersion	150 – one layer 300 – one layer	silicone II
U14	EPS TR 100	dispersion	150 – one layer 300 – one layer	silicone-silicate
U15	EPS TR 80	dispersion	150 – one layer	silicone I
U16	EPS TR 80	dispersion	150 – one layer	silicone II
U17	EPS TR 80	dispersion	150 – one layer	silicone-silicate
U18	EPS TR 80	dispersion	150 – two layers	silicone II
U19	EPS TR 80	dispersion	150 – two layers	silicone-silicate

2 models with the dimensions of 0.5×1.0 m, with the use of 0.05 m thick EPS, were prepared for each variant. The models were prepared in laboratory conditions (T23±2°C and RH 50±5%), according to the technology specified for individual solutions. They were seasoned for 28 days in the preparation conditions.

The following properties of the ETICS were determined:

- resistance to hard body impact,
- resistance to hail.

The resistance to hard body impact test was carried out for all the variants of the ETICS, whereas the resistance to hail for the systems with EPS 100, a single layer of 150 mesh and silicone finishing coats.

The resistance to hard body impact test was carried out on specimens with the dimensions of 0.5×0.5 m, whereas the resistance to hail test on specimens of 0.25×0.25 m, cut out from the models after completion of the seasoning. The edges of the specimens collected for the resistance to hard body impact test were covered with the epoxy resin for protection.

The resistance to hard body impact test was performed on the basis of ISO 7892. The steel balls with the weight ranging from 0.5 kg to 4.7 kg were used in the test. The balls were dropped from the height ranging from 0.61 m to 3.0 m, appropriately to the adopted impact energy level. Before the test, the specimens were immersed in water for 7 days and then seasoned for 7 days in laboratory conditions (T23±2°C and RH 50±5%). The specimens were placed on a stable surface with their facing towards the direction of the ball movement, while ensuring the stability of their position during the test. A steel ball was dropped freely from the height appropriate for the adopted impact energy level. The test was started with the impact with the energy of 3 J (the ball of 0.5 kg, the fall height of 0.61 m), and if the specimen was not damaged, the test was continued while increasing the impact energy (increasing the weight of the ball and/or the fall height) until the damage occurred. Any cracks, scaling, debonding between layers visible to a naked eye were considered as damage. Any possible permanent deformation that was not accompanied by the above-mentioned features, was not considered as damage. The test was carried out in the above-mentioned laboratory conditions. The result of the test is the maximum impact energy, at which no damage to the ETICS was found in five tests.

The resistance to hail test was made on the basis of PN-EN 13583. The hail impact was simulated with the use of a polyamide ball with a smooth surface, the weight of 38.5±0.5 g and the diameter of 40±0.5 mm. First the specimens were seasoned for 24 h at T23±2°C and RH 50±5%, then they were placed horizontally in the testing apparatus, with their facing towards the direction of the ball movement, while ensuring the stability of their position during the test. The polyamide ball was shot perpendicularly to the specimen surface at the speed of 2 m/s. If the specimen was not damaged at this speed, the test was continued while increasing the shot speed until the damage occurred. The same properties as during the hard body impact resistance test were considered as damage. The test was carried out in the above-mentioned laboratory conditions. The result of the test is the maximum speed at which the ball hits without causing any damage to the ETICS, confirmed in five tests.

3 Results and disscusion

The analysis of the results of the hard body impact resistance tests on the ETICS made with the use of EPS TR 100 and a single layer of glass fibre mesh 150 (Fig. 1) indicated that in the solutions where the mineral base coat was used, the type of finishing coat was the factor determining the resistance. The systems with the silicone I finishing coat were characterised by the resistance to impact with the energy of 20 J, while the remaining variants – with the silicone II finishing coat and the silicone-silicate finishing coat – had the resistance of 3 J. For the system with the dispersion base coat, the relations of the solution with both silicone finishing coat were characterised by the resistance of 20 J, whole the system with the silicate-silione finishing coat showed the resistance of 10 J. It should be also noticed that in the case of the solutions with the silicone II finishing coat and the silicone-silicate finishing coat, an increase in the impact resistance was observed when the mineral base coat was replaced with the dispersion base coat. There was no change for the systems with the silicone I finishing coat.

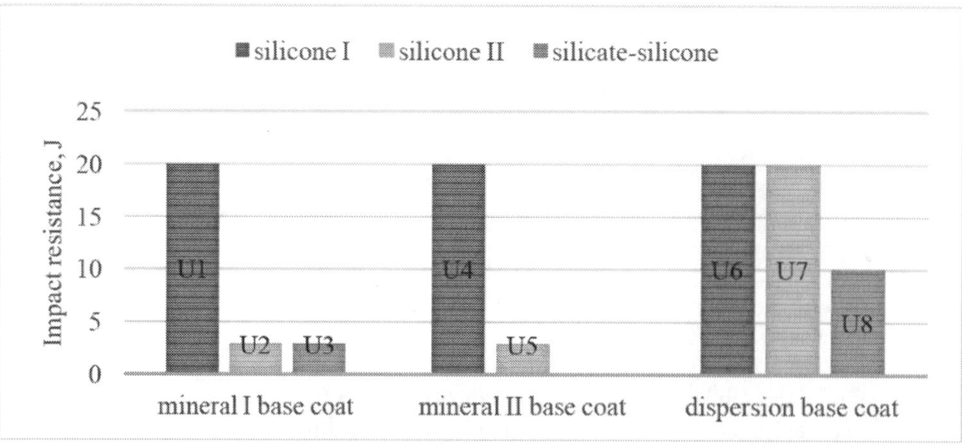

Fig. 1. The hard body impact resistance of the ETICS made with the use of EPS TR 100 and a single layer of glass fibre mesh 150 depending on the type of the base coat and finishing coat.

Selected samples after hard body impact resistance test are presented in Fig. 2.

a) b)

Fig. 2. Samples after hard body impact resistance test a) variant U1 - cracks at the circumference after impact with energy 30J, b) variant U4 - cracks at the impact point and at the circumference after impact with energy 50J.

The application of the reinforced mesh resulted, as expected, in a significant increase in the impact resistance of the system (Fig. 3). The systems made with the use of EPS TR 100 and the dispersion base coat, where a single layer of mesh 150 was used, were characterised by the resistance of 10 J – the systems with silicone finishing coats or 20 J – the system with silicone-silicate finishing coat. The application of the additional layer of the same mesh resulted in resistance increase to 30 J and to 70 J respectively. The use of the armored glass fibre meshes 300, together with the standard mesh 150, allowed to increase the resistance of the ETICS to as much as 120 J for the solutions with the silicone I finishing coat and silicone-silicate finishing coat and to 140 J for the solutions with silicone II finishing coat. The results are correlated with the tensile strength of the reinforced layer made with different types of reinforcement [6].

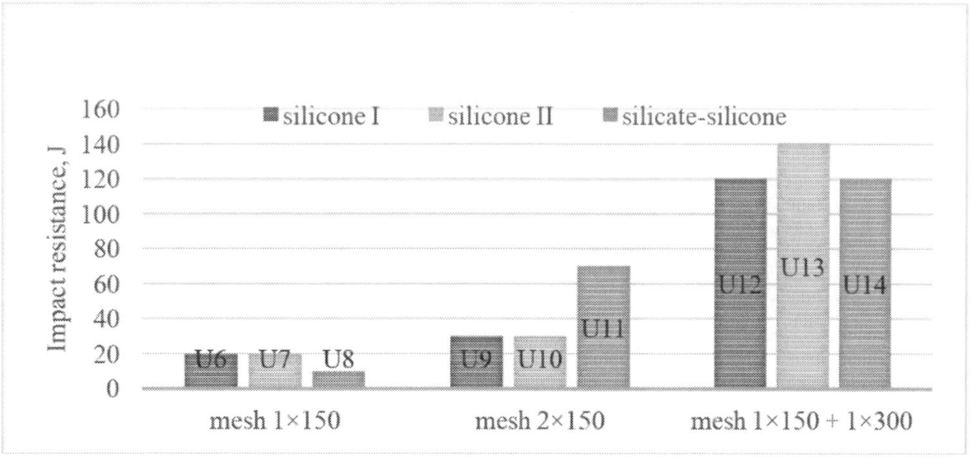

Fig. 3. The hard body impact resistance of the ETICS made with the use of EPS TR 100 and the dispersion base coat depending on the type of the glass fibre mesh and finishing coat.

When comparing the results of the impact resistance tests obtained for the systems made with the use of EPS TR 100 and EPS TR 80 it was found that the type of the expanded polystyrene – in view of the scope under consideration, typical for the ETICS – had an effect on the characteristics under consideration. The systems with the EPS TR 100 with silicone finishing coats were characterised by the resistance to impact with the energy of 20 J, while the systems with EPS TR 80 had the resistance of 10 J. The above was observed for the systems with the dispersion base coat, for all the types of finishing coat under consideration (Fig. 4).

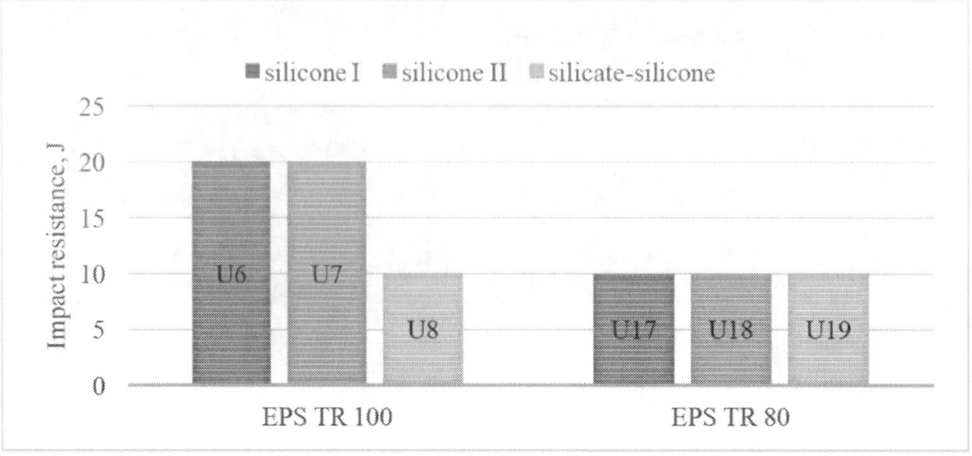

Fig. 4. The hard body impact resistance of the ETICS made with the use of the dispersion base coat and a single layer of glass fibre mesh 150, depending on the type of the finishing coat and insulation product

Similar relations were noted when analysing chosen systems with different reinforcement of the dispersion base coat – the solutions with the silicone II finishing coat and single layer of mesh 150 and (Fig. 5a) and the systems with the silicone-silicate finishing coat and double layer of mesh 150 (Fig. 5b). In other systems the impact resistance for both types of EPS was kept at the same level.

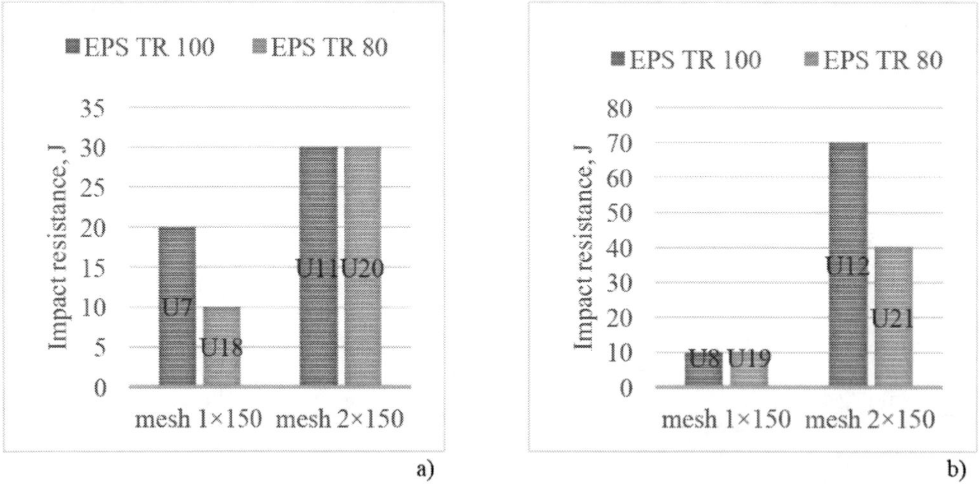

Fig. 5. The hard body impact resistance of the ETICS made with the use of the dispersion base coat and finishing coat a) silicone II, b) silicone-silicate depending on the type of the glass fibre mesh and insulation product

Summarising the results of the resistance to hard body impact tests it can be stated the majority of the tested ETICS solutions were characterised by the resistance of at least 10 J, which allows to classify them as appropriate for use in use category I, i.e. in the area that is readily accessible at ground level to the public and vulnerable to hard body impacts but not subjected to abnormally rough use. Only the solutions with the mineral base coat and the finishing coat other than silicone I, with the impact resistance of 3 J, can be classified as appropriate for use only in the use category II, that is in the area which is liable to impacts from thrown or kicked objects, but in public locations where the height of the ETICS will limit the size of the impact; or at lower levels where access to the building is primarily to those with some incentive to exercise care. At the same time, having in mind the extremely high values of the impact resistance of some of the solutions, even over 100 J, it can be assumed that they are appropriate for use even in the areas exposed to the above-average risk of a severe hard body impact, i.e. next to the sport infrastructure – located close to pitches and playgrounds, or around the entrances of staircases, on plinths, etc.

Analysis of the results of the tests of hail resistance of the ETICS carried out for the systems made with the use of EPS TR 100 and the weakest variant of the reinforcement in the form of a single layer of glass fibre mesh 150 (Fig. 6), shows that there is a slight effect of the finishing coat type on the characteristics under examinations in the case of the systems with the mineral base coat. The systems with the silicone I finishing coat achieved the resistance to hail of 10 m/s – mineral I base coat and 6 m/s – mineral II base coat, whereas those with silicone II finishing coat, a slightly smaller resistance – 7 m/s and 5 m/s respectively. These results correspond to the resistance to hard body impact (Fig. 1). In the case of the systems with the dispersion base coat, where the most advantageous characteristics were expected, a surprisingly low result was obtained for the silicone I finishing coat – 5 m/s. Obtaining such remarkably low result for this system (especially in comparison with hard body impact test results) may suggest some abnormality in the test, such as dissimilarity of the sample preparation or other issue, and additional investigation for that case is required. The hail resistance of the variant with the silicone II finishing coat was high – 22 m/s, which allows to consider this solution as resistant to impacts resulting from extreme weather conditions.

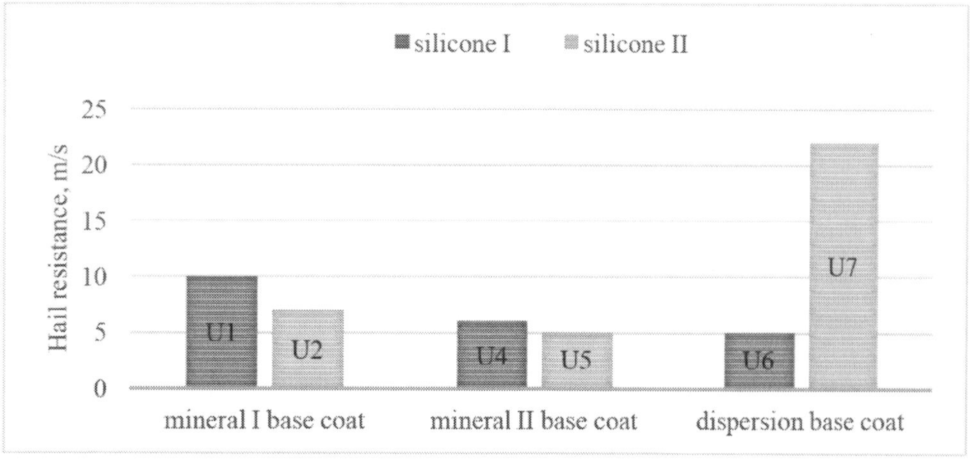

Fig. 6. The hail resistance of the ETICS made with the use of EPS TR100 and a single layer of glass fibre mesh 150 depending on the type of the base coat and finishing coat

4 Summary and conclusion

The research carried out on the External Thermal Insulation Composite Systems (ETICS) confirms that these solutions, with the appropriate system configuration, are characterised by very high resistance to hard body impact. Some of them achieved the resistance of over 100 J which allows to conclude that they are suitable for use even in the areas exposed to the above-average risk of a severe hard body impact. The tests showed the advantageous effect of the dispersion base coat and the armored mesh on the resistance to hard body impact. The advantage of the EPS TR 100 over EPS TR 80 was visible.

High results of the hail resistance of selected ETICS solutions indicate that they are resistant to impact resulting from weather conditions. However, this area of research needs to be expanded by carrying out further tests.

References

1. E. Barreira, V. P. de Freitas, *External Thermal Insulation Composite Systems (ETICS). An Evaluation of Hydrothermal Behavior* (Springer, 2015)
2. A. Wanat, Izolacje **22**, 78 (2017)
3. N. Dotzek, P. Groenemeijer, B. Feuerstein, A.M. Holzer, Atmos. Res. **93**, 575 (2009)
4. T. Buczek, Modern Eng. **1**, 16 (2016)
5. S. Zalewski, Izolacje **21**, 42 (2016)
6. P. Konca, M. Koniorczyk, A. Zawadowska, Izolacje **20**, 68 (2015)
7. ETAG 004 External Thermal Insulation Composite Systems (ETICS) with Rendering (EOTA, 2013)

MATEC Web of Conferences **163**, 08005 (2018) https://doi.org/10.1051/matecconf/201816308005
MATBUD'2018

Inverted roof insulation kits and their durability

Barbara Francke [1,*], and *Robert* Geryło[1]

[1] Building Research Institute, Filtrowa 1, 00-611 Warsaw, Poland

Abstract. In the paper there was analyzed a mechanism of loss of performance properties of inverted roof insulation kits resulted by moisture increase of the insulation layers. This problem is very important in such systems because the thermal insulation is placed above the waterproofing layer and is not protected against water absorption. Abovementioned problem is more and more common in the course of building's utilization in Polish climate conditions. Because inverted roof kits are based on extruded polystyrene (XPS) and expanded polystyrene (EPS) , those materials have been used in tested samples. The results of laboratory tests showed the phenomenon and its effect on durability and sustainability of roof covering performance. It was also found that the biggest influence on the increase of moisture in the thermal insulation layer has not the long term water absorption by immersion but freeze –thaw cycles.

1 Introduction

Roof and terrace are parts of external building envelope which close building or its fragment from above and which act as a ceiling and roof in structural and functional aspect, i.e. it transfers external loads and protects the building space against changes of microclimate, rainfall and weather influences. Roofs may be constructed in a traditional layout, in which the roof coverings are placed over thermal insulation, as well as using an inverted layer system, in which the layer sequence is reversed. In inverted roofing systems, the thermal insulation, loaded with surface protection layers (ballast or parking decks), is placed loosely on the waterproofing layer and is therefore exposed to weather, especially rainfall. That is the reason why the inverted roof kits require the use of insulation materials which[1,2]:

- do not absorb water,
- are resistant to repeated freeze and thaw cycles,
- have very good thermal resistance properties,
- have high compressive strength,
- are resistant to biological corrosion,

i.e. fulfill the basic role of thermal insulation, which is to reduce the loss of heat through building's envelope.

According to ETAG 031[2], materials suitable for use in inverted roof kits are both extruded polystyrene (XPS) and expanded polystyrene (EPS); using XPS is decadal tradition, while EPS has been utilized since recent years and publications point out insufficient experience in that regard [2,3].

* Corresponding author: b.francke@itb.pl

Recent inspections of inverted roofs show more and more frequent reveal of dampness of the thermal insulation made of materials the absorption of which is theoretically low (i.e. <3%). According to common knowledge of the civil engineering community, which is however unsupported by official requirements and normative documents, these products should guarantee the necessary thermal insulation of the roofing throughout the building's working life. These assumptions are not always reflected in real life conditions, where after only several years of usage, considerable dampness can be noticed in the insulating material, which adversely affects its basic function of providing heat insulation to the rooms under the roof [4,5,6,7].

Calculations of the thermal conductivity of construction barriers take into account modifiers due to air spaces in the insulation layer, mechanical fasteners that perforate the insulation material and, in case of inverted roofs, also influence of rainfall, as with this solution rainwater may flow under the insulation, causing periodic increase of heat transfer. The value of the index describing relative increase in thermal conductivity of materials used in inverted kits is shown in fig. 1[8], and values of the thermal conductivity in relation to moisture content – in fig. 2[8]. Graphs show that with excessive levels of moisture, the heat resistance properties drop considerably.

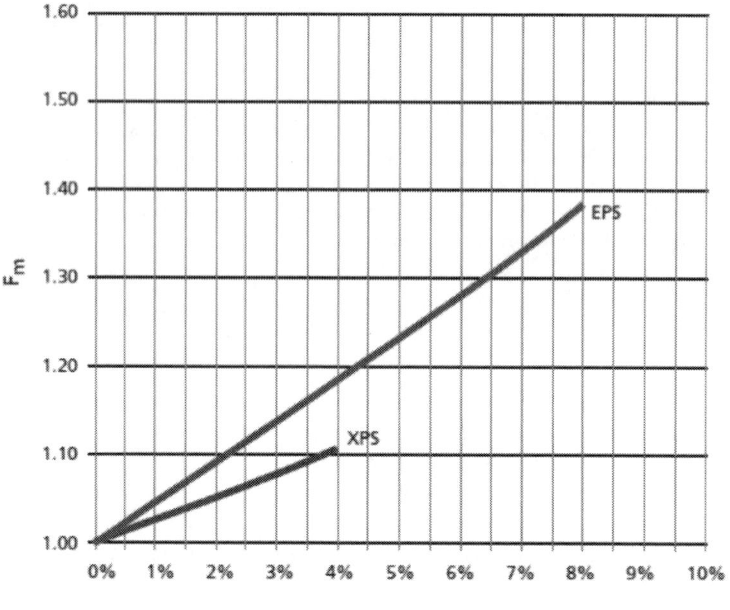

Fig. 1. Thermal conductivity change depending on product's moisture content [8].

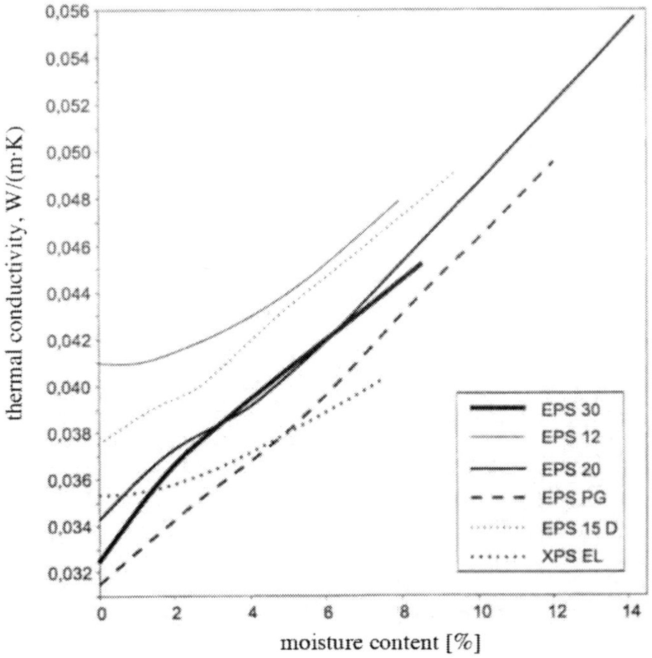

Fig. 2. Change of thermal conductivity depending on product's moisture content[8]

So far, technical literature does not report information supported by tests which would describe the mechanism of water absorption by thermal insulation products commonly of low absorption if they are used in inverse roofing. Another regarded as not profoundly explored issue is the problem of their resistance to frost, which results in increase of moisture content coupled with a small drop in strength, as well as the influence of these factors on durability of thermal resistance after freeze-thaw cycles.

2 Experimental program

The Building Research Institute started research of the above phenomenon in 2014-2017 in order to find an explanation for this issue. The research was carried out in two parallel paths, i.e.

- in laboratory conditions and on a test roof; tests used new materials purchased for the purposes of this research. As part of the laboratory tests an attempt was made to:
  - a) explain which service factor has the greatest influence on the rate of increase in moisture content of the insulation material, based on accelerated aging in laboratory conditions,
  - b) determine the relation between increase in moisture content of the insulation material and the thermal conductivity values,
- in buildings used for years, where dampness problems were diagnosed in insulation used for inverted roofs/terraces.

Materials used in the tests were:

- perimeter shape-moulded EPS boards with a profiled surface (compliant with standard PN-EN 13163),
- smooth-skin XPS boards (compliant with standard PN-EN 13164).

Their basic characteristics are given in table 1.

Table 1. Basic characteristics of EPS and XPS boards used in the research

Specimen label	Board type	Thickness, mm	Apparent density, kg/m^3	Compressive stress at 10% deformation, kPa
EPS 1	EPS 200	50	31	215
EPS 2	EPS 120	50	21	128
EPS 3	EPS 100	50	18	99
EPS 4	EPS 120	50	21	130.9
EPS 5	EPS 120	60	22	143.3
EPS 6	EPS 200	65	32	235.8
EPS 7	EPS 120	80	22	146.5
XPS I	XPS 300	40	33	453
XPS II	XPS 300	100	34	362
XPS III	XPS 500	100	34	533.3
XPS IV	XPS 500	100	32	503.9

3 Results

The research work started with long-term total immersion of the EPS and XPS boards in water for 28 days (in accordance with PN-EN 12087:2013-07). Wet samples were then subjected to 300 freeze-thaw cycles (in accordance with PN-EN 12091:2013-07). The results of these tests are presented in fig. 3 and 4. Indirect measurements of water absorption taken after 7 and 14 days of immersion showed that both in the case of XPS and EPS the most considerable increase of moisture content takes place in the first days. With time, the absorption rate decreases. Irrespective of the initial absorption after 28 days of immersion, after 300 freeze-thaw cycles the moisture content of the EPS and XPS rises significantly. With moisture levels recorded after immersion in water reaching: for EPS boards in the range of 0.2÷0.4% and for XPS boards 0.3÷0.5%, the final absorption resulting from the effect the freeze-thaw cycles had on the soaked samples was 7.3÷25.7% for EPS and 7.6÷21.3% for XPS. Test results confirmed that there is no direct relation between the absorption of insulation boards after long-term immersion in water and the effects of freeze-thaw cycles. Tested products show different ability to absorb water at over zero temperatures than in a situation of temperatures fluctuating from over to below zero.

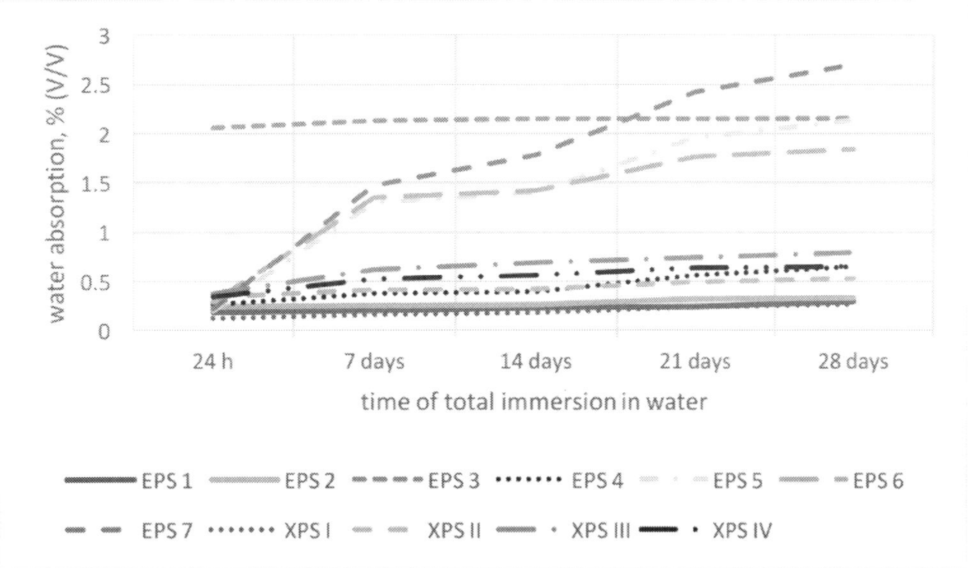

Fig. 3. Increase in water absorption as a result of long-term immersion of EPS and XPS boards over time.

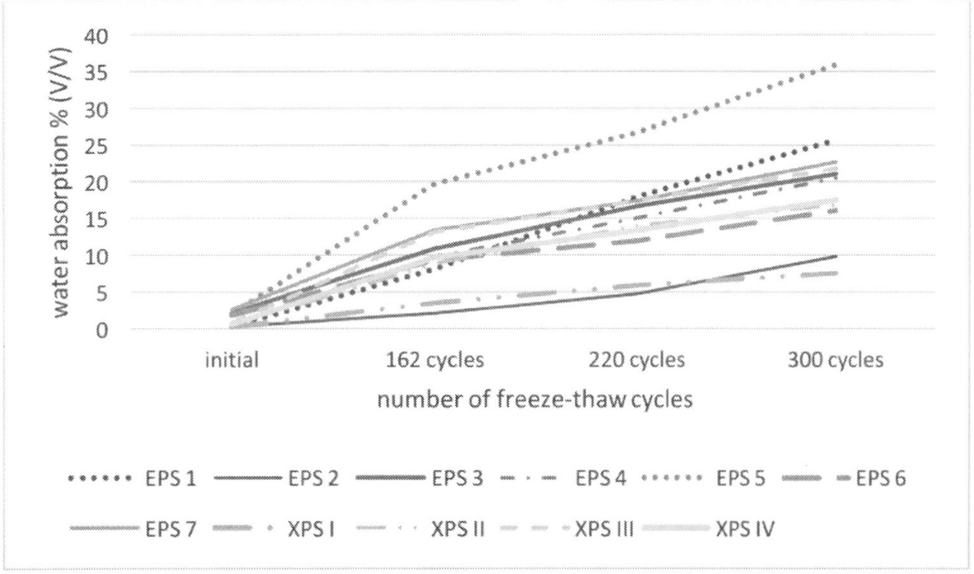

Fig. 4. The change of moisture content of EPS and XPS boards pre-soaked through immersion in water over 28 days, after subsequently being subjected to freeze-thaw cycles.

The observations so far allow to additionally conclude that the assessment of the boards' ability to absorb water only based on testing the long-term immersion in water does not yield a clear answer regarding preserving the characteristics of insulation materials used for roofs and terraces with inverted layering despite the fact that such assumption is adopted in current product standards, i.e. PN-EN 13163 and PN-EN 13164. According to assumptions in the above mentioned documents, only the change in compressive stress at 10% deformation is assessed after the freeze-thaw cycles, without evaluating the thermal aging effect on the boards' absorption. Due to the fact that the moisture content of the samples has direct impact on the material's thermal conductivity, this assumption does not allow to fully assess the product intended for use in roofs/terraces with an inverted roofing system. Fig. 5 shows the change in thermal conductivity for EPS and XPS boards subjected to accelerated aging in laboratory conditions (i.e. immersion in water over 28 days, then subjecting to 300 freeze-thaw cycles) as compared to the air-dry product (the new as well as the one after being subjected to the aforementioned aging processes and dried).

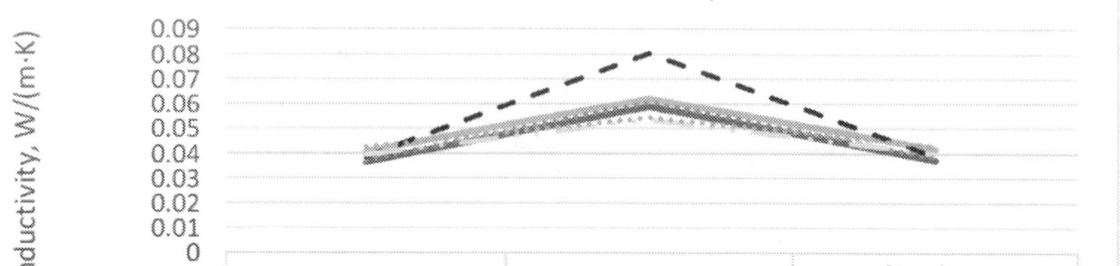

	air-dry state	after 300 freeze/thaw cycles	after drying specimens until permanent weight is reached
EPS 4	0.0362	0.0587	0.0373
EPS 5	0.0406	0.0619	0.0417
EPS 6	0.0369	0.0605	0.0382
EPS 7	0.0385	0.0801	0.0388
XPS 3	0.0393	0.0532	0.0399
XPS 4	0.0424	0.0543	0.0429

EPS 4 EPS 5 EPS 6 EPS 7 XPS 3 XPS 4

Fig. 5. Change in thermal conductivity for selected EPS and XPS boards after accelerated aging consisting in immersion in water over 28 days and 300 freeze-thaw cycles as compared to the air-dry product.

Obtained test results allow to conclude that moisture content has considerable impact on the λ coefficient and the increase of that coefficient caused by rising moisture content is several times higher than that resulting from ambient temperature change. However, the same material subjected to slow drying over a long time recovers heat conductivity equal to that of a new product. In standard laboratory conditions, i.e. at a temperature of (23±2)°C and relative humidity of (50±5)% drying process is very slow, showing noticeable effects only after 42 days, and final values after so long as 280 days. The rate of change in the thermal conductivity during the drying process for some EPS boards is shown in fig. 6 and the drying procedure for a selected EPS board in fig. 7. It should be emphasized that drying the product completely in use conditions is very hard.

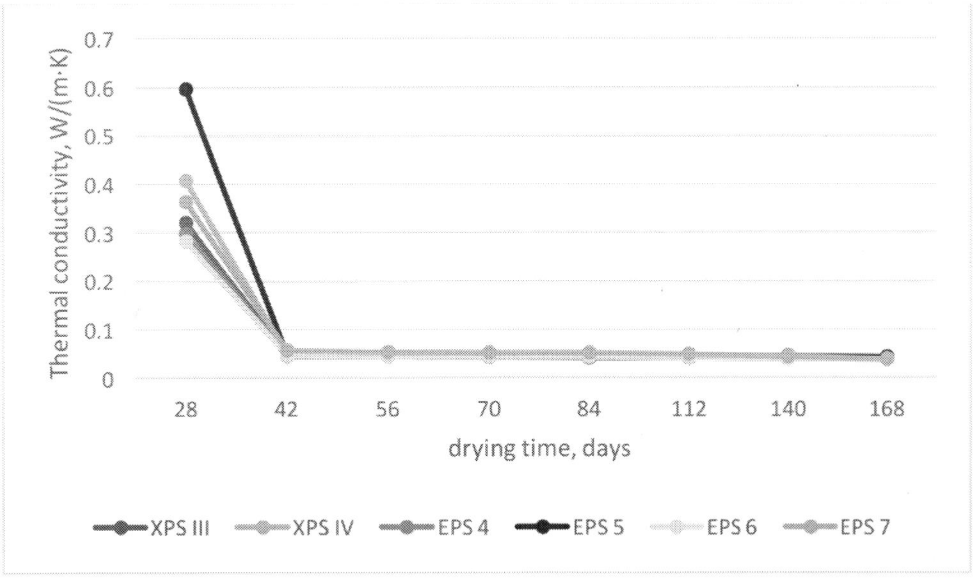

Fig. 6. Change of the thermal conductivity in the process of drying EPS boards in laboratory conditions after soaking and freeze-thaw cycles.

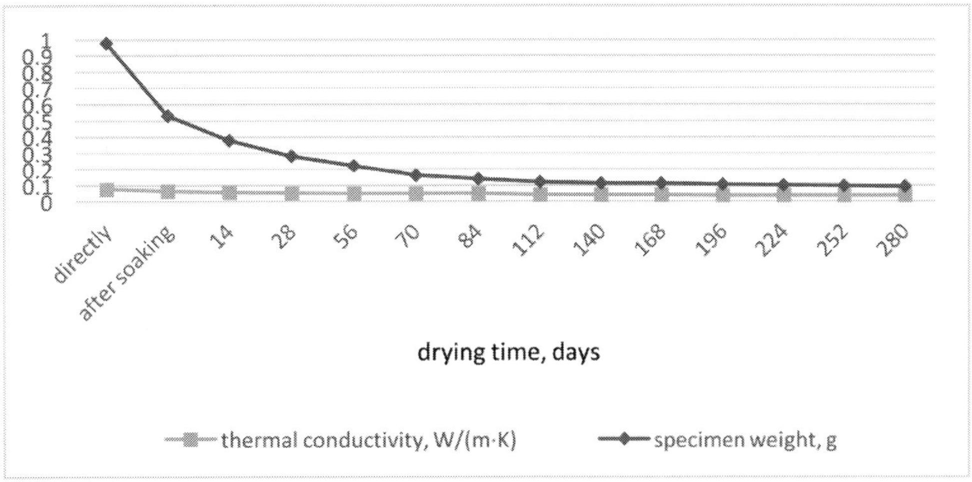

Fig. 7. Relation between change of mass in the drying process and the corresponding thermal conductivity for one of the samples (labelled as EPS 7).

Inspections of buildings suggest that the water absorbing process in thermal insulation layers of inverted roofs/terraces made of materials with properties compliant with ETAG 031[2] requirements typically starts after 5 to 10 years of use.

4 Conclusion

Test results discussed in this article suggest conclusions described below:

- following immersion of EPS and XPS boards in water, the most significant increase in moisture content takes place in first days of soaking and stays mostly consistent in subsequent days. Freezing and thawing the boards, irrespective of the initial water absorption, causes constant and considerable increase of moisture content,

- the tests did not show any relation between the initial water absorption following immersion
  in water and that reached after applying the freeze-thaw cycles. With the initial absorption being similar, the moisture content increase curves have different shapes,
- lack of relation between water absorption after 28 days of immersion and that caused by cyclic freezing and thawing limits options for assessment of fitness of polystyrene boards on inverted roofs/terraces based on declared water absorption in long-term total immersion,
- comparison of influence of moisture content on heat conductivity allows to conclude that moisture content has a considerable effect on the λ coefficient and the increase of that coefficient is tied to increase of moisture content. Wet insulation material left to dry on its own will recover its initial coefficient value.

The paper was prepared within the statutory research work at ITB (Warsaw - Poland), n° NZM-042/2017 *"Assessment of durability of inverted roof's and terrace's covering systems in the aspect of determining the optimal economic level of their application"* and with the technical support of Iwona Komosa and Aldona Wasilewska

References

1. K.Firkowicz-Pogorzelska, B.Francke, , *Projektowanie i wykonywanie stropodachów o odwróconym układzie warstw. Poradnik* (ITB, Warszawa, 2012)
2. Etag 031- European Organisation for Technical Approvals , *Guideline for European Technical Approval of Inverted Roof Insulation Kits Part 1: General* (2010)
3. J.A.Pogorzelski, K.Firkowicz-Pogorzelska, A.Bobociński, Prace Instytutu Techniki Budowlanej – Kwartalnik **3**, 139 (2006)
4. D.Zirkelbach, B.Schafaczek, H.Künzel, *XII DBMC International Conference on Durability of building Materials and Components* 1, (2011)
5. H.Künzel, K.Kiebl, *CIB W40 Proceedings Kyoto* 1 (1997)
6. H.Künzel, IBP-Mitteilung, *Bieten begrunte Umkehrdacher einen dauerhaften Warmeschutz* 271 (1995)
7. H.Künzel, IBP-Mitteilung, *Feuchteverhalten von Umkehrdachern mit massiven Deckschichten* 295 (1996)
8. J.Potter, H.Evans, *A review of recent guidance on inverted roof construction*, 11, 1 (2001)

Hygrothermal modelling of masonry blocks filled with thermal insulation

Balázs Nagy[1,*]

[1]Budapest University of Technology and Economics, Faculty of Civil Engineering, Department of Construction Materials and Technologies, Műegyetem rkp. 3. K.I.85., 1111 Budapest, Hungary

Abstract. Ceramic brick as building material has been used for thousands of years. Nowadays, the energy performance of new products has to meet rigorous requirements; therefore, in the design of new ceramic masonry blocks, building physical simulations are essential. The aim of this research is to evaluate existing masonry block shapes filled with different thermal insulation using conjugated heat and moisture transport finite element simulations with material properties measured in laboratory. The research compared four different internal structures: trapezoidal, triangular, rectangular, and with mixed shaped gaps according to existing masonry blocks. In the gaps, different thermal insulations were considered, such as mineral wool, expanded perlite and polyurethane foam. The research demonstrated that the perlite as filling material does not have a great effect on thermal conductivity comparing to unfilled blocks; however, polyurethane foam with an optimal internal structure can improve the thermal performance. Manufacturing inaccuracies in the materials' hygrothermal properties influences their performance, since a little difference in thermal conductivity has a noticeable impact on thermal transmittance, and it may result in underperformance according to regulations.

1 Introduction

In the 19th century, ceramic bricks overcome as the most popular building material of industrial and commercial building. In the 20th century, new masonry manufacturing procedures developed, such as the perforated fired clay bricks. The hollows in the masonry bricks reduced the amount of clay needed to manufacture the blocks, and improved the thermal performance too. Hollow bricks are lighter, therefore easier to handle, and usually used in single-wall constructions with finishes applied on the inside and outside. From the point of the present research, the development of hollow or filled masonry blocks is significant only after the 1960-s in Europe. Consequently the development of bricks occurred in two directions, after and regarding to the oil crisis in the 1970-s. The manufacturers started increasing the size of the holes, therefore decreasing the mass, as well as increasing the size of blocks to make the construction process more efficient and applying more porous material with smaller thickness [1]. In case of perforated bricks, the

* Corresponding author: nagy.balazs@epito.bme.hu

location of holes has an effect on the thermal performance. For example a zigzag design results in a longer route for the heat transfer, therefore increases the thermal resistance. Beside the location, the different shapes of the holes also influence the heat and moisture transfer of the masonry blocks, therefore it has to be investigated.

Nowadays, the energy performance requirements are more and more rigorous, so manufacturers increase the thermal resistance of blocks with different fillings in the holes. Since the air can move in the holes, most of the solid insulating materials have lower thermal conductivities than the moving air's equivalent value in the gaps of the building blocks. Using fillings in the holes can lead to better thermal insulating capability [2].

Fig. 1. Mineral wool filled masonry blocks [3].

Researchers in the past few years also started to investigate the thermal and hygrothermal behaviour of hollow and filled ceramic masonry blocks [1-2, 4-11] with different fillers and geometry. The presented research includes three of the most commonly used thermal insulation fillers in the industry, such as expanded perlite (Exp. Perlite), mineral wool (MW) and polyurethane foam (PUR foam). In this article four different fired clay masonry blocks with four different hollow shapes: rhombus, small rectangular, triangular and small rectangular, and big rectangular were investigated. The blocks are filled with the aforementioned three common thermal insulation materials.

2 Conjugated heat and moisture transport modelling

2.1 Masonry block geometry

Four different types of masonry blocks were modelled in 3D with different internal structures as shown in Fig. 2. a-d. Both geometry models represents an industry applied masonry block form. The dimensions of model '30HS' are 30 cm x 25 cm x 25 cm and this block have mostly rhombus based internal structure. The dimension of model '38A+' are 38 cm x 25 cm x 25 cm and this block have only small, rectangular based internal hollows. '44T' marked block's dimensions are 44 cm x 25 cm x 25 cm and this block have triangle and rectangle shaped internal structure. The fourth block is called '44T', and it's have the same size as '44K' however with only big rectangular hollows. In the industry, 44T is the most commonly used filled masonry block shape.

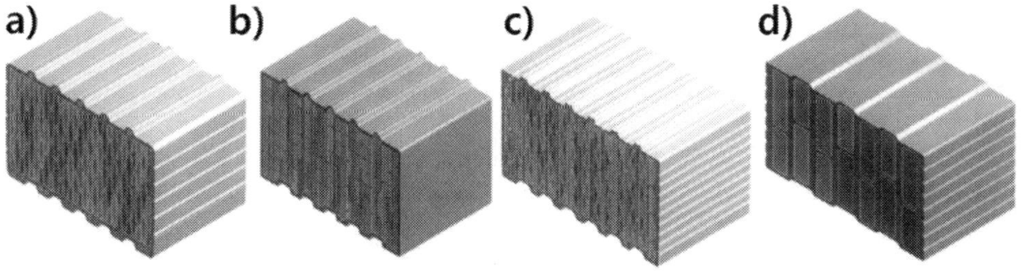

Fig. 2. Geometry models: a) 30HS; b) 38A+; c) 44K; d) 44T masonry blocks.

2.2 Partial differential equations

In this paper, the physics of steady-state conjugated heat and moisture transport in porous media was considered. The partial differential equations (PDE) based on Künzel [12] solved using Comsol Multiphysics finite element method (FEM) software [13]. The PDE for steady-state heat transfer is the following:

$$\nabla \cdot \mathrm{q} = \mathrm{Q} \tag{1}$$

$$\mathrm{q} = -(\lambda_{eff} \cdot \nabla \cdot \mathrm{T} + \mathrm{L}_v \cdot \delta_p \cdot \nabla \cdot (\phi \cdot \mathrm{p}_{sat})) \tag{2}$$

$$\lambda_{eff} = \lambda \cdot e^{f_T(T_2-T_1)} \cdot e^{f_\psi(\psi_2-\psi_1)} \tag{3}$$

where ∇ is nabla vectorial differential operator, q is heat flux [W/m^2], Q is heat source [W/m^3], λ_{eff} is temperature and moisture dependent effective thermal conductivity [W/mK] according to EN ISO 10456:2007 [14], T is temperature [K], L_v is latent heat of evaporation of water [J/kg], δ_p is vapour permeability [g/msPa], φ is relative humidity [-] and p_{sat} is the saturation pressure of water vapour [Pa]. f_T is the temperature conversion coefficient [1/K], f_ψ is the moisture conversion coefficient [m^3/m^3], when T_2 is the effective, T_1 is the declared temperature [K] and ψ_2 is the effective, ψ_1 is the declared volumetric moisture content. The PDE for moisture transfer is the following:

$$\nabla \cdot \mathrm{g} = \mathrm{G} \tag{4}$$

$$\mathrm{g} = -(\xi \cdot \mathrm{D}_w \cdot \nabla \cdot \phi + \delta_p \cdot \nabla \cdot (\phi \cdot \mathrm{p}_{sat})) \tag{5}$$

$$\xi = \frac{\partial \mathrm{w}}{\partial \phi} \tag{6}$$

where g is moisture flux [kg/m^{2}s], G is moisture source [kg/m^{3}s], ξ is differential moisture capacity [kg/m^3], D_w is moisture diffusivity [m^2/s] and w is the moisture content [kg/m^3] according to the hygroscopic sorption isotherms of the materials.

2.3 Boundary conditions

The boundary conditions for adiabatic conditions (eq. 7) and for heat flux (eq. 8) are the following:

$$-\mathbf{n} \cdot q_0 = 0 \tag{7}$$

$$-\mathbf{n} \cdot q_0 = (h_c + h_r)(T_{air} - T) \tag{8}$$

where q_0 is the inward heat flux [W/m^2], h_c is the wind speed dependent convective and h_r is the temperature dependent radiative heat transfer coefficient [W/m^{2}K] according to ISO 6946:2007 [15]. T_{air} is internal air temperature in steady-state conditions set to 293.15 K and external temperature set to 271.15 K according to MSZ 24140:2015 [16].
According to [1 and 13], the moisture transfer boundary conditions are the following:

$$-\mathbf{n} \cdot g_0 = 0 \tag{9}$$

$$-\mathbf{n} \cdot g_0 = \beta_p h_c (\phi_{air} \cdot p_{sat}(T_{air}) - \phi \cdot p_{sat}(T)) \tag{10}$$

where β_p constant is $7.7 \cdot 10^{-9}$ [kgK/WsPa] which if multiplied by h_c convective heat transfer coefficient gives the moisture transfer coefficient [kg/m^{2}sPa]. φ is the relative humidity, internally set to 0.65 and externally 0.9 was used respecting MSZ 24140:2015 [16].

2.4 Material properties

The measurements of the hygrothermal material properties were performed at the Budapest University of Technology and Economics Department of Construction Materials and Technologies Laboratory of Building Physics detailed in [1]. Each materials' ϱ density, λ thermal conductivity and δ vapour diffusion coefficient were measured, while c_p specific heat capacity, f_T and f_ψ conversion factors are from ISO 10456:2007 [14]. The hygrothermal properties listed in Table 1.

Table 1. Hygrothermal material properties.

Material	Hygrothermal properties					
	ρ [kg/m^3]	**λ [W/mK]**	**c_p [J/kgK]**	**f_T [1/K]**	**f_ψ [m^3/m^3]**	**δ [g/msMPa]**
Fired clay	1500	0.35	800	0.001	10	0.0133
Exp. Perlite	90	0.05	900	0.0035	3	0.1000
MW	75	0.031	1000	0.0045	4	0.1538
PUR foam	35	0.024	1400	0.0055	6	0.0033

Hygroscopic sorption properties were measured between the range of 0-97% relative humidity according to ISO 12571:2013 [17] detailed in [1]. Free water saturation point obtained by immersing the samples in water for 28 days. The results shown in Table 2.

Table 2. Water content [kg/m^3] according to relative humidity.

Material	Relative Humidity [1]				
	0.25	**0.5**	**0.75**	**0.97**	**~1**
Fired Clay	0.36	0.76	2.65	23.34	355.68
Exp. Perlite	0.05	0.08	0.10	0.38	1.42
MW	0.06	0.09	0.11	0.49	1.92
PUR foam	0.21	0.46	0.69	0.79	0.84

2.5 Calculating equivalent thermal conductivity

From the simulated results, the equivalent thermal conductivities of the masonry blocks were calculated using the following formula:

$$\lambda_{eq} = \frac{q \cdot A}{d \cdot \Delta T} \tag{11}$$

where q is the average heat flux measured on the internal side of the masonry block [W/m^2], A is the internal surface of the masonry block [m^2], d is the thickness of the masonry block [m] and ΔT is the temperature difference between the internal and external temperature [K].

3 Results

3.1 Equivalent thermal conductivity

The steady-state equivalent thermal conductivities of the thermal insulation filled masonry blocks are shown in Fig. 3. It is notable that both insulation fillers improved the thermal conductivity of the masonry blocks. Small hollows shows the same pattern in improvement due to better thermal insulation fillers. There is no significant difference in the thermal conductivities between '38A+' and '44K' models, when they were filled with insulations. However '44T' performed best when filled, this model improved the less with PUR foam filler compared to mineral wool filler, but also this model improved the most changing expanded perlite to mineral wool. It's observable that the shape of the hollows plays key role in the thermal insulation capability of the masonry blocks.

Masonry blocks can improve with PUR foam up to 31% compared to expanded perlite filler, and up to 19% compared to mineral wool. With expanded perlite and mineral wool fillers, masonry blocks containing smaller hollows performs almost the same thermal conductivity values. Comparing the 44 cm thick bricks, '44T', which have large rectangular hollows have lower equivalent thermal conductivity than '44K', which have small rectangular and triangular hollows.

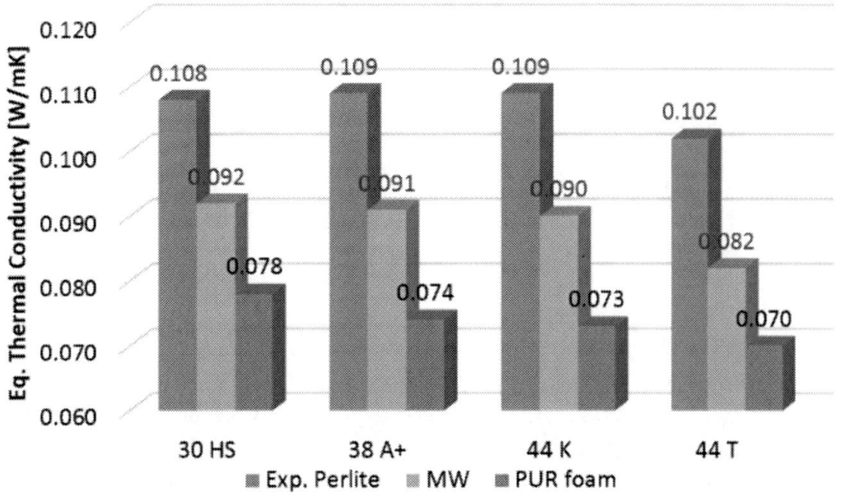

Fig. 3. Equivalent thermal conductivity of the filled masonry blocks.

A 3D surface coloured representation is obtained from each simulation showing the temperature and relative humidity distribution in the bricks. Mineral wool filled '44T' model shown in Fig. 4. The results were also obtained in the bisector of the masonry blocks.

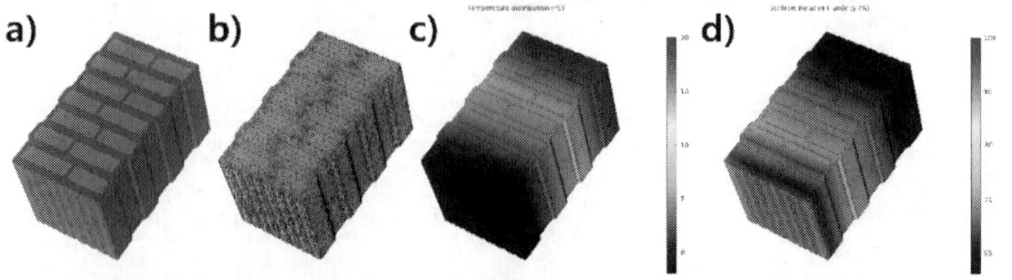

Fig. 4. Hygrothermal simulation of the '44T' model filled with mineral wool: a) geometry; b) Finite element mesh; c) Temperature distribution; d) Relative humidity distribution.

3.2 Temperature distribution

3D numerical simulated steady-state temperature results along the bisector line showed in Fig. 5. The graph also contains the simplified 1D calculated temperature distribution according to EN ISO 13788:2012, where one layer and the equivalent thermal conductivity showed in Fig. 3. was used. Using this simple one layer model calculation, the temperatures can calculated only on the surfaces. The difference between the temperatures calculated using simplified 1D calculation and 3D numerical simulation is very significant (see Fig. 5).

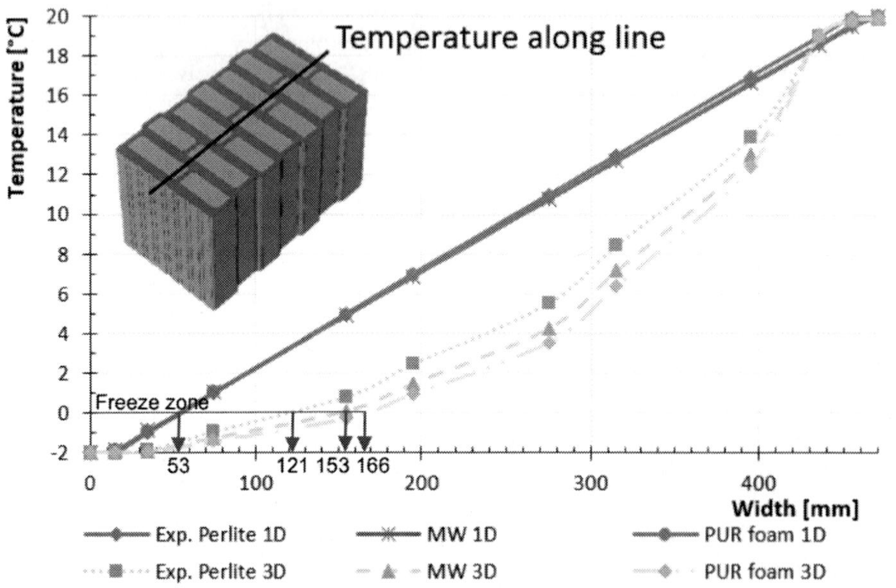

Fig. 5. Temperature along line comparison of 1D calculated and 3D simulated results.

If the temperature change assumed to be linear in the filled masonry blocks between the calculated surface temperatures, the simulated temperature distribution, in all cases, lower than the 1D obtained results. Simulated results shows remarkably thicker external side zone, 53 mm (All fillers) compared to 121 mm (Exp. perlite), 153 mm (MW) and 166 mm

(PUR foam) respectively, where the masonry block can freeze. There is also significant freeze zone increment due to different fillers, better insulating fillers have wider freeze zone. In the inside of the masonry blocks, simulated results can differ up to 7.2 °C. The shape of the graphs differs too compared to each other, 1D is linear, because both fired clay and the thermal insulation filler is calculated in one layer using the filled blocks' overall equivalent thermal conductivity, and also 1D calculation cannot take into account the geometric variability. Simulated results fits to an exponential curve.

3.3 Vapour pressure distribution

From the relative humidity along the bisector line, partial vapour pressure can be calculated using the water vapour saturation pressure depending on the temperature. Fig. 6. shows the partial vapour pressure calculated from the simulations and from 1D calculation (standard Glaser method) according to ISO 13788:2012 using each masonry blocks' equivalent vapour diffusion coefficient.

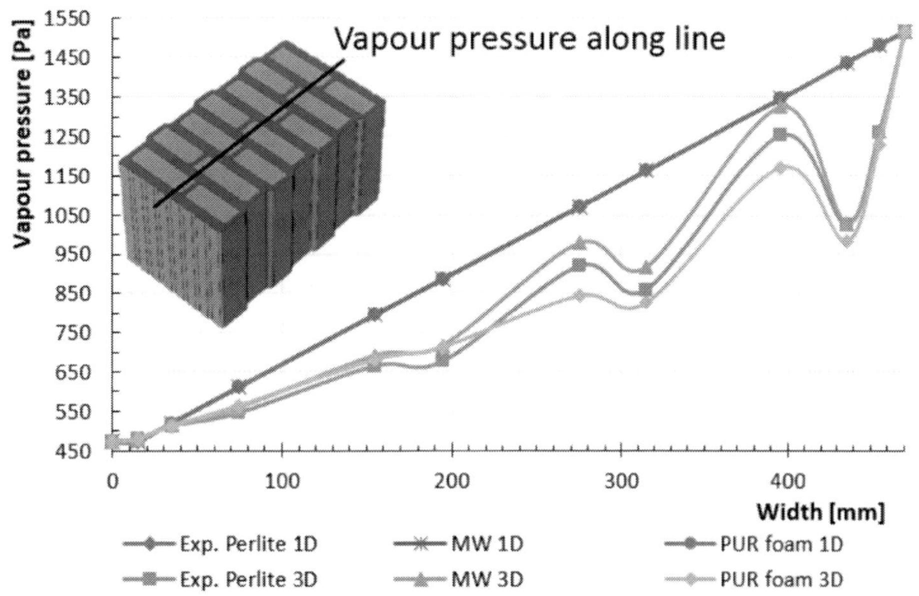

Fig. 6. Vapour pressure along line comparison of 1D calculated and 3D simulated results.

The shape of the curves are significantly different, while 1D calculated vapour pressure is linear in the one and only layer, while the simulated vapour pressure is not even decreasing steadily due to the changing materials along the line. It can be observed that expanded perlite's and mineral wool's vapour pressure curve remains the same distance to each other along the brick, but PUR foam filled brick's vapour pressure curve is going below the mineral wool's curve around the half of the block. This phenomena can be explained by the different hygroscopic sorption behaviour of the PUR foam, and the other two insulation, which is observable in Table 2 too. PUR foam is better thermal insulation, which results lower temperature distribution along line (see Fig. 5), but higher relative humidity, which means higher vapour pressure.

4 Conclusions

In this paper, four different geometries of masonry blocks with three different thermal insulation fillers were examined. The equivalent thermal conductivity of the building blocks shows that fillers with lower thermal conductivity results in lower overall results, however the improvement is not linear. Only the '44T' block is recommended to use with perlite filler, since there is not worth using fillers if they not lowering enough in the masonry blocks' thermal conductivity. PUR foam can significantly lower the values of all building blocks. Larger rectangular hollows perform better when they are filled with insulation than small rectangular, triangular or trapezoidal hollows. Filling smaller hollows is beneficial, when the filler have good thermal insulating capability and the shape of the hollows are well designed.

Steady-state condition simulations of temperature and vapour pressure distribution in the '44T' geometry along the bisector line showed that there can be up to 7.2 °C temperature and significant partial vapour pressure difference inside the blocks regarding the fillers compared to simplified one layer model results. The use of hydrophobic filler materials at least till 95 % relative humidity is essential. Simulations also showed that the outer one-third of the insulation filled masonry blocks' width is exposed to the risk of freezing, which includes two or three thermal insulation filled hollows depending on the filler material. With simplified one layer model, freeze zones can not be examined properly. Freezing can damage the not suitable filler materials and spoil their hygrothermal performance, therefore numerical simulation is advised, when complex inhomogeneous elements, such as insulation filled masonry blocks are investigated.

References

1. B. Nagy, E. Tóth, *RILEM Proceedings PRO 112*, 279-288 (2016)
2. B. Nagy, M. Orosz, Applied Mechanics and Materials **797**, 174-181 (2015)
3. http://clay-wienerberger.com (January, 2018)
4. M. Zukowski, G. Haese, Energy and Buildings **42**, 1402-1408 (2010)
5. J. Zach, J. Hroudova, M. Sedlmajer, *Proceedings of the 2nd Central European Symposium on Building Physics*, 169-174 (2013)
6. J. Kočí, J. Maděra, M. Jerman, R. Černý, *Proceedings of the 2nd Central European Symposium on Building Physics*, 635-639 (2013)
7. Y, Pavlík, M. Jerman, A. Trník, V. Kočí, R. Černý, J. Build.Physics **37**, 436-448 (2014)
8. A.B.T.P. Setty, S. Shaik, *Proceedings of the 4th International High Performance Buildings Conference at Purdue*, 3416 (2016)
9. J. Zach, V. Novák, Procedia Engineering **151**, 206-2013 (2016)
10. J. Wernery, A. Ben-Ishai, B. Binder, S. Brunner, Energy Procedia **134**, 490-498 (2017)
11. J. Li, X. Meng, Y. Gao, W. Mao, T. Luo, L. Zhang, Case Studies in Thermal Engineering **11**, 62-70 (2018)
12. H. Künzel, IRB Verlag, 102 (1995)
13. Comsol Multiphysics 5.3 Heat transfer module user's guide (2017)
14. ISO 10456 (2007)
15. ISO 6946 (2017)
16. MSZ 24140 (2015)
17. ISO 12571 (2013)
18. ISO 13788 (2012)

AUTHOR INDEX

AUTHOR INDEX